T0291802

CAMBRIDGE LIBRARY COLLECTION

Books of enduring scholarly value

Life Sciences

Until the nineteenth century, the various subjects now known as the life sciences were regarded either as arcane studies which had little impact on ordinary daily life, or as a genteel hobby for the leisured classes. The increasing academic rigour and systematisation brought to the study of botany, zoology and other disciplines, and their adoption in university curricula, are reflected in the books reissued in this series.

Algae

George Stephen West (1876–1919) was a prominent British botanist specialising in freshwater algae. In 1906 he became a lecturer in botany at the University of Birmingham and was later appointed the Mason Professor of Botany. This volume was first published in 1916 as the first of the Cambridge Botanical Handbooks series and provides a description of both marine and freshwater algae in the Myxophyceae, Peridinieae, Bacillarieae and Chlorophyceae classes. West describes the habitat, biological conditions, distribution, internal and external structures and life history of these algae in great detail, with a bibliography concluding each chapter. The book provided the first detailed description of the Myxophyceae (or blue-green) class of algae, and provides an insight into knowledge and classification of algae at the time of publication. A second volume containing a full taxonomic account of freshwater algae was planned, but not published owing to the author's death in 1919.

Cambridge University Press has long been a pioneer in the reissuing of out-of-print titles from its own backlist, producing digital reprints of books that are still sought after by scholars and students but could not be reprinted economically using traditional technology. The Cambridge Library Collection extends this activity to a wider range of books which are still of importance to researchers and professionals, either for the source material they contain, or as landmarks in the history of their academic discipline.

Drawing from the world-renowned collections in the Cambridge University Library, and guided by the advice of experts in each subject area, Cambridge University Press is using state-of-the-art scanning machines in its own Printing House to capture the content of each book selected for inclusion. The files are processed to give a consistently clear, crisp image, and the books finished to the high quality standard for which the Press is recognised around the world. The latest print-on-demand technology ensures that the books will remain available indefinitely, and that orders for single or multiple copies can quickly be supplied.

The Cambridge Library Collection will bring back to life books of enduring scholarly value (including out-of-copyright works originally issued by other publishers) across a wide range of disciplines in the humanities and social sciences and in science and technology.

Algae

Myxophyceae, Peridinieae,
Bacillarieae, Chlorphyceae

G.S. West

CAMBRIDGE UNIVERSITY PRESS

Cambridge, New York, Melbourne, Madrid, Cape Town, Singapore,
São Paolo, Delhi, Dubai, Tokyo, Mexico City

Published in the United States of America by Cambridge University Press, New York

www.cambridge.org
Information on this title: www.cambridge.org/9781108013222

This edition first published 1916
This digitally printed version 2010

ISBN 978-1-108-01322-2 Paperback

Cambridge Botanical Handbooks

Edited by A. C. Seward and A. G. Tansley

ALGÆ

VOLUME I

MYXOPHYCEÆ PERIDINIEÆ

BACILLARIEÆ CHLOROPHYCEÆ

CAMBRIDGE UNIVERSITY PRESS
C. F. CLAY, MANAGER
London: FETTER LANE, E.C.
Edinburgh: 100 PRINCES STREET

London: H. K. LEWIS AND CO., LTD., 136 GOWER STREET, W.C.
London: WILLIAM WESLEY AND SON, 28 ESSEX STREET, STRAND
New York: G. P. PUTNAM'S SONS
Bombay, Calcutta and Madras: MACMILLAN AND CO., LTD.
Toronto: J. M. DENT AND SONS, LTD.
Tokyo: THE MARUZEN-KABUSHIKI-KAISHA

ALGÆ

VOLUME I

MYXOPHYCEÆ, PERIDINIEÆ, BACILLARIEÆ, CHLOROPHYCEÆ, TOGETHER WITH A BRIEF SUMMARY OF THE OCCURRENCE AND DISTRIBUTION OF FRESHWATER ALGÆ

BY

G. S. WEST, M.A., D.Sc., A.R.C.S., F.L.S.

MASON PROFESSOR OF BOTANY IN THE UNIVERSITY OF BIRMINGHAM

Cambridge:

at the University Press

1916

Cambridge Botanical Handbooks

Edited by A. C. Seward and A. G. Tansley

THE rapid development of certain branches of botanical science in recent years has emphasised the need of books by specialists on different groups of the vegetable kingdom. After acquiring a superficial knowledge of the larger groups, the student who desires to pursue the subject beyond the limits of a general text-book receives but little assistance from existing treatises, at least as regards recent researches into the morphology and natural history of plants. The enormous output of original papers renders the task of keeping abreast of current work increasingly difficult: the accumulation of facts necessitates a periodic review from a broad standpoint of the results of recent work, more especially as they affect the actual problems of evolution presented by the various classes of plants. It is with a view of meeting this want that the present series is designed.

Professor West's book dealing with certain groups of Algæ is the first of the Handbooks to appear. The Volumes on Lichens, Fungi, and Gnetales by Miss Lorrain Smith, Dr Helen Gwynne-Vaughan, and Professor Pearson respectively are in an advanced state of preparation.

Botany School,
Cambridge.
August 1, 1916.

PREFACE

IT is now twelve years since the publication of the author's 'Treatise on British Freshwater Algæ' and several years since that work was sold out. Since the time of its publication great strides have been made in our knowledge of many groups of Algæ and it is now proposed to replace the 'Treatise' by two works of which this is one. The present volume, contributed to the series of Cambridge Botanical Handbooks, is a biological account of all the Algæ included in the Myxophyceæ, Peridinieæ, Bacillarieæ and Chlorophyceæ, both freshwater and marine, and therefore from the biological aspect more than covers the Algæ dealt with in the 'Treatise.' The author has also in preparation a distinct work which will be a complete systematic account, with illustrations, of all the Freshwater Algæ (with the exception of Desmids and Diatoms) which are known to occur in the British Islands. This is a task of some magnitude and will still take some time to complete.

An endeavour has been made to be impersonal throughout this volume, but the whole work must of necessity be largely the embodiment of the views of the author.

A chapter has been devoted to the Peridinieæ because these organisms are important as 'producers' of organic substance, especially in the marine plankton, and they store starch and oil as food-reserves; moreover, no comprehensive modern account of the group has previously been published in an English text-book. It might be suggested that the Flagellata should have been included in this volume, since the majority of them are also 'producers,' but the immense additions in recent years to our knowledge of these organisms certainly necessitates a separate volume to do them justice.

The greater part of this work deals with the Green Algæ. It is to this group that the author has devoted most of his investigations and a number of new details of classification are suggested. The treatment of this group is different from that of the Myxophyceæ, Peridinieæ and Bacillarieæ, being sectional in character. Generalizations are well nigh impossible in such a large group containing so many diverse types, and to set forth an intelligible and reliable account of the various forms in a manner which would not confuse the student, sectional treatment was deemed essential. It must always be remembered that in the Green Algæ taxonomy is intimately bound up with cytology and life-histories.

The bibliography of the various groups does not pretend to be complete, but consists only of the works which have been cited in the text. It will, however, be found to contain almost all the publications of importance which the student may wish to consult.

There are 271 illustrations comprising 1284 lettered or numbered figures, of which 681 are from original drawings by the author.

The author wishes to express thanks for permission to reproduce certain figures to Professor N. Wille of Christiania, to Professor C. A. Kofoid of Berkeley, California, to Dr F. Börgesen of Copenhagen, to Professor F. Oltmanns of Freiburg, and to Mr E. N. Transeau of Columbus, Ohio; his best thanks are also due to Mr W. B. Grove of Birmingham for kindly consenting to read through the whole of the final proofs.

There has been considerable delay in the publication of this volume, largely owing to a prolonged illness of the author in 1913–14 and partly owing to conditions which have arisen as a result of the present calamitous European upheaval.

G. S. WEST.

THE BOTANICAL LABORATORY,
THE UNIVERSITY,
BIRMINGHAM.
June, 1916.

CONTENTS

ERRATA

Page 96, line 36, *for* Mereschowsky *read* Mereschkowsky.

„ 142, to description of fig. 94 should be added 'from Oltmanns.'

„ 165, line 18, *for* Schwarda *read* Schmarda.

„ 168, in description of fig. 97 *I* and *J*, for *Pteromonas angulosa* (Carter) Lemm. read *Pteromonas Takedana* G. S. West.

„ 169, line 5, for *Chlamydomonas giganteus* read *Chl. gigantea.*

„ 194, line 33, for *Diplosiphon* read *Diplosphæra.*

„ 206, line 25, for *Ch. Nordstedtii* read *Ch. globosum.*

„ 217, in description of fig. 143 *I*, for *P. glandulifera* read *P. glanduliferum.*

„ 219, line 14, *for* '3 to 6 nuclei' *read* '4 to 8 nuclei.'

„ 425, line 23, for *E. majus* read *E. major.*

MYXOPHYCEÆ (OR CYANOPHYCEÆ)

(Blue-green Algæ)

THE Myxophyceæ (which are also known as the Cyanophyceæ[1] and as the Schizophyceæ[2]) are the lowest and most primitive of Algæ, having in their general organization a certain resemblance to the Bacteria. They combine two outstanding features: first, the copious secretion of mucilage, which often becomes very tough; and secondly, the characteristic blue-green colour of the cells. The latter feature may, however, be partially or wholly masked owing to the presence of other pigments, either in the cells or in coloured sheathing envelopes. Some are unicellular, some filamentous, and many are colonial. They occur everywhere in damp and wet situations; and quite a large number exist in subaërial habitats. Owing to the bright colours of the cells and sheaths, the Myxophyceæ furnish in moist climates many of the richest tints of the landscape, and as they occur in profusion on rocks, stones, and the trunks of trees, they sometimes impart a decided character to the country.

The Blue-green Algæ are mostly freshwater and subaërial, although a few inhabit brackish waters, and some are marine. Many forms exist in quantity in the freshwater plankton, and a few in the plankton of the warmer oceans. They also constitute the principal vegetation of hot springs.

[1] The group-name 'Cyanophyceæ' is a fairly good one, but it is being gradually superseded by the name 'Myxophyceæ,' which is perhaps a little better. Some 15 per cent. of the species of this group are not of a blue-green colour at any period of their life, and about 30 per cent. of the remainder have their cells so lodged within sheaths and gelatinous envelopes of a yellow, orange, red, brown, or purple colour that they also do not appear blue-green in the mass. The name 'Myxophyceæ' is now in general use except in the British Islands where it has been thought more advisable to retain a 'colour name.'

[2] The name 'Schizophyceæ' is a very inappropriate one, as 70 per cent. of the known species are propagated by hormogones and spores, and do not multiply by fission. Moreover, in the large group of the Hormogoneæ cell-division is accompanied by the laying down of a new transverse wall, and does not occur by 'fission' (consult p. 9).

Apart from their interest as very primitive organisms, with a world-wide distribution and a capacity for existence under the most varied conditions of environment, the Myxophyceæ present a cytological problem which as yet cannot be regarded as entirely solved. Much labour has been expended on the cytology of this group by many skilled investigators, but the conflicting opinions which have been expressed render it a matter of the greatest difficulty to give in a brief space a reasoned account of what might be regarded as our present knowledge of the minute structure of the Myxophycean cell.

Several summaries of this work were given during the years 1903—6, notably those of Kohl ('03), Phillips ('04), Olive ('05), and Guillermond ('06). Subsequently these four papers formed the subject of a critical article by Zacharias ('07). Since then many more investigations have been made with a view to elucidate the rather obscure structural details of the Blue-green Algæ, and some of these have been summarised by Pavillard ('10).

THE CELL-WALL. The cell-wall may be regarded as the definite layer immediately surrounding the protoplast. It is composite in character, and consists in its earlier stages largely of cellulose. Later, it ceases to give any cellulose reactions, offers much resistance to the penetration of reagents, and according to some authors has much in common with fungus-cellulose. Its resistance to reagents caused both Borzi and Hegler to state that there is much resemblance between it and the cuticle of higher plants. Hyams and Richards ('02) have shown that it sometimes contains silica.

In the adult cell of *Oscillatoria* and *Lyngbya*, and in the cells and spores of *Anabæna*, the cell-wall consists, according to Fritsch ('05), of an 'inner investment,' which is a modified plasmic membrane of a viscous gelatinous nature, and a 'cell-sheath,' which is probably the modified innermost layer of the external sheath of mucilage (fig. 1). Unlike the inner investment, the cell-sheath is soluble in chromic acid except in the mature spore. The cell-sheath would appear to form a coherent whole around the filaments of the Oscillatoriaceæ, but in the heterocystous forms it is split at each cell-division.

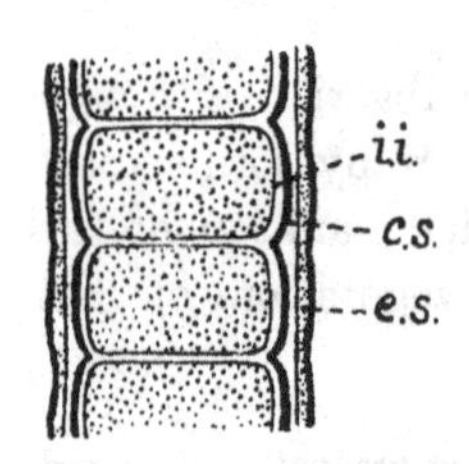

Fig. 1. Small portion of filament of *Tolypothrix* sp. after treatment with iodine. × about 1600 (slightly modified from Fritsch). *i. i.*, inner investment; *c.s.*, cell-sheath; *e.s.*, external sheath.

All Blue-green Algæ secrete mucilage to a greater or less extent. Most of the colonial unicells, and many of the filamentous forms, are embedded in a more or less extensive mass of mucus, the external surface of which may be covered with a thin cuticle. The gelatinous mass is frequently lamellose (fig. 2, *B*—*E*), and some or all of the layers may be coloured. In the filamentous forms the mucus is in the form of a sheath, which is often readily diffluent, as in *Anabæna*, many species of *Phormidium*, etc.; or it may attain various degrees of toughness and not infrequently become chitinized, as in *Lyngbya*, *Scytonema*, etc. Sheaths of this kind are the secretion of the enclosed cells of the filament, and all stages can be observed between a hyaline mucous investment and a tough, lamellose sheathing tube. The strong sheaths of some forms are therefore the homologues of the copious mucous integuments of others. All the gelatinous investments and sheaths originate as lamellæ, although this is sometimes scarcely evident, and they should really be regarded as portions of the cell-wall, in some cases undergoing gelatinization and in others a toughening to form a hard sheath. The gelatinous mass consists mostly of pectose compounds, and in certain species of *Glœocapsa* is brilliantly coloured. The investments of the unicells as well as those of the filamentous forms are often chitinized. In a few species of *Schizothrix* the sheath consists partly of cellulose and colours blue with chlor-zinc-iodine.

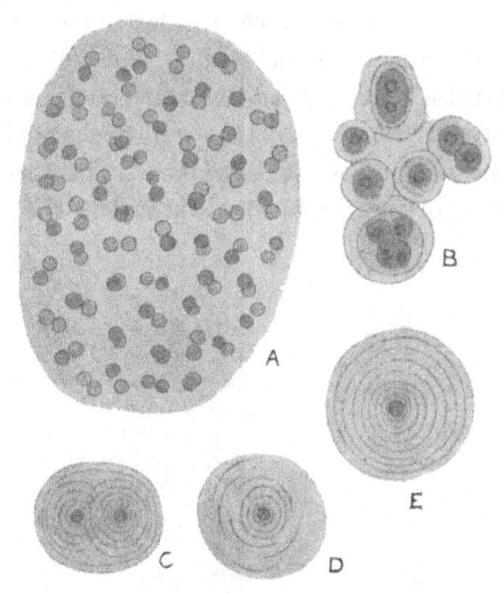

Fig. 2. Unicellular Blue-green Algæ embedded in mucus and showing the colonial habit. *A*, *Aphanocapsa Grevillei* (Hass.) Rabenh.; *B*, *Glœocapsa magma* (Bréb.) Kütz.; *C—E*, *Gl. montana* Kütz. All × 450.

The general function of the sheaths and gelatinous investments of the Myxophyceæ is undoubtedly to enable the plants to withstand periods of dryness, as the water is but slowly evaporated from such investments, and is very readily absorbed. They thus form a kind of water-reserve.

THE PROTOPLAST. Much controversy exists regarding the structure of the protoplast. It can readily be seen, sometimes even in the living cell, to consist of two parts, a peripheral coloured zone surrounding a colourless 'central body.' Both portions are generally granular, the granules of the central body being larger and as a rule only observable by appropriate staining. During the last thirty years this so-called *central body* has received much attention from cytologists, largely with the view of determining whether or not it should be considered as a true nucleus with functional activities of a similar nature to those exhibited by the nuclei of more highly organized cells. It occupies in most cases about one-quarter or one-third the volume of the cell, and its external form is largely determined by the shape of the cell; as for instance, in the numerous forms with disc-shaped cells where it is much compressed. The evidence brought forward concerning the precise nature of the central body is very conflicting, but most of the recent investigators agree that it differs considerably in its structural details from the nucleus of higher plants. The various authors come mostly within three categories: those who state that the central body is not a nucleus; those

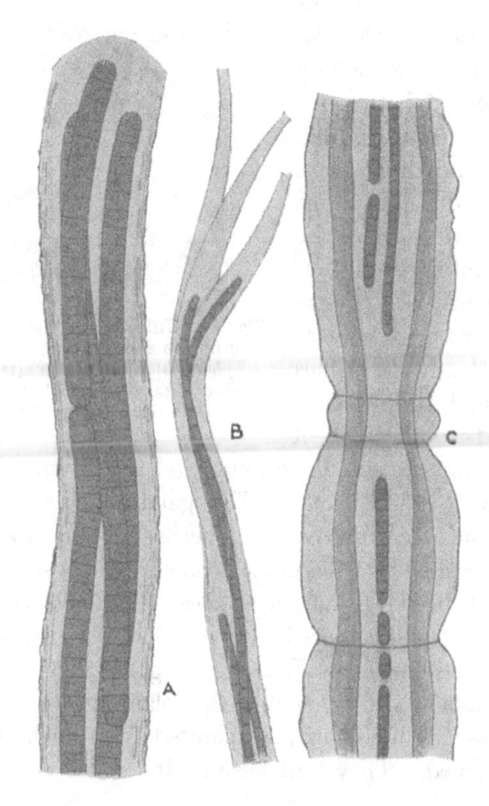

Fig. 3. Filamentous forms of Blue-green Algæ with conspicuous sheat
Mülleri Näg.; *B*, *S. lardacea* (Ces.) Gom.; *C*, *Dasyglœa amorpha* Berk. All ×460.

who regard the central body as a definite nucleus, comparable with that in higher types of cells; and those who hold the view that it is a nuclear structure of a somewhat primitive type.

I. Borzi ('86) failed to demonstrate the presence of a nucleus, and both Stockmayer ('94) and Zukal ('92) have declared that the central body has no relation to the nuclei of higher plants. Marx ('92), whose work was largely micro-chemical, concluded that it was not possible to demonstrate the existence of a nucleus. Zacharias ('90—'92) found that the central body differed in all respects from a true nucleus, and that there was no proof that it had the same function as the nucleus of higher plants. Deiniga ('91) did not assign to the central body the function of a nucleus, because, like Zacharias, he could not demonstrate its presence in all the cells of a filament. Palla ('93) found no chromatin in the central body, and regarded it as questionable whether it should be regarded as a nucleus or not; but he also added that the direct division may be more complicated than it appears to be. Massart ('02) affirmed that there was no reason to consider it as a nucleus owing to its vacuolation and indefinite outline. Macallum's investigations ('99), were mainly negative, and he believed that there was no cell-nucleus or any structure resembling one. Fischer ('05), in a very suggestive paper, also denies the existence of a nucleus. He describes, however, what he terms a 'carbohydrate mitosis' (or 'pseudomitosis of anabænin') which is an equal division of carbohydrate reserves by means of simple mitotic dispositions, but this interpretation of any of the structures described as occurring in the myxophycean cell has never been supported, whereas there is much evidence to the contrary.

II. In contrast to the opinions just enumerated, Wille ('83), Zacharias (in one of his earlier papers in 1885, the views in which he afterwards repudiated) and Scott demonstrated to their own satisfaction the presence of a body of a nuclear character. Hegler ('01) considered that his investigations conclusively proved that the central body was a nucleus. It consisted of a faintly stainable ground substance in which small granules were loosely embedded. The granules he regarded as chromatin because their behaviour during division, and towards stains and reagents, agreed in all respects with the behaviour of the chromatin of the nuclei of more highly organized cells. The nucleus, he stated, differs from that of higher types in the absence of a nuclear membrane and a nucleolus. Kohl ('03), as the result of some very careful work, states that the Blue-green Algæ possess a nucleus which differs from that of higher plants not only in the absence of a nuclear membrane and nucleolus, but also in its remarkable form. It possesses numerous radiating outgrowths of a pseudopodium-like character, which sometimes extend as far as the cell-wall and are retracted if the fixation is slow. It also contains a number of granules of albuminous material, the 'central granules,' and chromatin is always present distinct from other inclusions. Phillips ('04) regarded the central body as a true nucleus with chromatin in the form of hollow vesicles in the resting cell. Olive ('05) also stated that the central body was a true cell-nucleus.

The three last-named authors have each described a mitotic division of the declared nucleus during which a rudimentary spindle and rudimentary chromosomes can be demonstrated.

Gardner ('06) also considered the central body as a nucleus consisting of granules and chromatin embedded in an achromatic ground substance. He described three types of nuclear structure: (1) some forms (*e.g.* the large, short-celled species of *Oscillatoria*) in which the chromatin is disposed in the ground substance in the form of disconnected masses; (2) others (*e.g. Symploca muscorum*) in which it is partially united into a coarse, thread-like mass; and (3) still other cases (*e.g. Dermocarpa*) in which the chromatin is united into a definite network (fig. 4). The division of the nucleus was in all cases amitotic, with the possible exception of *Synechocystis* (fig. 8).

III. Between the extreme views of the above-mentioned authors there are the carefully considered opinions of those who regard the central body as a nucleus of a primitive or rudimentary kind. Hieronymus ('92) demonstrated the presence of granules in the central body, which, although not identical with chromatin, performed the functions of chromatin, and he also described the presence of vacuoles. He was the first to point out the absence of a nuclear membrane, and he suggested that the central body might well be regarded as an 'open nucleus' in contradistinction to the 'closed nucleus' of higher plants. Nadson ('95) considered the central body only as an aggregation of the middle alveoli of an alveolar protoplast, distinguished from the outer portion of the protoplast by the fact that it is the region where the so-called chromatin granules are exclusively or especially concentrated. He concluded that the central body corresponded to the nucleus of other organisms, but differed in its morphological peculiarities. Wager ('02) stated that the central body was vacuolated, and possessed granules which stained deeply with nuclear stains, resisted the action of digestive fluids, and gave strong reactions for phosphorus and masked iron. He considered that the granules had all the characters of nuclein and were comparable to the chromatin of a true nucleus.

Guillermond ('06) stated that the central body consists of a hyaloplasm in which there is an achromatic reticulum containing granules of chromatin. He regards it as a true chromatic network and compares it to the 'chromidial apparatus' described in

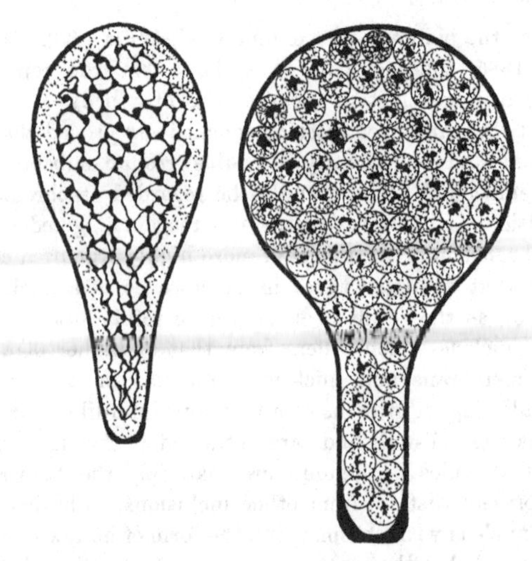

Fig. 4. *Dermocarpa fucicola* Saunders. Left figure, section of mature cell in vegetative condition, showing a thin layer of cytoplasm surrounding a definite network of 'chromatin.' Right figure, section of cell containing mature gonidia. Very highly magnified (after Gardner).

certain of the Protozoa. There is no spireme; there are no chromosomes, nor is there any division of the chromatin granules; and the division is amitotic. Swellengrebel ('10), from observations on only one species of *Calothrix*, found an alveolar achromatic ground substance in which were embedded granules and filaments of chromatin, with a more or less uneven distribution. He also stated that the distinction between the groundwork of the central body and the surrounding cytoplasm is somewhat slight, and sometimes the chromatin granules are diffused throughout the cell. The latter condition was observed by Guillermond, but only in vacuolated cells, and he therefore considered it merely as a pathological condition. Brown ('11) found in a species of *Lyngbya* a nuclear body consisting of a mesh of fine fibres embedded in a clear substance resembling nuclear sap. Small granules were scattered along the fibres, the latter staining like linin and the former like chromatin.

Still more recent work carried out by Miss Acton on various members of the Chroococcaceæ is largely confirmatory of Nadson's view that the protoplast is alveolar and that the 'central body' is a slight concentration of the alveoli in the meshes of which certain distinctive granules occur. This central concentration of the alveoli is really only apparent, as it is due to the absence of vacuoles from the 'central body,' whereas vacuoles cause a considerable breaking-down of the peripheral alveoli (consult fig. 5).

From this mass of conflicting evidence it is possible to sift out certain facts which hardly admit of dispute, and, if the whole question be carefully considered in the light of probable misinterpretations due to optical illusions and other causes, it is possible to arrive at conclusions which may not be far removed from the truth. The majority of recent workers agree that there is in the Myxophycean cell a structural differentiation (the so-called

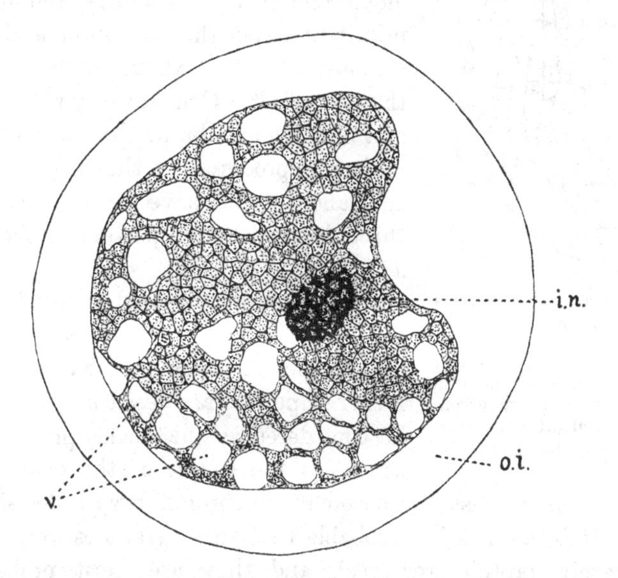

Fig. 5. Section through cell of *Chroococcus macrococcus* Rabenh. showing the incipient nucleus (*i.n.*) and the small vacuoles (*v.*) in the cytoplasm. The fine reticulation of the protoplast is also well shown. The outer line marks the limit of the outer integument or investment (*o.i.*) of the cell. ×1000 (after a drawing by Miss Acton).

'central body') which from its position and relationship to the rest of the protoplast is homologous with the nucleus. This body may be truly regarded as an *incipient nucleus*[1], and the disagreements as to the details of its structure are probably due to two causes: first, to morphological misinterpretations; and secondly, to the fact that it does not exhibit a uniformity of structure in the various members of the group. The Blue-green Algæ are undoubtedly very primitive organisms in which the protoplast shows the

[1] It would be preferable to discontinue the use of the term 'central body,' as there is no actual body with definite limitations. It is only a sort of concentration of varying degree of an alveolar protoplasmic network in which certain granules (possibly of one of the chromatin substances) are lodged, and would best be openly recognized as an *incipient nucleus*.

commencement of that differentiation which in higher types has resulted in the complete demarcation between nucleus and cytoplasm. Moreover, on considering the great differences in external morphological features which are exhibited by the various members of such a primitive group, it is not surprising that there should be different grades of incipient nuclear differentiation within the limits of the group.

This *incipient nucleus* consists of an achromatic ground substance occupying the alveoli of a reticulum in which are located minute granules, principally at the angles of the meshes. There is no limiting membrane to this structure, and in the more primitive forms the reticulum is directly continuous with that which occurs in the rest of the protoplast. Concurrently with the absence of a limiting membrane there are in some forms radiating processes of the achromatic ground substance, which have been shown to extend through the coloured portion of the protoplast as far as the cell-wall.

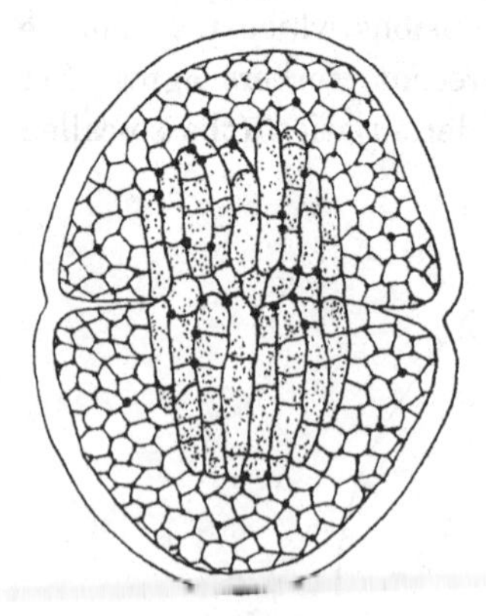

Fig. 6. Cell-division in *Chroococcus turgidus* (Kütz.) Näg. The drawing out in approximately parallel lines of the network of the incipient nucleus is well shown. Stained: iodine-green-fuchsin. ×1000 (after a drawing by Miss Acton).

The granules disposed along the threads of the mesh behave with stains somewhat after the manner of chromatin, but the staining is rather imperfect and less constant. They may be considered as the nucleo-protein substance which, in the course of the evolution of the higher types of protoplast, has become true chromatin with decisive staining properties. It is not at all improbable that these granules were in the first instance merely protein reserves[1], and they are most probably to be identified with Bütschli's 'red granules[2]' and Guillermond's 'corpuscules métachromatiques.'

No structures corresponding with nucleoli exist.

In some of the Blue-green Algæ (*e.g. Chroococcus macrococcus*) there is a clear differentiation of this incipient nucleus (fig. 5), whereas in contrast to this relatively advanced type, the most primitive of all the Myxophycean protoplasts is perhaps that of *Myxobactron* in which a differentiation has not yet been demonstrated (G. S. W., '12). The vacuoles described by certain

[1] The quantity of 'chromatin' described and figured by Gardner ('06) is so surprisingly large that one wonders whether it is really chromatin or merely the accumulation of protein reserves which behave very like chromatin with nuclear stains. Also, is not Gardner's 'chromatin' identical with Hegler's 'anabænin'? It seems probable that both observers were dealing with the same substance notwithstanding the wide difference of interpretation.

[2] These granules must not be confused with 'Bütschli's red corpuscles' in Diatoms.

authors as occurring in the 'central body' are probably due to pathological conditions and occur in cells which are undergoing degeneration.

Division of the protoplast. On the division of the cell there is much evidence to prove that the incipient nucleus divides without any mitosis such as is understood in fully differentiated nuclei. There is a constriction of the incipient nucleus, but the only suggestion of mitosis, even in the most highly developed species, is a slight tendency of the net-work to become drawn out in parallel lines, though the meshes are not broken except by the advancing constriction (fig. 6). The presumed chromatin granules do not divide during this amitotic division, but an approximately equal number is found in the two parts after division. There are in the numerous members of the Myxophyceæ various degrees of differentiation of the nuclear structure, which may account for the fact that the division has been interpreted in so many diverse ways. In some cases the constriction of the incipient nucleus takes place before the appearance of any division of the protoplast as a whole, and the view held by Macallum that the 'central body' initiates division is in these cases probably correct. In *Chroococcus macrococcus* there may be a twice-repeated division of the primitive nucleus before there is any sign of the division of the cell itself. In the Chroococcaceæ division of the cell occurs by a gradual constriction, either at the time of division of the primitive nucleus or subsequent to it. There is no septum laid down. On the other hand, in many of the Myxophyceæ with cylindrical filaments a transverse septum is formed on division, and the inward growth of this septum gradually divides both cytoplasm and primitive nucleus into two more or less equal parts.

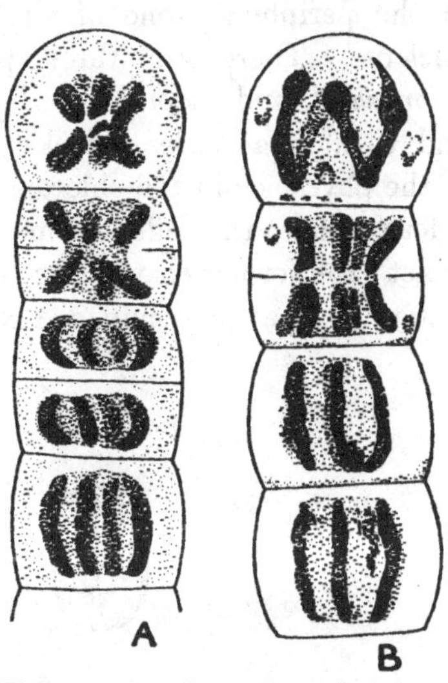

Fig. 7. *A* and *B*, the ends of two trichomes of *Tolypothrix lanata* (Desv.) Wartm., showing successive stages of supposed mitosis. Fixed in sulphurous acid and stained with iron-ammonia-alum hæmatoxylin. × 1160 (after Kohl).

The nuclear mitosis described by Kohl (consult fig. 7), Olive, and Phillips is probably of the nature of an illusion since it is quite easy to imagine the presence of mitotic figures in stained preparations of the dividing cells of many of the Blue-green Algæ. There is, however, such a mass of evidence to the contrary that the mitotic figures of supposed dividing nuclei published by these authors must be considered as the result of unconscious self-deception.

That the central differentiation of the Myxophycean cell can be regarded as an incipient nucleus is fairly clear, but it is not at all clear to what extent the nuclear functions are localized. The mere fact that division of the protoplast may begin by the ingrowth of a ring-shaped septum before there is any indication of division in the primitive nucleus seems to show that that structure has not yet acquired all the physiological attributes of a true cell-nucleus.

THE PIGMENTS AND THE CYTOPLASM. In the majority of the Myxophyceæ the cells are of a rich blue-green colour, the pigments being located in the peripheral zone of the protoplast which surrounds the incipient nucleus. There are three pigments of importance, viz. *phycocyanin, chlorophyll,* and *carotin,* of which the first is generally the most abundant and to a great extent masks the chlorophyll. It is the combined effect of the phycocyanin and chlorophyll which gives the characteristic 'blue-green' colouration to the cells. Some forms are reddish-pink in colour and others violet; the former colouration is due to the predominance of carotin, and the latter to the predominance of both phycocyanin and carotin over the

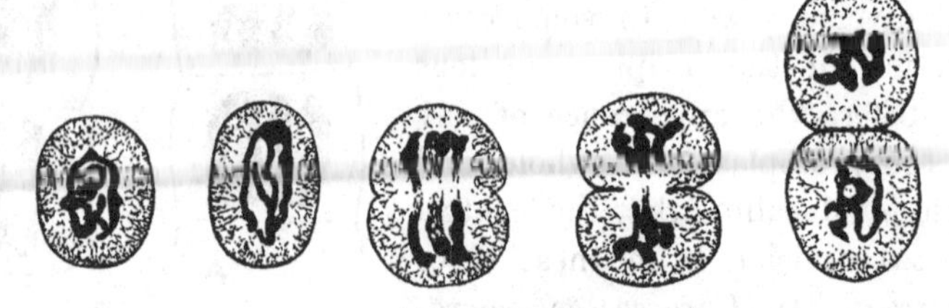

Fig. 8. *Synechocystis aquatilis* Sauvageau, showing various stages of cell-division. Very highly magnified (after Gardner).

chlorophyll. One form of carotin has been named *polycystin* by Zopf; and another form, which is identical with Sorby's *pink phycocyanin,* has been named *myxophycin* by Chodat. All these pigments can be extracted by appropriate methods (consult Lemmermann, '07, p. 9). The phycocyanin is soluble in water, and after rapidly killing the cells, may be easily obtained as a brilliant blue solution from which it can be crystallized.

It has long been known that the Myxophyceæ occurring in deep water are for the most part tinted red, and Gaidukov ('04—'06) has suggested that this is a complementary chromatic adaptation due to the fact that the quality of the light is affected by the depth of the water through which the sun's rays have to pass. He found that a species of *Oscillatoria* changed its colour when grown behind coloured glass or coloured solutions, and that the change was always in the direction of taking on the colour complementary to the light in which it was placed. This complementary colour was only assumed after a series of colour changes, and when once acquired was retained for months after the Alga had been restored to white light.

The changes of colour are due to modifications of the phycocyanin, which can be regarded as the 'adaptational pigment.' Gaidukov states that chlorophyll is not the only colouring matter useful in the photosynthetic process in those Algæ which grow in deep water. Sauvageau ('08) does not altogether agree with these views, and states that, although the rose-colour of certain Myxophyceæ is due to a transformation of the phycocyanin, such transformation is caused solely by the diminished intensity of the light. Most of the authors who have dealt with this question appear to have lost sight of the fact that red and rose-coloured Myxophyceæ are not confined to deep waters. For instance, *Oscillatoria rubescens,* which sometimes gives a red colouration to large lakes, is purely a plankton species, and so is *Trichodesmium erythræum,* which colours the Red Sea. Moreover, in fresh waters the red plankton species, which include certain of the Chroococcaceæ, are mixed with others of a rich blue-green colour. There are also quite a number of rose-coloured and purple filamentous forms which live in subaërial habitats.

The chlorophyll appears to be lodged in the cytoplasm in very minute granules, and the carotin, when present, appears to be similarly disposed; but some doubt has been expressed as to whether the phycocyanin exists in a diffuse state or is located in the same granules as the chlorophyll. It seems probable, however, that it exists in a diffuse state, since it is a water-soluble protein pigment of a similar nature to the red pigment of the Rhodophyceæ. In any case, the pigments are restricted to that portion of the protoplast surrounding the incipient nucleus, and the exact morphological interpretation of this pigmented zone has proved a matter of considerable difficulty. In *Glaucocystis* there are true chromatophores, but the presence or absence of such structures in the rest of the Blue-green Algæ has given rise to much discussion.

Borzi ('86) could not demonstrate the presence of a definite chromatophore, and both Stockmayer ('94) and Zacharias ('91; '00) have each declared that the pigmented part of the protoplast cannot be considered as a true chromatophore, but only as a coloured plasma. Macallum ('99) also concluded that there was no evidence of a special chromatophore.

In contrast to these statements, Deiniga ('91) found structures in the cells of *Nostoc* and *Aphanizomenon* which had the form of more or less reticulated plates in contact with the cell-wall, and he regarded them as true chromatophores. Zukal ('92) also considered that *Tolypothrix lanata* possessed a chromatophore, which consisted of a definite, demarcated part of the protoplasm, saturated with the characteristic colouring matter of the plant. Fischer ('97) stated that by means of hydrofluoric acid he could dissolve all the protoplast except a hollow, cylindrical or barrel-shaped structure containing colouring matter, and this he therefore considered as the chromatophore. Hegler ('01) found that the peripheral protoplasm was packed with granules, and he stated that each granule contained both chlorophyll and phycocyanin. He therefore regarded the granules as chromatophores and termed them *cyanoplasts.* Kohl ('03) also regarded these coloured granules as separate chromatophores, and stated that each granule may contain chlorophyll, carotin, and phycocyanin in variable proportions. Both Olive and Gardner concluded that there was no definitely organized chromatophore. Hieronymus ('92) described the cell as possessing a chromatophore consisting of small granules of chlorophyll strung like beads on a network of threads which extended through the peripheral part of the protoplast in which the phycocyanin was dissolved.

Nadson ('95) describes the peripheral cytoplasm as vesiculated, and states that the chlorophyll and phycocyanin are contained in the walls of the alveoli. The pigmented portion he regarded as functioning both as cytoplasm and chromatophore. Massart ('02) considered that the pigmented layer, although morphologically not a true plastid, yet functioned as such. Phillips ('04) also regarded the outer pigmented zone as one in which the functions of the chromatophore were not yet divorced from the functions of the cytoplasm.

The location of the pigment in the cells of the Myxophyceæ does not appear to be entirely uniform throughout the group. The pigmented zone is more indefinite in some cases than in others, and instances have been recorded in which the pigment has extended into the incipient nucleus. It is scarcely possible, in view of the evidence, to accept Hegler's and Kohl's ideas that the numerous minute coloured granules are each of them chromatophores. There is convincing evidence that the Myxophycean protoplast is of a primitive character, and most of the modern investigations show that nothing of the nature of a true chromatophore has yet been evolved, except in the puzzling genus *Glaucocystis*. The pigmented zone undoubtedly functions as a chromoplast and at the same time carries on the normal functions of cytoplasm, no complete differentiation having as yet taken place between the plastid and the cytoplasm.

Cell-sap vacuoles do not normally occur in the protoplasts of all Blue-green Algæ, and when present it is highly probable that they are often due to pathological conditions. In *Oscillatoria decolorata*, which has degenerated into a saprophyte and lost all trace of colour, the protoplast is very much vacuolated. In the cytoplasm of normal cells of *Chroococcus macrococcus*, Miss Acton has found that small vacuoles are not at all uncommon, and that vacuolation begins in this species near the periphery of the protoplast and proceeds towards the nuclear region.

Inclusions in the protoplast. Within the protoplast of the Blue-green Algæ are several kinds of granules, which have been described by various authors under different names. Many conflicting statements have been made concerning the staining of these granules, although much of the discrepancy may have been due to differences in fixation, and to the use sometimes of alkaline and sometimes of acid reagents. There are six inclusions deserving of special mention.

(1) *Central granules.* These are found scattered among the meshwork of the incipient nucleus (the so-called 'central body'), and they stain well

with methylene blue or with an aqueous solution of Bismarck brown, and very deeply with hæmatoxylin. They are identical with the 'α granules' of Gardner ('06) and appear to be equivalent to A. Mayer's red granules of volutin. They are of an albuminous nature and are insoluble in hydrochloric, nitric, or sulphuric acid; they are digested in about twenty-four hours with artificial gastric juice. They give a phosphorus reaction and contain masked iron. They are larger and more conspicuous than the nucleo-protein granules (presumed chromatin granules) in the threads of the meshwork of the primitive nucleus, and it is not unlikely that they also consist largely of some nucleo-protein.

Fischer ('05) has described the occurrence of what he terms 'anabænin' in the 'central body' of various filamentous forms of the Myxophyceæ. This substance he apparently

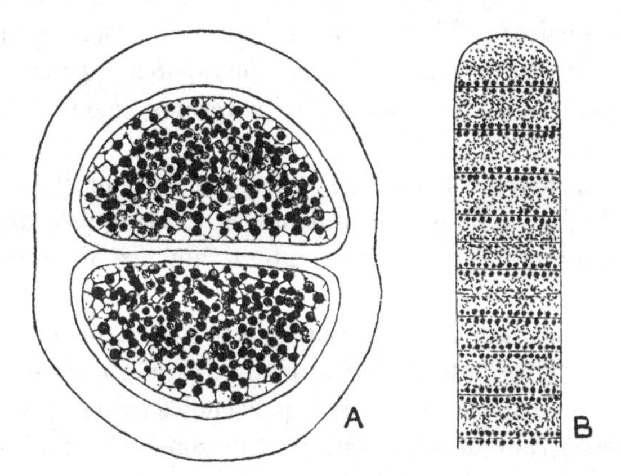

Fig. 9. *A*, *Chroococcus* sp. [? *Chr. minutus* (Kütz.) Näg.] stained with brilliant blue to show the cyanophycin granules. × 1200 (after a drawing by Miss Acton). *B*, *Oscillatoria limosa* Ag., end of a filament showing prominent cyanophycin granules on each side of the transverse walls. × 800.

identified with the central granules of certain other authors, but the identification is exceedingly doubtful as he states that anabænin is a carbohydrate and probably a transformation product of glycogen. Fischer's observations with regard to anabænin can scarcely be reconciled with the published statements of other investigators of the Blue-green Algæ. It is possible that anabænin is identical with the substance contained in the mucus-vacuoles, as that substance is of a carbohydrate nature and the vacuoles are often aggregated around the central body.

(2) *Cyanophycin granules.* These granules were so named by Borzi ('86), who considered them as a food reserve. They are identical with the 'reserve granules' of Bütschli and of Nadson, and with the 'β granules' of Gardner. They occur only in the cytoplasmic zone, more especially in its outer portion near the cell-wall, and in the Oscillatoriaceæ they are often very conspicuous on each side of the transverse walls of the filament (fig. 9,

A and *B*). They are small colourless granules, very refractive in appearance, and their outward form is either round or subangular. They give reactions for both nitrogen and phosphorus, and consist of albuminous protein in a crystalloidal form. They swell up and dissolve in 0·5 per cent. hydrochloric acid, and also in dilute nitric or sulphuric acids. In artificial gastric juice they are completely digested in about one hour. They give no reaction for iron. They have a special affinity for picro-carmine, stain blue-violet with Ehrlich's hæmatoxylin, a faint blue with Delafield's hæmatoxylin, and a bright red after a prolonged treatment with an alcoholic solution of eosin. Cyanophycin granules entirely disappear when the albumen-consumption of the protoplast is great, and their number and presence depend largely upon physiological conditions. They are absent in dividing cells, and they gradually disappear in the dark or when the vitality of the protoplast is impaired. They occur in large numbers in spores, but are consumed during the germination of the spores. There is little doubt that cyanophycin granules constitute a reserve food of an albuminous character.

(3) *Mucus-vacuoles.* These are the slime-balls of Palla and the slime-vacuoles of Hegler. They are difficult to distinguish from cyanophycin granules, but are larger, and to some extent they are differentiated by stains. It seems likely that they contain a substance allied to mucin, which is insoluble in hydrochloric acid, stains a red-violet[1] with hæmatoxylin, and red with eosin. This substance is apparently a glucoprotein and gives a carbohydrate chemical reaction. Mucus-vacuoles cannot always be detected, and when present they mostly occur in the zone of cytoplasm surrounding the primitive nucleus, although they may extend further outwards.

(4) *Glycogen and Sugar.* Glycogen was first mentioned by Bütschli as occurring in the Myxophycean cell, and it was shown to exist by both Hegler and Kohl. Gardner proved its existence experimentally, and also that sugar was present in many of the Blue-green Algæ. Aqueous extracts of many kinds of Myxophyceæ after pulverization with fine sand readily reduce Fehling's solution, and it seems probable, therefore, that sugar is the first product of photosynthesis in these plants, and that some of this sugar is converted into glycogen as a carbohydrate reserve.

Starch is entirely absent from the Myxophyceæ.

(5) *Oil.* Minute oil-drops may occur in the cytoplasm of many forms of the Blue-green Algæ, and can be detected by osmic acid. They have also

[1] Mucus-vacuoles are said by Phillips ('04) to stain *blue* with Delafield's hæmatoxylin, but it is necessary to point out that the colour of stained granules of this character is very misleading. A small refractive sphere of dense mucus stained with hæmatoxylin may appear of various shades of red, blue or violet according to the focus and the colouration of the protoplast in which it is embedded.

been observed in the heterocysts and in the germinating spores of *Glœotrichia* and *Anabæna*.

(6) *Pseudovacuoles.* Certain of the free-floating Myxophyceæ, more especially of the genera *Glœotrichia, Anabæna, Aphanizomenon, Cœlosphærium, Microcystis,* etc., which sometimes occur abundantly in the plankton, contain 'granules' of a more or less dark red colour scattered through the cytoplasm. These granules were originally discovered in *Glœotrichia echinulata* by P. Richter ('95), who at first described them as granules of amorphous sulphur, but afterwards regarded them as an optical illusion due to an alveolated protoplast. The red appearance, which is not very pronounced, is probably caused by diffraction. Klebahn ('95) and others have concluded that they are *gas vacuoles,* directly concerned with the floating capacity of the Algæ which possess them. They often appear almost black, and are removed by treatment with absolute alcohol or chromic acid. Klebahn, experimenting with *Glœotrichia,* and Molisch ('03) with *Aphanizomenon,* both state that the granules are not removed on placing the water containing the Algæ in a vacuum. Molisch, therefore, rightly concluded that they cannot be gas vacuoles, but expressed the opinion that they had a specific gravity less than that of water, and he termed them *suspensory bodies.* Lemmermann ('07), on the other hand, points out that those Blue-green Algæ containing 'pseudovacuoles' can be caused to sink by keeping them for a long time in a mixture of alcohol and glycerin, or in 2—4 per cent. formalin, without losing these bodies; and further, that some blue-green forms which are always fixed to a substratum also possess 'pseudovacuoles.' Fischer attributes these bodies to the optical effect of his so-called 'pseudomitosis of anabænin'; but if so, all Blue-green Algæ should possess 'pseudovacuoles,' which is not the case.

It is obvious from the foregoing remarks that our information concerning 'pseudovacuoles' is altogether insufficient, and that statements as to their nature are as yet largely conjectural. They are chiefly found in the Myxophyceæ of the plankton, organisms which are exposed to relatively intense light. They also appear in the hormogones of *Nostoc, Phormidium,* and *Lyngbya,* especially in those parts of the stratum exposed to the brightest light. In view of the fact that most Myxophyceæ without 'pseudovacuoles' live to a great extent in feeble light, either as shade plants or by reason of their strongly coloured sheaths and integuments, Lemmermann has suggested that the 'pseudovacuoles' are possibly protective bodies against light of too great intensity.

PROTOPLASMIC CONTINUITY. In some of the filamentous Myxophyceæ there exists a protoplasmic continuity between the cells. Wille ('84) was the first to point this out in *Stigonema compactum* var. *brasiliense.* Borzi ('86) described protoplasmic connections in species of *Nostoc* and *Anabæna*, and Nadson ('95) figured them in *Aphanizomenon* and *Tolypothrix*. Phillips ('04) stated that every multicellular blue-green organism studied by him was found to have intercellular protoplasmic continuity, although the protoplasmic connections were difficult to demonstrate in *Oscillatoria*. He considered that the fine protoplasmic threads passed through communicating pores in the cell-walls (fig. 10), and that there was usually one central pore, although other finer pores were sometimes present. The protoplasmic continuity described by Fritsch ('04) as occurring between the cells of *Anabæna*, was subsequently stated by that author ('05) to have been an error of interpretation, and that the connecting threads consisted only of the intercellular portion of the investment. This may also be true with regard to the supposed protoplasmic connections between the cells of *Nostoc* and other genera. Gardner ('06) entirely failed to demonstrate any protoplasmic continuity between the cells in the various forms he examined.

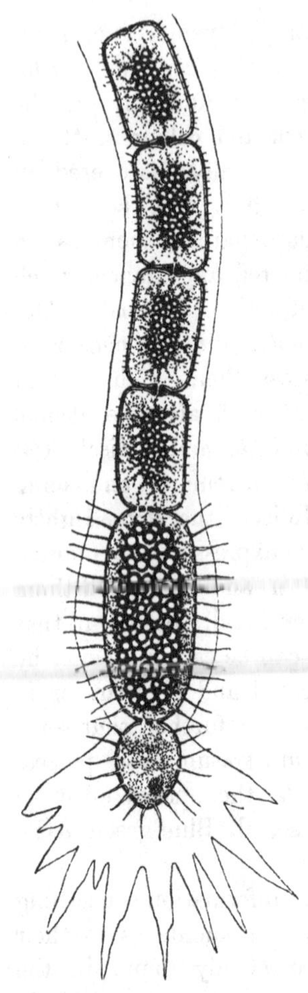

Fig. 10. End of trichome of *Cylindrospermum* sp. showing the terminal heterocyst and spore furnished with hair-like processes, and the gelatinous sheath frayed around the heterocyst. The protoplasmic connections are also shown through the communicating pores in the end-walls of the cells. Very highly magnified (after Phillips).

In certain species of *Stigonema* the protoplasmic connections are rather more conspicuous than in other blue-green forms, a condition being presented which is precisely similar to the protoplasmic continuity exhibited by the Rhodophyceæ (W. & G. S. W., '97 A, p. 242; G. S. W., '04). The continuity is effected by a single protoplasmic strand connecting the polar extremities of adjacent cells. The strands are by no means small and they pass through centrally-placed pores in the transverse walls of the filaments. The pores are best seen in the young branches of *Stigonema ocellatum*, particularly if plants previously dried are soaked in water, or if the protoplasts are caused to shrink by means of reagents.

Heterocysts. In the filamentous families of the Nostocaceæ, Scytonemaceæ, Stigonemaceæ, and Rivulariaceæ, there are certain differentiated cells known as *heterocysts*. They may be sparsely scattered in an intercalary manner between the vegetative cells of the filament (fig. 11 *C*), or they may be terminal (fig. 11 *F* and *G*). Sometimes they are seriate as in some species of *Tolypothrix*, *Calothrix*, *Nostoc*, etc. As a rule, they are a little larger than the vegetative cells, and in most cases stand out conspicuously among the other cells of the filament. There is almost

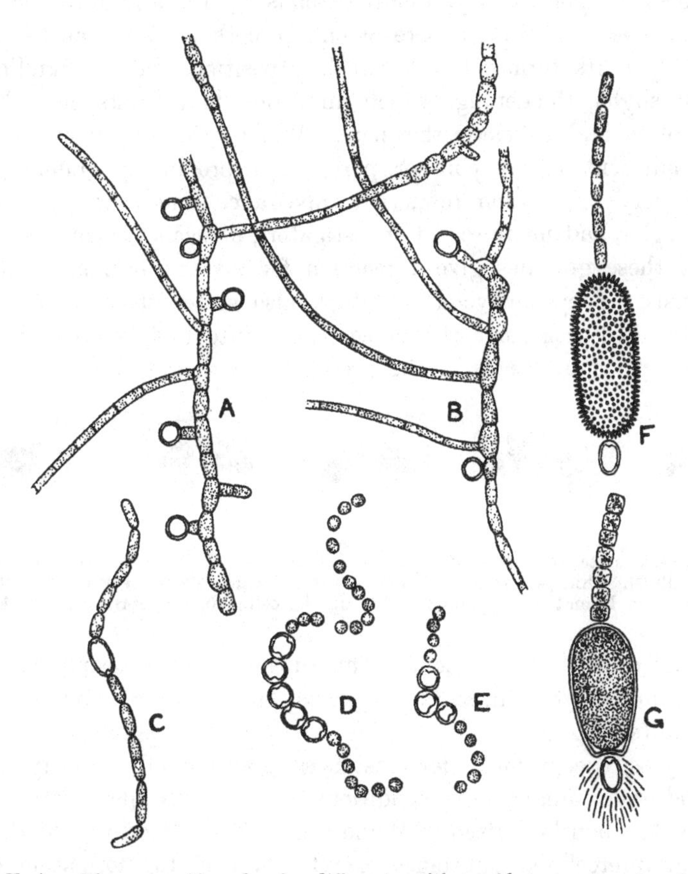

Fig. 11. Various Blue-green Algæ showing different positions of heterocysts. *A* and *B*, portions of two branched filaments of *Nostochopsis Goetzei* Schmidle, with stalked and sessile lateral heterocysts. *C*, fragment of filament of *Nostoc carneum* Ag., showing intercalary heterocyst. *D* and *E*, small portions of filaments of *Nostoc antarctica* W. & G. S. West, showing seriate intercalary heterocysts. *F*, *Cylindrospermum tropicum* W. & G. S. West with basal heterocyst and papillate spore. *G*, *Cylindrospermum indentatum* G. S. West, with basal heterocyst and smooth indented spore. In all the figures the pellucid heterocysts are represented without shading. The hair-like appendages around the heterocyst in *G* are slender bacteria radiating in the colourless mucilaginous envelope. *A*, *B* and *F*, ×520; *C*—*E*, and *G*, ×590.

a complete absence of pigment from the heterocysts, in consequence of which they present a very pellucid appearance, although they are often of a pale yellow-brown or yellow-green colour. The walls of fully-formed heterocysts are thicker than those of the ordinary vegetative cells and they are composed of cellulose.

Heterocysts are formed from young vegetative cells by the disintegration of the incipient nucleus and the gradual assumption of a homogeneous character by the whole of the cell-contents. The thickening of the wall commences at the poles and gradually extends equatorially until the wall is of equal thickness. There is a pore at one or both poles of the heterocyst, depending upon its terminal or intercalary position, and surrounding the pore is a slight thickening, which in older heterocysts may become a minute plate and so close the pore. Within the heterocyst and immediately adjacent to its pore or pores is a prominent granule, which presents a very bright and refractive appearance, and consists, according to Borzi, Hegler, and many recent investigators, of cyanophycin. Macallum states that these granules give a reaction for masked iron and therefore cannot consist of cyanophycin. Phillips also states that the heterocyst is gradually filled with some substance passed into it from the other cells through the pores by means of the protoplasmic threads which connect it

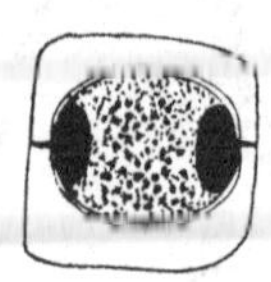
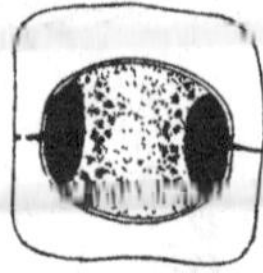
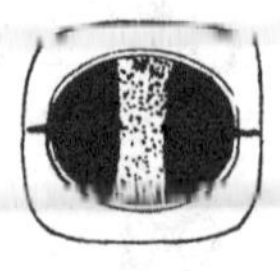

Fig. 12. Successive stages in the formation of a heterocyst of *Nostoc* sp. Greatly magnified (after Phillips). Phillips states that some substance which stains very deeply with iron-ammonia-alum hæmatoxylin gradually fills up the cell; this is represented black in the figures.

with adjacent cells of the filament. This substance stains deeply with iron-ammonia-alum hæmatoxylin and he suggests that it may possibly be related to chromatin (see fig. 12).

Heterocysts, except for a few instances, are normally solitary, but in cultures and under unfavourable conditions they may become seriate. Brand ('03), and subsequently Fritsch ('04) and others, have described the development of an intercellular substance excreted during the formation of the heterocysts. This substance, however, is not necessarily excreted by cells which are being transformed into heterocysts, as it can be observed remote from the heterocysts in *Scytonema*, and it is also excreted by the cells of certain species of the Oscillatoriaceæ, a family in which heterocysts do not exist.

The exact nature of heterocysts has long been a puzzle. They were thought by Borzi ('78) and Hansgirg ('87) to serve as limitations to the length of the filaments, and they are sometimes, more especially in young filaments and in cultures, concerned with the breaking of the filaments. In this connection, it should be remembered, however, that in normal plants of the genus *Anabæna* the filaments break readily at all points, and this fracture is in no way controlled by the heterocysts. Moreover, in *Nodularia*, the fracture of the filaments is almost invariably between the heterocysts. Neither does the structure of a filament of *Stigonema* or *Scytonema* support this view, although in the former genus heterocysts always limit the hormogones. Hieronymus ('92), Hegler ('01), and later, Fritsch ('04), have regarded the heterocysts as storehouses for reserve substances, the latter passing into the heterocyst along the protoplasmic threads which communicate with the adjoining cells. Phillips ('04) has suggested that as the heterocyst is usually next or near to the spores it might possibly be considered as a storehouse of food for them. He also states that the heterocyst of *Cylindrospermum* will develop into a spore if it gets sufficient nutriment and hereditary material passed into it from the other cells. The fact that spores occur abundantly quite away from the heterocysts, as in *Hapalosiphon*, *Scytonema*, and certain species of *Anabæna*, does not lend much support to this view; nor does the fact that in one filament there may be twenty, or even thirty, times as many spores as heterocysts.

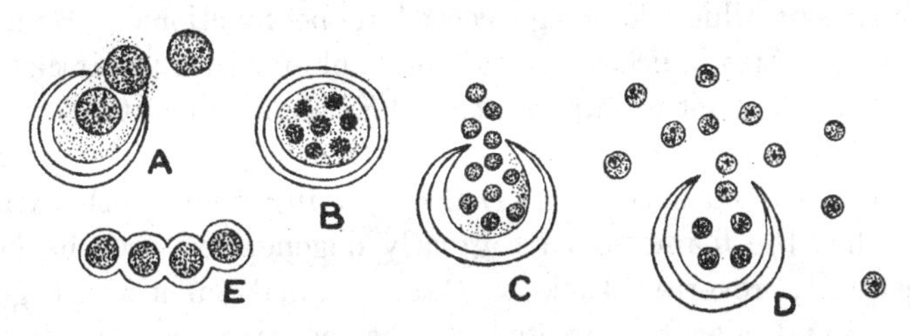

Fig. 13. *A—D*, formation of gonidia by the heterocysts of the aberrant *Anabæna cycadearum* Reinke. In *B—D*, a number of gonidia are being formed; in *A*, only three. *E*, young filament developed from a gonidium. All × 2200 (after Spratt).

The remark made by Spratt ('11) that in *Anabæna* detached heterocysts 'certainly occur under both natural and artificial conditions' is erroneous as a general statement. Detached heterocysts never occur in the heterocystous Myxophyceæ living under *normal* natural conditions. The species examined by Spratt was *Anabæna cycadearum* Reinke, which in addition to being the most specialized species of the genus, having adapted itself to conditions of environment totally unlike those under which species of *Anabæna* normally live, is also a degenerate form.

Brand ('01) observed the contents of the heterocysts in *Nostoc commune* and *N. microscopicum* set free as gonidia, which subsequently developed into new filaments. Fritsch has observed states which he thought might be due to arrest of gonidia-formation in the heterocysts of a species of *Anabæna*. These observations were made on material in *cultures*. Spratt ('11) has also observed the formation of gonidia in the heterocysts of *Anabæna cycadearum*, 'in material from old nodules and hanging-drop cultures.' The contents divide first into two, and eventually into a number of small spherical gonidia within the original walls of the heterocyst. These are ultimately liberated, and when set free they possess a very delicate membrane.

Each gonidium may rest for a time or germinate at once to form a new filament.

In the heterocystous Myxophyceæ of to-day that are living under normal conditions of environment, the heterocysts appear to have no function other than that of limiting the hormogones in certain types, and probably of limiting the filaments themselves in the younger or adult stages of certain other types[1].

It is not improbable that heterocysts sometimes act as storehouses for reserve materials of a protein nature, but there is no evidence to show that this normally occurs under natural conditions. It is mostly in cultures that such storage takes place. The position of the spores in the Rivulariaceæ and in certain species of the genus *Anabæna*, has led to the suggestion that heterocysts are in some way connected with spore-formation, but this is exceedingly doubtful. It has been suggested (Fritsch, '04; G. S. W., '04) that heterocysts are probably the lingering and abortive relics of a type of reproductive organ (gonidangium) once possessed by certain of the Myxophyceæ, but which long ago ceased to be functional. Brand's observations gave the first indication of this probable function; Fritsch's evidence, although slight, was not unimportant; and Spratt's observations on *Anabæna cycadearum* are particularly important in this respect. *A. cycadearum* is a profoundly modified species of *Anabæna*, living under such extraordinary conditions that the filaments have greatly degenerated and the heterocysts have apparently reverted back to their original function of gonidangia. Attention should also be directed to the position of the heterocysts in *Nostochopsis* (fig. 11, A and B). They are lateral, and either stalked or sessile, being strangely reminiscent of the position of gonidangia in many of the Green Algæ.

SPONTANEOUS MOVEMENTS IN THE OSCILLATORIACEÆ. Certain Blue-green Algæ of the family Oscillatoriaceæ exhibit spontaneous movements, generally of a slow oscillating, gliding, or rotatory character. It is in the genus *Oscillatoria* that these movements are most easily observed, but although they have been repeatedly investigated, no convincing explanation has yet been offered. The movements exhibited by some species of *Oscillatoria* are of three kinds: a gliding or creeping movement by which the whole filament travels slowly through the water in a serpentine manner, either forwards or backwards; a slow oscillation of a considerable portion of the apical part of the filament; and a rather more rapid bending of the extreme apex which generally closes each oscillation.

[1] Young *Nostocs* are always limited by heterocysts.

In the spirally twisted *Arthrospira Jenneri* the bending of the entire filaments is more pronounced, and the oscillation of their spiral extremities is more vigorous than in *Oscillatoria*, although spasmodic and jerky (G. S. W., '04). In *Spirulina*, in which the filaments are twisted into a very close spiral, there is a well-defined rotation around the axis of the spiral, accompanied by a decided propulsion of the whole filament. In the species with very long filaments the latter is often a steady and by no means a slow creeping movement. In *Spirulina turfosa* Bulnh., in which the filaments are relatively short, the rotatory motion is fairly rapid. The hormogones of many of the Hormogoneæ exhibit slow gliding or creeping movements. It has also been stated both by De Bary ('63) and Phillips ('04) that the filaments of *Cylindrospermum* are capable of similar movements. All these movements are quicker in bright illumination and at warm temperatures than they are in weak light and at cool temperatures.

The earliest attempts to explain the cause of the movement were by Nägeli ('49), who considered it to be due to osmotic currents between the cells and the surrounding water; and Siebold ('49), who regarded it as resulting from an extensive secretion of colourless mucus. Borzi ('86) offered no explanation, but pointed out that isolated cells were incapable of movement, and that no movements were exhibited by any form which possessed heterocysts. Cohn ('67), and also Correns ('96), have stated that the movements could only take place when the filaments were in contact with a substratum, and it is not improbable that in certain species contact with a solid body or with the surface-film of water is actually necessary. Hansgirg ('83) concluded that the movements were due to osmotic currents, and that they only took place within a very thin gelatinous sheath, which was secreted by the filament and attached to a substratum. Engelman ('79) found the thin mucous secretion described originally by Siebold, but interpreted it as a protoplasmic layer. He attributed the creeping movements of the filaments to this secretion, but the oscillating movements he regarded as due to the action of cilia similar to those which occur in the Bacteria.

The movements are in no way connected with growth, as was suggested by Wolle and others, because all the facts are entirely against such a conclusion (*vide* Phillips, '04, p. 316). Nor can a strict comparison be made between these movements and those exhibited by individual diatom-cells. The movements of the latter are now known to be connected with the raphe (*vide* p. 102), but no such structure exists in the much more primitive cells of the Myxophyceæ. Phillips ('04), after carefully reviewing the evidence and adding his own observations, suggests that there are only two plausible explanations of the movements under discussion: either (1) it is effected by a protoplasmic pellicle which creeps along on a substratum and acts in a peristaltic manner, or (2) the plant has some propelling organs such as flagella, cilia or pseudopodia, that either act upon the solid substratum or move freely through the water. One must agree with Phillips that the first of these possible explanations is quite inadequate to account for all the movements exhibited by the Oscillatoriaceæ. The second hypothesis is, however, at present unsupported by sufficient testimony for it to be accepted with any degree of confidence.

In certain species of *Oscillatoria* and *Phormidium* there have been described fine hair-like outgrowths from the terminal cells (see fig. 14). These have been regarded by many observers as parasitic or saprophytic Bacteria of the nature of *Ophiothrix*, but Phillips ('04, p. 320) asserts that those outgrowths observed by him were living portions of the algal filament which had grown out from the cell-protoplasm. He was unable to assign a definite function to these outgrowths, which he states at first stain like the protoplasm of the terminal cell, but afterwards react in the same manner as the cell-walls. He considered that they probably acted as tactile organs in piloting the trichome around obstacles, but that they were not the cause of the movements. These filaments were also found in *Cylindrospermum* (fig. 10).

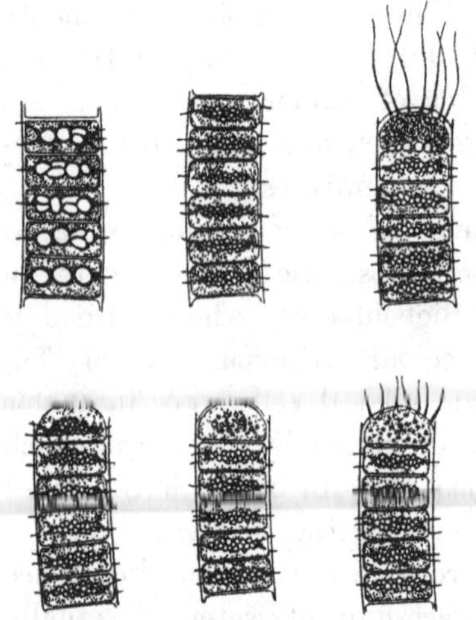

Fig. 14. Portions of filaments of *Oscillatoria* showing the hair-like outgrowths from the end-cells, and also the lateral cilia described by Phillips. Highly magnified (after Phillips).

Phillips states that by using Bunge's mordant, followed by carbol-fuchsin, a method which brings out the cilia of Bacteria very well, he was able to demonstrate the presence of fine cilia along the sides of the trichomes (see fig. 14). These cilia were very short, and could only be seen with extreme difficulty in the living organism. He considers that the reason they have been overlooked by other observers is that they 'mass down when placed in reagents and appear as granules of foreign substance on the exterior of the trichome.' He also concludes that the presence of these cilia 'explains more clearly the moving of the particles of indigo along the trichome as described by Schulze and others, and the massing down of the cilia will explain why the contour of the trichomes is so rough oftentimes, as is especially shown when stained with Heidenhain's iron-ammonia-alum hæmatoxylin, with but slight or no destaining.'

It seems scarcely possible to accept Phillips' conclusions that the creeping movements of the Myxophyceæ are caused by very delicate, short cilia without much confirmatory evidence. Such confirmation is essential if only because Phillips' statements regarding the structure and division of the 'central body' are not in agreement with the most reliable cytological work done on the protoplasts of these plants.

MULTIPLICATION. The general increase of the unicellular and colonial Blue-green Algæ is by simple cell-division, which may occur in every direction of space or in certain directions only. Indefinite or definite colonies are thus produced, and these are often increased by dissociation into smaller groups. In the filamentous forms, fragmentation of the filaments is a frequent method of multiplication in certain genera. The division of the protoplast has already been dealt with, and brief mention has been made of the fact that in many families, such as the Chroococcaceæ, Nostocaceæ and Stigonemaceæ, the cell often divides by a gradually deepening constriction, whereas in others a septum is formed during the division of the cell. The retention of the products of division by persistent envelopes, such as occurs in *Glœocapsa*, etc., results in the formation of colonies of considerable size, and often of an extended stratum.

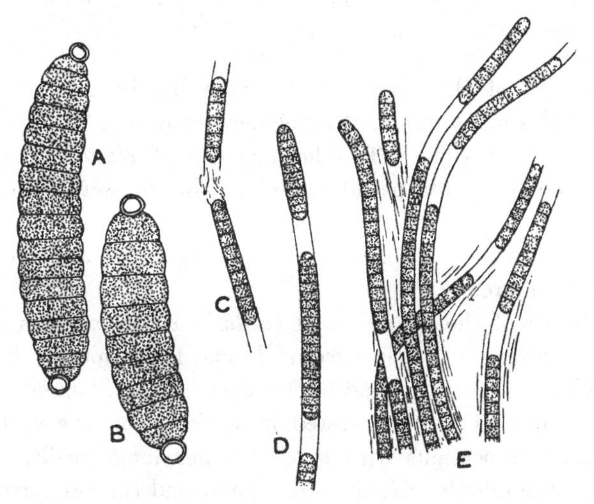

Fig. 15. *A* and *B*, hormogones of *Stigonema ocellatum* (Dillw.) Thur. *C*, hormogones of *Phormidium Corium* (Ag.) Gom. and *D* of *Ph. ambiguum* Gom. *E*, portion of extremity of erect tuft of filaments of *Symploca muralis* Kütz. showing formation of hormogones. *A*, *B* and *E*, ×520; *C* and *D*, ×500.

In the great section of the Hormogoneæ, the plants largely multiply by the formation of *hormogones*. These are short filaments of vegetative cells which are generally set free from the extremities of the main filaments and branches (fig. 15). They are capable of slow spontaneous movements; sometimes in a straight line, as in *Lyngbya, Nostoc, Scytonema, Rivularia* and *Stigonema*, or less often in a spiral manner, as in *Oscillatoria* and *Arthrospira*. The distance travelled by the hormogones of a species of *Phormidium* varied, according to Brand, from 2·8 μ to 34·8 μ in ten minutes. Hormogones can be considered as primitive multicellular gemmæ, and each one ultimately develops directly into a new plant by vegetative cell-division.

Sometimes the hormogone is terminated at each end by a heterocyst. Frequently entire filaments will break up into hormogones owing to the formation at intervals of biconcave discs of intercellular substance, each of which marks the limitations of the extremities of two consecutive hormogones. These *separation-discs* are much more frequent in some forms than in others. Each disc consists of an intercellular substance secreted by the two adjacent cells, and is often so thin in the middle as to appear almost like a ring. At first the discs are dark green, but they may become colourless. They are coloured yellow by chlor-zinc-iodine; they are not contracted by glycerin, nor do they stain with congo-red. (Consult Brand, '03; '05.)

Filaments are frequently broken, and thus multiplied, by the death of vegetative cells, which lose their turgor and are compressed by the neighbouring turgid cells. The dead cell finally decays, causing the dislocation of the filament. Such cells (called by Brand 'necridia') are strongly contracted by glycerin and stain with congo-red.

Quick growth of the filaments, accompanied by repeated hormogone-formation, enables many of the Hormogoneæ to form a considerable stratum in a relatively short time. The hormogones of *Nostoc, Phormidium* and *Lyngbya* frequently contain pseudovacuoles. In the developing hormogones of *Nostoc* the division of the cells produces a more or less spirally twisted filament within a firm and relatively wide integument, at each pole of which is a prominent heterocyst.

In two species of the Rivulariaceæ (*Calothrix adscendens* and *Glœotrichia natans*), Teodoresco ('07) has described the formation of hormogones of a peculiar kind. The upper piliferous part of the filament grows greatly in length and becomes differentiated into wider portions with thick, short cells, and narrower portions with much thinner, longer cells. A dislocation then occurs in the middle of each thin portion, and the two narrow dislocated ends grow past each other. In this way there may be formed several filaments, each gradually attenuated from the middle to the hair-like extremities, by which they are adherent. Soon the filaments completely separate and two adjacent vegetative cells in the thick middle part of each filament become transformed into heterocysts. The final separation occurs between these heterocysts, two complete *Calothrix*-filaments thus arising from each hormogone.

Asexual reproduction. Reproduction takes places asexually in many genera by the formation of gonidia or by resting-spores.

Gonidia occur in many genera of the Myxophyceæ, and are usually formed within gonidangia which have arisen from vegetative cells. They have long been known as the only method of increase in the various genera of the Chamæsiphoniaceæ, but their occurrence in certain genera of the Chroococcaceæ and the Hormogoneæ has only been demonstrated in relatively recent times. Apart from the Chamæsiphoniaceæ, they have been found in *Gomphosphæria aponina*, *Chroococcus macrococcus*, *Anabæna oscillarioides*, *Nostoc*, *Nostochopsis*, *Mastigocoleus*, *Leptochæte*, *Plectonema*, *Symploca*, *Lyngbya*, *Phormidium* and *Oscillatoria*.

In *Chamæsiphon* the gonidia arise by the formation of transverse walls beginning at the distal end of the elongated cell. The cells thus cut off are gonidia, which become rounded, and after they have burst through the

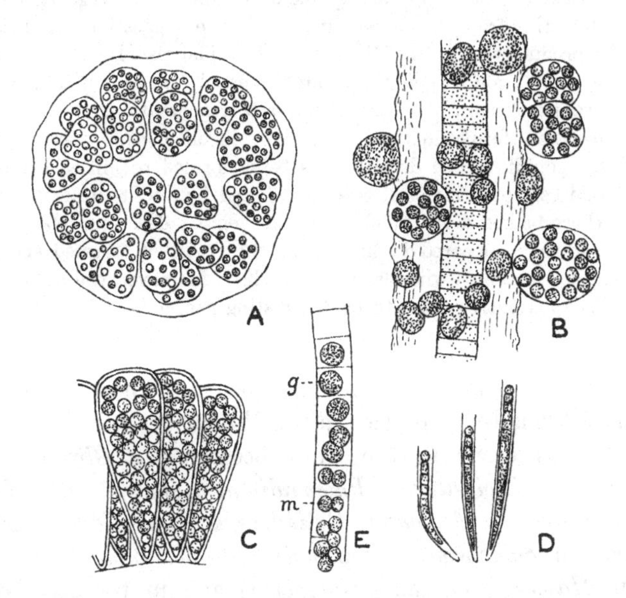

Fig. 16. *A*, colony of *Gomphosphæria aponina* Kütz. with minute gonidia in the cells; × about 500 (after Schmidle). *B*, *Xenococcus Schousbœi* Thur. attached to the rather wide gelatinous sheath of *Lyngbya semiplena* J. Ag.; × 1200. *C*, a few cells of an epiphytic colony of *Dermocarpa prasina* (Reinsch) Born. & Thur., showing gonidia; × 800. *D*, three cells of *Chamæsiphon gracilis* Rabenh.; × 500. *E*, part of filament of *Phormidium autumnale* (Ag.) Gom. showing gonidia (*g*) and microgonidia (*m*); × about 1000 (after Brand).

delicate sheath at the distal extremity they are gradually set free. The mother-cell may continue to grow, so that the gonidia are ultimately developed by abstriction from the free end of the protoplast[1]. In *Godlewskia* the gonidia arise by both transverse and longitudinal divisions of the protoplast of the mother-cell; and in *Pleurocapsa*, *Xenococcus*, *Hyella* and *Dermocarpa* the divisions are in three planes at right-angles or in many intersecting planes. A similar kind of division also occurs during the formation of gonidia in *Gomphosphæria* (Schmidle, '01).

[1] The cytology of this process still awaits investigation and should prove very interesting.

In *Phormidium*, Brand has described the formation, by further divisions, of microgonidia (fig. 16 *E*), and Lemmermann ('02) has observed similar microgonidia in *Plectonema capitatum*. Solitary gonidia have been described by Brand ('03) as occurring in *Phormidium autumnale* (= *Ph. uncinatum*).

The gonidia, such as those described above, usually germinate at once without any resting period.

Spratt ('11) found that the heterocysts of *Anabæna cycadearum* formed a variable number of gonidia by the rejuvenescence and subsequent division of their contents. The gonidia were spherical and each one was capable of forming a new vegetative plant. This observation is interesting from the support it lends to the suggestion that heterocysts are merely relics of reproductive organs of the nature of gonidangia. It must not be forgotten, however, that the production of gonidia is not a normal function of heterocysts, and in this particular case it may simply be reversion due to degeneration.

Fritsch has observed the production of 'gonidia' in old culture material of *Anabæna azollæ*. One large gonidium was described as being formed within each mother-cell by rejuvenescence, and before liberation it was surrounded by a well-marked membrane. It is evident that these bodies differ much from the gonidia described above, and Fritsch himself states that they appeared to him to be the result of arrested spore-formation, which is certainly not the case with ordinary gonidia. Moreover, the so-called 'gonidia' of *Anabæna azollæ* apparently passed through a resting period.

Reproduction by zoogonidia does not take place in the Myxophyceæ[1], and the few motile blue-green unicells which are known to exist appear to be Flagellate forms with no place in the Myxophyceæ.

Resting-spores occur in the Chroococcaceæ (in *Glœocapsa*), in the Oscillatoriaceæ (in *Oscillatoria*, *Phormidium*, *Lyngbya* and *Microcoleus*), in the Nostocaceæ (in *Anabæna*, *Cylindrospermum*, *Nostoc*, etc.), in the Scytonemaceæ (in *Scytonema*, *Tolypothrix* and *Microchæte*), in the Stigonemaceæ (in *Hapalosiphon* and *Stigonema*) and in the Rivulariaceæ (in *Calothrix* and *Glœotrichia*). In certain genera of the Nostocaceæ and Rivulariaceæ resting-spores are formed with sufficient regularity to become of specific importance.

The spores arise in all cases from vegetative cells, which generally increase in size, become largely filled with reserve materials, and develop two distinct membranes, the *endospore* and the *exospore*. The latter is the stronger membrane. It is often coloured yellow or brown, and sometimes, as in *Cylindrospermum majus* and *C. tropicum* (fig. 11 *F*), it has a

[1] Zukal ('94) has described the formation of small motile bodies of two sizes in *Tolypothrix lanata*, which he termed 'gametes,' since they associated themselves in pairs, a large one with a small one, although he was not certain that they fused. Goebel, also, has described zoogonidia in *Merismopedia*. Neither Zukal's nor Goebel's observations are in any way substantiated by the investigations of other authors, and for the present they must be regarded as exceedingly doubtful. It seems highly improbable that 'gametes' could occur in any group of organisms in which the nuclear differentiation is imperfect.

papillate exterior. The details of the development of the spores have been described by various authors in a number of genera. In the Hormogoneæ any vegetative cell of the filament may become a spore[1], or, as in the Rivulariaceæ, only the cell next the basal heterocyst may become thus differentiated. In some species (about 30 per cent.) of the genus *Anabæna* the spores arise in relation to the heterocysts, being developed from those cells adjacent to the heterocysts on either one or both sides[2]. In *Hapalosiphon* and *Scytonema*, in which the vegetative filaments are enclosed in strong sheaths, there is no enlargement of a vegetative cell on its conversion into a spore. The only obvious change is the development of a thick brown wall (W. & G. S. W., '97). In these genera spore-formation is in no way related to the heterocysts; nor, of course, can it be in the five genera of the Homocysteæ in which resting-spores are known to occur.

During the development of the spore, particularly if there is any considerable increase in size, much reserve material accumulates in it. According to Phillips ('04) this is largely passed into the spore from the adjacent 'nurse' cells by way of the intercellular pores. The two spore-membranes can be seen at a very early stage, and as development proceeds they gradually become more and more differentiated. The inner membrane (endospore), which from the first completely envelops the protoplast, remains colourless. The outer membrane (exospore) is at first only a cylindrical sheath, but subsequently it extends around the two ends of the spore and becomes a strong membrane. In the mature spores the endospore and exospore are separated by a very thin layer of mucilage. Fritsch ('05) states that in *Anabæna azollæ* the endospore and exospore are merely the fully-developed inner investment and cell-sheath respectively, both of which in the mature condition completely envelop the protoplast. The resting-spores vary much in shape in the different genera and species. They may be spherical, ellipsoid, cylindrical, or more rarely sublunate.

[1] Phillips describes the spores of *Oscillatoria* as often arising from more than one cell, sometimes as many as four cells fusing together by the absorption of the intervening cell-walls.

[2] This development has been referred to as 'centripetal' in contrast to the so-called 'centrifugal' development in which spore-formation begins in cells distant from the heterocysts and gradually advances towards them. It would, however, be wise completely to discard the use of these terms in relation to spore-formation in *Anabæna*, and in any case the application of the terms should be reversed since the only obvious fixed points are the heterocysts. Even in those species of the genus which were at one time referred to *Sphærozyga* (Agardh, 1827; Ralfs, 1850) spore-formation does not always begin in the cells adjacent to the heterocysts and gradually recede from them [the method which would be more correctly termed 'centrifugal'!], but is frequently mixed in the same spore-aggregate. The terms, as used, would be to some extent correct if the filaments of *Anabæna* consisted of a chain of vegetative cells terminated at each end by a heterocyst, but this is not the case except in certain very specialized species, such as the plankton-forms *Anabæna circularis* G. S. W. and *A. Tanganyikæ* G. S. W. (*vide* G. S. W. '07; and fig. 19 *A—E*).

They are sometimes solitary, but more often seriate, and the longest chains of spores are those found in *Nodularia, Scytonema* and *Hapalosiphon.* In some species the spores are enveloped in an external gelatinous investment, which may be very thin, as in certain species of *Nostoc,* or thick and firm, as in some of the plankton-species of *Anabæna.* In *Anabæna Lemmermanni* the spores agglutinate in masses, which float in such quantity in the surface layers of water as often to give a deep colouration to a whole lake.

Before germination the spores undergo a more or less prolonged rest, which in *Hapalosiphon, Scytonema,* and the plankton-species of *Anabæna* extends over many months[1].

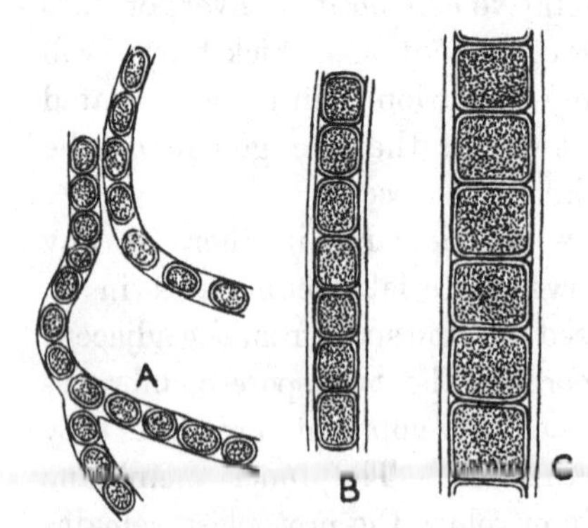

Fig. 17. *A*, resting-spores of *Hapalosiphon Welwitschii* W. & G. S. West, × 520; *B*, spores of *H. luteolus* W. & G. S. West, × 520; *C*, spores of *Scytonema coactile* Montagne, × 500.

The *germination of the resting-spores* has been followed out in a number of different forms and, although there is some variation in the actual mode of germination, in all cases it results in the more or less direct formation of the typical vegetative plant. This first-formed filament may in some cases break up into hormogones, each of which grows into a new plant. There are three main types of germination of the spores:

(1) The protoplast contracts from the wall of the spore and escapes either through a pore in the wall or by a lid-like portion of the wall becoming detached and so providing an orifice. The escape of the contents is a very gradual one and is primarily brought about by the secretion of mucilage, which forces the protoplast through

[1] Fritsch ('04) found that in *Anabæna azollæ* the spores can germinate as soon as they are fully formed, but this observation was made on material in cultures. It must also be borne in mind that *A. azollæ* is one of the greatly specialized forms of Myxophyceæ, and observations on this or any other equally aberrant species do not affect generalizations based upon normal members of the genus, or upon species of other genera. It is interesting to note in this connection that in another similar specialized species of *Anabæna* (*A. cycadearum*) the spores in culture-material are capable of immediate germination. The author has also found that spores of *Nodularia turicensis* which had been *developed in cultures* were able to germinate as soon as fully formed.

Concerning the retention of vitality by the resting-spores, Mr T. Goodey and Dr H. B. Hutchinson have recently made a remarkable discovery. Cultures made in 1912 of a Rothamstead soil which had been sealed up in a dry state since 1846 yielded a quantity of a form of *Nodularia turicensis.* The conditions under which the soil had been kept, and the cultures made, were such that no outside contamination could have taken place, and therefore the small spores of this species (only 7 μ by 8 μ) must have retained their vitality for a period of 66 years. Such spores were indeed resting-spores

a gradually widening aperture and then is itself exuded as an envelope for the escaped cell. The latter now begins to divide and forms a new plant. (Fig. 18, *A*, *B*, and *C*.)

(2) The first division of the protoplast occurs within the walls of the spore, and this may be followed by other divisions, so that from two to four cells may be formed before the spore-wall is burst open. Succeeding divisions then cause the free end of the young filament to extend further and further from the spore, although the other extremity may not free itself from the old spore-wall for some time. (Fig. 18 *A*, the lowest spore in the figure.)

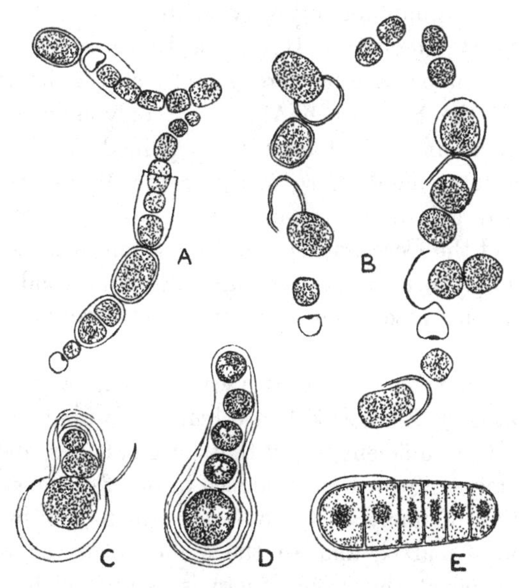

Fig. 18. *A* and *B*, germination of the spores of *Anabæna azollæ* Strasb.; *A*, ×680; *B*, ×850 (after Fritsch). In *A*, two spores show germination of the first type, and the lowest spore germination of the second type. In *B*, variations of the first type of germination are shown. *C* and *D*, germination of the spores of *Anabæna cycadearum* Reinke; *C* illustrates the first type of germination and *D* the third type; ×2200 (after Spratt). *E*, germination of a spore of *Oscillatoria* sp. (possibly *O. limosa* Ag.); ×about 700 (after Phillips).

(3) Both walls of the spore may swell up and become mucilaginous, forming a gelatinous envelope of considerable dimensions. The protoplast sooner or later begins its division within this wide coat of mucus. (Fig. 18 *D*.)

The first type of germination occurs more frequently than the other two. In *Anabæna* and *Nodularia* the spores may germinate while still attached end to end, but this is not usual in other genera.

Reproduction has been found to occur by Sauvageau ('97) in *Nostoc punctiforme* by the formation of '*cocci*,' which are small cells about the same size as the vegetative cells. They occur in large numbers, probably consisting of free vegetative cells, and were observed to form a scum on the surface of the water. The contents of each coccus divide into two and the original wall stretches out and becomes partially gelatinous. Then further divisions ensue, resulting in the formation of a coiled thread within the wide envelope. This thread may be set free by the complete gelatinization of the envelope.

POLYMORPHISM. At one time it was generally assumed that the Blue-green Algæ exhibited a considerable polymorphism, and that many of the described genera and species were merely stages in rather obscure life-histories. Modern investigations, and especially the methods of pure culture, have shown that this belief was a fallacious one, and that the amount of polymorphism in the group is really very small.

The sweeping statements of Itzigsohn, Hansgirg and Wolle that the members of the Chroococcaceæ were merely stages in the development of higher forms of the Blue-green Algæ, were only assumptions based upon crudities of observation. That such ideas gained credence was due very largely to the general habit of so many of the Myxophyceæ in occurring for the most part in gelatinous masses and strata. Thus, it is usual to find various members of the Coccogoneæ and Hormogoneæ as components of the same stratum, simply because they require the same conditions of environment and their mutual association more effectively results in the attainment of the necessary conditions.

No purpose would be gained by recapitulating any of the published statements of Hansgirg, Zopf, Zukal, or others concerning this supposed wide polymorphism. It is sufficient to say that pure cultures, and other carefully conducted experiments, furnish no evidence in proof of the suggested generic or specific identity of many of these forms which live intermingled in a common gelatinous matrix, and to the skilled observer with a thorough taxonomic knowledge of the group, there is rarely much difficulty in discriminating between the developmental stages of the higher types and the unicellular or colonial plants of a lower type (G. S. W., '04).

OCCURRENCE AND DISTRIBUTION. The Myxophyceæ have a world-wide distribution, occurring in abundance in all climates which are sufficiently moist. The majority of the known forms are either subaërial or freshwater, but some occur in saline or brackish areas, and other littoral forms are frequent on nearly all coasts. Many species are prominent constituents of the plankton of freshwater lakes, and a few occur in the marine plankton of the warmer seas. Some of the Chroococcaceæ are free-floating organisms in bogs and lakes, but many of them form gelatinous masses on wet and dripping rocks. Often the gelatinous mass is spread out as a stratum of a blue-green, violet, red-violet or grey colour. In parts of Western Scotland and the Hebrides *Glœocapsa magma* occurs in vast quantities on the ground and among wet stones, in the form of innumerable, small, brownish-purple, gelatinous patches, which are much lobed and curled[1].

The majority of the Hormogoneæ occur aggregated to form a stratum. This may be thin and papyraceous, but is more often slimy; or it may be tough and leathery, or even of a cartilaginous consistency. Sometimes the stratum is felt-like, and in *Scytonema Myochrous* it usually takes the form of a coarse mat. This felt may sometimes cover extensive areas of the ground, as in *Phormidium autumnale* and *Porphyrosiphon Notarisii*. The former is world-wide in distribution, but is most prolific in damp temperate areas, whereas the latter is mostly tropical and subtropical. In 1868 Welwitsch called attention to extensive sheets of a Blue-green Alga (since shown to be *Porphyrosiphon Notarisii*) in the damp sandy valley of the Cuanza River in Angola. The sheets were closely spread like a net over the soil, intergrown with small herbaceous plants and shrubs. The mat-like sheets of the Alga eagerly absorbed the atmospheric moisture during dewy nights, affording by this means a refreshing protection to the roots of many other and larger plants during the glowing heat of the following day. Welwitsch stated that the growth and thriving of the numerous small spermatophytes in these places was conditional on the co-presence of the Alga (W. & G. S. W., '97, p. 303)[2]. Some species of *Phormidium*, such as *Ph. purpurascens*, often cover the vertical faces of wet rocks with a brightly coloured stratum many square feet in extent.

One of the most frequent habitats of the Myxophyceæ is amongst Mosses and Hepatics in the deep gullies and glens of mountainous regions, and among similar luxuriant growths of Bryophytes on the tree-trunks of damp forests, more especially in the tropics. In these situations species of *Chroococcus, Glœocapsa, Scytonema, Stigonema,* and *Nostoc*, often occur in great quantity, together with various members of the Oscillatoriaceæ. It has been shown that where the climate is sufficiently moist Blue-green Algæ are often the first colonists of newly-bared rocks, just as they were the first colonists on the pumice and ash after the terrific volcanic outburst at Krakatau in the Straits of Sunda.

Quite a number of the Myxophyceæ have adapted themselves to a life in the freshwater plankton, and many of these forms have developed those dark-coloured pseudovacuoles which are regarded by some authors as special bodies for increasing the floating capacity of the Algæ which possess them. Species of *Anabæna* are among the most abundant of the plankton-Myxophyceæ, but the two species of *Oscillatoria*, *O. Agardhii* and *O. rubescens*, often occur in quantity, and a number of rather narrow species of *Lyngbya*

[1] These gelatinous masses are known to the local inhabitants as 'Mountain Dulse,' and in past times they were rubbed into a pulp and used for purging calves (Lightfoot, 1777).

[2] Mats of *Zygnema ericetorum* sometimes fulfil a similar function.

are not infrequent. In both *Lyngbya* and *Anabæna* certain of the plankton-species have become spirally coiled (fig. 19). In *Aphanizomenon* the filaments are straight and densely aggregated to form floating bundles, a habit which is also characteristic of the marine genus *Trichodesmium*, a pink species of which gives the colour to the Red Sea. Of the Coccogoneæ the principal genera found in the plankton are *Cœlosphærium*, *Gomphosphæria*, *Microcystis*, *Chroococcus*, and *Dactylococcopsis*, the two former genera sometimes occurring in such quantity as to dominate the entire plankton.

The phenomena of 'water-bloom' and 'the breaking of the meres' are due to the sudden and rather sporadic development of large quantities of

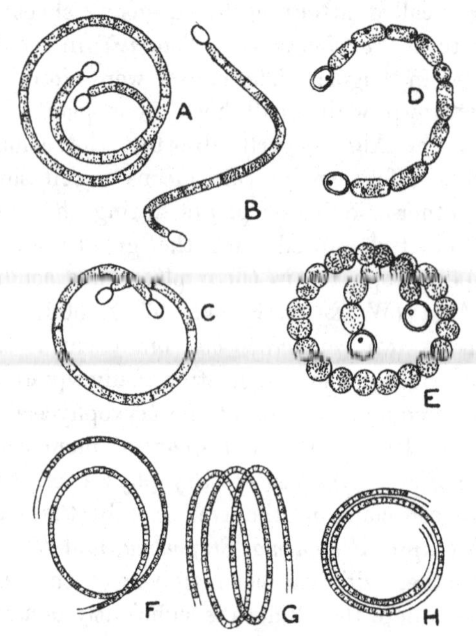

Fig. 19. Spirally coiled Myxophyceæ from the plankton. *A—C*, *Anabæna Tanganyikæ* G. S. West; *D* and *E*, *A. circularis* G. S. West; *F—H*, *Lyngbya circumcreta* G. S. West. All × 520.

a few species of Myxophyceæ, more especially certain of those which normally occur in the plankton of lakes and pools. The extraordinary rapidity with which these species multiply, with consequent discolouration of the water, is a fact which is not yet fully understood; but their disappearance, which is often equally rapid, may be due partly to exhaustion of available food-supplies and partly to the action of toxic substances secreted by themselves. It has been found by Nelson ('03) that water containing 'water-bloom' has often been fatal to cattle; and in this connection it is interesting to note that horses have frequently been killed in the Gulf of Manar by feeding upon *Lyngbya majuscula* (G. S. W., '04).

Many Myxophyceæ are epiphytic in habit, such as the genera *Chamæsiphon, Pleurocapsa, Dermocarpa* and *Oncobyrsa*; also certain species of *Calothrix, Rivularia, Lyngbya,* etc.

The Blue-green Algæ possess a remarkable capacity for withstanding adverse conditions of environment, and especially long periods of drought. It is not merely in moist climates that these plants occur: they are not uncommon in semi-arid districts in which the rainfall is relatively low. It is especially those members of the Oscillatoriaceæ which form slimy felt-like masses that are thus capable of withstanding considerable desiccation without the formation of resting-spores[1]. In Angola the prolific growth of *Scytonema Myochrous* var. *chorographicum* during the rainy season gives such a feature to the mountains of the Presidium that they are known as the *pedras negras* or 'black rocks.' At the end of May the Algæ begin to be discoloured with the intense heat, and the large patches begin to dry and peel off. Soon the rocks reappear in their natural grey or brownish-grey colour, and remain thus through the hot season until the beginning of the next rains (Welwitsch, 1868; W. & G. S. W., '97). In cold latitudes and at high altitudes Myxophyceæ of all kinds can withstand prolonged and repeated freezing. In the Antarctic in the vicinity of Ross Island and South Victoria Land (lat. 77° 32′ S. to lat. 77° 45′ S.) Blue-green Algæ form the dominant vegetation of the pools and lakes, which is remarkable considering the severity of the climate. The growing season only lasts for about four weeks, and during the remaining part of the year the Algæ are completely frozen. Sometimes they pass several years in this frozen state, without being once thawed, and yet

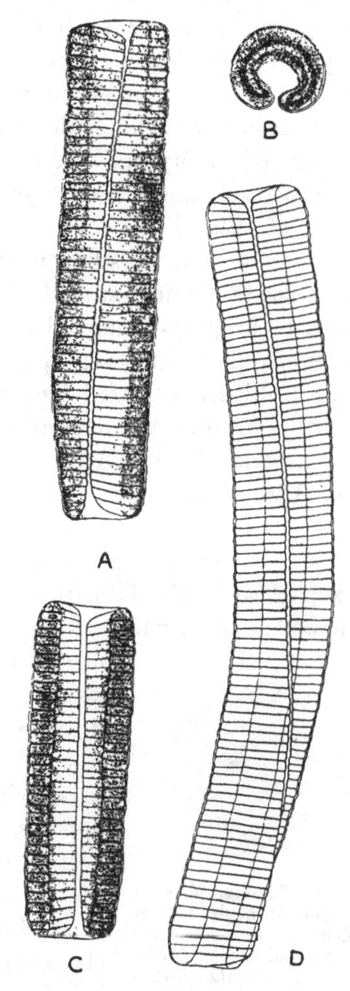

Fig. 20. *Gomontiella subtubulosa* Teodoresco. *A*, ventral view of filament; *B*, transverse section; *C*, optical longitudinal section; *D*, ventral view of a longer and more twisted filament. All × 580 (after Teodoresco).

[1] According to Phillips the spores of *Oscillatoria* will germinate after a resting-period of a year and a half.

the plants survive and resume activity as soon as circumstances are sufficiently favourable (W. & G. S. W., '11; and Fritsch, '12).

The genus *Gomontiella* exhibits a unique adaptation against desiccation. The plant is a member of the Oscillatoriaceæ in which the filaments are greatly flattened and at the same time rolled to form a capillary tube (fig. 20). The plants occur in shallow water-holes which are liable to become temporarily dry, but little evaporation can take place from the ends of the tube-like plant, and thus the filament retains its own water supply and very rarely becomes completely desiccated. (Teodoresco, '01.)

Some of the Myxophyceæ have become adapted to a life in hot water, and they constitute the principal vegetation of hot-springs. The part played by certain of these Algæ in the formation of rock-masses by the extraction of carbonate of lime or silica from the water of hot-springs is considerable. The deposits formed around the hot-springs in many parts of the world consist of brightly-coloured basins or terraces of travertine and sinter. All shades of yellow, orange-red, pink, blue, and blue-green occur, and are caused by the brilliantly-coloured Algæ included within the deposit. In the case of travertine deposits the deposition of the calcium carbonate is due largely to the extraction by the Algæ of the carbon dioxide dissolved in the water. That Algæ do actually cause the elimination of carbonate of lime from water in which it is contained in solution was first shown by Cohn ('62), and Weed ('88) has given a very able account of the assistance of the Myxophyceæ in the formation of the travertine and sinter deposits of the Yellowstone National Park in the United States. *Dichothrix gypsophila*, one of the Rivulariaceæ, is one of the most active species in building up calcareous deposits, and such deposits generally exhibit a well-marked zoning (fig. 21). The Algæ of hot-springs often grow as gelatinous masses in which a glassy form of silica gradually appears, and ultimately all except the peripheral portion becomes firmly silicified. Weed found that the thickness of travertine formed in three days varied from 1·25 to 1·5 mm. He also found that the character and colour of the deposit depended very largely upon the temperature of the water and the situation of the spring or geyser. The highest temperature at which Myxophyceæ have been found is 87·5° C., at which temperature *Phormidium laminosum* has been stated to occur. In the spray of a small geyser at Hveravellir in Iceland, *Phormidium angustissimum*, *Ph. tenue*, and *Mastigocladus laminosus* were thriving (G. S. W., '02). The temperature of this spray was 85° C., but it is highly probable that it would be so rapidly reduced that the wet stratum of the Algæ would be at

Fig. 21. Photograph of a section through a fragment of a calcareous nodule built up by *Dichothrix gypsophila* (Kütz.) Born. & Flah. Natural size. Note the well-marked zoning.

a temperature of possibly 10° less. Of the Chroococcaceæ, *Aphanothece thermalis* Brugger and *A. bullosa* Rabenh. have been observed living in water at a temperature of 68·75° C. On the whole, it may be said that no organisms, with the possible exception of certain Bacteria, possess such a capacity as the Myxophyceæ for withstanding extremes of temperature.

Myxophyceæ of the genera *Glœocapsa* and *Glœothece* are responsible for the formation of oolitic calcareous grains on the shores of the Great Salt Lake, Utah, and white grains of a similar kind are known from the Red Sea. On the bottom of Lough Belvedere, near Mullingar in Ireland, numerous spherical calcareous pebbles have been found, which have been gradually built up by the growth of *Schizothrix fasciculata* (Murray, '95) Similar pebbles are also known from other places, both in Europe and N. America. It is not improbable that the oolitic particles of rocks of various ages may have had a similar origin, and that the structure described as *Girvanella problematica* consists of the mineralised sheaths of an ancient genus of the Myxophyceæ. It is also possible that *Zonatrichites,* described by Bornemann from a breccia of the Keuper age, in Silesia, is a fossil member of the Rivulariaceæ which caused the construction of more or less hemispherical calcareous nodules (Seward, '98). When due consideration is given to the precisely similar nodules which are formed by *Dichothrix gypsophila* (*vide* fig. 21), and other living members of the Rivulariaceæ, the view that *Zonatrichites* was one of the Myxophyceæ is to some extent supported, notwithstanding the complete absence of cellular structure.

The living genus *Hyella* is a perforating Alga on the surface of the calcareous shells of Molluscs.

SYMBIOSIS. The curious endophytic *Anabæna Azollæ* Strasburger lives and grows in certain slime-filled hollows in the leaves of *Azolla,* into which cavities project hairs from the living layer of cells. It has been suggested that this is a case of symbiosis. The *Anabæna* is always present in the cavities and the terminal parts of the hairs become colourless and finally more or less disintegrated.

Anabæna Cycadearum Reinke is another endophytic species which lives within the modified tubercle-like roots of species of *Cycas.* The Alga occupies a well-marked intercellular-space-zone just within the cortex (fig. 22), and is associated with the two nitrogen-fixing Bacteria, *Pseudomonas radicicola* and *Azotobacter.*

Horejsi ('10) has given a good account of the distribution of the Alga in the cortex of the root, and he showed that by its penetration into the apical meristem it stopped normal growth and caused dichotomy. Thus the peculiar tubercle-like roots of *Cycas* and other Cycads are actually caused by the endophytic Alga. He found that the Alga maintained its existence in the soil, and that it entered the roots regularly at the beginning of every autumn through the lenticels. Spratt ('11) confirmed the fact that the Alga entered the roots through the lenticels, and showed that this was accomplished by means of gonidia which were developed from heterocysts and spores present in the soil. There is no doubt in this case of the symbiotic relationship between the Alga and the two nitrogen-fixing Bacteria, and there is every probability that the Cycads benefit to some extent by being able to absorb certain products of the metabolic activity of the lower organisms. Needless to say, both the above species of *Anabæna* are considerably different from the more normal species of the genus, and both have degenerated.

An endophytic member of the Nostocaceæ also lives in the hollow leaf-auricles of the Hepatic *Blasia*, but its exact relationships with its

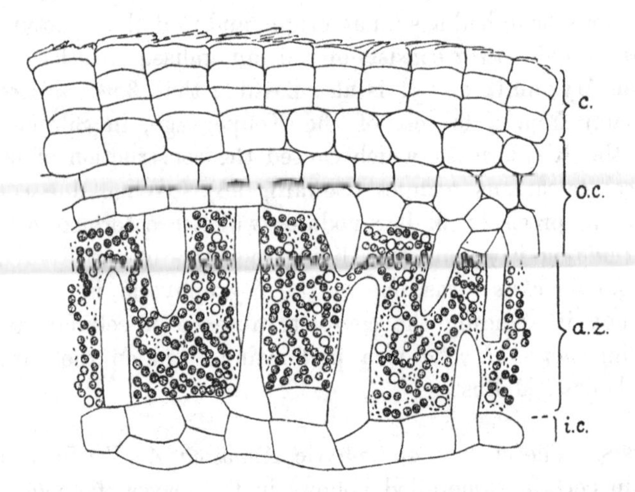

Fig. 22. Part of a transverse section of tubercle-like root of *Cycas* showing the algal zone (*a.z.*) containing *Anabæna Cycadearum* Reinke and certain Bacteria. *c.*, cork; *o.c.*, outer cortex; *i.c.*, inner cortex. ×400 (modified from Spratt).

peculiar environment have not yet been investigated. Another occurs in the stems of *Gunnera*.

Hugo Fischer ('04) has described a symbiotic relationship between certain of the Oscillatoriaceæ which carpet the ground (such as *Phormidium autumnale*) and *Azotobacter Chroococcum*. The nitrogen-fixing activity of the Bacterium enables the Blue-green Alga to make use of atmospheric nitrogen.

A still more remarkable member of the Nostocaceæ is *Richelia intracellularis* Johs. Schmidt, which lives within the cells of marine species of *Rhizosolenia* (fig. 23). As this Alga occurs within the living cells of the host it may be assumed that it is a partial parasite.

Some of the Myxophyceæ are regularly found associated with Fungi to form the dual organisms known as Lichens. The Algæ which have thus lost their individuality become considerably modified and generally lose almost all traces of their original specific characters. It has been customary to regard the thallus of the lichen as a case of commensal symbiosis, but in the great majority of Lichens the Alga is a very subordinate constituent, the fungus completely dominating it, even to the extent of curtailing and absorbing any undue increase in the algal cells. It seems to be much more in agreement with facts to look upon the Algæ as existing under conditions of beneficent slavery! The lichen-thallus thus provides an illustration of *helotism*, which is a form of symbiosis intermediate in character between commensalism and parasitism. The fungal hyphæ are partially parasitic on the algal cells, but at the same time they supply them with various excretory products, mineral salts, and water. It is true that inclusion within the lichen-thallus permits of the existence of the algal cells in situations in which they could not maintain themselves, but their multiplication is kept within restricted limits by the undoubted parasitism of the fungus.

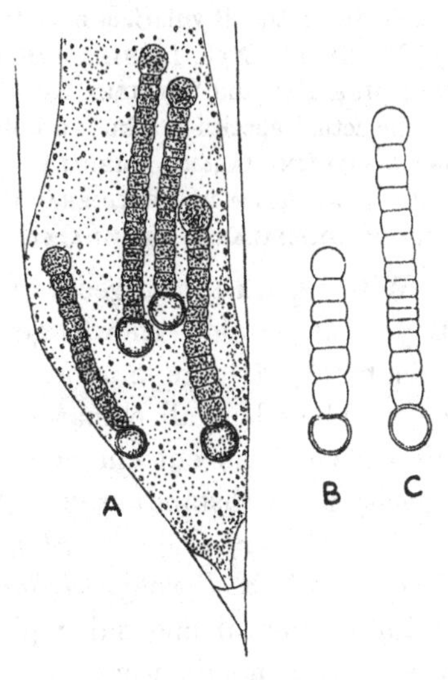

Fig. 23. *Richelia intracellularis* Johs. Schmidt. *A*, one extremity of cell of *Rhizosolenia styliformis* Btw. containing four filaments of the *Richelia*; *B* and *C*, two isolated filaments showing the basal position of the heterocyst. × about 450 (after Johs. Schmidt).

The following list shows some of the genera of Lichens which contain certain of the Blue-green Algæ:

Glœocapsa occurs in the thallus of ANEMA Nyl., COLLEMOPSIDIUM Nyl., CRYPTOTHELE (Th. Fr.) Forss., FORSSELLIA Zahlbr., GONOHYMENIA Stnr., JENMANIA Wächt., LEPTOGIOPSIS Nyl., PAULIA Fée, PECCANIA (Massal.) Forss., PHLŒOPECCANIA Stnr., PHYLLISCIDIUM Forss., PSOROTRICHIA (Massal.) Forss., PYRENOPSIS (Nyl.) Forss., and SYNALISSA E. Fr.

Chroococcus in PHYLLISCIUM Nyl. and PYRENOPSIDIUM (Nyl.) Forss.

Nostoc in ARCTOMIA Th. Fr., COLLEMA (Hill.) Zahlbr., DICHODIUM Nyl., HASSEA Zahlbr., HOMOTHECIUM Mont., HYDROTHYREA Russ., KOERBERIA Mass., LECIOPHYSMA Th. Fr., LEMMOPSIS (Wainio) Zahlbr., LEPIDOCOLLEMA Wainio, LEPROCOLLEMA Wainio, LEPTOGIUM (Ach.) S. Gray, LOBARIA (Schréb.) Hue [Section LOBARINA (Nyl.) Hue], NEPHROMA Ach. [in part], PANNARIA Del., PARMELIELLA Müll. Arg., PELTIGERA Willd., PHYSMA (Mass.) Zahlbr., PYRENIDIUM Nyl., PYRENOCOLLEMA Reinke, SCHIZOMA Nyl., SOLORINA Ach. [in part], and STICTA Schréb. [Section STICTINA (Nyl.) Hue].

Scytonema in COCCOCARPIA Pers., DICTYONEMA Ag., ERIODERMA Fée, HEPPIA Näg., LEPTODENDRISCUM Wainio, MASSALONGIA Koerb., PETRACTIS E. Fr., PLACOTHELIUM (Ach.) Harm., POROCYPHUS Koerb., STEREOCAULON Schréb. and THERMUTIS E. Fr.

Stigonema in EPHEBE E. Fr., LICHENOSPHÆRIA Born., PTERYGIOPSIS Wainio and SPILOMENA Born.

Genera of the Rivulariaceæ in CALOTRICHOPSIS Wainio, HOMOPSELLA Nyl., LICHINA Ag., LICHINELLA Nyl., LICHINODIUM Nyl., OMPHALARIA Gir. & Dum., POLYCHIDIUM Ach., PTERYGIUM Nyl. and STEINERA Zahlbr.

The actual species are known only in a few cases: *Glœocapsa montana* in BÆOMYCES ROSEUS and OMPHALARIA UMBELLA; *Nostoc sphæricum* in the Collemaceous HYDROTHYREA; *Nostoc punctiforme* in PELTIGERA, PANNARIA and STICTINA; *Rivularia nitida* in POLYCHIDIUM, OMPHALARIA, LICHINA, etc.; *Stigonema panniforme* in EPHEBE PUBESCENS.

With regard to the geographical distribution of the Myxophyceæ we have at present a wide knowledge but an insufficiency of detail. There are undoubtedly a large number of cosmopolitan species which occur all over the world and in all climates, a fact which is probably due in part to the primitive and ancient character of the organisms, and very largely to their capacity for withstanding drought and extremes of heat and cold.

Camptothrix, Chondrocystis, Katagnymene, Loefgrenia, Polychlamydum, Proterendothrix, Porphyrosiphon and *Trichodesmium* are, so far as is known, typically tropical and subtropical genera. A vast amount of detailed and *accurate* taxonomic work is necessary, however, before any reliable statements can be made concerning the geographical distribution of the Blue-green Algæ.

AFFINITIES. It has been customary to regard the Myxophyceæ as organisms of a primitive type closed allied to the Bacteria. This view received much encouragement by the general acceptance of Cohn's 'Schizophyta' as a group put forward to embrace the Schizomycetes (Fission Fungi or Bacteria) and the Schizophyceæ (Fission Algæ). Modern evidence, however, does not wholly support this view, and there is little doubt that the Bacteria and Myxophyceæ are separated by a gulf much wider than would be thus indicated. Cohn's name 'Schizophyceæ' should be entirely discarded as it is very misleading. Far from being the 'Fission Algæ,' only a small percentage of the known species divide by that gradual constriction which is so characteristic of the vast majority of Bacteria. Cell-division in the filamentous forms is generally by the gradual formation of a transverse septum.

At present it is necessary to make very guarded statements concerning the comparative cytology of the two groups. The investigation of that giant Bacterium *Hillhousia* has shown that the dry-staining methods so much in vogue in studying the cytology of both the Bacteria and the Myxophyceæ are to a considerable extent unreliable, and it is also possible that the bacterial protoplast may show various grades of simple structure. It is quite possible that true chromatin such as occurs in higher organisms does not exist in the bacterial cell, and the chromatin-substance which does occur is only present in the form of very minute granules. The chromatin-substance present in the protoplast of *Hillhousia* cannot be recognized by staining, and in the Blue-green Algæ it is exceedingly probable that the chromatin-substance which occurs in small particles and in small quantity is not of precisely the same nature as that which is present in higher types of nuclei.

The centrally differentiated incipient nucleus of the Myxophyceæ appears to have no exact counterpart in the Bacteria, and it furnishes clear evidence that the protoplast of the Blue-green Algæ is of a higher type. The cell-walls, also, in the young cells, and in the heterocysts throughout life, consist of cellulose, whereas the bacterial cell-walls never consist of cellulose. The presence of the pigments chlorophyll, carotin, and phycocyanin, enabling the myxophyceous protoplast to construct organic substance under the influence of light, is also a feature of great importance, and the fact that these pigments are restricted to the peripheral zone of the cell must not be lost sight of[1]. The resting-spores of the Myxophyceæ are *always* formed by the rejuvenescence of a vegetative cell, and are never of the nature of endospores such as are produced in the Bacteria. On the whole, the distinction between the Myxophyceæ and the Bacteria, although not absolutely sharp, is still sufficiently wide to make the use of such a term as the 'Schizophyta' rather unwise.

The group is primitive, and the living Myxophyceæ of to-day are the descendants, probably but little changed, of a group of organisms which were left aside very early in the evolution of plant-life. The complete absence of sexuality is undoubtedly to be associated with the incomplete differentiation of the protoplast and the consequent absence of a highly organized nucleus.

CLASSIFICATION. The different families of the Myxophyceæ are for the most part well-defined and easily diagnosed. The only controversial points have been those concerning the systematic position of the genera *Chroothece*, *Asterocystis* and *Glaucocystis*.

A careful examination of the original specimens of *Chroothece*[2] indicates that this genus would be best placed in the Chroococcaceæ. *Asterocystis* also appears to be a typical Chroococcaceous genus[3], but *Glaucocystis* is still

[1] There are no known saprophytic Myxophyceæ, with the exception of *Oscillatoria decolorata* G. S. West (*Journ. Bot.* 1899, p. 263; see also McKeever in *Trans. Edin. Field Nat. and Microscop. Soc.* Nov. 1911, p. 370).

[2] Some of the original material of *Chroothece Richterianum* Hansg. was issued in Wittrock & Nordstedt's *Alg. Exsic.* in 1884. The author has to thank Dr Borge of the Botanical Museum at Stockholm for kindly sending the specimens, and Miss Acton for making careful preparations of them in the botanical laboratory of the University of Birmingham.

[3] The cytological structure has been carefully examined by the author in the following species: *A. africana* G. S. West, *A. antarctica* W. & G. S. West and *A. smaragdina* (Reinsch) Forti, and found to be similar to that of some of the larger species of *Chroococcus*. The cells of *A. halophila* (Hansg.) Forti also appear in the living state to be like those of *Chroococcus*, but this species has not been examined cytologically.

something of a puzzle. The form of its cells, with strong cell-walls, and the method of formation of the daughter-cells are exactly as in the Chlorophyceous genus *Oocystis*. It has a definite nucleus with a nuclear membrane, and fully differentiated chromatophores of a rich blue-green colour (consult fig. 24). In the young cells the chromatophores are numerous and short, but in older cells there are from 10 to 20 of them, mostly curved, and more or less radiating from the central part of the cell.

Glaucocystis Nostochinearum is not uncommon in bogs in Western Europe, and a large form of it has been found in E. Africa. There is no doubt that the Alga differs from all the other Blue-green Algæ in the possession of a higher type of protoplast and true chromatophores, and it was this fact which led to the suggestion that the Myxophyceæ should be primarily divided into the Glaucocystideæ and the Archiplastideæ (G. S. W. '04). It is probable, however, that this Alga, notwithstanding its brilliant blue-green colour, may have to be removed entirely from the Myxophyceæ when more is known concerning the details of its cytology.

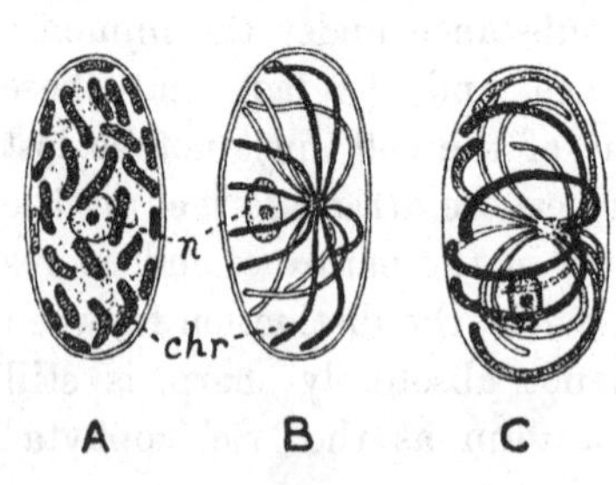

Fig. 24. *Glaucocystis Nostochinearum* Itzigsh. Three cells showing the nucleus (*n*) and the differences between the chromatophores (*chr*) of young and old cells. *A* is a young cell, *B* an older one, and *C* one which is still older. ×720 (after Hieronymus).

The curious Alga, *Porphyridium cruentum* (Ag.) Näg., has been recently investigated by Brand ('08 *A* and *B*), who brings forward strong evidence to show that it must be regarded as a rudimentary form of the Bangiaceæ, which has never got beyond the embryonic state.

Excluding *Glaucocystis* the Myxophyceæ are primarily divided into the two natural groups of the Coccogoneæ and Hormogoneæ.

Order I. COCCOGONEÆ. Unicellular or colonial plants, usually consisting of colonies of unicells surrounded by mucous investments. The cells vary much in form and in their disposition. The mucous investment is often structureless, but may be obviously lamellose. The normal method of multiplication is by simple cell-division, the larger colonies dissociating to form smaller ones. Rounded asexual gonidia are known to occur in some genera and species. The group includes the lowest of the Myxophyceæ and contains therefore the most primitive of the Algæ. They occur free-floating in bogs and lakes, as gelatinous masses on wet rocks, and more rarely as epiphytes.

Fam. **Chroococcaceæ.** This is the principal family of the Coccogoneæ and includes nearly all the unicellular and colonial Blue-green Algæ. In one genus (*Glœochæte* Lagerh.) the cells are epiphytic and dorsiventral, each cell being furnished with one or two bristles; but in all the other genera the cells or colonies are either free-floating or form a gelatinous stratum. The cells vary much in shape in the different genera, and with few exceptions they are provided with a copious mucous covering. In some the cells divide in every direction of space, producing an irregular colony, often of large size (*Aphanocapsa*, *Glœocapsa*, *Microcystis*, etc.). In others the cells divide only in two directions in the same plane, giving rise to a tabular colony (*Merismopedia*; fig. 26, *B* and *C*); and in others cell division takes place in one direction only (*Synechococcus*, *Glœothece*). In the genus *Tetrapedia* (fig. 26 *D*) the cells are compressed and very symmetrical. In *Chroococcus* (fig. 25) the mucous coats of the cells are often very firm and lamellose, and in *Glœocapsa* (fig. 2 *B*) the inner integuments often contain red, orange, or violet pigments. The largest colonies are macroscopic and may contain many thousands of cells, as in *Aphanothece prasina*. In *Cœlosphærium*, *Gomphosphæria* and *Merismopedia* the colonies are of limited size and definite shape, and in the two first-named genera a kind of budding takes place by means of which a new colony is developed from the side of the old one, ultimately becoming separated from it.

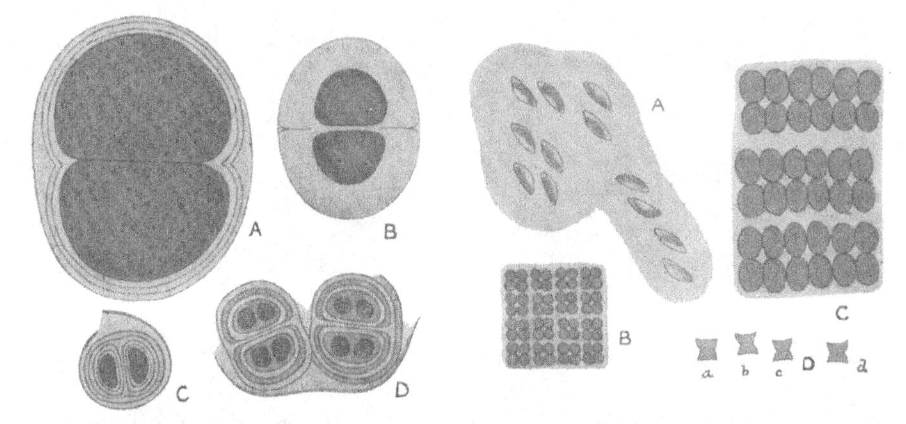

Fig. 25. *A*, *Chroococcus giganteus* W. West. *B*, *Chr. turgidus* (Kütz.) Näg. *C* and *D*, *Chr. schizodermaticus* W. West. All × 450.

Fig. 26. *A*, *Dactylococcopsis montana* W. & G. S. West. *B*, *Merismopedia glauca* (Ehrenb.) Näg. *C*, *M. elegans* A. Br. *D*, *Tetrapedia Reinschiana* Arch. All × 450.

Fam. **Chamæsiphonaceæ.** A family of epiphytes in which the cells are generally clustered in dense masses around the filaments of larger Algæ (fig. 27). The cells may be ovoid, pyriform, or cylindrical, and there is always a distinction between base and apex.

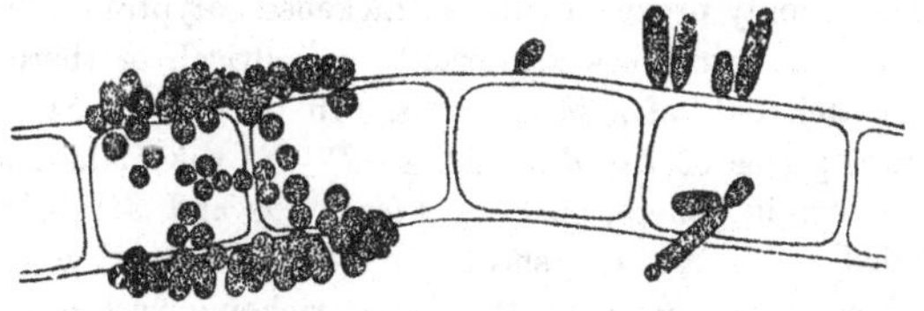

Fig. 27. *Chamæsiphon incrustans* Grun. epiphytic on a filament of *Rhizoclonium*. × 416.

Reproduction occurs by the formation of gonidia which are successively cut off from the upper part of the mother-cell (*Chamæsiphon*), or produced by the division of the contents of the mother-cell by three or more sets of division-planes (*Xenococcus*, *Hyella*, *Dermocarpa*).

Order II. Hormogoneæ. Includes all the filamentous Myxophyceæ. The filaments consist of a simple row of cells, very rarely naked, and usually enclosed within a sheath, which varies from a hyaline, indistinct envelope to a tough, coloured investment. In some instances several rows of cells occur within the same sheath. In most of the families heterocysts occur in more or less abundance, but in the Oscillatoriaceæ and Camptotrichaceæ they are absent. In some genera the filaments are branched, and in others a peculiar type of false branching occurs. Most of the Algæ in this group occur attached to a substratum. The plants multiply by means of hormogones, and more rarely by resting-spores.

A. *Psilonemateæ.* Trichomes[1] cylindrical, sometimes narrowed *at* the extremities, but never gradually attenuated *towards* the extremities.

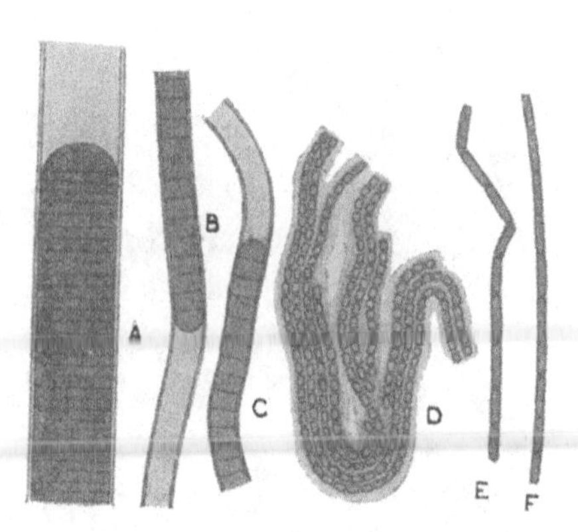

Fig. 28. *A, Lyngbya major* Menegh. *B* and *C, L. ærugineo-cœrulea* (Kütz.) Gom. *D, Phormidium molle* (Kütz.) Gom. *E* and *F, Ph. tenue* (Menegh.) Gom. All × 460.

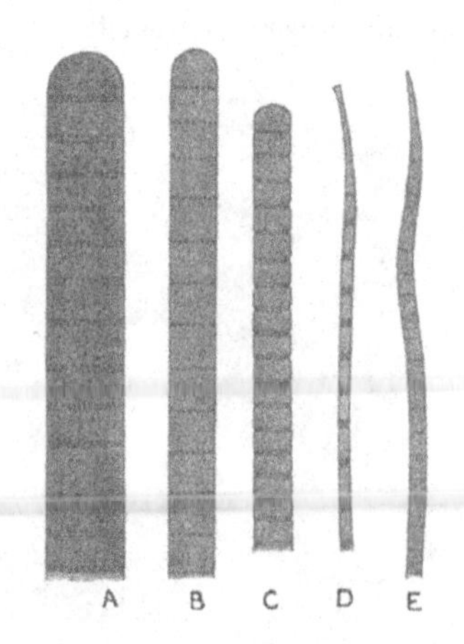

Fig. 29. *A, Oscillatoria limosa* Ag. *B, O. irrigua* Kütz. *C, O. tenuis* Ag. *D, O. splendida* Grev. *E, O. acuminata* Gom. All × 460.

Fam. **Oscillatoriaceæ.** A large family distinguished from the other families of the Psilonemateæ by the absence of heterocysts. The trichomes consist of cylindrical or disc-shaped cells, often somewhat modified at the extremities, the apical cell being frequently conical and not uncommonly provided with a thickened calyptra. The cells may be so compactly joined that the trichomes are exactly cylindrical, or there may be a slight constriction between each pair of adjacent cells. In the genera *Spirulina, Arthrospira* (fig. 30), and in some species of *Lyngbya* (fig. 19 *F—H*) the filaments are twisted into a regular spiral. In certain genera, such as *Microcoleus* and *Schizothrix* (fig. 3), several trichomes are included within a single sheath (sub-fam. Vaginarieæ), whereas in most of the other genera there is never more than one trichome within a sheath (sub-fam. Lyngbyeæ). The most abundant genus is *Oscillatoria* (fig. 29), in which the sheath is so close and delicate as to be easily overlooked. In *Lyngbya* (fig. 28 *A—C*) the trichomes possess a strong sheath which is often lamellose. *Phormidium* is almost as abundant as *Oscillatoria* and occupies a position half-way between that genus and *Lyngbya.* It is characterized by the sheaths becoming agglutinated or else entirely diffluent.

[1] The term 'trichome' is universally applied to the filament of a Blue-green Alga without its sheath, the term 'filament' being retained for the trichome together with its sheath.

In *Plectonema* the filaments are falsely branched, and in *Symploca* they form compact, erect, pointed tufts arising from a flat stratum. In some species of *Schizothrix* the sheaths are strongly lamellose and brilliantly coloured, but it is in the African genus *Polychlamydum* that the sheath exhibits its maximum differentiation.

The most extraordinary member of this family is *Gomontiella subtubulosa* (fig. 20) found by Teodoresco in Roumania. The flattened, *rolled* filaments are unique among the Hormogoneæ.

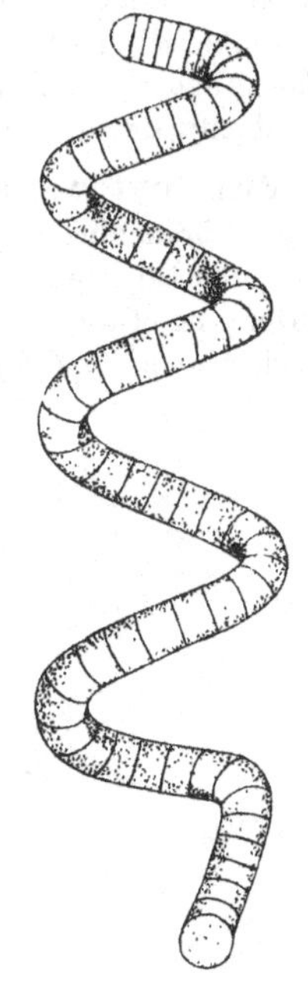

Fig. 30. *Arthrospira platensis* (Nordst.) Gom. Single filament, × 520.

Fam. **Nostocaceæ.** The filaments are simple, with delicate gelatinous sheaths. The trichomes consist of a single series of uniform and often torulose cells, amongst which are few or many heterocysts. In a few forms the sheaths are distinct, but in others they become confluent to form a mass of jelly enclosing a large number of trichomes. This jelly attains its maximum development in the genus *Nostoc* (fig. 31) in which the shape of the colony is largely determined by the toughness of the gelatinous mass. Thousands of trichomes occur in such a colony, and they are generally contorted and densely crowded. In *Anabæna* the trichomes are not associated to form definite colonies, as the gelatinous sheaths are much more fleeting and delicate, although they may be aggregated in masses. The trichomes may be rigid and straight or variously twisted and contorted (as in so many of the plankton-species).

Aphanizomenon occurs free-floating in the waters of pools and lakes, and the trichomes are often agglutinated to form spindle-shaped bundles.

In *Cylindrospermum* there is only one terminal heterocyst (fig. 32 *E—G*); and in *Nodularia* the heterocysts are rather numerous and occur at regular distances along the trichome (fig. 32 *H*).

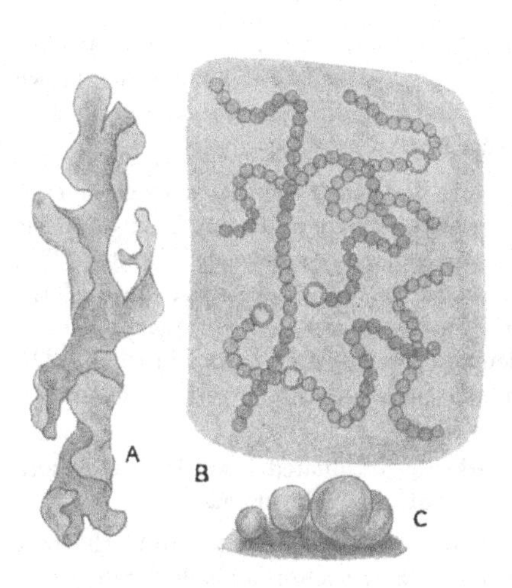

Fig. 31. *A* and *B*, *Nostoc Linckia* Bornet; *A*, nat. size; *B*, small portion of thallus, × 340. *C*, *N. cœruleum* Lyngbye, nat. size.

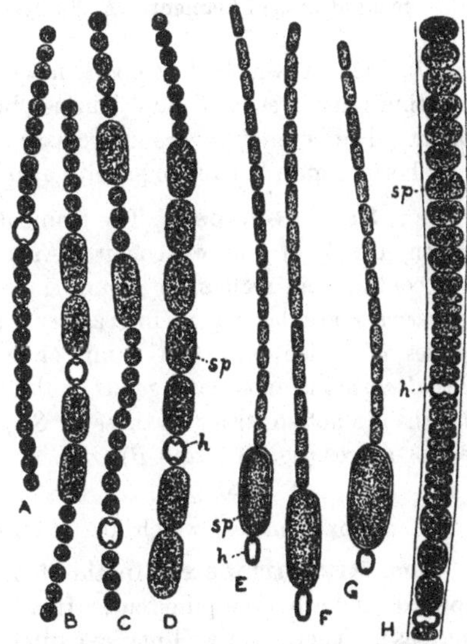

Fig. 32. *A—D*, *Anabæna inæqualis* (Kütz.) Born. & Flah. *E—G*, *Cylindrospermum stagnale* (Kütz.) Born. & Flah. *H*, *Nodularia sphærocarpa* Born & Flah. (All × 480.) *h*, heterocyst; *sp*, resting-spore.

Reproduction takes place by hormogones and by resting-spores. The latter are variable in shape, and may be solitary or seriate. In *Cylindrospermum* the spore arises from the cell next the terminal heterocyst (fig. 32 *E—G*). About half the species of *Nostoc* and *Cylindrospermum* occur in subaërial habitats, but all the other members of the family are aquatic. Species of *Nodularia* are mostly brackish-water forms.

Fam. **Scytonemaceæ.** The Algæ of this family are distinguished by their peculiar type of branching. As a rule there is only one trichome within a strong tubular sheath of regular thickness. The false branches arise either singly or in pairs, and are due to the perforation of the sheath of the primary filament by its trichome, which issues either as a single 'branch' (*Tolypothrix*; fig. 33 *E*) or as a pair of 'branches' (*Scytonema*; fig. 33

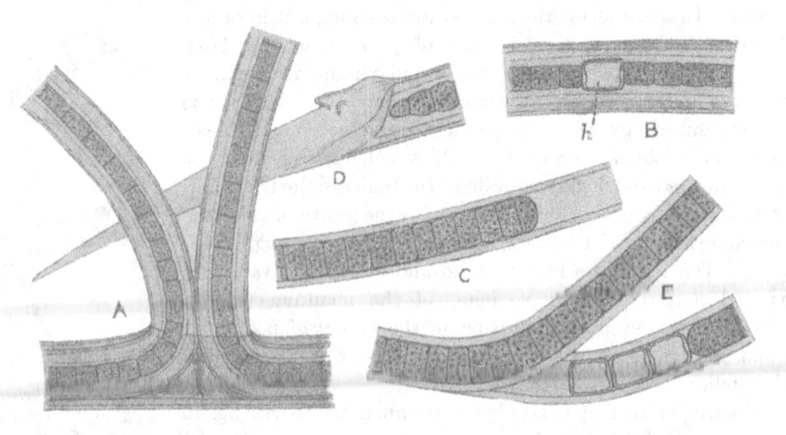

Fig. 33. *A—D*, *Scytonema mirabile* (Dillw.) Thuret. *A*, showing pair of false branches; *B*, part of filament with a heterocyst (*h*); *C*, apex of 'branch'; *D*, organ of attachment at base of filament. *E*, *Tolypothrix lanata* (Desv.) Wartm. All × 440.

A—D), each of which develops a new sheath. The sheaths are variable in character, sometimes homogeneous and colourless, but in other cases lamellose and of a golden-brown colour. The heterocysts are intercalary and rather infrequent. The normal reproduction is by hormogones, but resting-spores also occur (fig. 17 *C*).

Fam. **Stigonemaceæ.** The filaments are coarse and tough owing to a strong uneven sheath, usually of a brown colour. As a rule the filaments are much branched, and they may contain one regular, or several more or less irregular, series of cells (fig. 34). The heterocysts are lateral or intercalary. *Stigonema* is the principal genus, most of the species of which occur on damp or wet rocks, and are propagated by hormogones. *Hapalosiphon* is an aquatic genus, with more slender filaments and longer branches. The sheaths are not so strong as those of *Stigonema*, and the plants are largely reproduced by resting-spores (fig. 17 *A* and *B*).

B. *Trichophoreæ.* Trichomes markedly attenuated towards one or both extremities, which are often drawn out into hair-like points.

Fam. **Rivulariaceæ.** In this family the trichomes are attenuated from the base upwards and are often piliferous. In all except a few cases there is a single heterocyst at the base. There is a well-marked tubular sheath, gelatinous or membranous, and often strongly lamellose. There is a prominent false branching in the genus *Dichothrix*, accompanied by a conspicuous fusion of sheaths. The filaments of *Calothrix* generally form an expanded stratum on wet rocks and submerged stones (fig. 35). In *Rivularia* the filaments are disposed to form subglobose or hemispherical, attached masses, sometimes

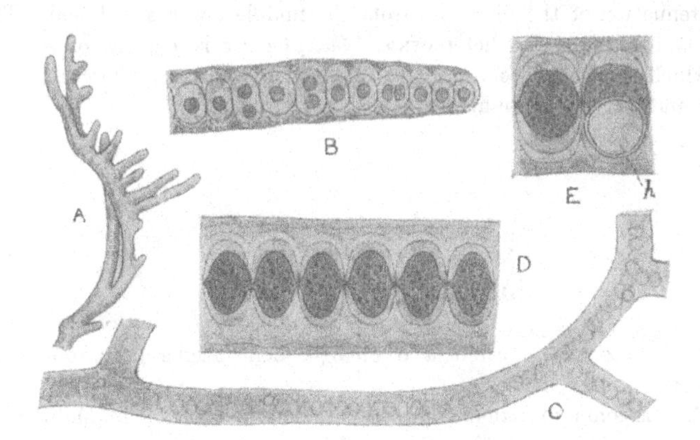

Fig. 34. *A* and *B*, *Stigonema minutum* Hass.; *A*, ×100; *B*, ×440. *C—E*, *St. ocellatum* (Dillw.) Thur.; *C*, ×100; *D* and *E*, ×400. *h*, heterocyst.

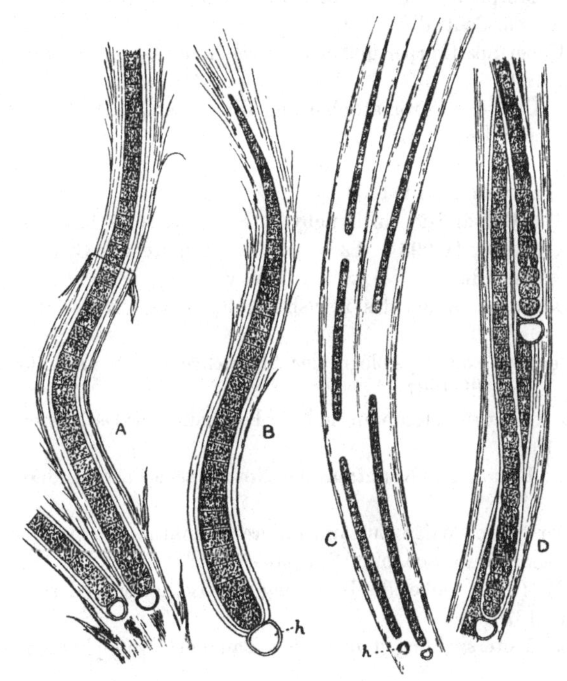

Fig. 35. *A* and *B*, *Calothrix parietina* (Näg.) Thur. *C*, *Dichothrix interrupta* W. & G. S. West. *D*, *D. Orsiniana* (Kütz.) Born. & Flah. All × 420.

gelatinous and sometimes incrusted with lime. In *Glœotrichia*, which occurs largely in the plankton, the gelatinous mass is usually free-floating. The genus scarcely differs from *Rivularia*, except in the formation of large cylindrical resting-spores. In *Homœothrix* and two other genera there are no heterocysts.

Fam. **Camptotrichaceæ.** This family was established in 1897 (W. & G. S. W. '97 A and B) to include certain Blue-green Algæ which differ from members of the Rivulariaceæ in the attenuation of the filaments from the middle towards each end. The plants are epiphytes and there are no heterocysts. *Camptothrix* is a genus of small tropical epiphytes in which the filaments terminate in blunt extremities, whereas *Hammatoidea* includes larger and more elongate filaments with piliferous apices.

LITERATURE CITED

BORZI, A. ('78). Note alla morfologia e biologia delle Alghe ficochromacee. Nuovo giornale Botan. Ital. x, 1878.

BORZI, A. ('86). Le communicazioni intercellulari delle Nostochinee. Malpighia, i, 1886.

BRAND, F. ('01). Bemerkungen über Grenzzellen und über spontan rote Inhaltskörper der Cyanophyceen. Ber. Deutsch. Bot. Ges. xix, 1901.

BRAND, F. ('03). Morphologisch-physiologische Betrachtungen über Cyanophyceen. Beihefte zum Botan. Centralbl. xv, 1903.

BRAND, F. ('05). Ueber Spaltkörper und Konkavzellen der Cyanophyceen. Ber. Deutsch. Bot. Ges. xxiii, 1905.

BRAND, F. ('08 A). Über das Chromatophor und die systematische Stellung der Blutalge (Porphyridium cruentum). Ber. Deutsch. Bot. Ges. xxvi a, 1908.

BRAND, F. ('08 B). Weitere Bemerkungen über Porphyridium cruentum (Ag.) Naeg. Ber. Deutsch. Bot. Ges. xxvi a, 1908.

BROWN, W. H. ('11). Cell-division in Lyngbya. Botan. Gazette, li, no. 5, 1911.

CHODAT, R. & MALINESCO, O. ('93). La structure cellulaire des Cyanophycées. Labor. botan. de l'université de Genève. Sér. 1, fasc. v, 1893.

COHN, F. ('62). Ueber die Algen des Karlsbader Sprudels. Abhandl. der Schles. Ges. Nat. 1862.

COHN, F. ('67). Beiträge zur Physiologie der Phycochromaceen und Florideen. Archiv für mikroscop. Anat. iii, 1867.

CORRENS, C. ('96). Ueber die Membran und die Bewegung der Oscillarien. Ber. Deutsch. Bot. Ges. xiv, 1896.

DE BARY, A. ('63). Beitrag zur Kenntniss der Nostocaceen, insbesondere der Rivularien. Flora, 1863.

DEINEGA ('91). Der gegenwärtige Zustand unserer Kenntnisse ueber den Zellinhalt der Phycochromaceen. Bull. Soc. imp. Nat. Moscou, 1891.

ENGELMANN, TH. W. ('79). Ueber die Bewegung der Oscillarien und Diatomeen. Botan. Zeitung, xxvii, 1879.

FISCHER, A. ('97). Untersuchungen über den Bau der Cyanophyceen und Bakterien. Jena, 1897.

FISCHER, A. ('05). Die Zelle der Cyanophyceen. Botan. Zeitung, lxiii, 1905.

FISCHER, HUGO ('04). Über Symbiose von *Azotobacter* mit Oscillarien. Centralbl. f. Bakt. xii, 1904.

FRITSCH, F. E. ('04). Studies on Cyanophyceæ. New Phytologist, iii, no. 4, 1904.

FRITSCH, F. E. ('04 B). Studies on Cyanophyceæ III. New Phytologist, iii, nos. 9 and 10, 1904.

FRITSCH, F. E. ('05). Studies on Cyanophyceæ II. Beihefte z. Botan. Centralbl. xviii, 1905.

FRITSCH, F. E. ('12). Freshwater Algæ of the National Antarctic Expedition. Report on Nat. Hist. vol. vi, 1912.

GAIDUKOV, N. ('04). Die Farbe der Algen und des Wassers. Hedwigia, xliii, 1904.

GARDNER, N. L. ('06). Cytological Studies in Cyanophyceæ. Univ. of California Publications. Botany, ii, no. 12, 1906.

GUILLERMOND, A. ('06) in Rev. Gén. Bot. xviii, 1906.

HANSGIRG, A. ('83). Bemerk. ueber die Bewegung der Oscillarien. Botan. Zeitung, xxxi, 1883.

HANSGIRG, A. ('87). Physiologische und algologische Studien. Prague, 1887.

HEGLER, R. ('01). Untersuchungen über die Organisation der Phycochromaceen-zelle. Pringsh. Jahrb. f. wiss. Bot. xxxvi, 1901.

HIERONYMUS, G. ('92). Beiträge zur Morphologie und Biologie der Algen. Cohn's Beitr. zur Biol. d. Pflanzen, v, 1892.

HOREJSI, J. ('10). Symbioticka rasa sina v Korenech u Cycas revoluta. Rozpr. ceske Akad. cis. Frantiska Josefa, Praze, xix, ii, 1910.

HYAMS & RICHARDS ('02). Notes on Oscillaria prolifica. Technol. Quarterly, Boston, xv, 1902.

KLEBAHN, H. ('95). Gasvacuolen, ein Bestandteil der Zellen der wasserblütebildenden Phycochromaceen. Flora, lxxx, 1895.

KLEBAHN, H. ('97) in Botan. Zeitung, lv, 1897.

KOHL, F. G. ('03). Über die Organisation und Physiologie der Cyanophyceenzelle und die mitotische Teilung ihres Kernes. Jena, 1903.

LEMMERMANN, E. ('02). Die Algenflora der Chatham Islds. Engl. Bot. Jahrb. xxxviii, 1902.

LEMMERMANN, E. ('07). Algen in Kryptogamenflora der Mark Brandenburg, Bd. iii, Heft 1, 1907.

MACALLUM ('98). On the Cytology of Non-nucleated Organisms. Trans. Canadian Institute, vi, 1898—9.

MARX ('92). Untersuchungen ueber die Zellen der Oscillarien. Inaug. Dissert. Erlangen, 1892.

MASSART ('02) in Recueil de l'Inst. Bot. Univ. de Bruxelles, v, 1902.

MOLISCH, H. ('03). Die sogenannten Gasvacuolen und das Schweben gewisser Phycochromaceen. Botan. Zeitung, 1903.

MURRAY, G. ('95). Calcareous Pebbles formed by Algæ. Phycological Memoirs, London, April, 1895.

NADSON, G. ('95). Ueber den Bau des Cyanophyceen-Protoplasts. Scripta bot. hort. Univ. Imp. Petropolit. iv, 1895.

NÄGELI, C. ('49). Gattungen einzelliger Algen. Zurich, 1849.

NELSON, N. P. B. ('03) in Minnesota Botan. Studies, iii, 1903.

OLIVE ('05). Mitotic division of the nuclei of the Cyanophyceæ. Beihefte Botan. Centralbl. xviii, 1905.

OSTENFELD, C. H. & JOHS. SCHMIDT ('01). Plankton fra det Röde Hav og Adenbugten. Vidensk. Medd. naturh. Foren. Kjöbenhavn, 1901.

PALLA, E. ('93). Beitrag zur Kenntnis des Cyanophyceen-Protoplastes. Pringsh. Botan. Jahrbuch. xxv, 1893.

PAVILLARD, J. ('10). État actuel de la Protistologie végétale. Progressus Rei Botanicæ, Bd. iii, Heft 3, 1910.

PHILLIPS, O. P. ('04). A Comparative Study of the Cytology and Movements of the Cyanophyceæ. Contrib. from the Botan. Lab. of the Univ. of Pennsylvania, ii, no. 3, 1904.

RICHTER, P. ('95). Scenedesmus und die roten Körner von Glœotrichia echinulata. Ber. d. naturf. Ges. zu Leipzig, 1895—6.

SAUVAGEAU, C. ('97). Sur la Nostoc punctiforme. Ann. des Sci. Nat. sér. 8, tom. iii, 1897.

SAUVAGEAU, C. ('08) in Compt. Rend. Soc. Biol. Paris, lxiv, 1908.

SCHMIDLE, W. ('01). Ueber drei Algengenera. Ber. Deutsch. Bot. Ges. xix, 1901.

SEWARD, A. C. ('98). Fossil Plants. Vol. I. Camb. Univ. Press, 1898.

SIEBOLD ('49). Ueber einzellige Pflanzen und Thiere. Zeitschr. für wiss. Zool. i, 1849.

SPRATT, E. R. ('11). Some Observations on the Life-history of Anabæna Cycadeæ. Ann. Bot. xxv, April, 1911.

STOCKMAYER ('94). Ueber Spaltalgen. Ber. Deutsch. Bot. Ges. xii, 1894.

SWELLENGREBEL, N. H. ('10) in Quart. Journ. Micr. Sci. new ser. liv, 1910.

TEODORESCO, E. C. ('01). Sur le *Gomontiella*, nouveau genre de Schizophyceæ. Verhandl. der k. k. zool.-bot. Ges. Wien, 1901.

TEODORESCO, E. C. ('07). Matériaux pour la flore algologique de la Roumanie. Beitr. z. Bot. Centralbl. xxi, 1907.

WAGER, H. ('03). The Cell-structure of the Cyanophyceæ. Proc. Roy. Soc. 1903. See also Report Brit. Assoc. 1901, p. 830 (1902).

WEED, W. H. ('87). The Formation of Travertine and Siliceous Sinter by the Vegetation of Hot Springs. U.S. Geol. Survey. Ann. Report, ix, 1887—8.

WEST, G. S. (G. S. W. '02). On Some Algæ from Hot Springs. Journ. Bot. July, 1902.

WEST, G. S. (G. S. W. '04). A Treatise on the British Freshwater Algæ. Camb. Univ. Press, 1904.

WEST, G. S. (G. S. W. '07). Report on the Freshwater Algæ, including Phytoplankton, of the Third Tanganyika Expedition. Journ. Linn. Soc. Bot. xxxviii, 1907.

WEST, G. S. (G. S. W. '12). Freshwater Algæ of the Percy Sladen Memorial Expedition in South-West Africa, 1908—11. Ann. South Afric. Mus. ix, part ii, 1912.

WEST, W. & WEST, G. S. (W. & G. S. W. '97 A). Welwitsch's African Freshwater Algæ. Journ. Bot. 1897.

WEST, W. & WEST, G. S. (W. & G. S. W. '97 B). A Contribution to the Freshwater Algæ of the South of England. Journ. Roy. Micr. Soc. 1897.

WEST, W. & WEST, G. S. (W. & G. S. W. '11). The Freshwater Algæ in Reports on Sci. Investig. Brit. Antarctic Expedit. 1907—9, vol. i, part vii, Dec. 1911.

WILLE, N. ('83). Ueber die Zellkerne und die Poren d. Wände bei den Phycochromaceen. Ber. Deutsch. Bot. Ges. i, 1883.

WILLE, N. ('84). Bidrag til Sydamerikas Algflora. Bih. till K. Sv. Vet.-Akad. Handl. Bd. 8, no. 18, 1884.

ZACHARIAS, E. ('90; '91; '92). Ueber die Zellen der Cyanophyceen. Botan. Zeitung, 1890; 1891; 1892.

ZACHARIAS, E. ('00) in Abhandl. a. d. Geb. Naturw. Ver. Hamburg, xvi, 1900.

ZACHARIAS, E. ('07) in Botan. Zeitung, lxv, 1907.

ZUKAL, H. ('92; '94). Ueber den Zellinhalt der Schizophyten. Ber. Deutsch. Bot. Ges. x, 1892; also xii, 1894.

PERIDINIEÆ

(Peridiniales or Dinoflagellata)

THE Peridinieæ form a very natural group of purely aquatic organisms embracing a large number of motile unicells furnished with two flagella and a cell-wall, which in the majority of forms is composed of a definite system of articulated plates. They are very largely organisms of the marine and freshwater plankton, and next to the diatoms they are the most important 'producers' of organic substance in the sea. Peridinians rarely become absolutely dominant in the freshwater plankton, but in the sea, and especially in the warmer oceans, they occur in vast numbers, and along with associated diatoms, constitute the principal food-supply of countless numbers of the smaller marine animals.

The marine genera and species are much more numerous than the freshwater ones, and they exhibit a much greater diversity of form, attaining a larger size and a more elaborate external configuration. The marine and freshwater species are quite distinct, but a few of the smaller marine forms occur in brackish waters.

The Peridinieæ are brown organisms, possessing as a rule a number of yellow-brown chromatophores. For the most part their nutrition is holophytic, starch and a fatty oil being stored as food-reserves. There are also a number of colourless, saprophytic forms and a few in which holozoic nutrition may sometimes occur. The cell-wall consists of cellulose, or more rarely has become converted into a wide gelatinous envelope. A few forms are naked—that is, they are without a cell-wall.

One of the leading features of the group is the presence of a transverse furrow, which divides each cell into a front *apical half* and a back *antapical half*. Crossing the transverse furrow is a much less evident longitudinal furrow of very variable extent.

The vast majority of the forms are motile during their active vegetative existence, swimming with some rapidity by means of two flagella, one of which lies and vibrates in the transverse furrow, while the other, which is responsible for the propulsion of the organism, passes from the longitudinal furrow obliquely outwards into the surrounding water. The transverse flagellum is in some forms a ribbon-like band of protoplasm encircling the body, and by its undulating movements causes the body to rotate upon its longitudinal axis and sometimes pursue a spiral course about the axis of progression. Its effect upon the general morphology of the Peridinieæ is fundamental and a transverse furrow for its lodgment is almost universally present throughout the group. The longitudinal flagellum also determines the direction of the main morphological axis. That part of the body which leads in locomotion is called the apical region. The side of the body on which the two ends of the girdle are found is called the ventral side.

Our knowledge of the life-history of the Peridinieæ is yet very deficient. It is definitely known that certain forms periodically enter into an encysted condition in the form of large resting-spores. Some become encysted in the winter, others in the spring and summer, and yet others in the autumn. Thus, in temperate climates some species are found in the active vegetative condition in every month of the year. In warmer climates it has been shown that the periodic forms of colder latitudes become perennial constituents of the plankton. The duration of vegetative and encysted states is very variable, depending largely upon temperature. Multiplication by cell-division is often very rapid, and has been shown to occur in both the motile vegetative condition and the encysted state.

The Peridinieæ, as generally understood, fall very naturally into three divisions, concerning which there has never been any controversy. To these it is apparently necessary to add a fourth group, the Pyrocystaceæ, in order to accommodate a group of organisms which cannot with justice be included in the other three. It will be convenient to consider these groups separately.

Class *PERIDINIEÆ.*

Order PERIDINIALES.

A. Cell-wall absent or consisting of a gelatinous covering. Vegetative condition motile. Fam. *Gymnodiniaceæ.*

B. Cell-wall consisting of a continuous cellulose membrane. Vegetative condition non-motile. Fam. *Pyrocystaceæ.*

C Cell-wall consisting of a cellulose membrane which is composed of definite articulated plates. Vegetative condition motile.

a. Cells with a girdle, usually in the form of a conspicuous groove; cell-wall composed of three or more plates. Fam. *Peridiniaceæ.*

This is the important group, the classification of which depends entirely upon the morphology of the plates composing the wall.

b. Cells without a girdle; cell-wall composed of two plates only. Fam. *Prorocentraceæ.*

To these four groups should probably be added the Phytodiniaceæ of Klebs (in part; see p. 57) and the Blastodiniaceæ of Chatton (see p. 53).

In the above arrangement the Peridiniales is the only order of the class Peridinieæ, but it may be expedient in the near future to establish a new order for some of the Phytodiniaceæ and perhaps for the Pyrocystaceæ.

Family **Gymnodiniaceæ.**

This group is distinguished by the absence from the motile stages of the life-history of anything of the nature of a strong cellulose wall, such as is possessed by other Peridinieæ. Most forms are furnished with a continuous plasma-membrane of great delicacy, and some are enveloped in a wide mucous investment. Sometimes the gelatinous investment is so loose and diffluent that the naked cell is able to move about within it. The vegetative, which is also the motile, state varies in shape from a rounded or egg-shaped cell to an elongate, spindle-shaped body.

The transverse furrow is a groove or trough in the peripheral plasma. It is ring-like in *Gymnodinium* (fig. 36 *A* and *B*), but in *Hemidinium* only the left half is developed. In *Spirodinium* (fig. 36 *C* and *D*) it takes the form of an ascending spiral which makes more than a complete turn, and

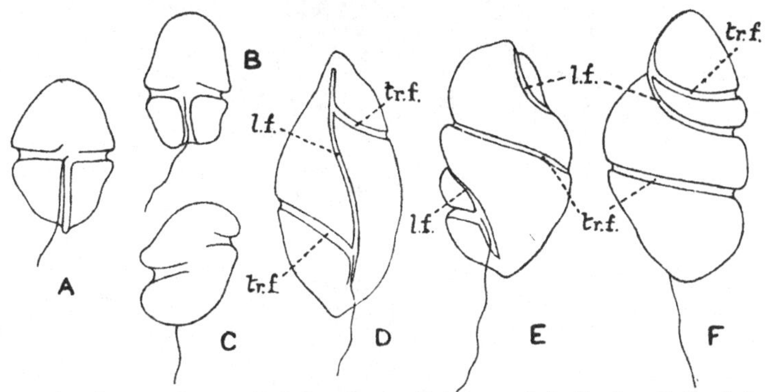

Fig. 36. Outline figures of several of the Gymnodiniaceæ. Only the longitudinal flagellum is shown. *A*, *Gymnodinium carinatum* Schilling, ×520 (after Schilling); *B*, *G. palustre* Schilling, ×750 (after Schilling); *C*, *Spirodinium hyalinum* (Schilling) Lemm., ×680 (after Schilling); *D*, *Sp. spirale* (Bergh) Schütt, ×500 (after Schütt); *E* and *F*, *Cochlodinium strangulatum* Schütt, ×160 (after Schütt); *E*, ventral view; *F*, dorsal view. *l.f.*, longitudinal furrow; *tr.f.*, transverse furrow.

in *Cochlodinium* (fig. 36 *E* and *F*) it makes several spiral turns. In *Amphidinium* the transverse furrow is so near the anterior end of the cell that the apical half appears almost as a lid to the rest of the cell (fig. 37). The longitudinal furrow is much less prominent than the transverse furrow, which it crosses. In *Gymnodinium paradoxum* Schill. and *G. pulvisculus* Klebs it is practically absent, and in a number of other forms it is only feebly developed, extending but a little way on either side of the transverse furrow. When the latter is spirally disposed, as in *Cochlodinium*, the longitudinal furrow has also a spiral twist. In the genus *Spirodinium* there are often two longitudinal flagella in place of the more normal solitary one.

The protoplast is colourless, or, as in *Cochlodinium archimedes*, uniformly tinted; and in *Pouchetia* there are numerous red droplets in the peripheral region. Vacuoles of considerable size may be present in the cytoplasm, through which branched protoplasmic threads pass to the peripheral part of the protoplast. The nucleus, which is of comparatively large size and somewhat variable in shape, may be anterior in position (as in *Gymnodinium fuscum*) or posterior (as in *Hemidinium*). The chromatin granules are scattered, and on division there is a primitive mitosis.

A red *pigment-spot* is present in some forms, but wanting in others. In the genus *Pouchetia* there is a lens-like body associated with the pigment-spot, and it is assumed that the concentration of light upon the pigment-spot by this body largely influences the orientation of the cell. Trichocyst-like

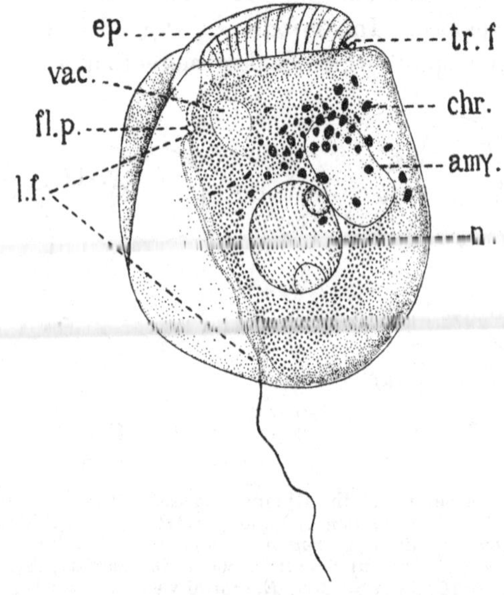

Fig. 37. Lateral view of *Amphidinium sulcatum* Kofoid. *amy.*, amyloid body; *chr.*, chromatophore; *ep.*, anterior part of cell in front of transverse furrow; *fl.p.*, flagellar pore; *l.f.*, longitudinal furrow; *n.*, nucleus; *tr.f.*, transverse furrow (with flagellum); *vac.*, vacuole. × 740 (after Kofoid).

structures occur in certain species of *Gymnodinium* and *Spirodinium*, sometimes forming a more or less continuous peripheral zone. Some species of these genera throw out numerous threads of mucus, either just previous to entering the resting state or on the advent of unfavourable external conditions: these threads rapidly swell up and form a colourless gelatinous envelope.

The chromatophores are disc-shaped, rod-like, or band-like, and are mostly of a golden-brown colour, although *Gymnodinium viride* is green, *G. æruginosum* is blue-green, and *G. cœruleum* blue. The two latter organisms contain phycocyanin in addition to the green and brown pigments. The nutrition of the coloured forms is normally holophytic, but in certain cases there appears to be in addition a truly animal-like method of taking up solid food by means of pseudopodia protruded from the region of the transverse furrow. This occurs, according to Zacharias, in *Gymnodinium Zachariasi* Lemm. Some forms, such as *Gymnodinium fucorum*, *G. spirale*, etc., are without chromatophores and lead a saprophytic existence. A few parasitic forms have also been described, notably *Gymnodinium parasiticum* Dogiel, *G. Pouchetii* Lemm., and *G. roseum* Dogiel. The first-named lives during part of its life-history within the eggs of Copepods[1].

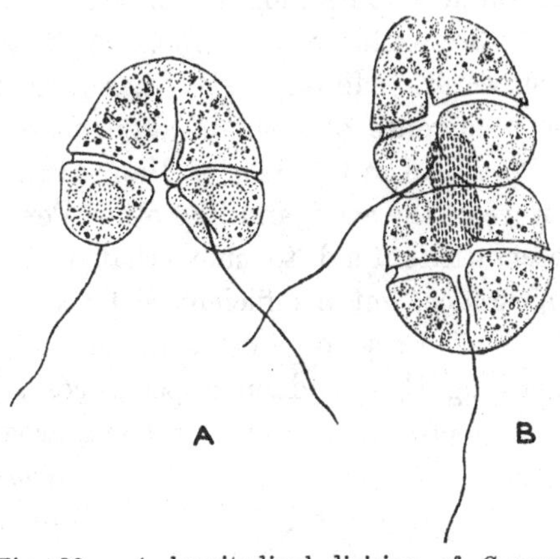

Fig. 38. *A*, longitudinal division of *Gymnodinium viride* Penard, the fission almost completed; *B*, transverse fission of *G. rufescens* (Penard) Lemm. showing the dividing nucleus (after Penard). Only the longitudinal flagella are shown. × about 450.

In the colourless forms *Gymnodinium Vorticella* Stein, *G. helveticum* Penard, and *Spirodinium hyalinum* (Schill.) Lemm., the nutrition is stated to be completely holozoic, but this assertion requires very careful confirmation. In the last-named, the organism loses its flagella and becomes amœboid before the ingestion of solid food-substances. In *Gymnodinium Vorticella* Stein, the ingestion of food takes place while the organism is active, and the food-remains are thrown out when in the resting-state (Dangeard, '92).

Multiplication occurs in most cases by cell-division while the organisms are in the motile state. The division is usually longitudinal, although in a few cases, such as in *Gymnodinium rufescens*, it is transverse (fig. 38 *B*). Unfavourable conditions of existence rapidly bring about a quiescent, encysted state, in which multiplication may sometimes occur. In this non-motile state the cell may be surrounded by a thick gelatinous envelope (as in *Gymnodinium Zachariasi*) or only by a thin membrane (as in *Hemidinium* spp. and *Gymnodinium fucorum*). After the division of the parent-cell the two daughter-cells are set free either by dissolution of the gelatinous envelope or by the bursting of the thin membrane. A resting-state with thick cellulose walls has occasionally been observed in some forms.

[1] E. Chatton (*Comptes Rendus Acad. Sc. Paris*, cxliii, 1906; *ibid.* cxliv, 1907) has also described several parasitic species of the genus *Blastodinium*, which he regards as the type of a new family of the Peridinieæ. These organisms are parasites upon various pelagic Copepods and Appendicularians.

Most of the Gymnodiniaceæ are marine organisms, although many species of the genus *Gymnodinium* occur in fresh waters, generally in small pools and ditches containing much vegetation. *Spirodinium*, *Cochlodinium*, and others are exclusively marine, but the species of *Hemidinium* chiefly inhabit fresh water. Herdman ('11) has furnished some interesting observations on the occurrence of *Amphidinium operculatum* Clap. & Lachm. on the beach at Port Erin in the Isle of Man. It is evidently a shore-form, living on the surface of the sand, and like other Peridinians, sensitive to light. Herdman's observations tend to show that it is either periodic or spasmodic in its appearance, but insufficient is known concerning the physiological requirements of these organisms to enable any definite statements to be made regarding their sudden appearances and disappearances.

A member of the Gymnodiniaceæ which requires special mention is *Polykrikos* (fig. 39). Originally described by Bütschli ('73) as an Infusorian and afterwards referred by Bergh ('81) to the so-called Cilioflagellata, this plankton-organism has recently been carefully examined by Kofoid ('07 A), who has shown that it is merely a member of the Gymnodiniaceæ with a permanent colonial organization. There are two, four, or eight transverse groves, each with its flagellum and each corresponding to the equatorial region of a single individual. The longitudinal furrows are all continuous, but there are as many longitudinal flagella as there are transverse flagella. The hindmost flagellum is however somewhat longer and acts as the propelling organ. Kofoid also found that the nuclear division was not always coincident with, but often lagged behind the multiplication of superficial structures such as furrows, flagella, etc. *Polykrikos* is therefore a permanent colonial member of the Gymnodiniaceæ consisting of two, four, or eight individuals which are only incompletely separated. Temporary colonies occur in *Spirodinium geminatum* and in the active *Gymnodinium*-stages of some of the Peridiniaceæ (*vide* fig. 51 *K* and *L*).

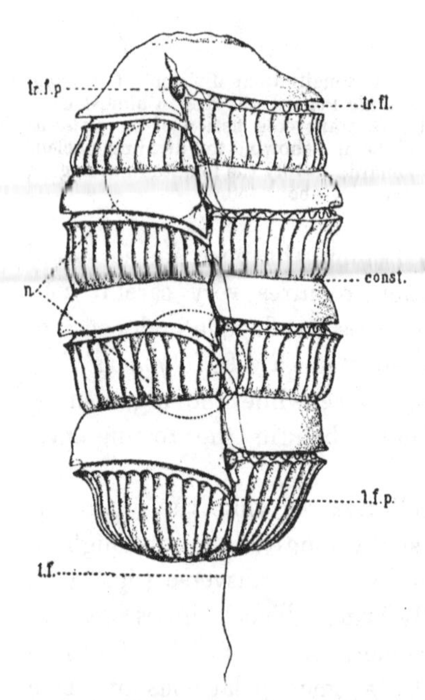

Fig. 39. *Polykrikos Schwartzii* Bütschli. Ventral view showing surface markings, flagella, and flagellar pores in a somewhat contracted individual. *Const.*, constriction between individuals of colony; *l.f.*, longitudinal flagellum; *l.f.p.*, pore of longitudinal flagellum; *n.*, nuclei; *tr. fl.*, transverse flagellum; *tr. f. p.*, pore of transverse flagellum. × 300 (after Kofoid).

In the peripheral zone of the protoplast, *Polykrikos Schwartzii* Bütschli and *Pouchetia armata* Dogiel are provided with nematocysts. It is by no means easy to explain how these structures are functional in unicellular organisms of such a type. It is easy to state that they play a part in nutrition, but there is no evidence to show that the nutrition of these organisms is other than saprophytic. Possibly they act purely as protective structures.

The organism known for a long time as *Pyrocystis lunula* (fig. 40 *B* and *C*) has quite recently been transferred by Klebs ('12) to the Gymnodiniaceæ and made the type of a new genus, *Diplodinium*. This change is made as the result of the recent work of Apstein ('06) and Dogiel ('06), the genus being based upon the fact that there are two successive encysted stages in the life-history. The primary cysts are large and globose, the contents dividing to form two, four, or eight secondary cysts of a crescentic or lunate form, which are set free by the rupture of the old wall. The secondary cysts now produce by successive divisions of the protoplast either 8 or 16 flagellated organisms of the *Gymnodinium*-type. After 'swarming' for a time within the wall of the lunate cyst, they become quiescent and surround themselves with a membrane; but nothing is known of the development of the primary cysts from these small flagellated cells.

Klebs has placed those Gymnodiniums which produce thick-walled, sublunate, and often horned resting-cysts in a genus which he describes as *Cystodinium*. He has also established a further genus, *Hypnodinium*, in which the division of the encysted protoplast produces two naked cells with the *Gymnodinium*-like furrows, but no flagella. These are set free, and grow into new cysts without ever becoming motile.

The present author does not agree with Klebs in the transference of *Glenodinium* to the Gymnodiniaceæ (see foot-note on p. 61).

Family **Pyrocystaceæ.**

The Pyrocystaceæ is represented by the solitary genus *Pyrocystis*, which was originally discovered during the voyage of the "Challenger." The organisms often occur in great abundance in the pelagic plankton of tropical and subtropical seas, wherever the temperature of the water is over 68° or 70° F. They are strongly luminous, and there is little doubt that the diffuse luminosity of the open sea in warmer regions is mainly due to species of *Pyrocystis*. When floating quietly they give out no light, but agitation of the surface water causes them to shine out as points of bluish light. It is probable that the luminosity is due to the oxidation of certain substances, such as lecithin, cholesterin, ethereal oils, etc., which are contained within the cell. Agitation of the water increases the rapidity of oxidation, either by augmenting the supply of oxygen or by raising the irritability of the protoplasm. Blackman ('02) has shown that dilute acetic acid and also absolute alcohol will so raise the irritability of the protoplasm that the organisms react more vigorously to the stimulus of shaking and thus glow with a brighter light.

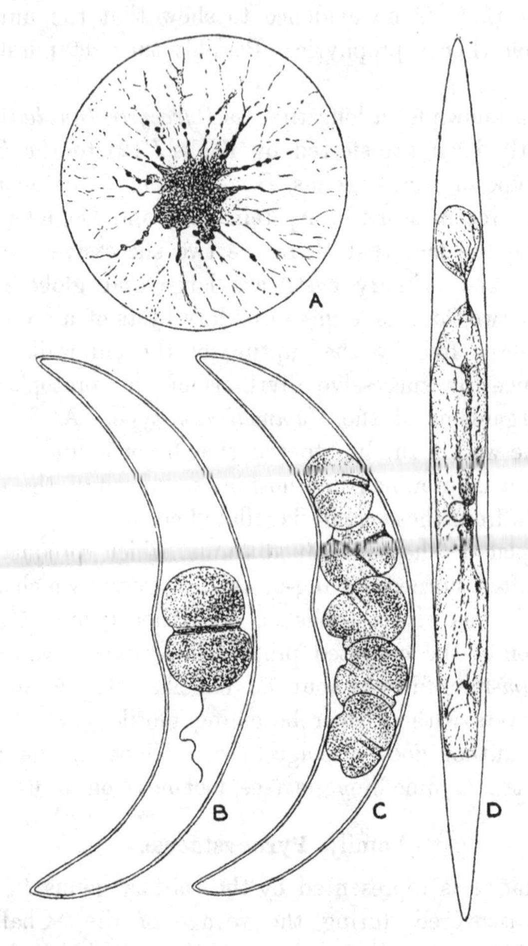

Fig. 40. *A*, *Pyrocystis pseudonoctiluca* Wyv. Thoms., a normal uncontracted specimen after treatment with osmic acid, × 150; *B* and *C*, *Diplodinium lunula* (Schütt) Klebs (= *Pyrocystis lunula* Schütt), showing flagellated spores of the *Gymnodinium*-type, × 475; *D*, *Pyrocystis fusiformis* Wyv. Thoms., a slightly contracted specimen, × 75. (*A* and *D* after Blackman; *B* and *C*, after Schütt).

Each individual is furnished with a strong wall of cellulose, about 5 μ in thickness (Blackman, '02), which was at first erroneously regarded as siliceous owing to its glassy appearance. The nucleus is large and may be centrally placed, as in the elongated *Pyrocystis fusiformis* Wyv. Thoms., or laterally situated, as in the spherical *P. pseudonoctiluca* Wyv. Thoms., but in all cases it is located in a concentration of the cytoplasm from which numerous protoplasmic strands stretch out through the large vacuole which occupies the greater part of the interior of the cell. Numerous small, oval chromatophores give the cells a brown or golden-brown colour. There is also an abundance of small oil globules, and larger refractive amylum-bodies, the latter sometimes almost filling the whole cell. The normal vegetative condition in the Pyrocystaceæ appears to combine features usually exhibited by the resting and the encysted states of other Peridinieæ. The cell possesses the thick cell-wall characteristic of the resting-state, but undergoes repeated divisions such as occur in the more usual encysted condition.

The systematic position was discussed by Blackman ('02), who stated that 'but for the somewhat meagre evidence provided by these figures of Schütt, the Pyrocysteæ would have to be considered as an algal group of quite unknown affinities.' The flagellated spores observed by Schütt (in '*Pyrocystis lunula*') were of the *Gymnodinium*-type, with a transverse furrow, and it was for this reason that he placed the Pyrocysteæ in the Gymnodiniaceæ. It is, however, the recent investigations of Apstein ('06) and Dogiel ('06) which have furnished the strongest evidence of the Peridinian nature of the Pyrocystaceæ. Even though the old '*Pyrocystis lunula*' has been transferred to the Gymnodiniaceæ as *Diplodinium lunula*, yet its encysted states present no essential differences from the cells of *Pyrocystis*. In both cases the walls are of cellulose, the nucleus is large, there are numerous yellow-brown chromatophores, and oil-globules and amylum bodies are found to be the food-reserves. Since *Diplodinium lunula* is now definitely known to form small *Gymnodinium*-like flagellated cells in one stage of its life-history, it is not unreasonable to suppose that species of *Pyrocystis* are closely allied organisms in which the encysted state has become the normal vegetative phase and the motile state has been suppressed.

Family **Phytodiniaceæ** [Klebs in part].

Klebs ('12) established this family of the Peridinieæ to include four newly described genera (*Phytodinium*, *Tetradinium*, *Stylodinium* and *Glœodinium*) and the genus *Pyrocystis*. Since, however, the relationships between these new genera are not at all clear, and their affinities with *Pyrocystis* are possibly somewhat remote, the present author prefers to retain *Pyrocystis* in the separate family Pyrocystaceæ (as above). All the new genera described by Klebs include comparatively small fresh-water forms, whereas *Pyrocystis* is a genus of much larger marine forms in which all the species are luminous. It is probable that the phylogenetic history of *Pyrocystis* is very different from that of *Phytodinium*, *Tetradinium*, etc. Until more observations have been made, or further allied forms discovered, it is difficult to discuss the new genera proposed by Klebs. For instance, *Tetradinium javanicum* Klebs is so like the encysted state of certain fresh-water species of *Ceratium* that one may be forgiven for suggesting that there is, as yet, no absolute proof that it is not such a state, especially in view of the fact that divisions may occur in cysts of the Peridiniaceæ with the formation of new cysts (G. S. W. '09). It may, of course, be an encysted state which has become permanent, never reverting to the motile condition of the form from which it originated. *Glœodinium montanum* Klebs bears a striking resemblance to *Chroococcus macrococcus* Rabenh. (= *Urococcus insignis* of most authors, but probably not *Ourococcus insignis* Hassall), and is moreover found in precisely similar habitats. As *Chroococcus macrococcus* is a true member of the Myxophyceæ, with only an incipient nuclear development (see fig. 5) it would be of interest to see detailed figures of the nuclear structure of *Glœodinium montanum.*

Family **Peridiniaceæ.**

In the Peridiniaceæ the vegetative condition is motile. The cell-wall is firm and hard, being built up of at least three, and usually of a number of articulated plates. There is a well-marked transverse furrow composed of one or more girdle plates. The external form is very varied; the cell may be globose, ovoid, rhomboidal, or even flat and leaf-like, and in many kinds (chiefly marine) various horn-like and wing-shaped expansions are developed, thus greatly increasing the floating capacity of the cell.

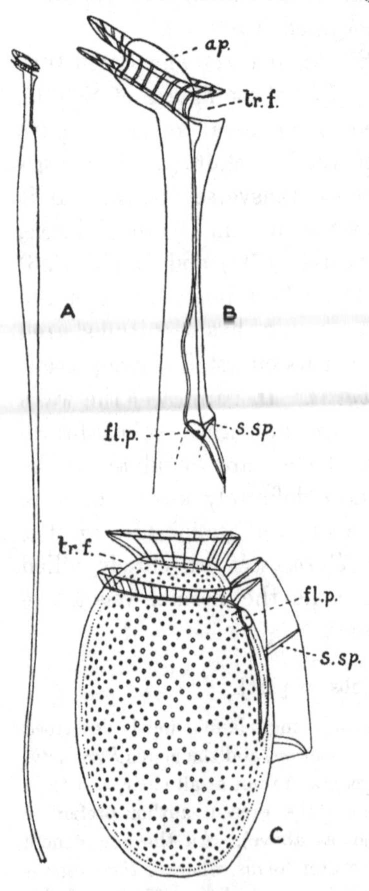

Fig. 41. *A* and *B*, *Amphisolenia spinulosa* Kofoid. *A*, single cell viewed from right side, × 140; *B*, anterior extremity of cell, × 912; *C*, *Dinophysis ellipsoides* Kofoid, view of right side, × 850. *ap.*, apical portion of cell; *fl. p.*, flagellar pore; *s. sp.*, suture spine; *tr. f.*, transverse furrow. (All after Kofoid.)

TRANSVERSE AND LONGITUDINAL FURROWS. In all the Peridiniaceæ except *Podolampas* and *Blepharocysta*, the *transverse furrow* is a conspicuous feature of the cell. The individual cell is thus plainly divided into an apical and an antapical half, the latter being usually the larger. In *Peridinium bipes* Stein and most of the freshwater species the apical half is the larger, whereas in *Oxytoxum*, *Ceratocorys*, *Dinophysis*, *Amphisolenia*, etc., the apical half is only feebly developed. The transverse furrow is approximately equatorial in *Peridinium africanum* and in *P. minutum*, but in most forms it takes a slightly spiral course around the cell, either to the left or to the right, one end of the furrow being thus displaced anteriorly. It may also be somewhat deepened by the development of prominent lips upon either margin.

In *Podolampas*, in which there is no true transverse furrow, 'the missing girdle is represented by a narrow band fused to the lower ends of the precingular plates. On the surface of this band, which is in the place along which the transverse flagellum passes, is a very shallow furrow' (Kofoid, '09 B).

Crossing the transverse furrow on the ventral side of the organism is a rather more open *longitudinal furrow*, which is best seen from an anterior or posterior view of the organism. This furrow usually extends but a little way into the apical half of the cell, but it may reach to the extreme apex, as in *Gonyaulax apiculata* (Pen.) Entz. In many cases it extends only across the median part of the cell, but it may sometimes be confined to the antapical half, as in *Peridinium Penardii* Lemm. The left margin of the furrow is sometimes provided with a wing-like expansion, and if both margins are thus extended the left one is the more strongly developed. In some cases the furrow is so expanded as to form a *ventral area*.

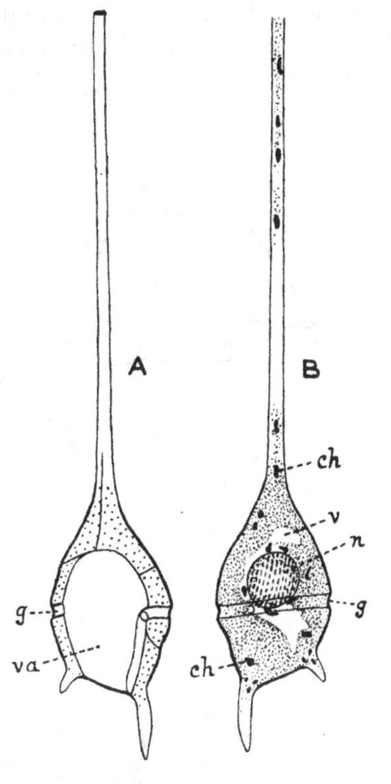

Fig. 42. *Ceratium teres* Kofoid. *A*, ventral view of empty cell. *B*, dorsal view of cell showing contents. × 400. *ch*, chromatophores; *g*, girdle; *n*, nucleus; *v*, vacuole; *va*, ventral area. (After Kofoid.)

STRUCTURE OF CELL-WALL. The firm cell-wall is composed of definitely articulated plates which consist largely of cellulose and to a lesser extent of callose and pectose (Mangin, '07). The whole outer covering thus forms a sort of exoskeleton which is divided by the transverse furrow (or girdle) into an *epivalve* (or epitheca) and a *hypovalve* (or hypotheca). These two halves, except in the Podolampinæ, are separated by the girdle, which may or may not be subdivided into several plates.

It is, of course, the disposition of the transverse flagellum and its location in a groove which in all the Peridiniaceæ is the primary cause of the division of the cell into an apical half covered by the epivalve and an antapical half covered by the hypovalve.

Several proposals have been made with regard to the naming of the plates composing the cell-wall Stein ('83), Bütschli ('85), Schütt ('96), Paulsen ('06), Fauré-Fremiet ('08), and Kofoid ('09) have all put forward systems of nomenclature, but that proposed by Kofoid has so many advantages over the other systems that it will without doubt be generally adopted. It is the simplest and the most easily applied, having a definite morphological basis which admits of an accurate comparative treatment of all the genera of the Peridiniaceæ. It recognizes that all the plates are in transverse series, and that the girdle (or transverse furrow) rather

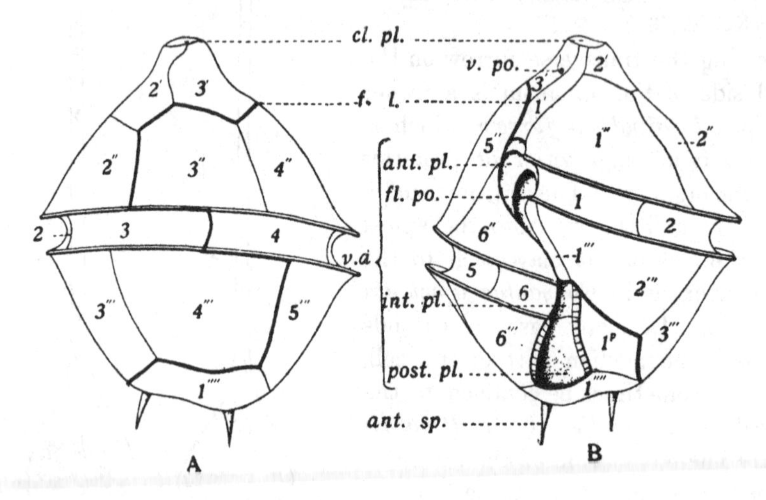

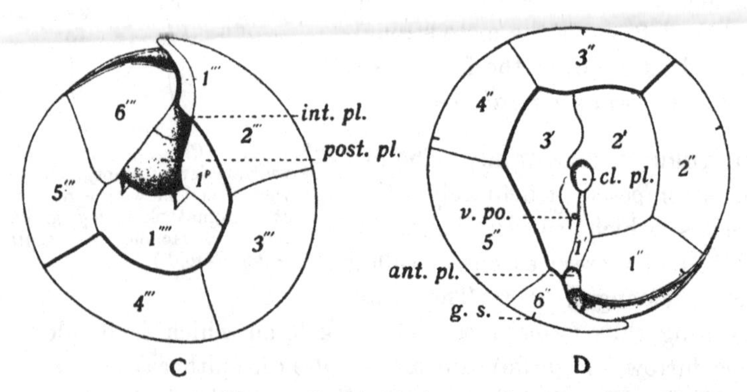

Fig. 43. *Gonyaulax spinifera* (Clap. & Lach.) Diesing, to show the plates of the cell-wall. *A*, dorsal view; *B*, ventral view; *C*, antapical view; *D*, apical view; 1—6, girdle series of plates; 1′—3′, apical series; 1″—6″, precingular series; 1‴—6‴, postcingular series; 1ᵖ, posterior intercalary; 1⁗, antapical plate; *ant. pl.*, anterior plate of ventral area; *ant. sp.*, antapical spine; *cl. pl.*, closing plate of apex; *f.l.*, fission line along which the wall parts in cell-division; *fl. po.*, flagellar pore; *int. pl.*, intermediate plates (usually four in number, hidden in constricted region of ventral area); *post. pl.*, posterior plate of ventral area; *v.a.*, ventral area (=longitudinal furrow); *v. po.*, ventral pore. ×1000 (after Kofoid).

than the equator should be used as a basis, since the plates are invariably arranged with reference to the girdle, which is not necessarily equatorial in position.

The girdle (or *cingulum*) separates the epivalve from the hypovalve, and in all the genera except *Glenodinium* and *Ptychodiscus*, these two halves of the cell-wall are composed of definite series of plates[1]. There are two transverse encircling series of plates in each half; that is, on each side of the girdle. The apical part of the epivalve is composed of the *apical plates*, either arranged as a cluster (*Oxytoxum*, *Amphidoma*) or in a series grouped about an open apical pore (*Ceratium* and some species of *Peridinium*). Behind these plates, and bordering the girdle, is the second series, generally more numerous and of larger size, the *precingular plates.* In a few genera a third series occurs between these two upon the dorsal side and often asymmetrically displaced to the left; these are the *anterior intercalary plates*, which never form a complete encircling series. There are two or three of them in *Peridinium*, and one in *Heterodinium*. In the hypovalve there is a series of *postcingular plates* immediately behind the girdle, usually fewer than in the precingular series and sometimes of larger size. The remaining part of the hypovalve, which is the posterior extremity of the cell, consists of one or two *antapical plates.* A solitary *posterior intercalary plate* may also occur.

The series of plates in both the epivalve and the hypovalve may be interrupted by a forward or backward extension of the longitudinal furrow or ventral area. This interruption in the mid-ventral line forms a convenient starting-point for numbering the plates of each series, the order being continued in the direction taken by the transverse flagellum; that is, from left to right. Kofoid's method of designating the various series is by using acute accent marks from $'$ to $''''$ for the four series, and *a* and *p* respectively for the anterior and posterior intercalaries. Thus, the cell-wall of one of the Peridiniaceæ, which contained the maximum number of plates, would consist of:—Apicals $1'$—$4'$; anterior intercalaries 1^a—3^a; precingulars $1''$—$7''$; postcingulars $1'''$—$5'''$; posterior intercalary 1^p; antapicals $1''''$ and $2''''$.

So little is known of the plates which form the girdle, and of those of the ventral area, that they have not been included in this general system of nomenclature. There are six girdle plates in *Gonyaulax* and three in *Peridinium.*

Schütt ('95) has shown that the plates can only be accurately determined by following the sutures, and that the superficial ridges, etc., frequently offer

[1] The present author (G. S. W. '09) has observed a delicate tracery on the two valves of *Glenodinium uliginosum* exactly resembling the suture-lines of certain of the smaller species of *Peridinium.* Schilling ('91) also records such an instance. This fact, along with its thick wall of three distinct parts and its method of cyst-formation, indicates that *Glenodinium*, although intermediate between *Gymnodinium* and *Peridinium*, is much more closely allied to the latter than the former and is rightly placed in the Peridiniaceæ. Klebs has recently placed *Glenodinium* in the Gymnodiniaceæ, but on grounds which do not appear to be sufficient. Following up Klebs' suggestion, Schilling ('Dinoflagellatae' in *Die Süssw.-fl. Deutschl. Österr. u. der Schweiz*, Jena, 1913) has replaced the names Gymnodiniaceæ and Peridiniaceæ by 'Kyrtodiniaceæ' and 'Krossodiniaceæ' respectively.

no clue to the sutures. This has been still further emphasized by Kofoid ('07 c), who states that the 'actual separation of the plates *in situ* is the only safe guide to an analysis of the thecal wall.'

The plates are often covered with minute spines, or with a close network of ridges giving them an areolated appearance. Both pores and poroids occur on the plates, only the former being real perforations through the wall. The wing-like expansions which occur in some of the marine forms are supported by variously arranged radial and transverse horns and ribs. These expansions do not act merely as parachute-like structures: their disposition is such that the sinking organism always assumes a definite position. The expansions of *Ceratocorys*, *Phalocroma*, *Ornithocercus*, etc., and the apical

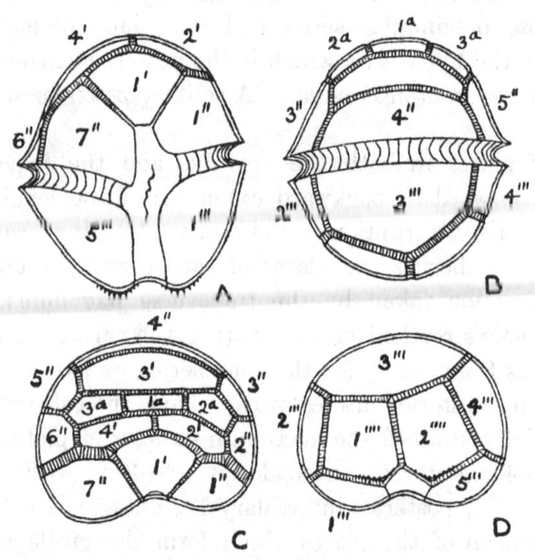

Fig. 44. ***Peridinium Willei*** **Huitf.-Kaas, to show the plates composing the wall. *A*, ventral view; *B*, dorsal view; *C*, apical view; *D*, antapical view. 1′—4′, apical plates; 1″—7″, precingular plates; 1^a—3^a, anterior intercalary plates; 1‴—5‴, postcingular plates; 1⁗ and 2⁗, antapical plates. The sculpture of the plates is not shown in the figures. × 500.**

and antapical horns of *Ceratium*, are always asymmetrically disposed. In some species Kofoid ('10) has demonstrated that in the passive sinking of the organisms, this fundamental asymmetry of horns and expansions causes the body of the cell to turn on to its ventral face with the maximum exposure of body and horns to the direction of sinking, thus delaying the descent.

The sutures of the plates are sometimes scarcely visible, as in *Peridinium berolinense* Lemm., but more often the lines of junction are marked by *intercalary bands*, which not infrequently possess prominent cross-striations (figs. 44 and 45). These attain their greatest development in *Peridinium multistriatum* Kofoid ('07 B), in which species they occupy a bigger area of the cell-wall than the plates themselves (consult fig. 45). It has been noticed in many species of *Peridinium* that the intercalary bands are wider and the striations stronger in old than in young cells, and it is probable that this increase in width of the intercalary bands is a means whereby the armour-plated cell may grow in size. Adjacent plates are joined by overlapping margins, which are often grooved and firmly cemented. It is by an extension of the surface of these thin outer flanges of the plates that the increase in width of the intercalary bands is accomplished. Kofoid ('09 A) states that the increased width of the intercalary bands is associated with the thickening of the cell-wall and the reduction in the porulate areas of the plates, and he suggests that these bands may be structural adaptations facilitating communication between the exterior and the interior of the cell,

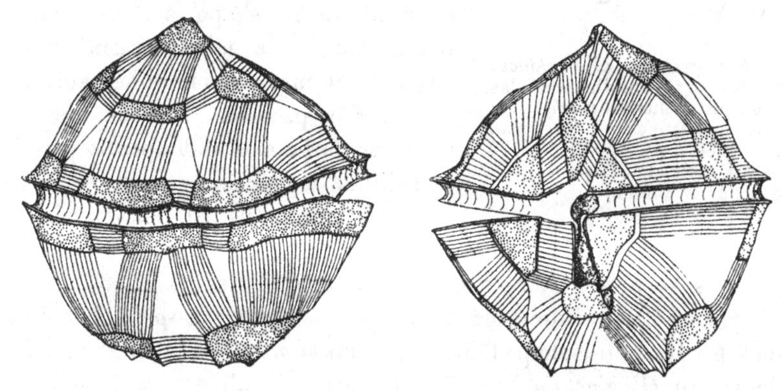

Fig. 45. *Peridinium multistriatum* Kofoid. Dorsal (left figure) and ventral (right figure) views of cell, showing very wide striated intercalary bands. × 530 (after Kofoid). In both figures the cell-wall is slightly dislocated immediately behind the girdle.

and increasing the strength of the union between the plates. The striations are oblique structures within the wall or on the oblique faces of the overlapping margins of the plates. They are most probably of the nature of modified pores or canals, and are sometimes continued beyond the edges of the overlapping margins.

In some of the marine forms there is an intracellular skeleton consisting of curved rods, five-rayed stars, or perforated basket-like structures (fig. 46 *A* and *B*).

In some part of the longitudinal furrow or of the ventral area there is an aperture through the cell-wall known as the *flagellar pore*, through which the two flagella pass as they leave the protoplast. This pore may be cleft-like, round, or oval, and its outer margin is sometimes provided with small teeth.

THE PROTOPLAST. The protoplast is colourless in most forms, but in its outer portion it may be faintly tinged with red. The peripheral region is of a denser character than the more central part, and as a rule contains many large granules. The central region, on the other hand is much more finely granulate, and contains the nucleus, the cell-sap vacuoles, and a contractile vacuole. Sometimes the distinction between these zones is not very distinct, but in other species the layers are sharply separated. There is also a thin layer of cytoplasm on the exterior of the cell-wall which is directly connected with the inner cytoplasm by the fine strands passing through the pores of the wall. The *nucleus* is spherical, oval, or reniform in shape, and is located in most forms in a fairly central position, although it may be sometimes anterior or posterior in position. On the whole, it is relatively large and undoubtedly primitive. It exhibits a striation due to the presence of rather numerous parallel threads. The latter are sometimes rather stout and they consist of nuclear rods enclosed in nuclear tubes. There may be one or more nucleoli present.

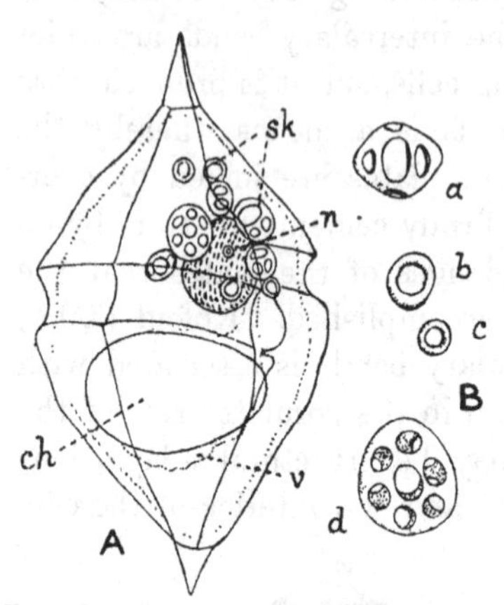

Fig. 46. *A*, ventral view of *Amphidoma biconica* Kofoid to show the protoplast and skeletal inclusions; *ch*, chromatophore; *n*, nucleus; *sk*, intracellular skeleton; *v*, vacuole. ×750. *B*, isolated skeletal inclusions, *a*–*d*, ×1650. (After Kofoid.)

In some of the Peridiniaceæ there is a red *pigment-spot* (the so-called 'eye-spot'). It is disc-shaped in *Glenodinium neglectum* Schütt and horseshoe-shaped in *Gl. cinctum* Ehrenb. It agrees in all essential details with the pigment-spot of the Flagellata, consisting of a protoplasmic groundsubstance in which are embedded small rod-shaped masses of hæmatochrome. In the genus *Peridinium* the pigment-spot has only been observed in the two species, *P. quadridens* Stein and *P. balticum* (Lev.) Lemm. Zacharias has stated that *P. quadridens* possesses two pigment-spots, one in each half of the cell.

The *longitudinal flagellum* is generally longer than the cell, and is directed backwards, corresponding to the trailing flagellum of many of the Flagellata. Sometimes it lies in the longitudinal furrow, but more often it stands out obliquely from the cell. Its movements are usually whip-like, but it may spirally contract and then suddenly unroll itself. In certain species of *Ceratium* and *Podolampas* it has also been observed to withdraw itself entirely within the flagellar pore, and occasionally in *Ceratium tripos* and *C. cornutum* two longitudinal flagella have been seen.

The *transverse flagellum*, which is at all times difficult to demonstrate, lies concealed in the transverse furrow, and in some forms the concealment is all the more complete by the development of ridge-like expansions at the margins of the furrow. Only in the Podolampinæ is the transverse flagellum exposed on the surface of the wall and not hidden in a groove. It is always turned to the left side of the cell-body. It is in some forms a thread-like flagellum and in others a band-like structure. Its nature is not easily

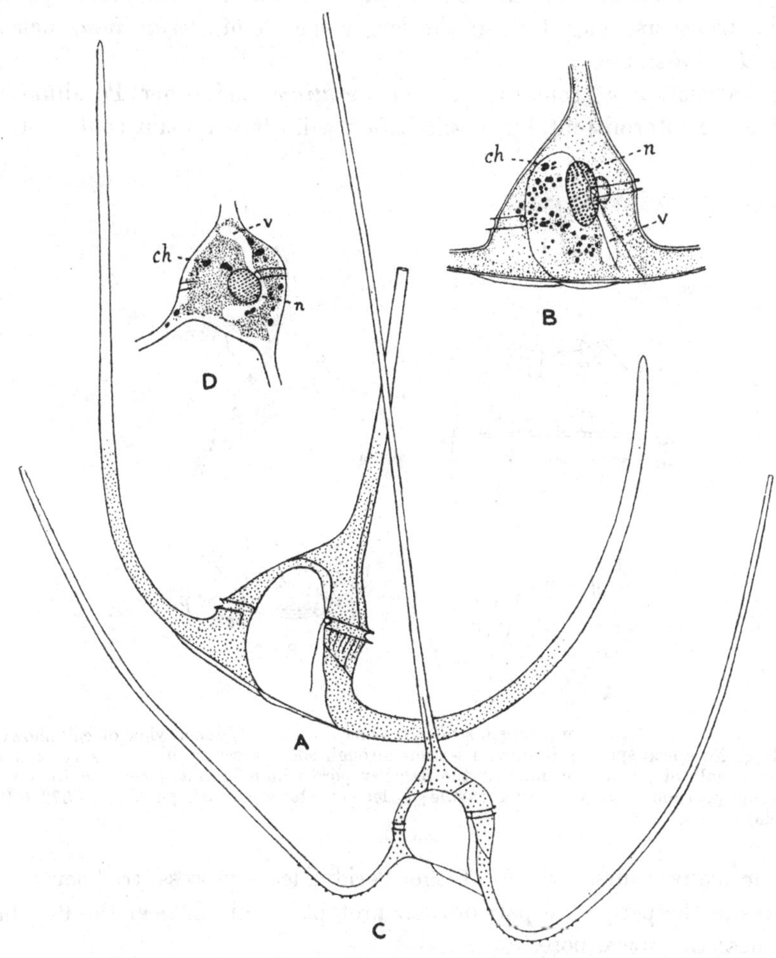

Fig. 47. *A* and *B*, *Ceratium Schrankii* Kofoid, ×310; *A*, ventral view; *B*, ventral view of central part of cell to show contents. *C* and *D*, *C. gallicum* Kofoid, ×250; *C*, ventral view; *D*, central part of cell to show contents. *ch*, chromatophores; *n*, nucleus; *v*, vacuoles. (After Kofoid.)

determined owing to the difficulty in observing the flagellum in the living cell[1]. The transverse flagellum of *Pyrodinium bahamense* is described by

[1] In some species of *Gymnodinium* and *Peridinium* the thread-like transverse flagellum can be seen vibrating by gently running in a small quantity of iodine solution or a 2 per cent. solution of osmic acid at one side of the cover-glass. Whenever the living organisms come

Plate ('06) as a fixed band running round the transverse furrow and with both ends passing into the flagellar pore. The inner fixed border of the band is thickened, whereas the outer border is very delicate, and the movement consists of wave-like undulations of this delicate free edge. Lemmermann ('10) also states that he has on several occasions observed a similar band-like transverse flagellum in the larger species of *Peridinium*, such as *P. bipes*, *P. Willei*, etc.

The locomotion of some species of *Ceratium* and other Peridiniaceæ appears to be intermittent, but *Glenodinium* will often remain continuously

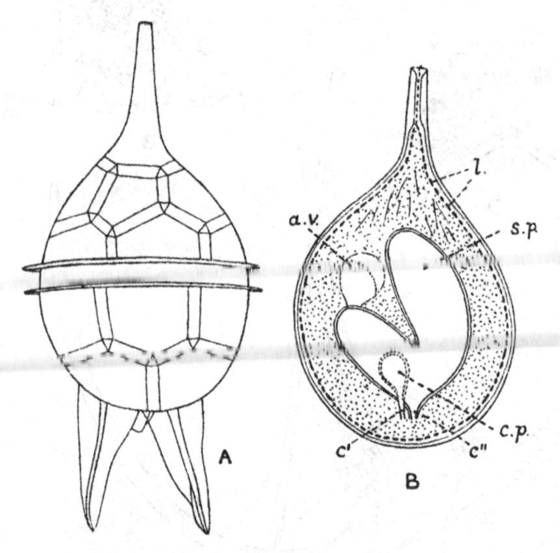

Fig. 48. *Peridinium Steinii* Jörg. subsp. *mediterraneum* Kofoid. *A*, dorsal view of cell showing the large antapical spines; *B*, optical section through body of cell; *a.v.*, accessory vacuole; *c'*, *c''*, canals of pusules opening at the flagellar pore which in this species is in a very posterior position; *c.p.*, collecting pusule; *l*, leucoplasts; *s.p.*, sack pusule. ×570 (after Kofoid).

active for many hours. Some of the Peridiniaceæ possess trichocyst-like structures in the peripheral part of their protoplasts, either near the flagellar pore or near the apical pore.

In some forms, and particularly in *Peridinium Steinii*, there is a well-marked 'pusule apparatus,' which was first described by Schütt ('92; '95). This consists of certain peculiar vacuoles in the protoplast, and it is doubtful if they are present in more than a few forms. In *Peridinium Steinii* (fig. 48) there is a large bilobed 'sack pusule' which opens by a short funnel-shaped canal leading to the flagellar pore. On the ventral

within range of these reagents their movements immediately slow down and in many cases the undulating transverse flagellum is thrown right out of the transverse furrow.

side of this 'pusule,' and in a median position, is another smaller spherical 'pusule,' which also has a duct leading to the flagellar pore. Schütt termed this the 'collecting pusule.' Other much smaller 'pusules' are aggregated about the larger ones. Schütt apparently regarded the canals as ducts for the discharge of the fluid contents of the 'pusules' through the flagellar pore. Kofoid ('09 A) has, however, produced evidence to show that the small 'pusules' form on the walls of the main sacs and pass peripherally, 'discharging their contents, perhaps by osmosis, at the surface.' His observations all indicate that the movement of fluid is inwards 'through the flagellar pore into the sack and collecting "pusules," and thence into the daughter and accessory vacuoles, perhaps also into typical plasma vacuoles, and ultimately out through the apical and other pores of the thecal wall.' He also points out that the maximum development of the 'pusule apparatus' occurs in those colourless forms in which the chromatophores are replaced by leucoplasts, and suggests that perhaps this system of cell-structures is in some way concerned with a saprophytic mode of nutrition such as might occur in zones of decaying plankton.

THE CHROMATOPHORES AND NUTRITION. In the Peridiniaceæ the chromatophores are so delicate that they can only be satisfactorily observed in the living organisms or by the most careful fixation and staining. In outward form they vary considerably; in some they are disc shaped, in others rod-like or band-like, and they may be much lobed. The disc-shaped and band-like chromatophores are usually thick in the middle and thin at the margin, and are disposed in a parietal manner. On the other hand, the rod-like chromatophores are usually arranged radially from the periphery inwards. The chromatophores are embedded in the peripheral part of the cytoplasm, but their distribution is by no means uniform in different parts of the cell. In *Glenodinium uliginosum,* and in the encysted state of *Peridinium anglicum,* the cell may contain only one large parietal, lobed chromatophore (G. S. W., '09). In colour they vary from yellow to brown, but in *Peridinium herbaceum* Schütt they are green. The usual yellow-brown chromatophores are stated to possess three pigments: *phycopyrrin, peridinin,* and a peculiar form of *chlorophyll,* but further investigation of these pigments is desirable before any definite statements can be made concerning them. It is not improbable that the different shades of colour are due to varying proportions of chlorophyll, xanthophyll and carotin, as in the case of Diatoms. The depth of colour varies according to the environment. In some of the freshwater forms the chromatophores are paler in summer than in spring or autumn.

Some of the Peridiniaceæ, such as *Glenodinium apiculatum, Peridinium achromaticum, P. Steinii,* etc., are destitute of chromatophores.

Nutrition. The great majority of the Peridiniaceæ possess chromatophores and have a typical holophytic nutrition. The colourless forms are mostly saprophytes. An animal-like nutrition has been either observed or suspected to occur in a number of genera and species, even in certain forms which possess definite chromatophores. Both Entz and Schütt have stated that the ball-like structures found within the cells of certain species of *Ceratium, Gonyaulax, Goniodoma, Peridinium, Dinophysis, Oxytoxum,* and other genera, are the remains of ingested food-materials, but in no single instance has the actual ingestion been observed. Zacharias found the valves of minute diatoms of the genera *Cyclotella* and *Navicula* within the cells of *Glenodinium apiculatum,* and Schilling has also seen the valves of small diatoms within the protoplast of *Glenodinium edax.* The natural assumption is that these diatoms were picked up and engulfed as food. There is, however, no proof of this; neither does it seem probable that an organism completely enveloped in an armature of strong plates, the slightest dislocation of which usually causes its death, is capable of engulfing solid bodies such as diatom-cells.

The *food-reserves* are starch and a fatty oil. Starch is found in grains of variable size, which sometimes show a stratification; it is not confined to the forms which possess chromatophores, but occurs in certain of the colourless species, the grains possibly arising by the activity of leucoplasts. Fatty oil is stored in small globules or as irregular masses. It may accumulate in any part of the cell or be more or less confined to certain regions, and its accumulation is brought about by the agency of lipoplasts which are surrounded by a special membrane.

Cell-division and multiplication. The multiplication of the Peridiniaceæ is effected almost entirely by vegetative cell-division. Such divisions occur mostly during the night, but only under favourable conditions, and generally while the organisms are motile. The plane of division is longitudinal in *Phalacroma,* and obliquely longitudinal in many genera, such as *Ceratium.* The nucleus first divides, the parallel threads (*vide* page 64) dividing across the middle, whereupon the whole nucleus constricts into two halves, each of which becomes rounded. In *Ceratium,* division of the whole protoplast follows immediately after the division of the nucleus, and then the hard cell-wall begins to separate into two pieces along certain of the existing sutures between the plates. Lauterborn ('95) first described this line of separation in *Ceratium hirundinella,* but Kofoid ('07 c and d) has given a much better account of the fission-plane in a number of other species of the genus (figs. 43 and 49).

Starting from the ventral area in the apical half of the cell, the line of fission passes between 4″ and 4′, 4″ and 3′, 3″ and 3′, 3″ and 2′, 3″ and 2″; then across the girdle into the antapical half between 4‴ and 3‴, 3‴ and 2⁗ 2‴ and 2⁗, 2‴ and 1⁗, 1‴ and 1⁗, and so back again to the ventral area (see fig. 49). The so-called ventral plate, which forms the ventral area, is thus obliquely divided by the fission-plane into an anterior and a posterior portion. The position of the fission-plane with reference to the skeletal plates was correctly given by both Bütschli ('85) and Bergh ('86), but only Kofoid ('07 C and D) has given a complete and correct account of the number and disposition of the plates.

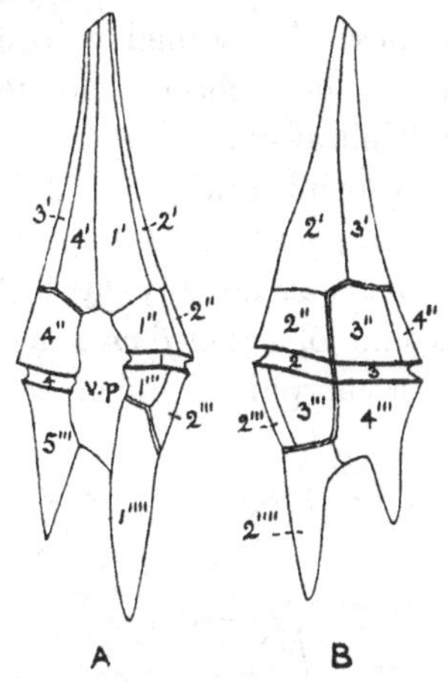

Fig. 49. *Ceratium furca* (Ehrenb.) Duj. × 350. *A*, ventral view; *B*, dorsal view. 1—4, girdle plates; 1'—4', apical plates; 1"—4", precingular plates; 1‴—5‴, postcingular plates; 1⁗ and 2⁗, antapical plates; *v.p.*, ventral plate. The sutures shown by the double line indicate the *line of fission*. (After Kofoid.)

As the daughter-cells separate after the division is completed, their exposed protoplasmic surfaces become so moulded as to complete the normal outward form of the organism. They are at the same time clothed with a thin wall which soon shows the plates and sutures characteristic of the species.

In *Peridinium*, and other allied genera, the old thick wall is frequently thrown off, and the protoplast escapes in a new thin wall. In this condition cell-division may occur (*Gonyaulax*), often in a transverse plane. Sometimes the protoplast on its escape becomes enveloped in a wide gelatinous envelope, under cover of which the division takes place (fig. 50 *G*). The ecdysis of the old cell-wall (figs. 50 *D* and 51 *I*) often results from the bursting of the girdle, the two valves falling apart (*Glenodinium* spp., *Peridinium aciculiferum*); by the loosening of certain plates only, as in *Peridinium Willei*, where the apical and some of the precingular plates are set free; or quite irregularly, as in several species of *Peridinium*.

A method of multiplication has been observed in *Peridinium anglicum* by the production of thin-walled, non-resting 'cysts' (G. S. W., '09). The protoplast becomes rounded, develops a new cell-wall, and then throws off the old articulated wall. The development now proceeds in one of two ways: (1) The rounded 'cyst' may divide at once into two cells each of which rapidly becomes an adult individual with a wall of typical plates. Sometimes each cell divides again before the formation of the tabulated wall, so that four daughter-cysts are formed from the original 'cyst.' (2) The 'cyst' may give rise to a motile cell, with a thin plasma-membrane, which at once escapes from the cyst-wall and begins to divide while slowly swimming about. Such cells generally divide transversely several times, and the final generation of daughter-cells become typical adult individuals. During these divisions the organism is in the '*Gymnodinium*-stage,' and curious states can be observed before the complete separation of the daughter-cells (fig. 51 *K* and *L*). Thus, in *Peridinium anglicum*, a species which attains its maximum vegetative activity in the early spring and then rapidly disappears, the formation of these thin-walled non-resting 'cysts,' with the resulting divisions both in the non-motile and motile states, provides a means of very rapid multiplication.

Multiplication by cell-division frequently results in chains of individuals in the genus *Ceratium*. Such chains have been studied by many investigators, but notably by Kofoid ('09 c), who has supplied the most detailed account of *chain-formation* and has shown that sometimes the chains are heteromorphic in character. He states that 'the morphology of chain-

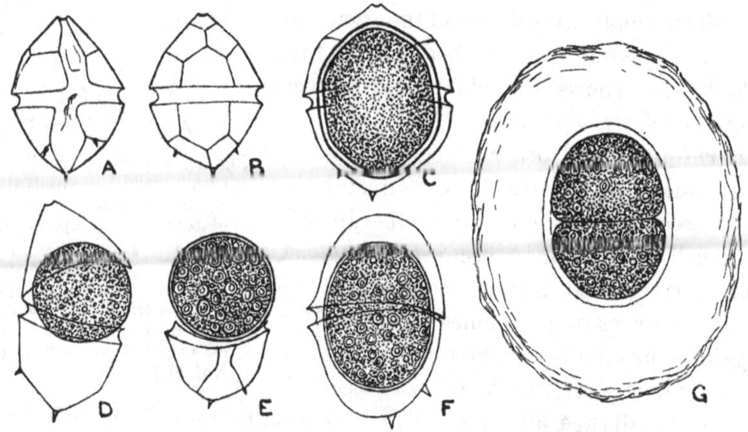

Fig. 50. *Peridinium aciculiferum* Lemm. *A*, ventral view; *B*, dorsal view; *C*, thick-walled resting-spore; *D*—*F*, thin-walled resting-spores; *G*, division of cell in encysted state. All × 500.

formation is correlated with the presence of an apical pore at the end of an apical horn. As the new skeletal moieties are formed respectively on the posterior and anterior regions of the diverging schizonts, the plasma of the posterior member is drawn out in a long strand which becomes the apical horn. Its tip rests immediately upon the distal end of the newly forming girdle, at which point the plasma of the two individuals remains in continuity without interference by the forming skeleton. As the newly forming skeletons are completed, the apical pore of the posterior schizont is set under the anterior shelf or list of the distal end of the girdle at the margin of the ventral plate of the anterior schizont. The posterior list of the girdle is not formed at this point, and the apical horn as it passes posteriorly lies in a channel or depression on the ventral face of the mid-body along the right margin of the ventral plate.' Kofoid designates the place of junction on the anterior schizont 'the attachment area,' and the depression 'the chain channel' (consult fig. 53).

Lohmann observed heteromorphic chains of *Ceratium tripos*, and Kofoid has given some interesting information on two cases of mutation, one of *Ceratium tripos* to *C. californiense*, and the other of *C. Ostenfeldii* to *C. californiense*. In the rare instances in which mutations occur, Kofoid

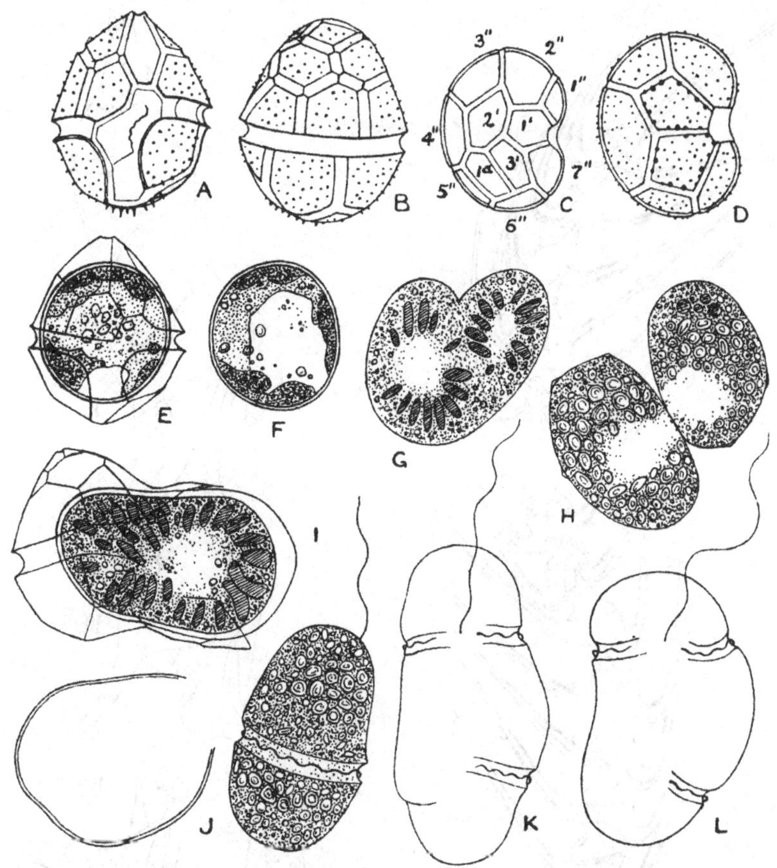

Fig. 51. *Peridinium anglicum* G. S. West. *A*, ventral view; *B*, dorsal view; *C*, apical view; *D*, antapical view; *E*, thin-walled 'cyst' within old wall; *F*, escaped 'cyst'; *G* and *H*, division of escaped 'cyst'; *I*, escape of an elongated 'cyst' from old wall; *J*, escape of naked motile individual from 'cyst'; *K* and *L*, divisions of naked motile state. All × 500. In fig. *C*: 1′—3′, apical plates; 1″—7″, precingular plates; 1^a, anterior intercalary plate.

finds very abrupt changes involving a whole complex of fundamental characters, so that in one, or at most in two generations of asexually-produced individuals the descendants differ profoundly from the ancestral type. The fact that the mutating chains have been seen so rarely in comparison with the vast numbers of normal chains which have been observed, coincides with observations on mutations in other organisms, and

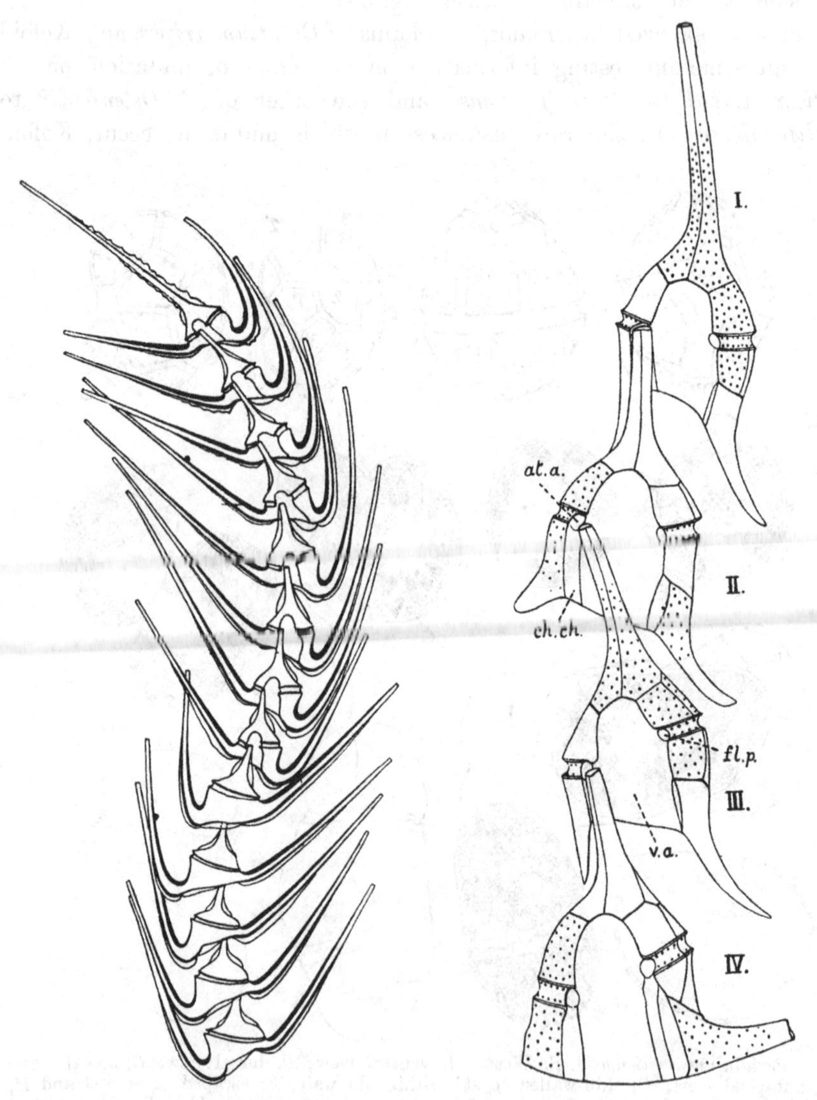

Fig. 52. Normal chain of *Ceratium vultur* Cleve, twisted at the fifth cell from the rear, showing ventral view anteriorly, and dorsal view posteriorly. ×100 (after Kofoid).

Fig. 53. Mutating chain of *Ceratium*. The posterior cell (IV) is *C. tripos* (O. F. M.) Nitzsch, and cells I—III are *C. californiense* Kofoid. *at.a.*, attachment area; *ch.ch.*, chain channel; *fl.p.*, flagellar pore; *v.a.*, ventral area covered by so-called 'ventral plate.' ×430 (modified from Kofoid).

the instances of their occurrence suggest the influence of environmental factors in producing the mutations.

If unfavourable conditions arise during cell-division, the process may be suddenly arrested, and undetached division-stages may be formed. These are not unlike certain of the presumed conjugation-states, and have sometimes been mistaken for such.

RESTING-SPORES. At the close of the active vegetative period many of the freshwater Peridiniaceæ are known to enter into an encysted condition by the production of thick-walled resting-spores (figs. 50 *C* and 54 *E*). Only one resting-spore is formed from a single individual. When the period of rest is of some length, as in *Peridinium aciculiferum*, where it extends over nine or ten months (G. S. W., '09), the wall of the resting-spore is very thick. Similar resting-spores are formed at any time in some species when the environmental conditions become unfavourable, such as by a sudden change

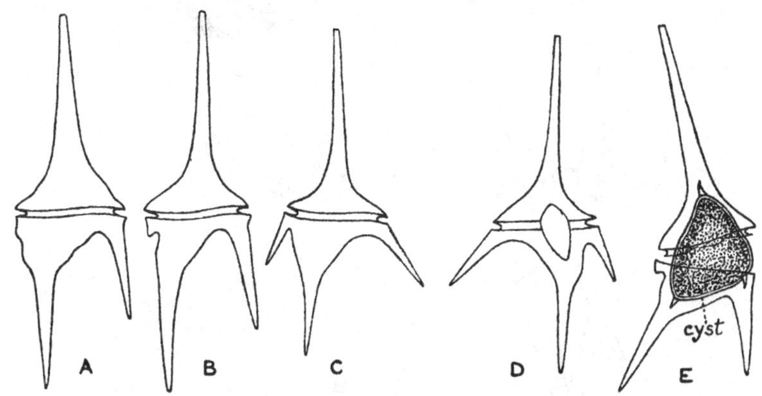

Fig. 54. *Ceratium hirundinella* O. F. Müll. *A* and *B*, three-horned forms; *C* and *D*, four-horned forms; *E*, three-horned form with resting-cyst. *D* is a ventral view; *A—C*, and *E* are dorsal views. All × 200.

of temperature or by alteration in the chemical constituents in solution in the water. In most species the resting-spores are ellipsoid or ovoid, but in *Ceratium hirundinella* they are usually three-angled or four-angled and somewhat twisted, and each angle is furnished with a short spine (fig. 54 *E*). It is most probable that these resting-spores remain in the mud at the bottom of the water until the conditions are again favourable for the resumption of the active vegetative phase, but their development has not yet been observed.

In some of the marine forms the protoplast becomes rounded off, secretes a wide gelatinous coat, to which the remains of the tabulated wall often stick; and it either rests in this condition as a 'gelatinous spore' or divides into a number (up to 128) of smaller cells which remain within the same envelope. The further fate of these spores has not been traced.

The occurrence of 'swarm-spores' of the *Gymnodinium*-type is known in certain genera of the Peridiniaceæ, but the details of their formation and their ultimate fate require further investigation.

Zederbauer ('04) has described a presumed sexual reproduction in *Ceratium hirundinella* by the conjugation of two individuals and the formation of a zygote (fig. 55). He states that the two cells become attached by their ventral areas in such a manner that their respective anterior and posterior extremities point in reverse directions. Then from each individual a conjugation-vesicle is extruded through the flagellar-pore, into which the protoplast gradually passes. The fusion of the two proto-

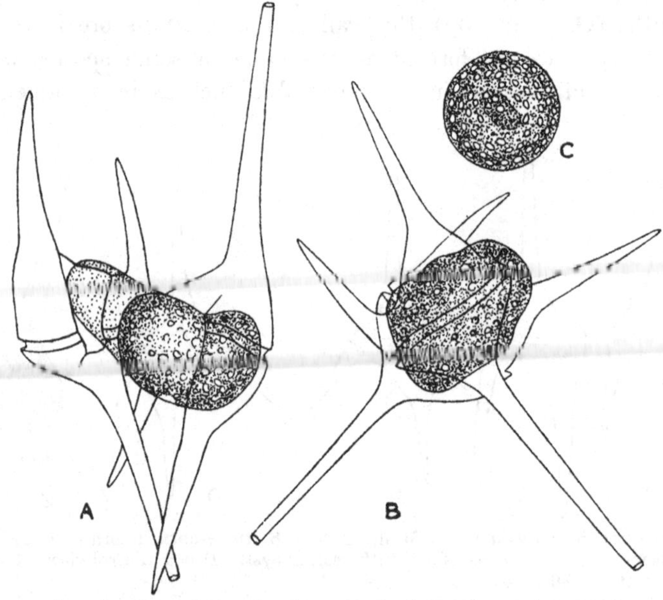

Fig. 55. Presumed conjugation of *Ceratium hirundinella* O. F. M. *A* and *B* showing two positions of conjugating (?) cells; *C*, supposed zygote (probably a cyst). ×660 (after Zederbauer).

plasts (gametes) results in a rounded zygospore, which becomes transformed into a three-horned resting-spore. In contrast to these statements, Entz ('07) describes the zygote as being formed within one of the old walls, which is in consequence burst open. Moreover, he states that only the nucleus of the one gamete passes into the other, the cytoplasm of the first gamete being lost. Neither of the nuclei undergoes any change, nor is there any nuclear fusion. The observations of Zederbauer and Entz are somewhat contradictory, and if conjugation does occur in the Peridiniaceæ it is of very rare occurrence, and still requires full investigation[1]. At present there is not sufficient proof that conjugation does exist in the Peridiniaceæ.

[1] Two doubtful cases of conjugation have been recorded in marine species of *Ceratium*: one by Pouchet in *Ceratium fusus*, and the other by Kofoid in *C. biceps*. There are also several records of conjugation, all very doubtful, in *Glenodinium* and a marine species of *Peridinium*.

OCCURRENCE AND DISTRIBUTION. The great majority of the Peridiniaceæ are constituents of the marine plankton, but there are quite a number of forms in the plankton of rivers and lakes, and also in the more stagnant waters of bogs and ditches. One form, *Glenodinium foliaceum* Stein, is apparently restricted to brackish water, and a few others occur both in brackish and fresh water (*e.g. Hemidinium nasutum* Stein, *Gymnodinium æruginosum* Stein, *Ceratium hirundinella* O. F. M., *Gonyaulax apiculata* (Pen.) Entz and *Peridinium Willei* Huitf.-Kaas). Although the genera and species of the Peridiniaceæ are not numerous in fresh water, certain forms, such as *Ceratium hirundinella* and species of *Peridinium*, often occur in prodigious abundance and are sometimes largely the cause of the 'water-bloom' previously described (p. 32). In the sea, a distinct colouration of the water may be caused by the abundance of Peridinians. In the Indian Ocean vast quantities of *Ceratium volans* Cleve sometimes give the water a brownish-purple colour; in the Japanese seas *Gonyaulax polygramma* Stein is frequently the cause of a brown colouration of the water; and in the vicinity of Bombay *Peridinium sanguineum* Carter has been known to colour the sea red. *Gonyaulax polyedra* Stein causes a red colouration of large areas of the sea off the coast of California during the summer months; it also exhibits a luminosity at night. In describing the occurrence of this organism Kofoid states that 'the decay of countless millions of these organisms in the water and upon the beaches where they are continually stranded by the receding waves, creates a nauseous and penetrating stench of a most disagreeable nature. The products of decay (and metabolism ?) of these organisms are toxic to many marine organisms, which die in great numbers and are cast up by the tide upon the beaches.' The organisms affected are mainly bottom-forms, such as Holothurians, Sipunculids, littoral Crustaceans, and bottom-feeding fish, such as the Sting Ray and the Guitar Fish (Kofoid, '11).

Kofoid ('10) finds that the more common plankton-species live within 100 fathoms of the surface of the ocean, and that below 50 fathoms a large percentage of the individuals show degenerative changes in the chromatophores and become moribund.

Some members of the Peridiniaceæ, more especially some of the species of *Ceratium*, are distinctly luminous, and to some extent they contribute to that luminosity of the sea of which the Pyrocystaceæ are the principal cause.

Of the freshwater forms, some attain their maximum vegetative abundance in the warmer months and others in the colder months. Also, certain species which in more northern latitudes are periodic in their assumption of the motile vegetative phase, and therefore periodic constituents of the plankton, are perennial constituents of the plankton of more southern latitudes. Such are *Peridinium Willei* and *Ceratium hirundinella*.

One of the great difficulties in the biological investigation of the organisms embraced in the Peridinieæ is their extreme sensitiveness to even trifling alterations of environment. The slight changes to which they are necessarily subjected during collection and microscopical examination are sufficient to cause almost immediate death and degeneration.

It is not possible to make any definite statements on the geographical distribution of the Peridiniaceæ with our present imperfect knowledge of the group, but there is much evidence to show that many forms are only found in the warmer oceans and that others are similarly restricted to temperate areas. Among the freshwater forms, *Peridinium aciculiferum* appears to be a northern type, and *Ceratium hirundinella* is without doubt the most ubiquitous.

A few fossil representatives of the Peridiniaceæ are known. The species found by Ehrenberg ('36) in a siliceous rock of Cretaceous age from Delitzsch in Saxony, and described as *Peridinium pyrophorum*, bears a striking resemblance to the recent *Peridinium divergens* Ehrenb.

Family **Prorocentraceæ.**

This small group of the Peridinieæ is characterized by the entire absence of the transverse and longitudinal furrows. According to the morphological interpretation put forward by Bütschli, Schütt, and others, the cell is compressed from the poles so that the longitudinal axis is the shortest one. It is ellipsoid, egg-shaped, or from the dorsal view the outline may be lanceolate (*Prorocentrum micans* Ehrenb.). In all except *Haplodinium*, in which the outer cellulose covering is structureless, the cell-wall is composed of two watch-glass-like plates which cover respectively the anterior and posterior halves of the protoplast. These two plates are precisely similar and, as the girdle is absent, their edges are directly joined along the median equatorial line, which is also the greatest circumference of the cell. In most of the forms the plates are furnished with distinct pores, which may be distributed over the whole of the wall or restricted to definite regions. In the middle of the ventral surface is the flagellar pore, which in *Cenchridium* projects inwardly into the cell-cavity as a tube. In the disposition of the longitudinal flagellum the Prorocentraceæ differ from all the other Peridinieæ. This flagellum is carried in front of the cell so that during the progression of the organism through the water the body of the cell is dragged after the longitudinal flagellum. The transverse flagellum is sometimes outstanding, but more often vibrates close to the cell-wall along the equatorial line. It is of about the same length as the longitudinal flagellum and usually swings round its base, but does not extend for more than about one fourth of the circumference of the cell. Close to the flagellar pore there are in some forms minute ridge-like projections of the cell, or in others more or less prominent tooth-like structures, which may possibly be the rudiments of the expanded borders of the transverse furrow.

The protoplast is similar to that in the Peridiniaceæ, consisting of an outer granular zone containing the chromatophores, and an inner more hyaline mass with cell-sap vacuoles and a nucleus. In the inner part of the protoplast there are also pusules (consult p. 66). The nucleus lies as a rule in that part of the cell towards the flagellar pore. The chromatophores are two in number, of a golden-brown colour, disc-shaped or expanded and lobed, sometimes reticulated, and they may overlap so as to give the appearance of one large parietal plate. In the larger chromatophores a pyrenoid may be present, as in *Exuviaella lævis.*

Multiplication takes place by division in the plane of the equatorial joint, that is, at right angles to the longitudinal axis of the cell. Each daughter-

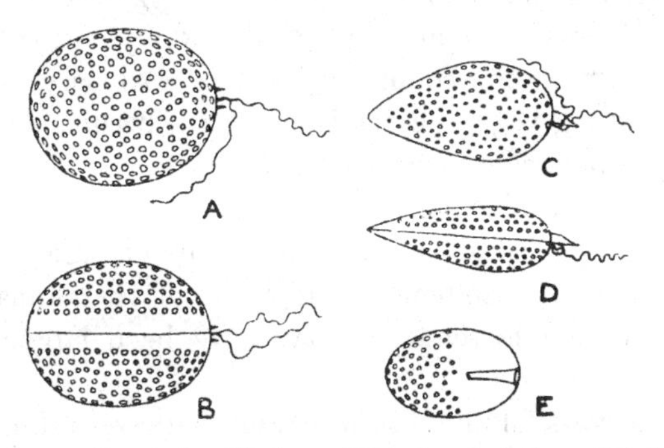

Fig. 56. *A* and *B*, *Exuviaella marina* Cienk., ×400; *A*, apical view; *B*, side view. *C* and *D*, *Prorocentrum micans* Ehrenb., ×400; *C*, apical view; *D*, side view. *E*, *Cenchridium globosum* (Williams) Stein, apical view, ×300. (After Schütt.)

cell retains one of the two old wall-plates and becomes provided with a new half-wall which is soon firmly joined to the old one. Resting-spores have only been observed by Cienkowski in *Exuviaella.* They are described as pyriform and attached by the pointed end to various Algæ.

There are only four genera of the Prorocentraceæ, *Cenchridium*, *Exuviaella*, *Prorocentrum*, and the recently described *Haplodinium*, all of which occur in the marine plankton, although a few forms have been observed in brackish water.

Klebs ('12) has suggested that the orientation of the cell is not as described by Schütt, but that the suture may be longitudinal, the flagella being anterior in their attachment. This interpretation certainly emphasizes the remarkable similarity between *Prorocentrum* and some of the Cryptomonads, with *Haplodinium* as a form more or less intermediate in character.

Thus, the Prorocentraceæ may be quite as primitive as the Gymnodiniaceæ, and not degenerate, reduced forms of the Peridiniaceæ as suggested by some authors.

NATURE AND AFFINITIES OF THE PERIDINIEÆ. The question may be asked why the Peridinieæ (or Dinoflagellata) should be included in a botanical text-book, and the answer would be that the balance of evidence indicates that in the evolution of these organisms from more primitive Flagellates the vegetable tendencies have so far become dominant that over 90 per cent. of them are true vegetable organisms with a holophytic nutrition.

The immense advance in the knowledge of the biology of unicellular organisms during the past twenty years has shown that in many cases there are no real distinctions between the animal and vegetable unicell. There can be no doubt in the mind of any biologist who has made a special study of the Flagellata, that this group of primitive organisms, exhibiting as it does such great diversity in morphological and cytological structure, has played a leading part in the commencement of various evolutionary series, in some of which strong vegetable tendencies have become dominant, whereas in others animal tendencies have come to the front. The more primitive coloured Flagellates must be regarded as fundamentally vegetable organisms with certain animal potentialities, and in groups of organisms which have evolved from them the subsequent development of the animal tendency and inhibition of the vegetable tendency may have been largely a question of environment.

The only sound basis for the discrimination between animal and vegetable organisms is nutrition. It must be borne in mind that all protoplasts, be they animal or vegetable, require practically the same classes of food-substances, and, moreover, they assimilate them in precisely the same way. The vegetable protoplast has, however, acquired the power of constructing its own organic food-substances. In contrast, therefore, to the animal protoplast, which requires its organic food presented to it in an available and assimilable form, the vegetable protoplast is capable of performing the preliminary synthetic work of constructing complex food-substances from raw materials. This constructive work is carried out in the normal plant by means of chromatophores, and is dependent upon light, and hence the photosynthetic activity of the typical vegetable organism is its fundamental characteristic. The raw materials enter the protoplast in a state of solution, and the elaborated materials which are the final products of photosynthetic activity are from the beginning *within the protoplast* ready for immediate assimilation whenever the action of enzymes renders them thus available. Such nutrition is *holophytic.*

The animal organism, on the other hand, has to take in *from the outside* elaborated materials insoluble in water, and for the necessary enzyme-action to work efficiently during the assimilation of these substances, they must be confined within a space which is more or less limited. Consequently, the normal animal organism has had perforce to adopt a method of ingestion of solid food-substances. This type of nutrition is *holozoic.*

Other methods of nutrition have arisen by modifications of the holophytic or holozoic types. The saprophytic Peridinieæ are probably mostly degenerate forms.

The ancient idea that the power of independent locomotion was the chief criterion of the animal nature of an organism has had to be entirely discarded in the light of modern knowledge[1]. It was this belief which for so long stood in the way of the full and proper investigation of the various groups of Flagellates.

The Peridinieæ are particularly interesting for the reason that in a few forms it seems probable that there has not been that complete divorce between the holophytic and holozoic methods of nutrition which is as a rule evident at the very inception of an evolutionary series.

There are a few observations, most of them rather old, on the ingestion of solid food by certain of the Peridinieæ by means of pseudopodia extruded from the protoplast; and there is good evidence that one or two members of the Gymnodiniaceæ, the most primitive group of the Peridinieæ, become amœboid in one stage of their life-history and take in solid food-particles. Should these observations receive further confirmation they must be regarded as actual facts, and their explanation is fairly clear. The action of the environmental influences on the Flagellates from which the more primitive Peridinieæ were evolved, was at first insufficient to bring about the complete inhibition of whatever tendencies these organisms possessed towards holozoic nutrition. Thus, there are a few examples within the Peridinieæ of holophytic, or degenerate saprophytic, organisms which for a brief period, either spasmodically or at some definite stage in their life-history, assume a holozoic method of nutrition.

The occurrence of trichocysts in several widely different forms of the Gymnodiniaceæ and Peridiniaceæ, and especially the presence of nematocysts in *Polykrikos*, is yet a further proof that the animal attributes were not completely eliminated during the evolution of the Dinoflagellates.

The affinities of the Peridinieæ are somewhat obscure, but that the group is not far in advance of certain sections of the Flagellata is fairly certain, and it may be that they originated far back among the free-swimming Cryptomonadineæ as suggested by Pascher ('11). There really seems to be every

[1] Consult footnote on p. 162.

possibility that the Prorocentraceæ have been derived from forms similar to *Wysotzkia* and allied genera; and there is much to be said in favour of Pascher's suggestion that some members of the Gymnodiniaceæ may have originated from Cryptomonads not far removed from *Protochrysis Phæophycearum.* It is also possible, as suggested by Bergh, and further emphasized by Klebs, that the Dinophyseæ (of Schütt) may have arisen from the Prorocentraceæ. Since it is not improbable that the origin of the Gymnodiniaceæ in the Cryptomonads was somewhat different from that of the Prorocentraceæ, and as there is no doubt that from the Gymnodiniaceæ the characteristic group of the Ceratieæ has arisen through such forms as *Glenodinium,* it is quite likely that the Peridiniaceæ as at present understood consists of two distinct series of forms, the Ceratieæ and the Dinophyseæ, and is therefore diphyletic. The possible evolution of the Peridinieæ is represented in the accompanying table.

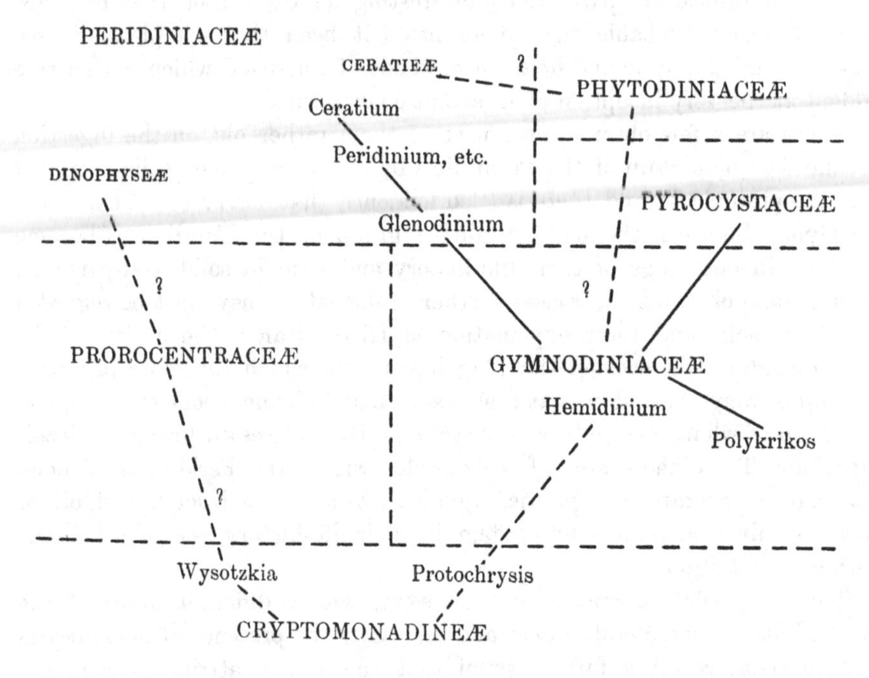

It is quite possible that the Bacillarieæ are remotely related to the Peridinieæ. In both groups the cell-wall is composed of two distinct halves, and in cell-division (excluding the Gymnodiniaceæ and certain other low types) each daughter-cell must acquire a new half-wall. The pigment of the chromatophores is similar, and the food-reserves are the same in each case. On the other hand, the longitudinal furrow with its flagellar pore, through which the two flagella pass, indicates that the locomotion of the Peridinieæ is a retention of an ancestral character, whereas in the Bacillarieæ all traces of a flagellate type of locomotion have vanished, certain of the more advanced forms having acquired a new method of locomotion through the agency of the raphe.

LITERATURE CITED

APSTEIN, C. ('06). Pyrocystis lunula und ihre Fortpflanzung. Wiss. Meersunters. N.F., Kiel, ix, 1906.

BERGH, R. S. ('81). Der Organismus der Cilioflagellaten. Morph. Jahrb. vii, 1881.

BERGH, R. S. ('86). Über den Theilungsvorgang bei den Dinoflagellaten. Zool. Jahrb. Bd. ii, 1886.

BLACKMAN, V. H. ('02). Observations on the Pyrocysteæ. New Phytologist, i, 1902.

BÜTSCHLI, O. ('73). Einiges über Infusorien. Archiv für. mikr. Anat. ix, 1873.

BÜTSCHLI, O. ('85). Dinoflagellata in Bronn's Kl. u. Ord. des Tier-Reiches. Bd. I, Abh. II, 1885.

CAVERS, F. ('13). The Inter-relationships of Flagellata and Primitive Algæ. New Phytologist, xii, 1913.

DANGEARD, P. A. ('92). La nutrition animale des Péridiniens. Le Botaniste, sér. 3, 1892.

DOGIEL, V. ('06). Beiträge zur Kenntnis der Peridineen. Mitt. Zool. St Neapel, xviii, 1906—1908.

EHRENBERG, C. G. ('36) in Abh. k. Akad. wiss. Berlin, 1836.

ENTZ, G. jun. ('07). A Peridineak Szervezetéröl. Különlenyomat az allattani Közlemények VI. [With German abstract.]

FAURÉ-FREMIET, E. ('08). Étude descriptive des Péridiniens et des Infusoires ciliés du plankton de la Baie de la Hougue. Ann. des Sci. Nat. Zool. sér. 9, vol. 7, 1908.

HERDMAN, W. A. ('11). On the Occurrence of *Amphidinium operculatum* Clap. et Loch. in vast Quantity at Port Erin (Isle of Man). Journ. Linn. Soc. Zool. xxii, 1911.

KLEBS, G. ('12). Über Flagellaten- und Algen-ähnliche Peridineen. Verhandl. d. naturhist.-med. Vereines zu Heidelberg, xi, Heft 4, Juli 1912.

KOFOID, C. A. ('07 A). The Structure and Systematic Position of *Polykrikos* Bütsch. Zool. Anzeiger, xxxi, Mar. 1907.

KOFOID, C. A. ('07 B). Dinoflagellata of the San Diego Region, III. Descriptions of New Species. Univ. of California Publications. Zool. vol. iii, no. 13, 1907.

KOFOID, C. A. ('07 C). The plates of *Ceratium* with a note on the unity of the genus. Zool. Anzeig. xxxii, Oct. 1907.

KOFOID, C. A. ('07 D). On *Ceratium eugrammum* and its related species. Zool. Anzeig. xxxii, Juli, 1907.

KOFOID, C. A. ('09 A). On *Peridinium Steini* Jörgensen, with a note on the nomenclature of the skeleton of the Peridinidæ. Archiv für Protistenkunde, xvi, 1909.

KOFOID, C. A. ('09 B). The morphology of the skeleton of *Podolampas*. *Ibid.* xvi, 1909.

KOFOID, C. A. ('09 C). Mutations in *Ceratium*. Bull. Mus. Comp. Zool. Harvard, lii, no. 13, 1909.

KOFOID, C. A. ('10). Significance of Certain Forms of Asymmetry of the Dinoflagellates. Advance print from Proc. Seventh Internat. Zool. Congress (Boston Meeting, Aug. 1907). Cambridge, Mass. 1910.

Kofoid, C. A. ('11). Dinoflagellata of the San Diego Region, IV. The Genus *Gonyaulax*. Univ. of California Publ. Zool. viii, no. 4, Sept. 1911.

Lauterborn, R. ('95). Protozoenstudien. I. Kern- und Zellteilung von Ceratium hirundinella O.F.M. Zeitschr. für wiss. Zool. Bd. 59, 1895.

Lemmermann, E. ('10). Peridiniales in Kryptogamenflora der Mark Brandenburg. Algen I. Juli, 1910.

Mangin, L. ('07). Observations sur la constitution de la membrane des Péridiniens. Compt. Rendus Acad. Sci. Paris, t. 144, 1907.

Pascher, A. ('11). Über die Beziehungen der Cryptomonaden zu den Algen. Ber. Deutsch. Bot. Ges. xxix, 1911.

Paulsen, O. ('06). On some Peridinidæ and Plankton Diatoms. Med. Komm. Havundorsög. vol. i, no. 3, 1906.

Plate, L. ('06) in Archiv f. Protistenkunde, vii, 1906.

Schilling, A. J. ('91). Die Süsswasser-Peridineen. Flora, 1891.

Schütt, F. ('92). Über Organisationsverhältnisse des Plasmaleibes der Peridineen. Sitz.-Ber. d. Akad. Berlin, 1892.

Schütt, F. ('93). Das Pflanzenleben der Hochsee. Kiel und Leipzig. 1893.

Schütt, F. ('95). Die Peridineen der Plankton Expedition. Teil I. Kiel und Leipzig. 1895.

Schütt, F. ('96). Peridiniales in Engler und Prantl, Die natürlichen Pflanzenfamilien. Teil I, Abt. I b. 1896.

Stein, F. R. v. ('83). Der Organismus der arthrodelen Flagellaten nach eigenen Forschungen in systematischer Reihenfolge bearbeitet. Der Organismus der Infusionsthiere, III Abt. 2 Hälfte. 1883.

West, G. S. (G. S. W., '09). A Biological Investigation of the Peridinieæ of Sutton Park, Warwickshire. New Phytologist, viii, 1909.

Zederbauer, E. ('04). Geschlechtliche und ungeschlechtliche Fortpflanzung von Ceratium hirundinella. Ber. Deutsch. Bot. Ges. xxii, 1904.

BACILLARIEÆ

(Diatoms)

THE Bacillarieæ include a very large number of unicellular Algæ commonly known as Diatoms. They are mostly of minute size, and owing to the beautiful sculpture of their cell-walls they very early attracted the attention of microscopists. Investigations extending over the greater part of a century, and in almost all countries of the world, have resulted in the description of some 12,000 species; but it is only during recent times that these plants have been studied from a biological standpoint.

In such a large group of unicells it is not surprising to meet with a great variety of external form, and the sculpture of the walls is no less varied; but at the same time all diatoms possess such salient features that the observant student can scarcely fail to recognize a member of this group.

Diatoms are equally abundant in both fresh and salt water, and numerous fossil forms are known. As a rule, the species of marine diatoms are entirely different from those which inhabit fresh water, but there are a number of typically brackish-water species, found principally in estuaries of rivers and in brackish marshes, which sometimes occur in marine, and more rarely in freshwater, situations.

One of the leading characteristics of the group is the siliceous nature of the cell-wall, which consists of an organic matrix combined more or less closely with silica. The durable nature of this cell-wall accounts for the fact that so many forms have been preserved in a fossil state, there being far more fossil representatives of the Bacillarieæ than of any other group of algæ. The silica can be removed by the action of hydrofluoric acid leaving the organic matrix behind; or the organic matrix, which according to Mangin ('08) consists of pectic compounds, can be removed by calcination leaving behind the siliceous constituent.

The individual diatom-cell, which in the language of the diatomologist has long been known as a *frustule*, possesses a segmented cell-wall of an almost unique character. It consists of four or more segments, the two largest of which are exactly opposite each other and are called the *valves*. In most cases the characteristic sculpture of the wall is confined to the valves, each of which forms one entire face of the diatom. No matter what the external shape of the valve, its edges are always bent at right angles to the general plane of the valve-face, and it thus possesses a flange like that on the lid of a box. Closely joined to the flange of each valve is a *connecting band*, so that the wall of the diatom consists of two halves, each half being constituted by a valve and its corresponding connecting band. Owing to the method of cell-division one half is older than the other, and as the older valve is very slightly larger than the younger valve the connecting band of the older half fits over that of the younger like the lid of a cardboard box. The valve of the older half is known as the *epivalve*; that of the younger half as the *hypovalve*. (Consult figs. 57 and 65 *5*.)

The two connecting bands, which in many cases are not very firmly united with their respective valves, together form the *girdle*. The connecting bands are not closed hoops, but as shown by Palmer & Keeley ('00) and others each is a two-ended strip with the ends overlapping; and although in the majority of diatoms this is not apparent by mere inspection it can readily be seen on boiling rather fiercely large diatoms, such as *Surirella nobilis*, *Navicula major*, etc., in fuming nitric acid, as under such circumstances the connecting bands become loose and their free ends can be easily observed.

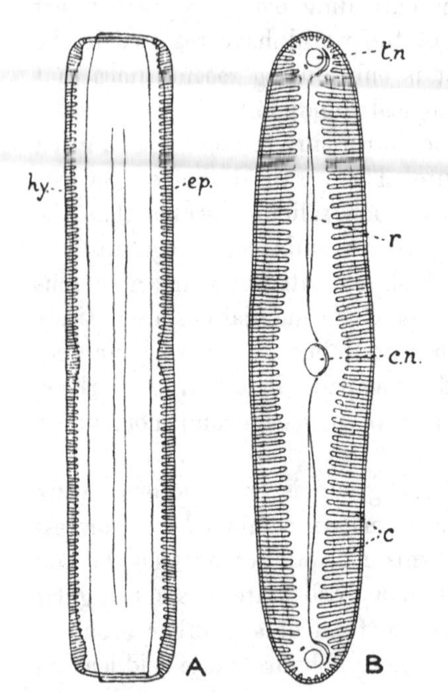

Fig. 57. *Navicula* (§ *Pinnularia*) *viridis* Kütz. *A*, girdle view; *B*, valve view. *ep.*, epivalve; *hy.*, hypovalve; *c.*, costæ; *c.n.*, central nodule; *t.n.*, terminal nodule; *r.*, raphe. × 800 (after Pfitzer).

The aspect in which the girdle of a diatom is exposed to view is known as the *girdle view*, and that in which the face of the valve is exposed to view as the *valve view*. (Fig. 57.)

In the average diatom just discussed the cell-wall consists of four segments, but sometimes the girdle is more complicated in structure owing to the interpolation of *intercalary bands* between the valves and connecting bands (fig. 58 *A*, *B*, *E* and *F*). One, two, or more of these intercalary bands may be present, the cell being correspondingly widened when seen in girdle view. The greatest number of such bands occurs in *Rhizosolenia*, in which genus the girdle view is the normal aspect presented by the diatom. In fact, the great elongation of the girdle renders it almost impossible to obtain a view of the valve-face because the diatom will not rest on that face. Similar elongated girdles occur in the genera *Guinardia, Attheya, Peragallia*, and other centric diatoms.

The girdle is also complex in the freshwater *Eupodiscus lacustris* described by Wille ('03) from the Zambesi, and in some of the naviculoid diatoms, such as *Stauroneis Biblos* Cleve ('92). Intercalary bands arise during the process of cell-division, being formed immediately after the new valves, the connecting band being the last-formed segment of the new half-cell. Like the

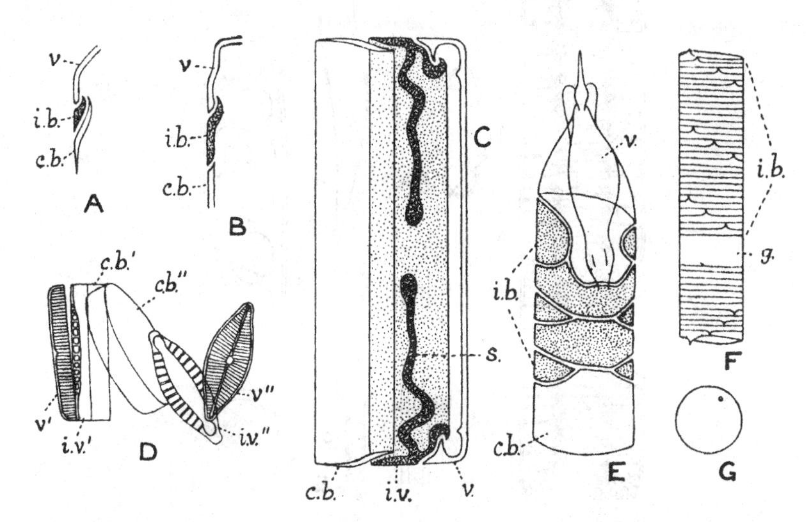

Fig. 58. *A*, section of edge of valve of *Epithemia turgida* (Ehrenb.) Kütz. to show attachment of connecting band (*c.b.*), intercalary band (*i.b.*) and valve (*v*). *B*, similar section of *Climacosphenia moniligera* Ehrenb. *C*, longitudinal section of the half-cell of *Grammatophora maxima* Grun. showing the intercalary valve (*i.v.*), which is an intercalary band furnished with a longitudinal septum (*s.*). *D*, single cell of *Mastogloia Smithii* Thwaites, disjointed so as to show the two valves, two intercalary valves, and two connecting bands. *E*, half-cell of *Rhizosolenia styliformis* Brightw., showing valve, connecting band, and intervening intercalary bands. *F* and *G*, girdle and valve views respectively of *Guinardia flaccida* (Castr.) Perag. *g.*, the two connecting-bands forming the girdle; *i.b.*, the numerous intercalary bands. (*A*, *B*, *C* and *E*, after O. Müller; *D*, after W. Smith; *F* and *G*, after Van Heurck.)

connecting bands, the intercalary bands are in many cases but imperfectly closed hoops. The openings in adjacent bands, however, are not in the same line, and are invariably covered by some portion of a neighbouring band. The valves, connecting bands, and intercalary bands fit against one another by bevelled and generally curved edges (consult fig. 58 *A*—*C*), but the connection is in some cases much looser than in others.

Some diatoms possess *longitudinal septa*, which are invariably ingrowths from the intercalary bands. Such septa are always more or less considerably perforated and approximately parallel to the valve-faces. The number of longitudinal septa depends upon the number of intercalary bands. In *Mastogloia* each cell has two longitudinal septa, in *Tabellaria* there are from two to twelve or more, and in *Tetracyclus* and *Rhabdonema* there may be as many as twenty. These septa are generally plane, although they may be undulated, as in *Grammatophora.* Karsten has distinguished those intercalary bands which possess longitudinal septa as *intercalary valves* (fig. 58 *C* and *D*). The perforations are large but variable. Sometimes one large window-like foramen is situated in the median part of the septum,

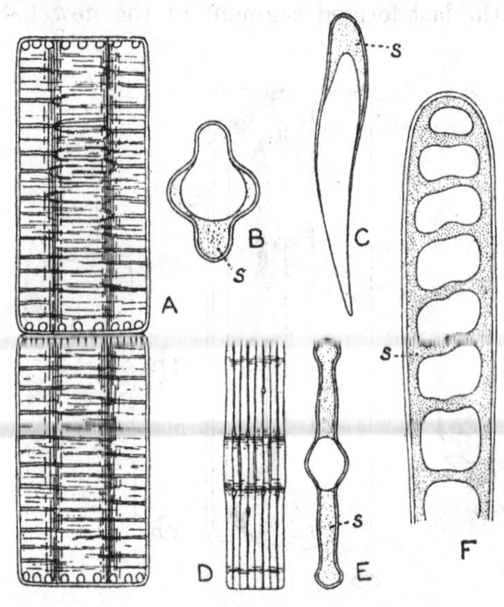

Fig. 59. *A* and *B*, *Tetracyclus lacustris* Ralfs; *B* is a longitudinal section of cell showing one of the partial septa. *C*, longitudinal section of cell of *Licmophora Lyngbyei* (Kütz.) Grun, showing partial septum. *D* and *E*, *Tabellaria fenestrata* (Lyngb.) Kütz.; *E* is a longitudinal section of cell showing the extent of the septa. *F*, *Climacosphenia moniligera* Ehrenb., section of upper portion of cell to show the perforated septum. *s*, septum. *C* and *F* are after O. Müller.

and at other times the septum is only partial, the foramen occupying most of the median part and the whole of one end (fig. 59 *A*—*E*). In *Climacosphenia* each septum has a series of numerous perforations (fig. 59 *F*). In *Denticula* the partial longitudinal septa are fused with numerous imperfect transverse septa which pass inwards from the valves.

Thus the interior of the diatom-cell may become chambered in various ways, and as will be noted subsequently these chambers very largely accommodate lobes of the chromatophores.

SYMMETRY OF THE DIATOM-CELL. The diatom-cell may be perfectly symmetrical or it may be completely asymmetrical, and the various degrees of symmetry or asymmetry have been brought about by unequal growth of either the valves or the girdle. This question of symmetry was regarded as of primary importance from a classificatory point of view by both Heiberg and Pfitzer, but unless taken in conjunction with other characteristics it is of little real importance.

O. Müller ('95) has given an excellent synopsis of the different kinds of external symmetry observed in various types of diatoms. Some are zygomorphic in one plane only, some in three planes at right-angles, and others exhibit a radial symmetry (fig. 60). In the majority of diatoms

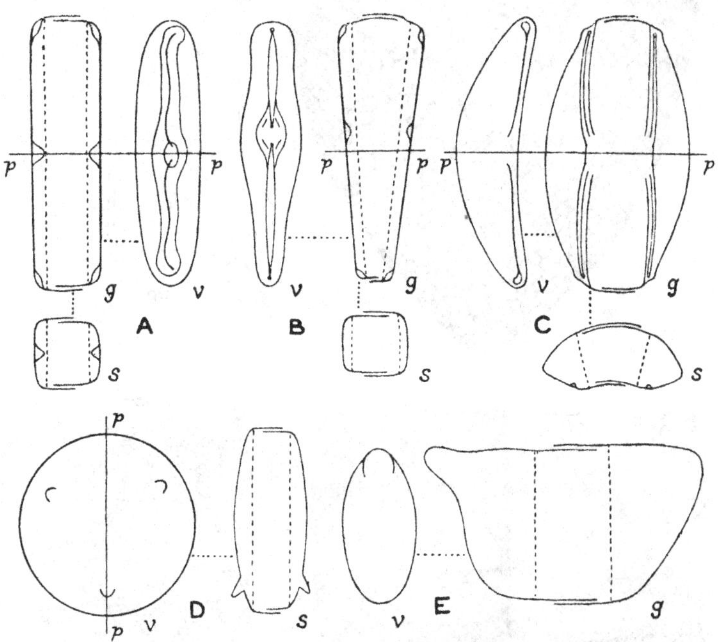

Fig. 60. Diagrams to illustrate the symmetry of the diatom-cell. *A*, *Navicula viridis*; *B*, *Gomphonema elegans*; *C*, *Amphora ovalis*; *D*, *Eupodiscus Argus*; *E*, *Isthmia enervis*. *A—C* are pennate diatoms, *D* and *E* are centric diatoms. *v*, valve view; *g*, girdle view; *s*, section of cell; *pp*, plane of section. (All the figures somewhat modified from O. Müller.)

the sculptures on the valves are also arranged in relation to the external symmetry, a fact of such importance from a classificatory standpoint that it allows the division of diatoms into the two well-marked primary groups of the Centricæ and Pennatæ.

STRUCTURE OF THE CELL-WALL. The cell-wall of diatoms consists of an organic ground substance which is more or less strongly silicified. The girdle is usually relatively thin, but the valves vary much in strength and thickness. In the great majority of species they are more or less symmetrically sculptured, such markings generally having the appearance of striations or areolations. A few diatoms are known in which the valves are smooth, no trace of striations or other markings having been discovered. Such diatoms, in which the wall is very thin and apparently structureless, are exemplified by *Navicula perlepida*, *N. glaberrima*, and *Tropidoneis lævissima*.

Detailed microscopical investigation has shown that the markings consist of small cavities within the siliceous cell-wall which are in most cases

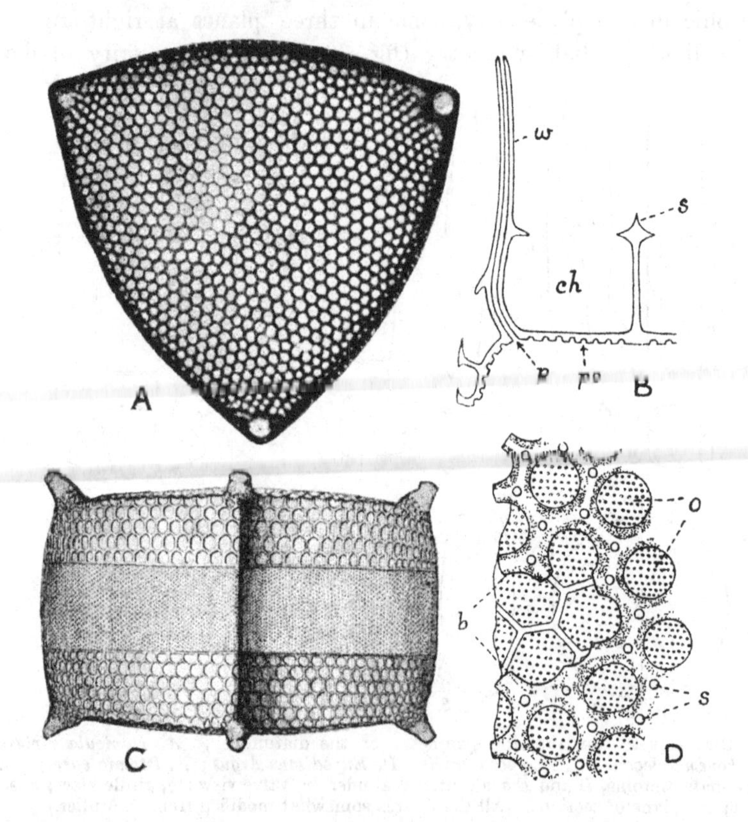

Fig. 61. *Triceratium Favus* Ehrenb. *A*, valve view; *B*, transverse section across the rim of the valve; *C*, girdle view; *D*, surface view of portion of valve to show details of structure. *b*, limitations of broken part of valve to show the hexagonal chambers (*ch*); *o*, external openings of the large chambers; *p*, pore; *po*, poroids or dots; *s*, spines at the corners of the honeycomb-like chambers; *w*, external wing at the rim of the valve. (*B* after O. Müller; *C* after W. Smith; *D* after Pfitzer.)

arranged in regular rows, thus giving the appearance of striations. In some cases the marks are due to ridges, and both the cavities and ridges may be either upon the inner or the outer side of the cell-wall. When the cavities giving rise to the appearance of striations are very small they are known as *punctæ*. The striæ vary in strength from conspicuous ribs or *costæ* to lines so fine as readily to escape detection, and species with the most delicate striation (*e.g. Amphipleura pellucida,* various species of *Gyrosigma* (= *Pleurosigma*) are frequently used as test objects for the definition and angular aperture of the object-lenses of microscopes. During more recent years our knowledge of the minute structure of the wall of diatoms has been much increased, largely by the investigations of Pfitzer ('71), Lauterborn ('96), Schütt and O. Müller ('98—'01).

In the centric diatoms, which are for the most part marine, there is generally a radial disposition of the markings on the valve-faces. In some

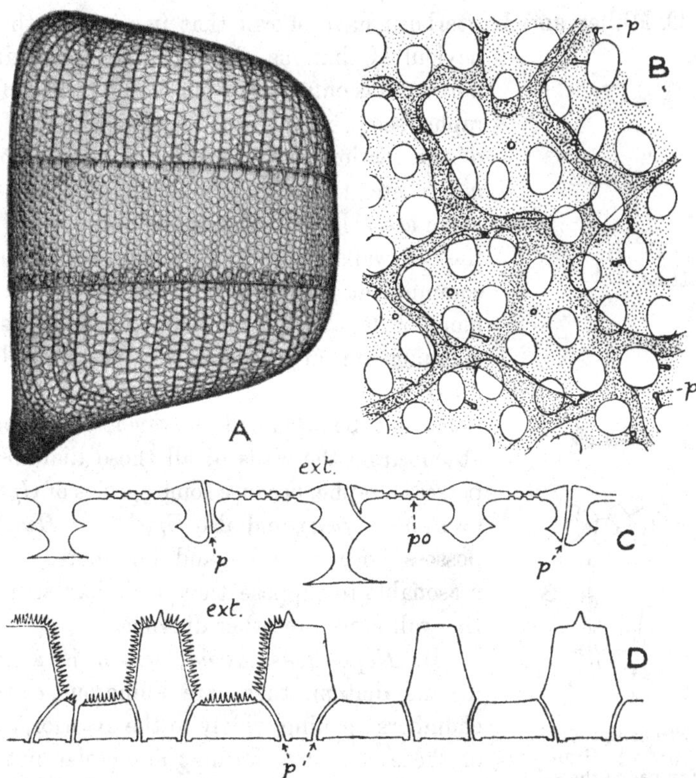

Fig. 62. *A—C, Isthmia nervosa* Kütz. *A*, girdle view; *B*, part of the end of a valve, seen from the inner side, showing the primary and secondary chambers in the wall; *C*, ideal section through the primary and secondary chambers of the wall. *D*, ideal section through the chambered wall of *Eupodiscus Argus* Ehrenb. *p*, pores; *po*, poroids or dots; *ext.*, outer side of cell-wall. (*A*, after W. Smith; *B—D*, after O. Müller.)

of the larger species the valves exhibit a beautiful areolation, due to numerous, closely adjacent chambers in the siliceous wall. The chambers may be open to the exterior or they may be covered by a thin siliceous membrane, and their inner walls are pierced by exceedingly minute canals which pass right through to the interior of the cell. These minute canals, which are termed "pores," are not present in all diatoms, as some species which have been most carefully examined with this object in view have failed to reveal any trace of true perforations through the cell-wall. They occur beyond question, however, in many species, and Schütt states that it is highly probable that they exist in a large number of others. Schütt instituted a distinction between pores and dots; and O. Müller ('00) has still further emphasized this distinction by suggesting that all the minute circular dots over 0·6 μ in diameter, which are tiny cavities resembling pores but not actual perforations, should be termed 'poroids' in contrast to the true pores which vary from 0·1 μ to 0·5 μ in diameter.

Both O. Müller and Lauterborn have shown that in some of the larger naviculoid diatoms the cell-wall is destitute of pores and is only broken through by the cleft of the raphe (see p. 92). In the large species of *Navicula*, of the section *Pinnularia*, such as *N. nobilis*, *N. major*, etc., the bilaterally arranged costæ on each valve are elongated chambers on the inner side of the cell-wall. Each of these furrow-like chambers communicates with the interior of the cell by a more or less wide opening, and the edges of the openings appear in the valve-view as fine longitudinal lines crossing the costæ (fig. 57 *B*). It must not be assumed, however, that pores are absent from the walls of all those diatoms which possess a raphe, because some species of *Gyrosigma* (=*Pleurosigma*), and also *Epithemia Hyndmanni*, possess both a raphe and fine pores, and it is reasonable to suppose that a similar structure of the wall occurs in other diatoms.

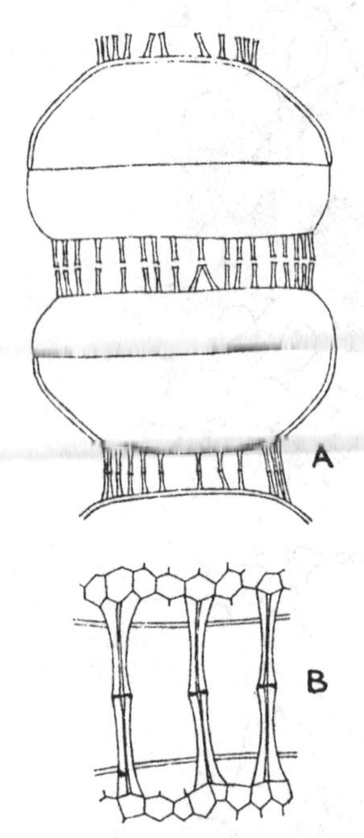

Fig. 63. *Stephanopyxis Palmeriana* (Grev.) Grun. to show the spines on the valves by which the cells are united to form chains. B is an enlargement (× 1000) of three pairs of these spines to show their tubular character. (After O. Müller.)

In *Eupodiscus Argus*, which is a frequent marine diatom, there are numerous cup-shaped chambers opening widely to the exterior, the walls of these chambers forming an areolar network of ridges on the surface of the valve. The walls of the chambers are finely papillate, and several minute, obliquely sloping canals perforate the base of each chamber (fig. 62 *D*).

The structure of *Triceratium Favus* is somewhat similar, only the chambers opening to the exterior are polygonal and their walls are smooth. Minute poroids occur on the inner side of the inner walls of these chambers, and pore-canals pass through the flange of the valve as minute tubes (fig. 61 *B* and *D*).

In *Isthmia nervosa* there are larger primary and smaller secondary chambers. The inner walls of the secondary chambers possess a number of poroids, and pore-canals are present here and there, passing rather obliquely through the separating ridges (fig. 62 *B* and *C*).

In some diatoms, such as *Stephanopyxis Palmeriana* (fig. 63), *Sceletonema costatum*, and *Lauderia annulata*, O. Müller ('01) has shown that the spines by which the cells are united to form chains are really tubes, through which there is a protoplasmic continuity from cell to cell.

The most numerous and best-developed pores are found in the centric diatoms. O. Müller has suggested that in certain species they are in relation to an extracellular layer of cytoplasm which brings about a centrifugal thickening of the valve, but it is by no means certain that they always serve for the passage of protoplasm. Karsten has arrived at the conclusion that the extracellular layer of cytoplasm has a morphogenic activity only in certain of the pelagic genera, such as *Coscinodiscus*, *Planktoniella*, *Valdiviella*, etc. Perhaps the comparative absence of pores from the pennate diatoms is to be associated with the presence of the raphe, which in some of these forms attains a high development.

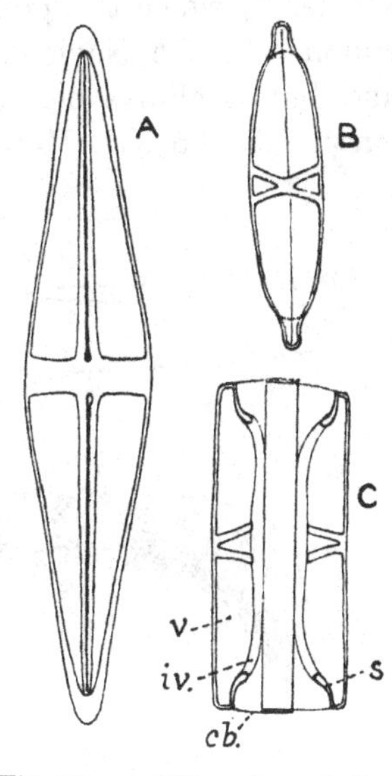

Fig. 64. *A*, Valve view of *Stauroneis acuta* W. Sm. showing stauros. *B* and *C*, valve and girdle views respectively of *Schizostauron Crucicula* Grun. showing the forked stauros. All × 500. *v*, valve; *iv.*, intercalary valve with short septa (*s*); *cb.*, connecting band. The fine striations on the valves are not shown in the figures.

For the careful study of the finer structure of the cell-wall O. Müller recommends treatment with hot sodium carbonate and potassium hydrate.

The valves of many pennate diatoms, especially those of the Naviculaceæ, possess small internal thickenings which are of a somewhat rounded or conical shape, generally contain a cavity, and are known as *nodules*. They occur at each extremity of the valve and in the centre, and in many of the pennate diatoms are connected by a median line known as the *raphe*. The central nodule is sometimes expanded laterally to form a *stauros*, which may be simple (*Stauroneis*; fig. 64 *A*) or forked (*Schizostauron*; fig. 64 *B* and *C*); in other cases it is prolonged into paired horns (as in *Diploneis*). In *Amphipleura* it is greatly extended in a longitudinal direction. Frequently there are smooth areas on a valve which is otherwise striated. Such hyaline areas are generally round the central nodule and on each side of the raphe. The former, which is known as the *central area*, often extends across the median portion of the valve from one margin to the other; the latter is termed the *axial area* and varies much in width. Less often, smooth areas occur parallel to the axial area but nearer the margin of the valve; these are known as *lateral areas.*

The *raphe* (or median line) may be straight, undulating, or sigmoid. In its most highly developed state it is a fissure of a more or less complicated character by which the protoplast is placed in contact with the surrounding medium. In the Naviculaceæ the cleft of the raphe is not in a vertical plane, but is always bent, and not infrequently V-shaped in cross-section. (Consult fig. 65 *5*). Moreover, the cleft is not wholly open, being in many

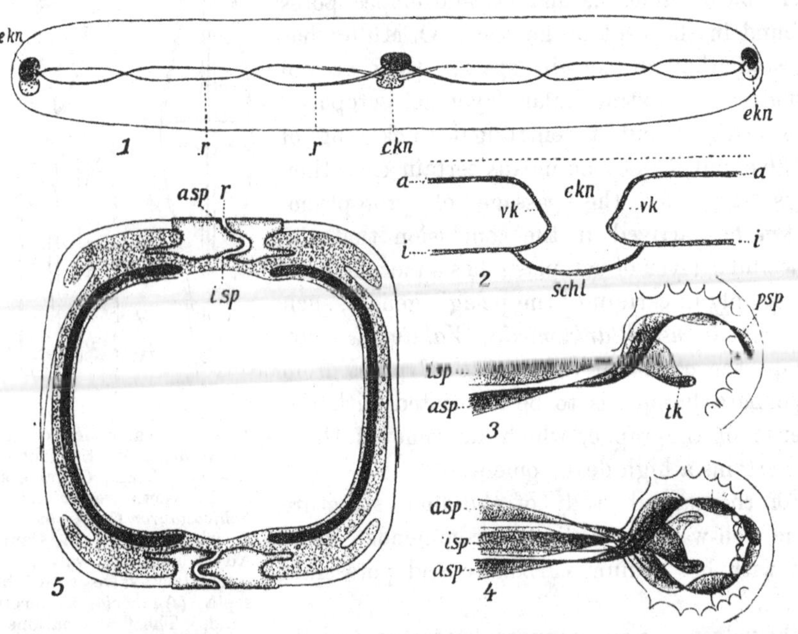

Fig. 65. *Navicula* (§ *Pinnularia*) *viridis* Kütz. *1*, diagram with the raphe of one valve almost superimposed upon the other; *ekn*, terminal nodule; *ckn*, central nodule. *2*, longitudinal section through the central nodule showing the two canals (*vk*) joining the outer (*a*) and inner (*i*) clefts of each half of the raphe; *schl*, the canal which joins both halves of the inner fissure. *3*, terminal nodule showing ends of the two fissures of the raphe; *isp*, inner fissure terminating in the funnel-shaped body (*tk*); *asp*, outer fissure ending in the polar cleft (*psp*). *4*, superimposed terminal nodules of epivalve and hypovalve. *5*, transverse section across the cell, showing the inner (*isp*) and outer (*asp*) fissures of the raphe (*r*), the nature of the costæ in the valves, and the disposition of the two chromatophores. (From Oltmanns; *1—4*, after O. Müller; *5*, after Lauterborn.)

forms closed in its middle region; that is, along the bend (or in cross-section about the point of the 'V'). Thus, in reality there are two cleft-like fissures, one on the inner side and one on the outer side of each valve.

The raphe is interrupted by the central nodule which is perforated by two canals, each of which joins together the outer and inner fissures of one half of the valve. There is also a canal running along the inner side of the central nodule joining together both halves of the inner fissure (fig. 65 *2 schl*).

The inner fissure terminates at each end of the valve in a funnel-shaped structure which projects into the cavity of the terminal nodule (fig. 65 *3 tk*). The outer fissure terminates in a definite polar cleft, which is a curved slit in the terminal nodule (fig. 65 *3 psp*). The polar clefts (often called terminal fissures) at the extremities of the same valve are usually curved in the same direction, but a number of species are known in which they are curved in opposite directions. The corresponding clefts of the epivalve and hypovalve of a single individual are invariably curved in opposite directions (fig. 65 *4*).

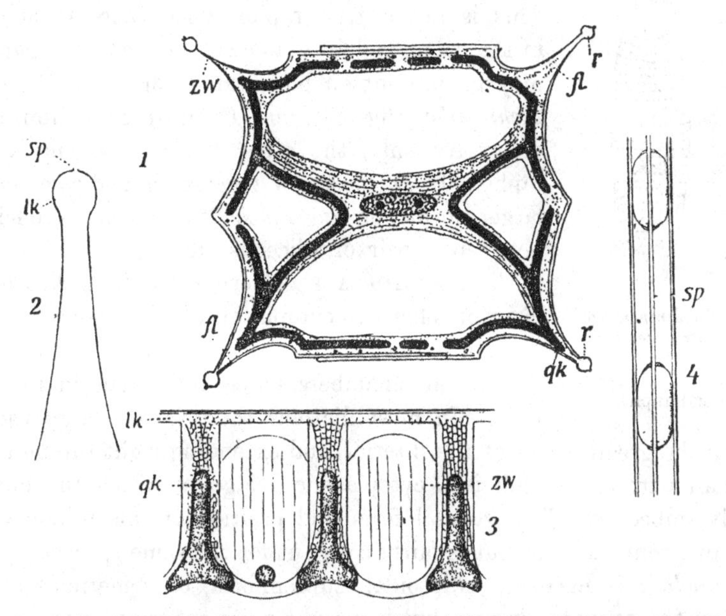

Fig. 66. *Surirella Capronii* Bréb. var. *calcarata* (Pfitzer) Hustedt. *1*, transverse section across the cell showing the four wing-like expansions (*fl*), each with its raphe-canal (*r*). *2*, transverse section of wing, showing raphe-canal and outer cleft. *3*, longitudinal section of wing. *4*, wing viewed from the edge. *qk*, cross-canals; *zw*, thin intercalary pieces of wing; *lk*, raphe-canal (or plasma-canal); *sp*, cleft of raphe. From Oltmanns (after Lauterborn).

The somewhat complex system of clefts and canals described above is concerned with streaming movements of the cytoplasm, the latter being brought into direct contact with the surrounding medium by means of the outer fissure of the raphe. This is stated by O. Müller to be the direct cause of the curious movements of those diatoms which possess a true raphe. (Consult page 102.)

The raphe of other pennate diatoms is essentially different. In *Surirella* the lateral margins of each valve are produced into wing-like expansions, of which there are therefore four to each cell. Near the free edge of each wing is a fine raphe-canal, with a longitudinal fissure extending its whole length and placing it in communication with the exterior, while it is connected with the interior of the cell by cross-canals (fig. 66). The free edge of each wing is thus somewhat thickened to accommodate the raphe-canal, and its flat portion consists of alternating cross-canals (fig. 66 *qk*) and thin intercalary pieces (fig. 66 *zw*).

In *Nitzschia* there is a raphe-canal with a longitudinal fissure similar to that of *Surirella*, extending along the whole length of the keel of each valve.

In some diatoms there is a narrow, hyaline, axial area without a central nodule, either median or submedian in position. This is not a true raphe, since there is no cleft in the valve, and it is known as a *pseudoraphe.* It may be present on both valves, or, as in *Cocconeis*, *Achnanthes* (fig. 67), and *Rhoicosphenia,* limited to one valve only, the other valve possessing a true raphe. In *Dictyoneis, Vanheurckia,* and some of the larger forms of *Stauroneis,* the raphe is enclosed between a pair of siliceous ribs.

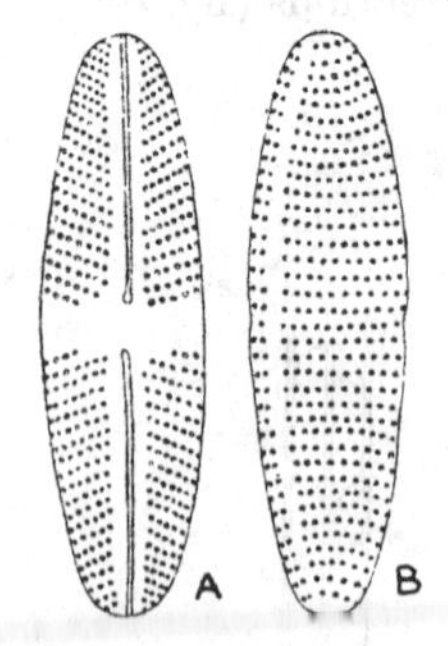

Fig. 67. *Achnanthes brevipes* Ag. var. *intermedia* (Kütz.) Cleve. Two valves of one cell; *A* with raphe, *B* with pseudoraphe.

THE PROTOPLAST. There is a well-marked lining layer of colourless cytoplasm, which extends into the various extensions of the cell-cavity, and also into the various chambers, pores, and canals in the cell-wall. In most diatoms there is one large vacuole occupying the greater part of the interior, and extending right through it in the median part of the cell is a cytoplasmic bridge in which the nucleus is usually embedded. The general form and position of this bridge varies much in different diatoms, but is fairly constant for the same species.

The *nucleus* is narrowly ellipsoidal, subreniform, or sometimes almost fusiform, in the pennate diatoms, but it may be quite spherical in the centric species. It contains one or more fairly conspicuous nucleoli, and Lauterborn ('96) has demonstrated the presence of a centrosome in certain of the larger species of *Navicula* and *Surirella.* Other authors have described the presence of a macro- and a micronucleus in certain diatoms. Lauterborn attempts to reconcile the various interpretations of nuclear structures said to be observed in diatoms, and states that it is highly probable that at one time the cell possessed two equally constructed nuclear bodies, but that gradually a division of labour has brought about a greater and a greater differentiation until one has acquired a purely morphological and the other a purely physiological function. The macro- and micronucleus of certain authors are possibly equivalent to the nucleus and centrosome of Lauterborn. This author has also followed out the mitotic division of the nucleus of several of the larger diatoms, in which there is apparently a relatively large number of chromosomes (consult fig. 68). After division, and on the reconstruction of the daughter-nuclei, the centrosome lies in a slight hollow of the nucleus. Heinzerling states that nuclear division is always mitotic.

In a somewhat denser mantle of cytoplasm surrounding the nucleus are a number of curious, small, bar-shaped structures disposed in pairs, the function of which is unknown.

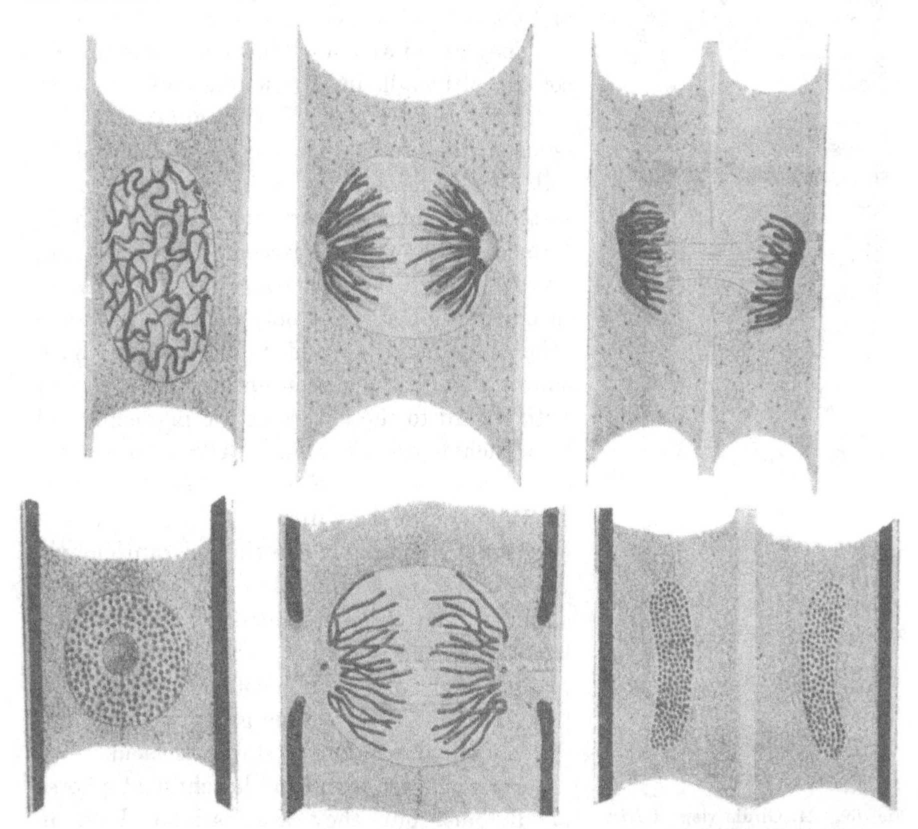

Fig. 68. Stages in the mitosis of two diatoms. Upper three figures represent three stages of nuclear division in *Nitzschia sigmoidea* (Ehrenb.) W. Sm. Lower figures are of *Navicula* (§ *Pinnularia*) *oblonga* Kütz.; the left-hand figure is of a resting nucleus with a large nucleolus. (After Lauterborn.)

THE CHROMATOPHORES. The chromatophores of diatoms vary much in the different tribes and families. One or many may be present in each cell; they may be small and discoidal, large and plate-like, or extensive anastomosing structures occupying a large part of the living layer of cytoplasm. They are frequently very irregular in form, and are sometimes band-like, much lobed, or they present the appearance of perforated plates. In cases where the interior of the cell is divided up by incomplete partitions lobes of the chromatophores generally extend into the chambers.

Plate-like chromatophores occur in nearly all the naviculoid diatoms, and they have approximately the same form and disposition in groups of allied species. In most of the centric diatoms the chromatophores take the form of rounded or lobed discs and are more or less numerous in each cell (fig. 69 *B* and *C*).

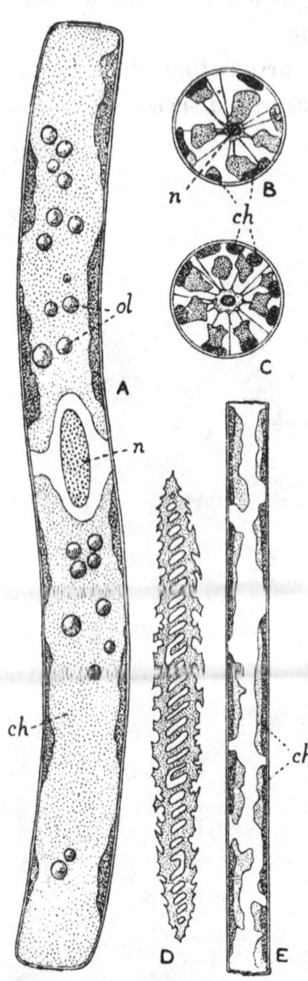

Fig. 69. *A*, Girdle view of *Nitzschia sigmoidea* (Ehrenb.) W. Sm. showing two large chromatophores with oil globules, and also the nucleus. *B* and *C*, Valve views of *Cyclotella compta* (Ehrenb.) Kütz. showing numerous discoidal chromatophores. *D*, Isolated perforated chromatophore of *Gyrosigma balticum* (Ehrenb.) (after O. Müller). *E*, Girdle view of *Synedra Ulna* (Nitzsch) Ehrenb. showing the irregular chromatophores. *ch*, chromatophores; *n*, nucleus; *ol*, oil globules.

They are of a yellow-brown or golden-brown colour, although in a few diatoms, such as *Navicula viridis* and *N. cuspidata*, they are occasionally green. They contain chlorophyll, but this is masked by a brown colour often stated to be due to a pigment which has been termed 'diatomin.' Recent investigations, however, into the colouring matters present in chromatophores, not only of diatoms, but of other algæ and also of higher plants, have shown that many misconceptions have existed with regard to the nature of the pigments, and that such names as 'diatomin' are to a great extent meaningless. Kohl ('06) has shown that the pigment in the chromatophores of diatoms consists of chlorophyll, carotin and xanthophyll. Molisch ('05) asserts that leucocyanin also occurs in the chromatophores of diatoms, and that it is this pigment which causes these plants (and also an alcoholic solution of 'diatomin') to turn blue-green on the addition of hydrochloric or sulphuric acid.

Pyrenoids are found in the chromatophores of diatoms, but they are variable both in number and disposition. They consist of an albuminous protein and are usually lens-shaped. Not infrequently they are grouped in clusters, and they often project internally from the chromatophores, Mereschowsky ('03) having recorded instances in which they have partially or entirely emerged from the chromatophores, appearing as free, colourless bodies on their inner surfaces.

In most diatom cells there are drops of a *fatty oil* which are generally more conspicuous than the pyrenoids. This oil is easily soluble in ether and is blackened in osmic acid. It can be shown to be a food-reserve by keeping diatoms for a considerable period in tap-water in closed vessels, in which case it is completely used up. There are also so-called 'oil-drops' of another nature. These are often of larger size, and they do not dissolve in ether, neither do they become black with osmic acid. Their function is obscure and they are known as *Bütschli's red corpuscles.*

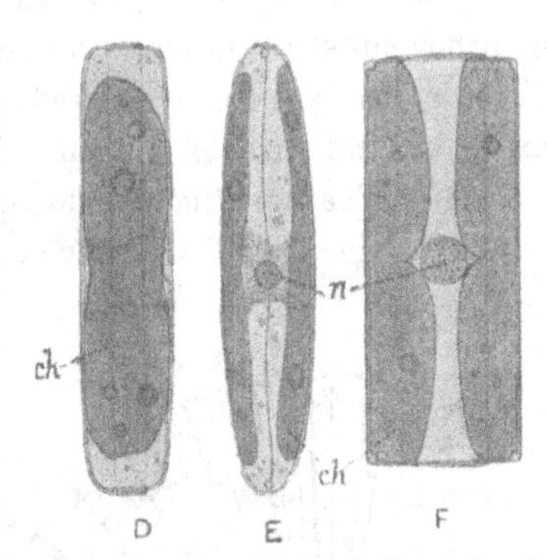

Fig. 70. *D* and *E*, girdle and valve view of *Navicula viridis* Kütz. showing chromatophores (*ch*) and nucleus (*n*); *F*, *Eunotia gracilis* (Ehrenb.) Rabenh. girdle view showing chromatophores and nucleus. (× 400.)

The normal *nutrition of diatoms* is holophytic, but several saprophytic forms have been described. In the latter there is a complete absence of pigment, and such forms occur for the most part where there is an abundance of decaying organic matter in the water. Benecke ('00), who obtained the colourless diatoms *Nitzschia putrida* and *N. Leucosigma* from Kiel Harbour, affirms that there are no transitional states between brown and colourless diatoms. Most of the known saprophytic forms are extremely motile. Karsten ('01) succeeded in producing a saprophytic form of *Nitzschia palea* by cultivating it in favourable nutritive media such as glycerin or grape-sugar. *Nitzschia putrida* has become so far saprophytic that all attempts to induce the formation of chromatophores in this diatom have failed. *Synedra hyalina* Provazek ('00) is another colourless diatom.

There is evidence to show that some diatoms do not require calcium, but that sodium is essential.

THE BUILDING OF COLONIES AND THE SECRETION OF MUCUS. Diatoms are frequently solitary and free-floating, but they are also united in various ways to form colonies, generally of small size. In some of these colonies the individuals are united by their valve-faces, in which case they form ribbon-like (*Eunotia, Tetracyclus, Fragilaria, Rhabdonema*) or thread-like (*Melosira, Bacteriastrum, Chætoceras*) chains. In others the cells are joined in an irregularly zigzag manner at the corners of the valves (*Tabellaria*, some species of *Diatoma, Triceratium*, and *Biddulphia*). In all these cases the attachment is by mucus secreted by the cells themselves. The cells of the ribbon-shaped and thread-like colonies are joined either by a layer of mucus or by mucous strands between the closely applied valve-faces, and the cells of the zigzag colonies are united by small mucous cushions at the corners of the valves.

Some diatoms are stipitate, secreting gelatinous stalks, either simple or branched, by which they are attached to larger aquatic plants or to some other substratum (*Gomphonema, Achnanthes,* some species of *Cocconeis*); and there are yet others which secrete a copious mucus so that a large number of individuals become embedded in a common gelatinous mass (*Epithemia alpestris, Cocconema* spp., *Vanheurckia rhomboides* var. *saxonica*). Less often the cells are enclosed in simple or branched mucous tubes

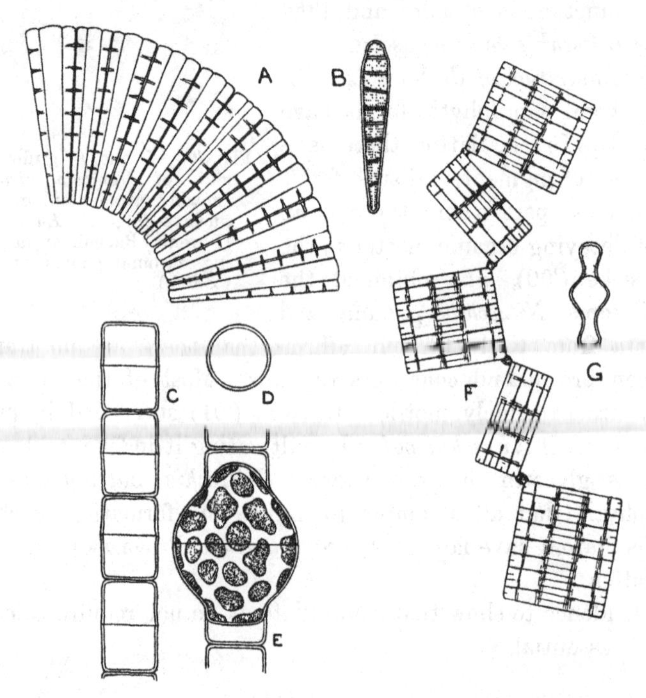

Fig. 71. *A* and *B*, *Meridion circulare* (Grev.) Ag.; *A*, part of spirally disposed colony; *B*, valve view of individual cell. *C—E*, *Melosira varians* Ag.; *C*, part of a long filament of cells; *D*, valve view of individual cell; *E*, cell of filament showing formation of auxospore. *F* and *G*, *Tabellaria flocculosa* (Roth) Kütz.; *F*, part of a zigzag colony seen from the girdle view; *G*, longitudinal section of cell showing partial septum.

[*Cocconema* (§ *Encyonema*) spp., *Navicula* (§ *Schizonema*) spp.], a condition which is more frequently met with in marine than in freshwater diatoms.

The mucus is often secreted through special pores, with definite localized positions in the valves of different genera. In *Tabellaria* such secretory pores occur near the middle of the valve (fig. 72 *F*), in *Diatoma* there is always one pore towards the end of the valve in a somewhat asymmetrical position (fig. 72 *A*). In *Synedra* and *Fragilaria* the pores are also near the extremities of the valves (fig. 72 *D* and *E*). The different forms assumed by diatom colonies are to be accounted for very largely by the varying position of the secretory pores in the different genera and species.

The colonies of *Asterionella*, *Meridion*, and *Eucampia* are particularly noteworthy for the beautiful disposition of the cells composing the colony. In the first-named genus the elongated frustules are united by their corners in such a manner that they radiate from a central circle like the spokes of a wheel (fig. 73 *A*); in the other genera an unequal growth of the girdle of each cell results in the development of spiral ribbons. In the freshwater plankton one of the abundant forms of *Tabellaria fenestrata* assumes this habit and has been named *T. fenestrata* var. *asterionelloides* (fig. 73 *B*).

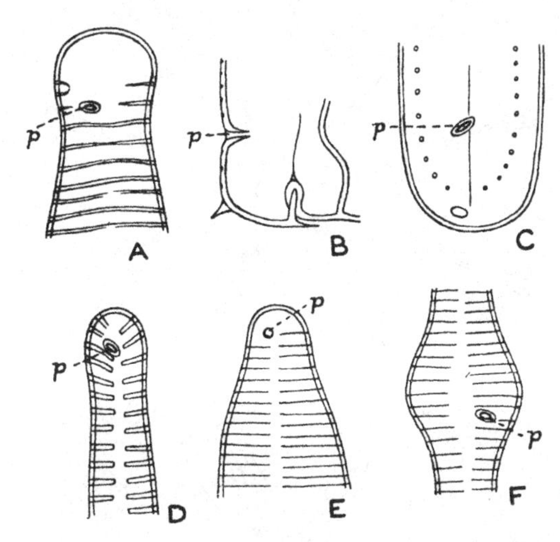

Fig. 72. Parts of the valves of various diatoms showing the pores through which mucus is secreted. *A*, *Diatoma grande* W. Sm. *B* and *C*, *Grammatophora serpentina* Kütz., girdle and valve views respectively. *D*, *Synedra Ulna* (Nitzsch) Ehrenb. var. *splendens* (Kütz.) V. Heurck. *E*, *Fragilaria virescens* Ralfs. *F*, *Tabellaria fenestrata* (Lyngb.) Kütz. *p*, pore. All × 2200. (After O. Müller.)

In this diatom, and in the plankton-species of *Asterionella*, Voigt ('01) states that a gelatinous membrane can be detected stretched between the radiating cells of the colony, thus considerably augmenting their floating capacity.

THE MOVEMENTS OF DIATOMS. Many of the free, unattached diatoms exhibit movements which even yet are but little understood. In some species the movements are relatively slow, but in others they are much quicker, and the individuals can be seen propelling themselves backwards and forwards in the direction of their longer axis. This movement is generally of a jerky character, although more rarely creeping and steady. It has been suggested that it only takes place when one valve-face of the diatom is in contact with some kind of substratum, but it must be admitted that there is much evidence to show that movements do take place when the active diatom is only in contact with the surrounding water. The power of locomotion is particularly manifested in those diatoms of a naviculoid form, and various explanations have at different times been put forward to account for it. The following historical survey, although brief and rather imperfect, will give some idea of the suggestions which have been made concerning this interesting biological problem.

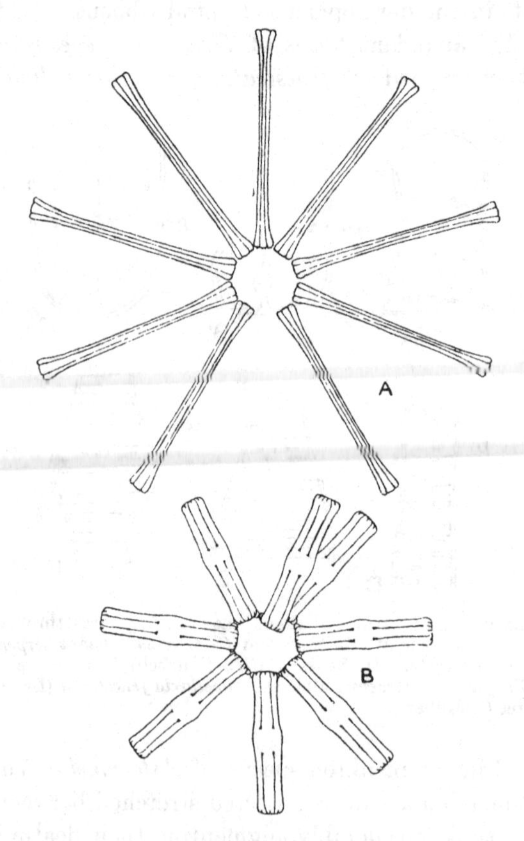

Fig. 73. *A*, *Asterionella formosa* Hass. *B*, *Tabellaria fenestrata* (Lyngb.) Kütz. var. *asterionelloides* Grun.

Ehrenberg (1838) imagined the movement to be due to the protrusion of cilia or of a pseudopodium through the raphe of the valve, whereas Nägeli (1849) attributed it to the passage of osmotic currents through the cell-wall. Max Schultze (1865), who observed the movements of minute foreign particles down the length of the raphe, attributed the locomotion to the contractility of a small portion of the protoplasm which was protruded through the raphe. Hallier (1880) considered it to be due to a contractile layer of protoplasm, and Onderdonk (1885) also regarded it as due to an external movement of protoplasm, but Mereschkowsky (1880) concluded that the evidence was in favour of Nägeli's theory of osmotic currents through the cell-wall. It was O. Müller in 1889 who first realised the significance of the pores in the central nodule and the polar clefts at the ends of the valves, and who showed that the raphe in the larger species of *Navicula* (§ *Pinnularia*) was really a rather complex cleft. He also demonstrated a streaming movement of the protoplasm in the raphe, and ascribed the movements of the cell to the reaction of the motive forces of this living stream of protoplasm upon the surrounding water. Schilberszky (1891), from observations on *Synedra*, agreed with Pfitzer that the movement was due to an outer coating of protoplasm which escapes from the raphe and is in a state of vibratile motion. He believed that the currents along the raphe were usually interrupted jerking or pulsating movements.

Cox (1890) revived the idea of a line of cilia along the raphe, and suggested that the absence of silica along this line could be accounted for by the obstruction of the moving cilia. Bütschli (1892) also imagined that the presence of a cilium or a fine flagellum would explain the phenomenon, but such structures have never been demonstrated.

The movements of some of the larger species of *Navicula* (§ *Pinnularia*) have been stated by Bütschli (1892) and by Lauterborn (1894) to be due to the production of a delicate mucilaginous filament which is protruded from the raphe at some point close to the central nodule. The whole cell is enveloped in a distinct mucilaginous mantle, and both this and the protruded filament are colourless and transparent. The filament is described as elongating by a series of jerks, and Bütschli put this forward as the explanation of the jerky movement of diatoms, the cell being pushed backward by the elongation of the filament, the distal end of which is fixed to the substratum.

In 1893, O. Müller again emphasized his previous statements relative to the movements of diatoms, and contested the views put forward by Bütschli and Lauterborn. In 1894, Lauterborn affirmed that the production of motility by the streaming of protoplasmic currents would be an isolated phenomenon in either the vegetable or animal kingdom, whereas movements are known to occur in the Desmidiaceæ and Oscillatoriaceæ[1] as a result of the excretion of mucilage. Müller replied in 1894 to the criticism of his hypothesis, and stated that the analogy which had been drawn between the movements of diatoms and of desmids was a false one. The cytoplasmic stream described by Müller is partly indicated in the accompanying diagram (fig. 74). Supposing the diatom to be moving in one definite direction,

[1] The movements exhibited by various genera of the Oscillatoriaceæ may not be due to the excretion of mucilage; see page 21.

there is a current passing along the outer cleft of the raphe from the anterior polar cleft to the anterior pore of the central nodule, and a corresponding current in the opposite direction in the inner cleft of the raphe. Similarly, in the posterior half of the diatom there is a current flowing from the posterior pore of the central nodule along the outer cleft of the raphe to the posterior polar cleft, and a corresponding current in the opposite direction in the inner cleft. These protoplasmic streams are connected by the joining canal of the central nodule (fig. 65 *2 schl*), and it should be remembered that the cytoplasm in the inner cleft of the raphe is really part of the lining layer of protoplasm of the cell.

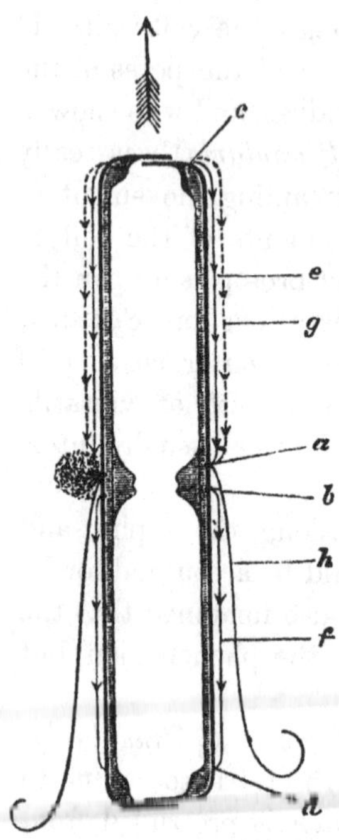

Fig. 74. Girdle view of a large *Navicula*. *a*, anterior opening of canal in central nodule; *b*, posterior opening of same; *c*, polar cleft in the front terminal nodule; *d*, opening in posterior terminal nodule; *e*, anterior plasma-stream in outer cleft of raphe; *f*, posterior plasma-stream; *g*, stream of granules when Indian ink is added to the water; *h*, slender mucous thread with granules attached. The arrow indicates the direction of movement of the diatom. (After O. Müller.)

Continued investigations by O. Müller, from 1896 to 1909, on the structure of the raphe in many of the larger diatoms, and on their movements, indicate that in those forms with a high development of the raphe the locomotion is due primarily to the protoplasmic currents which are circulating along this complicated cleft. It has been shown that the pressure within the cell is often as much as four or five atmospheres, and therefore there is probably sufficient friction between these circulating streams and the surrounding water to bring about the movements of the cells.

The observations of Palmer ('10) on *Surirella elegans* are certainly confirmatory of Müller's investigations. Palmer has shown that it is the actual protoplasm which circulates along the raphe-clefts in the keels of *Surirella*, and that the protoplasmic streams reacting upon the surrounding medium result in those peculiar rolling and turning movements which are so well exhibited in this genus.

Lauterborn has described and figured the formation of slender mucous threads by *Navicula major*, which can be best seen when the diatom is immersed in Indian ink. The thread apparently issues from the anterior pore of the central nodule of each valve, extending in a direction opposite to that in which the diatom is moving (fig. 75 *2* and *3*). The mucous threads are only discernible owing to the particles of ink, and it is not improbable that they merely indicate the direction of a slow current in the surrounding medium which originated in the outer cleft of the anterior half of the raphe, such a current carrying with it a small quantity of mucus secreted by the exposed protoplasm. Whatever mucus may be secreted in such a thread, the amount is much too small to cause the movements of a diatom as large as *Navicula major*. O. Müller has also carefully studied these threads and states that the movements of the diatoms which form them cannot be caused by the backward thrust of such a thread.

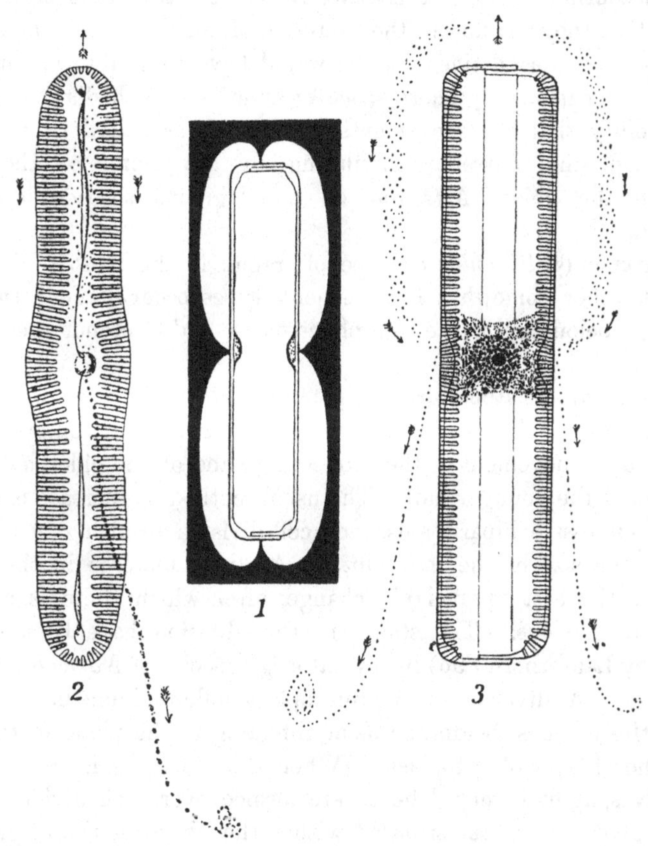

Fig. 75. *Navicula* (§ *Pinnularia*) *major* Kütz. *1*, individual cell immersed in Indian ink; *2* and *3*, valve and girdle views respectively, showing the narrow threads to which are attached the particles of ink. (From Oltmanns, after Lauterborn.)

A band-like thread of much larger size has been described and figured by Schröder ('02) as being formed by *Amphipleura pellucida*. This is a sluggish diatom with scarcely appreciable changes of position, and the secretion of such a mucous band might be sufficient to account for the feeble movements which occur.

The movements of diatoms are not all of a uniform character, and it is quite conceivable that they may not all be due to the same cause. The slow rolling and turning movements exhibited by some forms are very different from the rapid forward and backward propulsion so conspicuously shown by others. Those diatoms with the highest development of the raphe show the quickest movements, generally a jerky propulsion which enables them to move with more or less rapidity from place to place. The movements of some of the smaller species of *Navicula*, and also of *Nitzschia*, are so rapid and continuous that it is scarcely possible to accept Lauterborn's explanation that they are due to the excretion of mucilage. If that were the case, some of these very active diatoms would have to secrete an incredible amount of mucus in a very short space of time. In such diatoms there is every probability that the movements are caused, as stated by O. Müller, by the protoplasmic currents circulating in the clefts of the raphe. Moreover, in the genus *Nitzschia* no mucilaginous threads have been demonstrated.

The *Navicula* (§ *Pinnularia*) type of raphe is the most perfectly developed, but other somewhat less efficient types occur among the naviculoid diatoms belonging to the Gomphonemaceæ and Cocconemaceæ.

CELL-DIVISION. Diatoms are incapable of growth in size owing to the siliceous nature of their cell-walls, but slight alterations in volume can take place by a sliding movement of the connecting band of the older half of the cell over that of the younger half. The usual method of increase is a multiplication by successive bipartitions, each cell-division resulting in a gradual reduction in the size of the individuals. A slight increase in the volume of the cell is the first appreciable change, after which there is a mitotic division of the nucleus. The stages in this division have been carefully worked out by Lauterborn ('96) in several large species of *Navicula*, *Surirella* and *Nitzschia*. A division of the protoplast follows immediately on the division of the nucleus, beginning as an infolding in the plane of the girdle of the peripheral layer of cytoplasm. When this infolding is complete, new siliceous valves, at first very delicate, are formed over each divided surface. These new valves are at first situated within the girdle of the original cell, but with the growth of the new valves and the development of their connecting bands, the two connecting bands of the original girdle become separated, each forming one half of the girdle of the daughter-cells. Thus, each individual produced by division consists of a new half and an old half, and the connecting band of the old valve overlaps that of the new valve. Owing to the formation of a pair of new valves within the girdle of the old ones, and since the cells when once formed are incapable of growth, the newer half of every successive generation becomes reduced in size by the double thickness of a connecting band. This statement, however, is not altogether true, as it has been shown in some of the filamentous species, and is possibly the case in many other solitary forms, that daughter-cells are often produced of larger size than the parent-cells. Such individuals are recognized by the thickened rim of the valves. It is very probable that in many cases the new valves are only feebly silicified until they have been extruded beyond the connecting-band of the old girdle. They are thus capable of slight extension before their size becomes fixed by strong silicification. Such an increase in the size of the valves, however slight, would have a retarding influence on the diminution in size of the cells, the reduction in size not

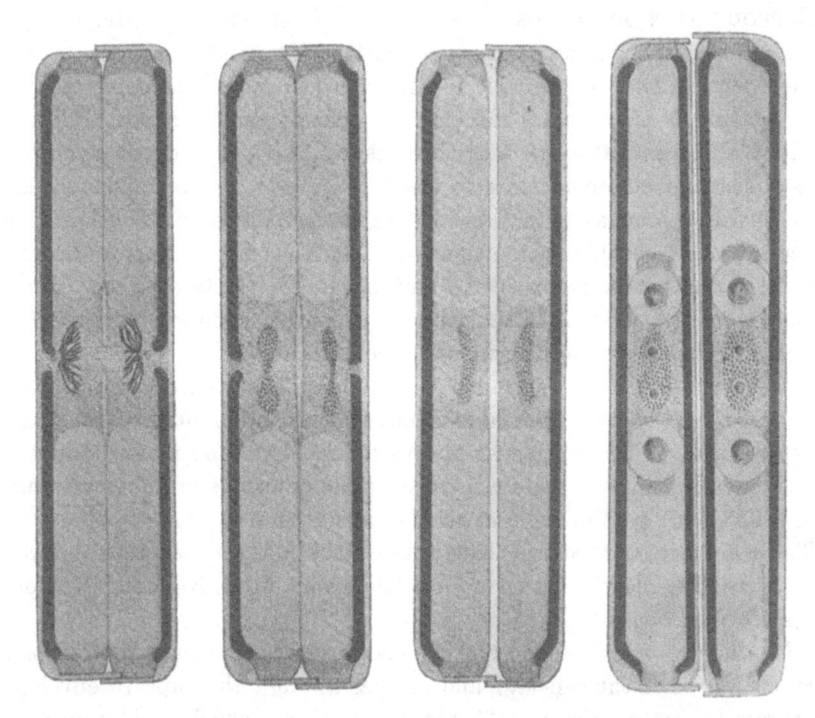

Fig. 76. Four stages in the cell-division of *Navicula* (§ *Pinnularia*) *oblonga* Kütz. The pair of conspicuous rounded bodies in each daughter-cell of the right-hand figure are Bütschli's red corpuscles. (After Lauterborn.)

being proportionate to the number of bipartitions. In the genus *Melosira*, O. Müller has shown that multiplication of the cells takes place in such a manner as to prevent as much as possible the division of the smaller daughter-cells.

Allen and Nelson ('00) have shown by series of pure cultures that cell-division in *Nitzschia Closterium* forma *minutissima* resulted in no appreciable reduction in size even though the number of generations in two years was incalculably great.

In the filamentous colonies of the genus *Melosira*, O. Müller ('03) has observed cells with thick walls and coarse markings, and other thin-walled cells with much finer and differently disposed markings. In some cases, indeed, one valve of a cell would be of the first type and the other of the second type. Müller's first interpretation of these differences was that he was dealing with mutations, although later he changed his view to agree with that expressed by Gran that it was a case of polymorphism.

Gran ('04) has recorded what he considers to be seasonal dimorphism in *Rhizosolenia hebetata*, with a winter form, forma *hiemalis*, and a summer one, forma *semispina*; but it seems probable, as pointed out by Kofoid, that there are really two distinct species which under the influence of certain environmental factors mutate the one to the other.

Reproduction by auxospores. Diatoms reproduce themselves by a type of spore known as an *auxospore*. Although formed frequently in some centric forms, in the vast majority of diatoms auxospores do not appear to be of common occurrence. The normal auxospore can be regarded as produced by the conjugation of two gametes, but differing from a zygospore by its almost immediate increase in size. There is a sort of rejuvenescence of the spore as soon as it is formed. Those auxospores produced without conjugation are probably parthenogenetic. It has been commonly accepted that the formation of auxospores counteracts the results of repeated cell-division, whereby the individuals have been greatly reduced in size. This view regards auxospore-formation as a method of regaining the maximum size of the species. It is probable that the rejuvenescence is much more important than the mere re-establishment of size, and it must be noted that auxospores are not by any means always formed from the most diminutive cells. In nearly all cases there is a considerable mucous secretion enveloping those cells taking part in the formation of auxospores.

There are five methods of auxospore-formation, the distinctions between which were first clearly drawn up by Klebahn ('96). Karsten ('00) only recognizes four types.

(1) The protoplast of one of the smaller, reduced cells swells up, forces apart the halves of the cell-wall, and escapes through the rupture enveloped in a thin cellulose membrane. This is the auxospore, which rapidly increases in size and assumes an outward shape more or less like the original cell. The wall soon becomes silicified and sometimes the markings characteristic of the species are at once acquired (fig. 78 *C* and *D*). In other cases the individual organized from the auxospore remains somewhat irregular, but very soon undergoes cell-division. The new valves are much more perfect in shape and sculpture, and the individuals of each succeeding generation rapidly regain their characteristic form and elegance. Miquel ('92—'98), from numerous experimental cultures of various species of diatoms, states that the re-establishment of the maximum size is habitually brought about by the formation of this simple type of auxospore. It is merely the rejuvenescence of a single cell accompanied by an increase in size.

(2) Two auxospores may be produced from a single cell by the division of the protoplast. Each portion increases in size, emerges from the old wall, and finally develops as in the first method. This type of auxospore-formation has only been observed in *Achnanthes longipes* (fig. 77 *F*) and *Rhabdonema arcuatum*.

(3) An auxospore may be formed by true conjugation between two

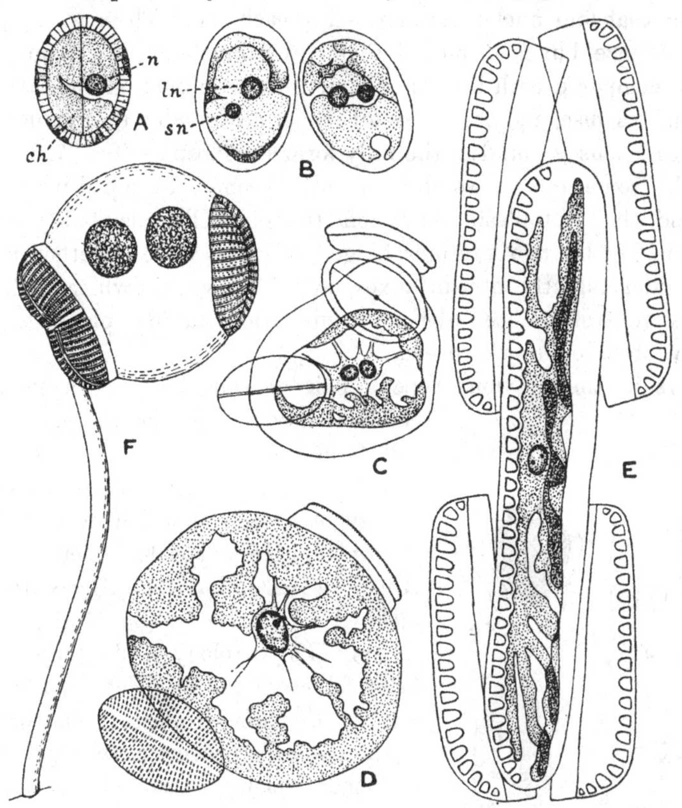

Fig. 77. *A—D*, *Cocconeis Placentula* Ehrenb.; *A*, vegetative cell showing chromatophore (*ch*) and nucleus (*n*); *B—D*, conjugation to form auxospore; *ln*, large nucleus; *sn*, small nucleus; *B*, two cells of *Cocconeis* before conjugation but after the division of the nuclei; in *C* the large nuclei are lying side by side, the smaller nuclei having disappeared, and in *D* the large nuclei have fused. *F*, *Achnanthes longipes* Ag. showing formation of two parthenogenetic auxospores from one protoplast. *E*, *Surirella saxonica* showing auxospore formed by conjugation. (*A—D*, and *E*, after Karsten; *F*, after W. Smith.)

individual diatoms. A large non-motile gamete (aplanogamete) emerges from each cell by the rupture of the girdle, and the two gametes fuse to form a zygote, which, as a rule, increases considerably in size, as in the case of those auxospores formed without conjugation. The cytology of this conjugation has been worked out by Karsten ('00) in the two species *Surirella saxonica* and *Cocconeis Placentula*. In the latter, division of the nucleus takes place before conjugation (consult fig. 77 *B*—*D*), so that the conjugating cells have two nuclei, one of which is large and one small. During conjugation the large nuclei fuse whereas the smaller ones gradually disappear. Conjugating cells 18—19 μ in length and 12—13 μ in breadth produced auxospores 40—41 μ in length by 28—33 μ in breadth. In *Surirella saxonica*, Karsten found that division of the nucleus again occurred previous to conjugation, but that four nuclei were formed in each cell. These were at first all of the same size, but soon one of them grew much in size, forming a large nucleus as compared with the other three. On conjugation the large nuclei of the gametes fuse, and the six small nuclei presumably disorganize, as only the fusion-nucleus is found in the fully-formed auxospore (fig. 77 *E*).

(4) In some cases two diatom-cells become enveloped in a common mucus and the protoplasts throw off the old cell-walls, but there is no conjugation. The two protoplasts generally lie close together and each develops independently into an auxospore. This type, in which each gamete undergoes rejuvenescence without conjugation taking place, is not infrequently observed in *Cocconema*.

(5) As in the previous type a pair of cells become enveloped in a common mucus, and the protoplast of each cell forms two gametes by division. The gametes of the two cells fuse in pairs forming two zygotes, each of which develops into an auxospore. This type has been observed in *Amphora ovalis*, *Epithemia Argus*, *Rhopalodia gibba*, *Navicula limosa* (fig. 78 *A*), and a few others. The cytology of the conjugation has been followed out by Klebahn ('96) in the case of *Rhopalodia gibba*. The protoplasts of the two cells become closely approximated, and the nucleus of each protoplast divides first into two and then into four. Two of these nuclei soon become reduced in size while the other two are conspicuous by their larger size. Each protoplast now divides into two portions, each of which contains two nuclei, one large one and one small one. The halves of the approximated protoplasts now coalesce, the larger nuclei fusing while the smaller nuclei gradually disorganize (fig. 79). The important feature thus brought to light by Klebahn is the division of the nucleus of the original protoplasts into four,

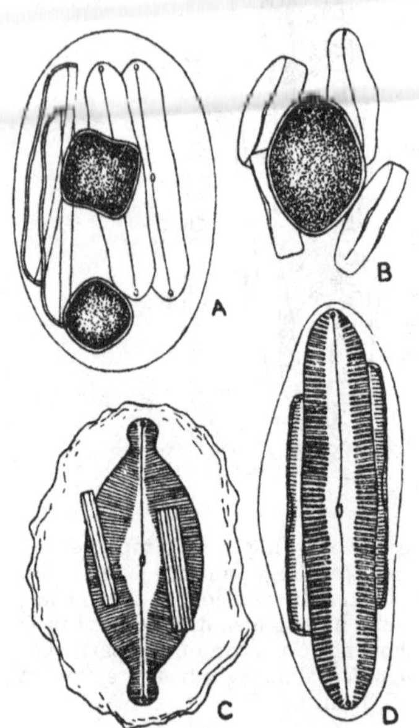

Fig. 78. *A*, *Navicula limosa* Kütz. (× 450). *B*, *Achnanthes flexella* (Kütz.) Bréb. (× 450). *C*, *Navicula Amphisbæna* Bory (× 450). *D*, *N. viridis* Kütz. (× 350). *C* and *D* illustrate the first method of auxospore-formation, *B* the third method, and *A* the fifth method.

and the significance of this tetrad-division, which precedes conjugation, may lie in the reduction of chromosomes.

There yet remains much work to be done on the cytology of auxospore-formation, as most of our present knowledge is based upon a few isolated cases.

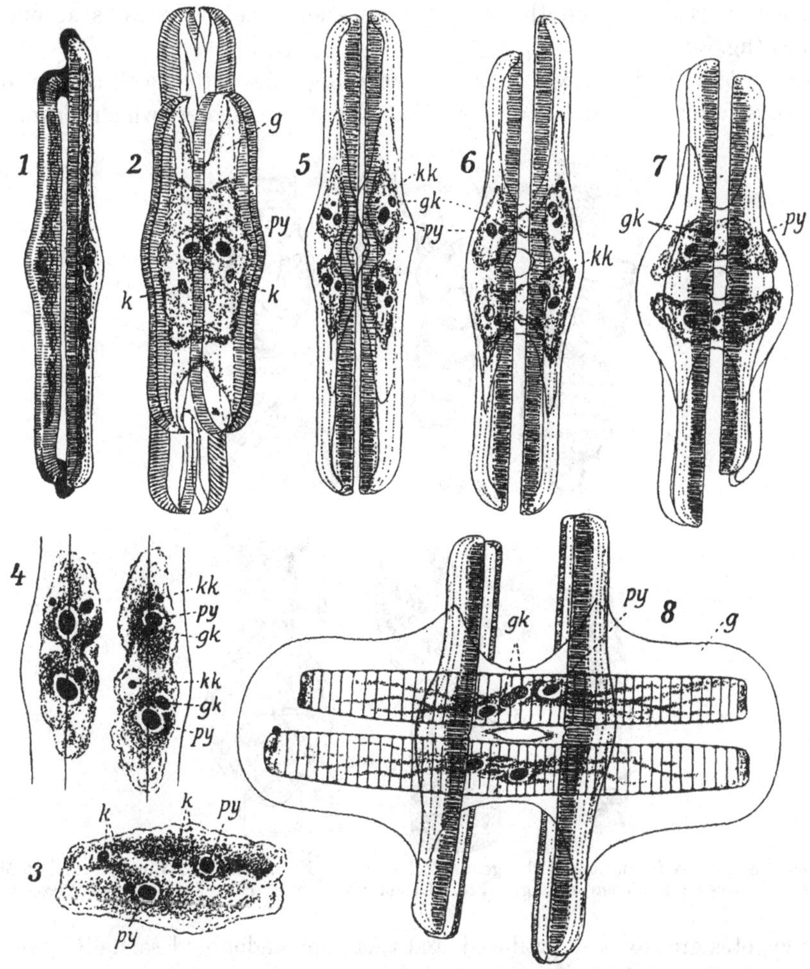

Fig. 79. Conjugation of *Rhopalodia gibba* (Kütz.) O. Müll. The conjugating cells are seen in the valve view except in fig. *2*, in which they are seen in an oblique girdle view. *k*, nucleus; *kk*, small nucleus; *gk*, large nucleus; *py*, pyrenoid; *g*, gelatinous investment. (From Oltmanns, after Klebahn.)

Karsten has stated that the type of auxospore-formation which occurs in the greater number of ground species is that in which two auxospores are formed by conjugation (the fifth type described above), but the observations of those who have spent many years in the investigation of freshwater and littoral diatoms scarcely support this statement, whereas there is much evidence to show that the first and fourth types are probably of the most frequent occurrence. The most essential feature in the formation of an auxospore is the increase in size of the cell. The first method of auxospore-formation is common in certain species of *Melosira,* and as the auxospores grow into filaments of cells while still attached to the old threads, the discrepancy in size between the new filaments and the old ones is at once obvious (fig. 80 *1*).

The little evidence we possess concerning the cytological details of auxospore-formation tends to show that those auxospores which develop

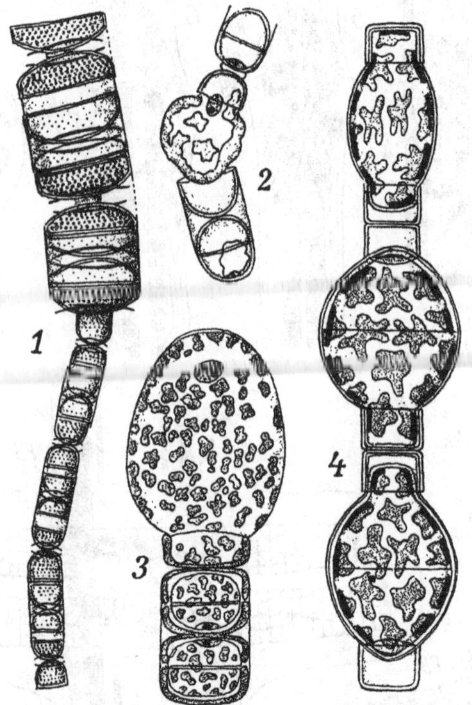

Fig. 80. Auxospore-formation in the genus *Melosira*. *1, M. nummuloides* Borr.; *2* and *3, M. Borreri* Grev.; *4, M. varians* Ag. (From Oltmanns, after W. Sm., Karsten, and Pfitzer.)

from zygotes are sexually produced, and therefore undoubted sexuality exists among diatoms. Such spores can be compared with the zygotes of the Desmidiaceæ, but whereas the latter remain without rejuvenescence as zygospores, the zygotes formed in the Bacillarieæ undergo an immediate rejuvenescence to form auxospores. Moreover, if those auxospores produced without conjugation are to be regarded as parthenospores, then the production of parthenogenetic spores is much more frequent than the production of zygotes in the Bacillarieæ, whereas parthenospores are of very rare occurrence in the Desmidiaceæ.

MICROSPORES. Reproduction by the formation of small spores termed *microspores* has been described as occurring in a number of diatoms, The existence of minute spores was suspected by Kitton ('85), who stated that they were so small as to readily pass through filter papers. Castracane ('97) has gone so far as to state that the normal method of reproduction of diatoms is by microspores, and that multiplication by division, although very common, is the exception rather than the rule. The latter statement, however, cannot be accepted in view of the fact that all observations tend to show that *the normal method of multiplication of diatoms is by cell-division.* This is still further emphasized by the numerous culture-experiments of Miquel ('92—'98), and of Allen & Nelson ('00), in which the diatoms multiplied *only* by cell-division. Gran has described the formation of microspores in *Rhizosolenia styliformis* ('02) and *Chætoceras decipiens* ('04) by the successive division of the protoplast to form sixteen spores. Ostenfeld ('10) has confirmed Gran's observations in the latter species, although he found in the samples he examined that division of the nuclei may not be at once followed by division of the cytoplasm.

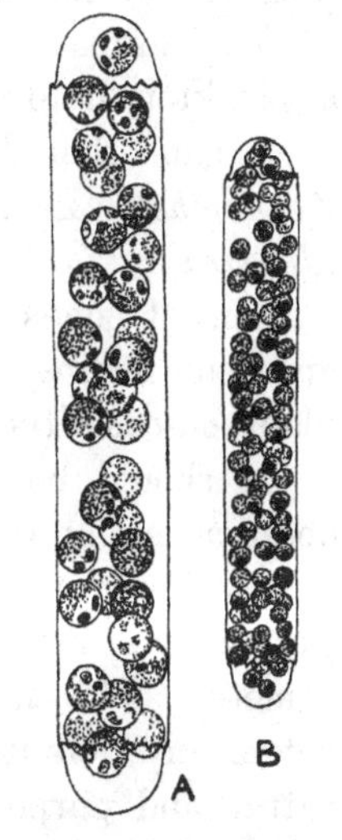

Fig. 81. Formation of microspores in *Corethron Valdiviæ* Karsten. *A*, division of protoplast into 32; *B*, further stage of division into 128 small rounded portions, each of which becomes a microspore (in this case a gamete). × 334. (Somewhat modified from Karsten.)

Karsten ('04) has given some interesting observations on the formation of microspores in *Corethron Valdiviæ*, one of the centric diatoms of the antarctic seas. In this case the protoplast divided by successive binary fissions to form as many as 128 globular cells each surrounded by a protoplasmic membrane (fig. 81 *A* and *B*). Simultaneous division of the nuclei occurred during the formation of these small cells, and in the division from 16 to 32 nuclear spindles were found, thus indicating that all the divisions are probably mitotic. Eventually the small cells escape and hang in clusters entangled among the spines of mature individuals. These cells are *gametes*, which fuse in pairs with those from other cells to form *zygotes*. The latter increase in size and divide into two daughter-cells, each of which possesses two nuclei. Only one nucleus remains as the daughter-cell becomes further organized into the diatom-cell, the other disappearing. The protoplast of the developing cell escapes from its first wall, and then expands to form a complete *Corethron*. The new siliceous wall develops gradually and the normal length is attained by the elongation of the girdle-bands. The 'tetrad-division' here *follows* conjugation (as in the Desmidiaceæ) and is a twice-repeated mitosis of the zygote-nucleus, in contrast to the 'tetrad-division' which *preceded* conjugation in the formation of the auxospores of *Rhopalodia gibba*.

Schiller ('09) has observed the formation of microspores in *Chætoceras Lorenzianum*, a species occurring in the plankton of the Adriatic. He found two types of microspores, one in which the cells were quite round, and another of more oval form, one end being more or less acute. These spores varied from 2·7—5 μ in diameter, and were provided with a wall of hardened protoplasm. Neither cilia nor any active movement could be observed. The differences between the two types are regarded by Schiller as sexual, but there is no proof of this, nor was it found out what became of these spores. Selk ('12) has also found microspores in *Coscinodiscus biconicus* in the plankton of the Elbe.

The formation of microspores just described was in each case in a plankton-diatom, but it would appear that such spores are sometimes formed in other diatoms. Kitton's suspicions concerned a small species of *Achnanthes*. Lemmermann has observed the production of microspores in *Melosira varians*, a species which is often most abundant at the weedy margins of ponds and ditches. Hustedt ('11) has also described and figured apparent microspores in *Eunotia lunaris*, a diatom which is never a true plankton-constituent.

It should be clearly understood that with the exception of Karsten's observations on *Corethron Valdiviæ*, the fate of these presumed 'microspores' has not been traced. Karsten's small spores were apparently non-motile gametes, and it is possible that those observed by Schiller were also passive gametes. The 'microspores' of diatoms require much further investigation before any definite statements can be made with regard to their general nature and purpose.

The observations of Bergon, and of Peragallo, in which motile spores or zoospores are recorded in several species of diatoms[1], require considerable confirmation before they can be accepted. It is only too well remembered how the presence of an internal parasite caused that most acute observer, Archer, to describe zoospore-formation in the Desmidiaceæ.

Murray ('97) has described a method of reproduction in certain marine plankton-diatoms by the formation of spores of the nature of gonidia. By successive divisions of the original protoplast as many as eight or sixteen rounded gonidia were observed in *Coscinodiscus concinnus*. Other aggregates of small *Coscinodiscus*-cells indicated that these gonidia probably developed

[1] Bergon (in *Le Micrographe Préparateur*, xiii, 1905) has described the development of 32 or 64 spores within the cells of *Biddulphia mobiliensis*. He states that these spores are motile, being furnished with two long cilia, and that his observations are confirmatory of Rabenhorst's discovery in 1853 of the reproduction of diatoms by zoospores. After further investigations Bergon has given fuller particulars of the development of the motile spores, but has not succeeded in tracing their fate. Peragallo (in *Proc. Stat. Biol. Soc. Sci. d'Arcachon*, viii, 1904—5 (1906), p. 127) has also described the transformation of a diatom-cell into a sporangium, from which zoospores ultimately escaped.

into normal frustules. Similar divisions of the protoplast were observed in *Chætoceras borealis.* This sporulation is not essentially different from the microspore-formation described by Gran in *Chætoceras decipiens* and by Schiller in *Chætoceras Lorenzianum.* Murray also observed the formation of one or two smaller frustules within the mother-cell both in *Coscinodiscus concinnus* and in *Biddulphia mobiliensis.* These observations are interesting if only because they indicate a possible means of rapid multiplication of plankton-diatoms, but further comment is not possible until the matter has been re-investigated.

RESTING-SPORES. In a few species of plankton-diatoms thick-walled resting-spores have been found. They occur in at least one marine species of the genus *Rhizosolenia,* and in the two freshwater species, *R. setigera* and *R. morsa* (W. & G. S. W., '09). They are also known in *Chætoceras, Bacteriastrum* and *Attheya.* In these genera only one resting-spore is formed within the mother-cell. The protoplast shrinks until it occupies only a small part of the original cell, after which it surrounds itself with a thick siliceous wall, which often develops spines or processes. In *Melosira italica,* O. Müller has observed the formation of pairs of resting-spores. Resting-spores of this nature appear to be formed at the end of the vegetative season, and up to the present time their germination has not been observed. Resting-spores have also been observed in *Surirella spiralis* (G. S. W., '12), a diatom which occurs in boggy marshes. In this case eight thick-walled spores were formed within the mother-cell (fig. 83).

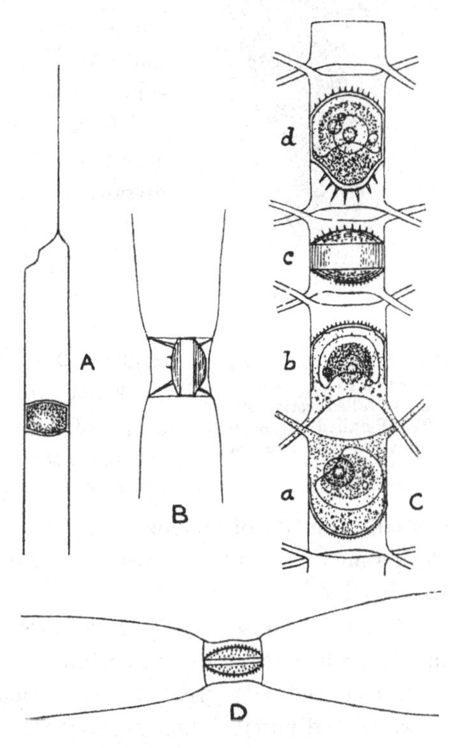

Fig. 82. *A,* Part of cell of *Rhizosolenia morsa* W. & G. S. West with resting-spore. *B,* Resting-spore of *Chætoceras ceratosporum* Ostenf. within the old mother-cell-wall. *D,* Similar state of *Ch. gracile* Schütt. *C,* Part of filament of *Ch. paradoxum* Schütt showing stages *a—d* in formation of resting-spores; *c* is a fully-formed resting-spore in surface view; *d,* a similar fully-grown spore in optical section. (*B* and *D,* after Ostenfeld; *C,* after Schütt.)

In some diatoms a resting condition has been noticed in which the protoplast has shrunk in size and a pair of new valves has been formed within the old ones. The resting-spores of diatoms are really aplanospores, in every way comparable with those which occur in many groups of the Chlorophyceæ.

CULTURES. The most extensive culture experiments have been in connection with plankton-diatoms. Miquel ('92—'98) conducted a large series of experiments with a view to obtaining cultures of single species of diatoms and obtaining information concerning the conditions requisite for their rapid increase. His methods were devised in the first instance for freshwater diatoms, but were afterwards found by him to succeed with marine littoral species. For the culture of marine diatoms Miquel prepared a special medium. Two solutions were made up as follows:

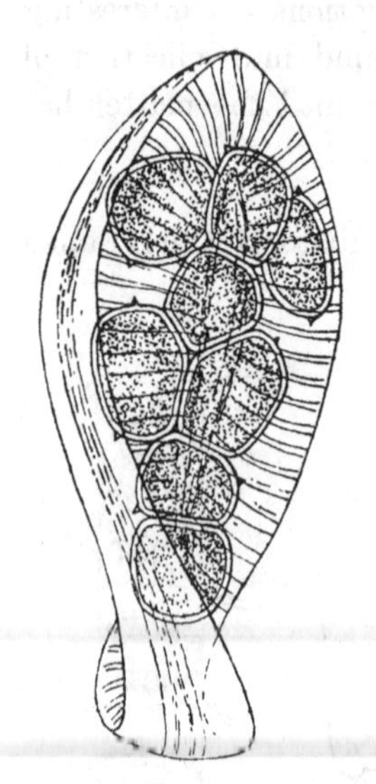

Fig. 83. A cell of *Surirella spiralis* Kütz. with eight thick-walled resting-spores. The detailed structure of the valve is not indicated. × 400.

Solution A. Magnesium sulphate 10 grm., Sodium chloride 10 grm., Sodium sulphate 5 grm., Ammonium nitrate 1 grm., Potassium nitrate 2 grm., Sodium nitrate 2 grm., Potassium bromide 0·2 grm., Potassium iodide 0·1 grm., Water 100 grm.

Solution B. Sodium phosphate 4 grm., Calcium chloride (dry) 4 grm., Hydrochloric acid 2 cc., Ferric chloride 2 cc., Water 80 cc.

Forty drops of solution *A* and from 10 to 20 drops of solution *B* were added to every 1000 cc. of sea-water which had been sterilized by keeping it at 70° C. for about 20 minutes. Miquel also added a small quantity of sterilized organic matter. Good cultures were obtained in this medium, and cultures of single species of diatoms were obtained by fractional subdivision.

Allen & Nelson ('00) have also obtained excellent results by this method, which they say is certain and gives good cultures. These authors have also used with great success a modification of Miquel's culture medium. It was found possible to reduce the first solution to one of potassium nitrate without detriment, the two solutions being as follows:

Solution A. Potassium nitrate 20·2 grm., Distilled water 100 grm.

Solution B. Sodium phosphate ($Na_2HPO_4 . 12H_2O$) 4 grm., Calcium chloride ($CaCl_2 6H_2 0$) 4 grm., Ferric chloride (melted) 2 cc., Hydrochloric acid (pure, concentrated) 2 cc., Distilled water 80 cc.

2 cc. of solution *A* and 1 cc. of solution *B* are added to each 1000 cc. of sea-water, and the whole sterilized by heating to 70° C. When cool the clear liquid is decanted from the precipitate which forms when solution *B*

is added to the sea-water. This medium was found to give constantly satisfactory results without the addition of any sterilized organic matter.

The best method of obtaining a culture was to add one or two drops of plankton to 250 cc. of the sterilized medium, which was then poured into Petri dishes. These should be left undisturbed and exposed to moderately bright diffuse light. The temperature should be kept as constant as possible and at about 15° C. In a few days colonies of different species of diatoms will be observed on the bottom of the Petri dishes. These can be removed by means of a fine pipette and transferred to flasks containing fresh culture medium. Successful persistent cultures of single species may in this way be obtained, but a word of warning must be given to the enthusiast who imagines that in this way he may reap a rich harvest of a beautiful diatom. On the contrary, deformed and distorted individuals are the rule in the earlier stages of the culture, and the wall is often so feebly silicified that the characteristic markings are not present. In older cultures the true form of the species is to a great extent regained, but the frustules are small.

Mention has already been made of certain saprophytic forms of diatoms which have been produced by cultivating normal brown forms in suitable media. Many diatoms can be cultivated on nutrient gelatin and agar, and some have been definitely shown by Richter ('03) to liquify gelatin and to dissolve grooves in the agar into which they sink. Miquel found the yellow rays of light the most favourable and obtained marked results by placing cultures under yellow glass. In the case of plankton-diatoms, however, Allen & Nelson were unable to obtain satisfactory cultures under yellow light.

Diatoms will thrive in water at freezing point, but on the whole they cannot withstand much freezing. There is evidence to show that the vitality of many species is destroyed about −15° C., although it is reasonable to suppose that some of the arctic and antarctic diatoms, more especially the freshwater ones, are able to survive a much lower temperature. The optimum temperature for cultures has been found to be about 15° C., and the maximum, which is rapidly fatal, lies between 35° and 40° C.

OCCURRENCE AND DISTRIBUTION. Diatoms are amongst the commonest of microscopic objects and they are ubiquitous in all kinds of damp and wet situations. They occur in great quantity in both fresh and salt water, and some species, chiefly of marine origin, occur habitually in brackish water. There is experimental evidence to show that apart from the quantities of available nutrient materials, the salinity of the water can be varied within large limits without appreciable effect on marine diatoms. On the other hand, a relatively small degree of salinity is fatal to the majority of freshwater diatoms. These facts largely account for the preponderance of marine species in brackish situations, and the almost complete absence of species common to both fresh waters and the sea.

Marine diatoms are most abundant in cold latitudes, having a decided preference for cold water. Freshwater diatoms are abundant all the world over, but they are more numerous and show greater diversity in temperate and arctic latitudes. Mountain ranges always yield a good diatom-flora. A few diatoms are inhabitants of warm and hot springs, and have been obtained in a living state at a temperature of certainly as high as 55° C. Some species are cosmopolitan, occurring in all parts of the world, but there are on the contrary many species and genera which occur only in certain seas and climates. The same is true to a lesser extent of freshwater forms, and when these forms are more fully and accurately investigated from a systematic point of view, many marked geographical peculiarities will doubtless be brought to light.

The free and unattached diatoms often form a yellow-brown scum at the surface of the water or on the sediment at the bottom of shallow pools. They also occur in abundance among the leaves of aquatic phanerogams or among the branches of larger algæ. Many species are epiphytic, being attached to the thallus of larger algæ. In fresh waters the filaments of various species of *Cladophora*, *Rhizoclonium*, and *Vaucheria* are frequently thickly covered with *Cocconeis Placentula* or *C. Pediculus* (fig. 84), and often

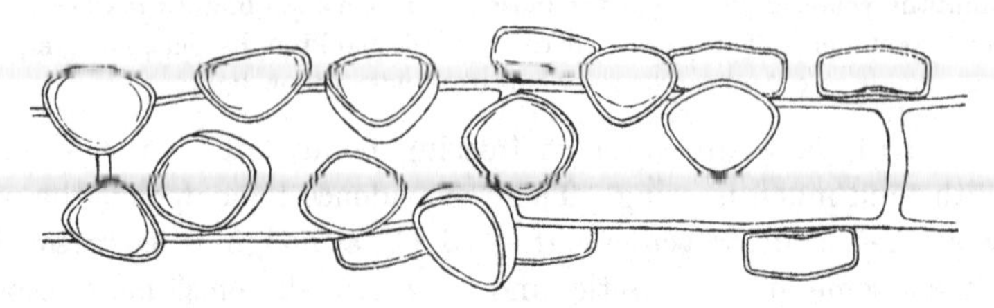

Fig. 84. A small portion of a filament of *Rhizoclonium hieroglyphicum* Kütz. largely covered with the epiphytic diatom *Cocconeis Pediculus* Ehrenb. ×375. Only the outlines of the cells are shown.

with *Epithemia turgida*. Several minute species of *Achnanthes* are also epiphytes of this nature. In the sea the genera *Isthmia*, *Grammatophora*, *Rhabdonema*, *Licmophora*, and others are epiphytes on the smaller seaweeds.

Numerous diatoms pass the whole of their existence free-floating in the surface-waters of the sea, or of lakes and large rivers. These are the plankton-diatoms which occur in prodigious quantity in the cold waters of the Arctic and Antarctic Oceans, and to a lesser degree in the warmer oceans. The marine plankton-diatoms are almost exclusively centric, and some of the principal genera are *Chætoceras*, *Biddulphia*, *Thalassiosira*, *Coscinodiscus*, *Rhizosolenia*, and *Dytilum*. The freshwater plankton-diatoms, on the other hand, are mostly pennate in character, consisting largely of species of *Tabellaria*, *Asterionella*, *Synedra*, *Fragilaria*, *Nitzschia*, and *Surirella*. The only centric freshwater genera of importance are *Melosira*, *Cyclotella*, and *Rhizosolenia*. The plankton-diatoms form the principal part of the food of countless freshwater and marine animals. They are the most important 'producers' of organic substance in the sea, and the life in the sea is therefore very largely dependent upon the inconceivable numbers of plankton-diatoms. In many of them the cell-walls are thin and rather delicate, with only a slight impregnation of silica.

The valves of diatoms of both the plankton and the benthos are found in quantity in the alimentary tracts of Molluscs, Tunicates and Fishes. They are also found in abundance in Guano, having passed through the digestive tracts of birds which feed on marine animals.

Some of the freshwater species, notably *Asterionella*, are sometimes the cause of foulness of drinking water (Whipple & Jackson, '99). This is due to the escape from the dead frustules of the oily products of metabolism. Such foulness can be obviated by storing water in the dark, or by treatment with copper sulphate in the proportion of not more than one part in two million parts of water.

FOSSIL DIATOMS. Great accumulations of diatoms are now being formed on the floor of both the Arctic and Antarctic Oceans, and beds are also accumulating in many of the larger freshwater lakes. Not only are these minute plants actively engaged at the present time, however, in forming oceanic and lake deposits, but the numerous *Diatomaceous Earths* bear testimony to their activity in former ages. These earths are mostly of a white or grey colour, sometimes hard, but more often so friable as to crumble readily between the fingers, and they are composed almost entirely of the siliceous valves of diatoms. Thus, large numbers of *fossil diatoms* are known. They may have had a marine or a freshwater origin, and most of the forms contained in the deposits belong to genera, and many of them to species, now living. The deposits are for the most part relatively recent, being principally associated with rocks belonging to the Tertiary formations. Some of them are of economic importance, being used as polishing powders ('Tripoli'), as non-conducting materials in the manufacture of fire-proof and sound-proof partitions, as absorbents for nitro-glycerin in the manufacture of dynamite ('Kieselguhr'), as constituents of dentifrices, and for other purposes. In deposits of this kind there are usually a number of species, but one or more may be dominant, forming the great bulk of the material.

Diatomaceous Earths have been found in many parts of the world, notably in Hungary, Bohemia, the United States, Barbados, and Trinidad. In the British Islands the best-known deposits are those at Dolgelly in Wales, and at Toome Bridge in Antrim, Ireland. The famous deposit at Biln, in Bohemia, which averages 14 feet in thickness, was estimated by Ehrenberg to contain some 40,000,000 frustules of diatoms in every cubic inch. It attains in places a thickness of 50 feet and is hard and flinty, the cementing silica having been formed by the solution of some of the diatom valves. The well-known deposit at Richmond, Virginia, U.S.A., is very extensive and attains a thickness of 30 feet, while on some of the geological surveys in the western states of America beds have been discovered no less than 300 feet in thickness and containing 80 per cent. of silica in the form of the frustules of diatoms. Diatoms have been found in the London Clay of the Lower Eocene, and are abundant in the Cretaceous rocks of the Paris basin.

All the Tertiary and Cretaceous diatoms show a close resemblance to existing species, and in some instances the species are identical. This is particularly the case with the pennate diatoms, but many of the fossil

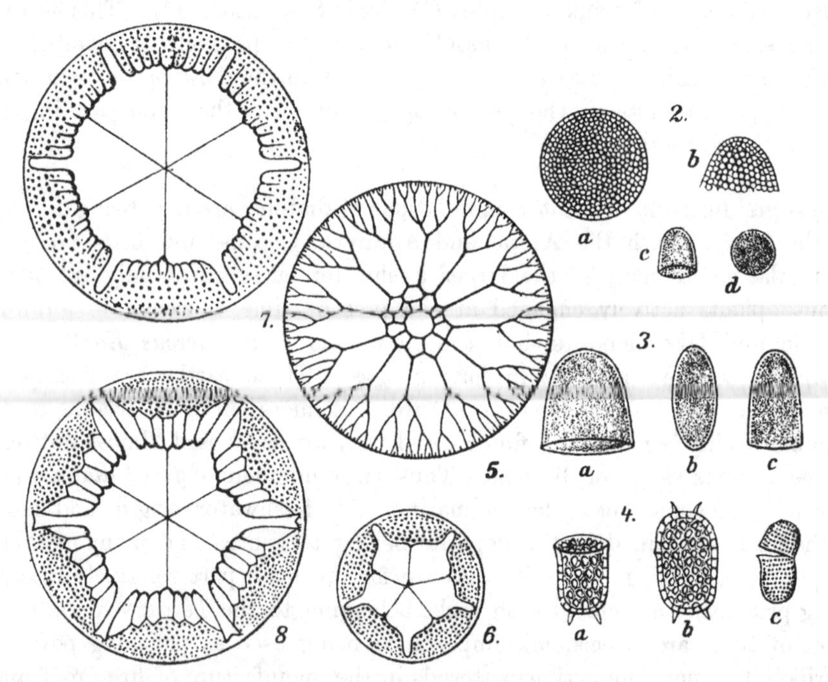

Fig. 85. A few fossil diatoms. *2 a—d*, *Pyxidicula bollensis* Rothpl. and *3 a—c*, *P. liasica* Rothpl., both from the Upper Lias. *4 a* and *b*, *Stephanopyxis* sp. from the Oligocene. *4 c*, *Pyxidicula* sp. from the Miocene. *5*, *Asterolampra Ralfsiana* Grev.; *6*, *A. marylandica* Ehrenb.; *7*, *A. crenata* Grev.; *8*, *A. decorata* Grev. All four from the Barbados deposit. All × 500. (After O. Müller.)

species of centric diatoms are distinct from the recent ones, and in this section considerably more than twenty exclusively fossil genera are known. Some of the more important are *Porodiscus, Actinodiscus, Brunia, Pyrgodiscus,* etc. *Actinoptychus,* although represented by some living marine species, embraces a large number of fossil forms, a fact which is also true of the genus *Asterolampra.*

It is probable, as believed by Pantocsek, that the deposit at Kusnetzk in Hungary belongs to the Trias, in which case this is the oldest known diatomaceous earth. The statement made by Castracane that he had found the valves of diatoms in coal from the English Carboniferous strata, was apparently made in error, and has never been verified (Cleve, '94; Seward, '98). Edwards has stated that he has found the valves of diatoms in still older rocks in New Jersey, but this observation also lacks verification. Cleve has examined a large number of Silurian clays and limestones, and also Rhætic and Cretaceous rocks, without finding a trace of any diatom. He also remarks upon the richness of the post-glacial strata of Sweden in both freshwater and brackish diatoms.

Cleve ('94) has emphasized the importance of the study of fossil diatoms to geologists. He states that microscopical examination of the pre- and inter-glacial deposits of northern Germany and Denmark has furnished evidence that these strata were formed in inlets from the North Sea and not from the Arctic Sea, and that accurate investigation of the geographical distribution of the living freshwater forms[1] will enable the geologist to ascertain the climate of the periods in which the numerous freshwater deposits were accumulated.

AFFINITIES. The Bacillarieæ were at one time included within the Phæophyceæ, but modern knowledge of their structure and life-history shows them to be widely removed from the brown seaweeds. Certain authors have placed them in close proximity to the Conjugatæ, and in 1904 Oltmanns so far extended the scope of the 'Akontæ' as to embrace the Bacillarieæ as well as the Conjugatæ.

Diatoms agree with Desmids in their unicellular and colonial habit, and in the fact that each cell consists of a newer and an older half. They multiply by cell-division in much the same way, but the comparison instituted by Oltmanns between the intercalary bands of the diatom-girdle and the intercalary pieces of cell-wall developed in certain species of *Closterium* and *Penium* scarcely holds good. The intercalary bands of the diatom are formed in definite order during cell-division, and they are developed before the actual connecting band of each new half is laid down. On the other hand, the so-called 'girdle-band' possessed by a few desmids is not a paired structure, even though two segments of it may be present in an old cell; but it is a cylindrical piece of cell-wall, arising subsequent to cell-division, and intercalated between the new and the old half-cells. Moreover, less than one-half per cent. of the known species of desmids ever develop such a structure, and in those in which it occurs only one 'girdle-band' may be formed during a period covering upwards of twenty divisions.

The cell-wall of most desmids is furnished with pores, but in contrast

[1] It is a surprising fact that the existing freshwater diatoms are not so well known as the fossil forms. Our knowledge of them is very incomplete and far from accurate.

to the wall of diatoms it consists mostly of cellulose and is not silicified. In both groups the chromatophores exhibit great diversity in form, but their normal disposition in diatoms is parietal, whereas in the vast majority of Conjugates they are axile. Moreover, no desmids possess the small parietal, disc-like chromatophores which are characteristic of such immense numbers of centric diatoms. The difference of pigment is also important, and so is the general storage of oil as a reserve of food by diatoms. The details of conjugation have been argued as a proof of affinity, but while conjugation of similar aplanogametes results in each case in a zygote, the latter remains as a resting zygospore in the Conjugatæ, whereas in diatoms there is an immediate rejuvenescence to form an auxospore. In the Conjugatæ conjugation is both usual and normal, and except in a few instances the spore is always a zygospore; but in diatoms the majority of auxospores are parthenogenetic, and nothing comparable with microspores exists at all in the Conjugatæ. The fact that Conjugates have never been able to adapt themselves to a marine life is also significant. On the whole, there is no satisfactory evidence that the Bacillarieæ and Conjugatæ are in any way nearly related. One must regard the similarity of the unicellular and colonial habit of diatoms and desmids, and also the resemblance between their cell-division, merely as a parallelism of modification brought about by the adaptation of two phylogenetically distinct groups to similar conditions of environment.

Up to the present time little if any light has been thrown on the affinities of diatoms. Their characters are so distinctive, and the group as a whole exhibits such a uniformity of structure, that there is every justification for treating them as a distinct class, the affinities of which are very obscure.

Classification. Space does not allow even an outline of the various proposals which have been put forward for the classification of the numerous genera and species of diatoms. Those systems based upon the disposition and mode of division of the chromatophores, which have been proposed by Pfitzer ('71), Petit ('77), Pelletan ('92), and Ott ('00), are impracticable owing to the fact that so many genera and species are unknown in the living state, and that considerable diversity in the structure of chromatophores has been found within the limits of a single genus. The classification which was suggested by H. L. Smith, and subsequently adopted by Van Heurck ('85) in his classical systematic work on the Diatomaceæ, was based upon the presence or absence of a raphe or pseudoraphe; and all diatoms were grouped under the three divisions of the Raphideæ, Pseudoraphideæ, and Cryptoraphideæ. More recently, Van Heurck ('09) has remodelled this classification, but he has most inaptly used two of H. L. Smith's divisional names in quite a different sense. As a classification it fails because of the admittedly artificial group of the Cryptoraphideæ.

There are two classificatory schemes which have the advantage of being much more natural than any of the others, and are therefore deserving of more special mention. These are the schemes of Schütt (1896) and Forti (1912).

The classification proposed by Schütt ('96) has been adopted with slight modifications by most investigators of the biology and taxonomy of diatoms, and its great merit lies in the fact that all the principal groups are to a great extent undoubtedly natural assemblages of forms. Its fundamental division, separating all those diatoms with a radial symmetry of the valves (Centric Diatoms) from those with zygomorphic or irregular valves in which the structure is in relation to a longitudinal line (Pennate Diatoms), is sharp and distinctive. It is briefly as follows:

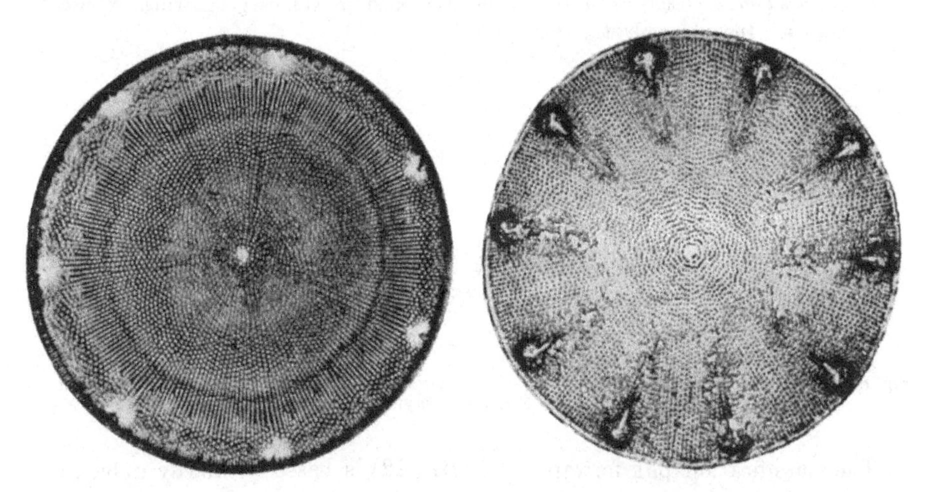

Fig. 86. Two centric diatoms. The left-hand figure is the valve view of *Aulacodiscus multipedex*, × 150; the right-hand figure is the valve view of *Aulacodiscus Crux*, × 120.

A. **Centricæ.** Valves with a concentric or radiating structure around a central point; without a raphe or pseudoraphe; valve view circular, polygonal, or broadly elliptical, rarely boat-shaped or irregular.

a. *Discoideæ.* Cells shortly cylindrical or disc-shaped, in valve view circular; hyaline or with radiating or areolated markings. Includes the subdivisions Coscinodisceæ, Actinodisceæ, and Eupodisceæ.

b. *Solenoideæ.* Cells elongate, cylindrical or compressed, circular or elliptic in cross-section (or in valve view); valves often conical and furnished with a stout or slender spine; girdle complex, with a considerable development of intercalary bands. Includes the Lauderiinæ and the Rhizosolenieæ.

c. *Biddulphioideæ.* Cells short, often a little longer than broad, box-like; valves generally with two, or sometimes with more, poles, each pole with a hump-like protuberance or a horn; cells in cross-section mostly elliptic, more rarely polygonal or circular. Includes the Chætocereæ, Biddulphieæ, Anauleæ, and Euodieæ.

d. *Rutilarioideæ.* Cells boat-shaped, with an irregular or radiating structure. Includes the Rutilarieæ.

B. **Pennatæ.** Valves zygomorphic, or less frequently irregular, never centric; mostly boat-shaped or needle-shaped, with a structure arranged in relation to a raphe or pseudoraphe.

e. *Fragilarioideæ.* Cells mostly straight, rod-shaped or lanceolate, without a raphe, but sometimes with a pseudoraphe or with indications of the commencement of a raphe. Includes the Tabellarieæ, Meridioneæ, and Fragilarieæ.

f. *Achnanthoideæ.* Cells crooked or suddenly bent, with a raphe on one valve and a pseudoraphe on the other. Includes the Achnantheæ and the Cocconeideæ.

g. *Naviculoideæ.* Cells with a raphe on each valve, with or without a keel in the line of the raphe (sagittal line). Includes the Naviculeæ and Nitzschieæ.

h. *Surirelloideæ.* Cells with the raphe concealed in submarginal wings of each valve. Includes only the Surirelleæ.

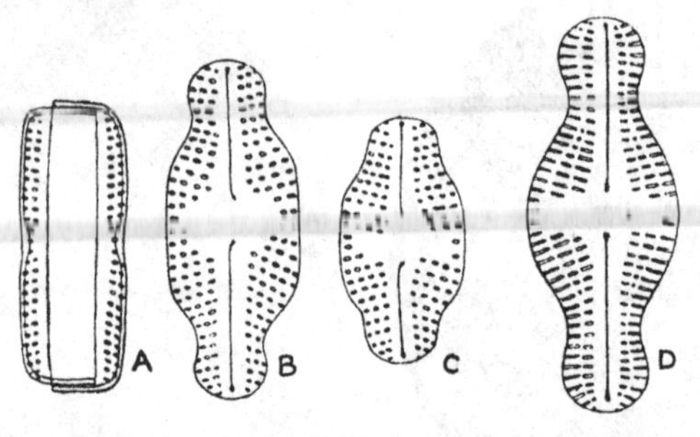

Fig. 87. Two small pennate diatoms. *A—C*, *Navicula muticopsis* Van Heurck. *D*, *Navicula globiceps* Greg. All ×1500. *A* is a girdle view, *B—D* are valve views.

The classification put forward by Forti ('12) is based primarily upon the fact that some diatoms are capable of spontaneous movements, whereas others are not. Forti accepts O. Müller's interpretation of these spontaneous movements as being due entirely to protoplasmic currents circulating in the raphe, and therefore the primary division separates all those diatoms which possess a raphe (whether perfectly or imperfectly developed) from those which do not. One improvement upon Schütt's arrangement is the more definite separation of the naviculoid diatoms from the Nitzschieæ. The scheme is briefly as follows:

I. **Immobiles.** Diatoms incapable of spontaneous movement. Conjugation unknown (perhaps non-existent) or possibly by the fusion of microspores.

A. Valves usually circular, more rarely elliptic or reniform. Includes Melosireæ, Coscinodisceæ, Asterolampreæ, Heliopelteæ, and Eupodisceæ.

B. Valves usually elliptic or lanceolate, sometimes linear, often regularly polymerous and stellate or stellate-sinuate; girdle generally complex, with numerous intercalary bands which are sometimes scale-like. Includes Biddulphieæ, Tabellarieæ, Licmophoreæ, and Entopyleæ.

C. Valves usually linear or lanceolate; girdle simple; generally with a pseudoraphe in each valve. Includes the Diatomeæ, Fragilarieæ, Synedreæ, Raphoneideæ, and Plagiogrammeæ.

II. **Mobiles.** Diatoms which exhibit spontaneous movements. Conjugation known in all the families.

A. Raphe imperfect or keeled (carinate). Includes Eunotieæ, Epithemieæ, Nitzschieæ and Surirelleæ.

B. Raphe perfectly formed, and interrupted in the middle of the valve, which is also furnished with central and terminal nodules; more or less extensive central and lateral areas occur in relation to the raphe. Includes the Heteroideæ, Tropidoideæ, and Naviculoideæ.

Forti has given the details of the genera of the section 'Immobiles' and of division A of the 'Mobiles.' Division B of the latter section, which includes all the naviculoid diatoms, was very carefully and comparatively synopsized by Cleve in 1894–5.

* * * * * * * * * *

It seems very desirable to combine the most important features of both the above classifications. Schütt's divisions of the Centricæ and Pennatæ are so fundamental, probably representing two distinct lines of descent, that they cannot be discarded. Also, in view of the fact that all the centric diatoms are non-motile, it might be much wiser to restrict the differentiation between the motile and non-motile forms entirely to the Pennatæ. This differentiation would then quite coincide with the degree of development of the raphe, which is without doubt the most important morphological structure of the pennate diatom.

The author would therefore suggest the following classification of diatoms:

Class *BACILLARIEÆ.*

Order I. CENTRICÆ.

[Includes divisions A and B of the 'Immobiles' of Forti.]

Order II. PENNATÆ.

Sub-order 1. *Non-labiles.*

[Includes division C of the 'Immobiles' of Forti.]

Sub-order 2. *Labiles.*

[Synonymous with the 'Mobiles' of Forti.]

A. Raphe imperfect or keeled.

B. Raphe perfectly formed.

LITERATURE CITED

ALLEN, E. J. & NELSON, E. W. ('00). On the Artificial Culture of Marine Plankton Organisms. Journ. Marine Biol. Assoc. viii, no. 5, 1900.

BENECKE, W. ('00). Über farblose Diatomeen der Kieler Föhrde. Pringsh. Jahrb. xxxv, 1900.

BÜTSCHLI, O. ('92). Mitteilungen über die Bewegung der Diatomeen. Verhandl. des Naturhist.-med. Vereins zu Heidelberg. Bd. iv, Heft 5, 1891 (1892).

CASTRACANE, TH. ('97) in Annales de Micrographie, ix, 1897.

CLEVE, P. T. ('92). Diatomées rares ou nouvelles. Le Diatomiste, i, March 1892.

CLEVE, P. T. ('94 ; '95). Synopsis of Naviculoid Diatoms. Kongl. Sv. Vet.-Akad. Handl. Bd. xxvi, no. 2, 1894 ; Bd. xxvii, no. 3, 1895.

FORTI, A. ('12). Contribuzioni Diatomologiche, XII. Atti del Reale Istituto Veneto di Scienze, Lettere ed Arti. Tom. lxxi, parte 2, 1912.

GRAN, H. H. ('02). Das Plankton des nordischen Nordmeeres. Report of the Norwegian Fishery and Marine Investigations, ii, no. 5, 1902.

GRAN, H. H. ('04). Fauna Arctica, III, Heft 3, 1904.

HUSTEDT, F. ('11). Beiträge zur Algenflora von Bremen. IV—Bacillariaceen aus der Wumme. Sonder-Abdr. Abh. Nat. Ver. Bremen, xx, 1911.

KARSTEN, G. ('00). Die Auxosporenbildung der Gattungen Cocconeis, Surirella, und Cymatopleura. Flora, lxxxvii, 1900.

KARSTEN, G. ('01). Über farblose Diatomeen. Flora, lxxxix, 1901.

KARSTEN, G. ('04). Die sogenannten 'Microsporen' Der Plankton Diatomeen und ihre Entwicklung beobachtet an *Corethron Valdiviæ* n. sp. Ber. Deutsch. Bot. Ges. xxii, 1904.

KITTON, F. ('85). On the Mysterious Appearance of a Diatom. Journ. Quekett Micr. Club, ser. 2, ii, 1885.

KLEBAHN, H. ('96). Zur Kenntniss der Auxosporenbildung. Pringsh. Jahrb. xxix, 1896.

KOHL, F. G. ('06) in Ber. Deutsch. Bot. Ges. xxiv, 1906.

LAUTERBORN, R. ('94). Zur Frage nach der Ortsbewegung der Diatomeen. Ber. Deutsch. Bot. Ges. xii, 1894.

LAUTERBORN, R. ('96). Untersuchungen über Bau, Kernteilung und Bewegung der Diatomeen. Zool. Instit. Univ. Heidelberg, Leipzig, 1896.

MANGIN, L. ('08) in Ann. Sci. Nat. sér. 9, viii, 1908.

MERESCHKOWSKY, C. ('80) in Botan. Zeitung, 1880.

MERESCHKOWSKY, C. ('03) in Flora, xcii, 1903.

MIQUEL, P. ('92 ; '93 ; '98 ; '03 ; '04). Recherches expérimentales sur la physiologie, la morphologie et la pathologie des Diatomées. Annales de Micrographie, iv, 1892 ; v, 1893 ; x, 1898. Also in Le MicrVSgr. Préparateur, xi, 1903 ; xii, 1904.

MOLISCH, H. ('05) in Botan. Zeitung, lxiii, 1905.

MÜLLER, O. ('83). Die Zellhaut und die Gesetze der Zelltheilungsfolge von Melosira arenaria Moore. Berlin, 1883.

MÜLLER, O. ('89) in Ber. Deutsch. Bot. Ges. vii, 1889.

MÜLLER, O. ('93). Die Ortsbewegung der Bacillariaceen betreffend. Ber. Deutsch. Bot. Ges. xi, 1893.

MÜLLER, O. ('95). Über Achsen, Orientirungs- und Symmetrieebenen bei den Bacillariaceen. Ber. Deutsch. Bot. Ges. xiii, 1895.

MÜLLER, O. ('94 ; '96; '97; '08 ; '09). Die Ortsbewegung der Bacillariaceen. II. Ber.

Deutsch. Bot. Ges. xii, 1894; III, *l.c.* 1896; IV, *l.c.* 1896; V, *l.c.* 1897; VI, *l.c.* 1908; VII, *l.c.* 1909.

MÜLLER, O. ('98; '99; '00; '01). Kammern und Poren in der Zellwand der Bacillariaceen. Ber. Deutsch. Bot. Ges. xvi, 1898; II, xvii, 1899; III, xviii, 1900; IV, xix, 1901.

MÜLLER, O. ('03). Sprungweise Mutation bei Melosireen. Ber. Deutsch. Bot. Ges. xxi, 1903.

MURRAY, G. ('97). On the Reproduction of some Marine Diatoms. Proc. Roy. Soc. Edinburgh, xxi, 1897.

OLTMANNS, F. ('04—5). Morphologie und Biologie der Algen. 2 vols. Jena. 1904—5.

OSTENFELD, C. H. ('10). Marine Plankton from the East Greenland Sea. I.—List of Diatoms and Flagellates. Danmark-Ekspeditionen til Grönlands Nordöstkyst 1906—1908. Bd. III, no. 11. Köbenhavn, 1910.

OTT, E. ('00). Untersuchungen über den Chromatophorenbau der Süsswasserdiatomaceen usw. Sitz.-Ber. d. Akad. Wiss. Wien, cix, 1900.

PALMER, T. C. ('10). Locomotion in the genus *Surirella.* Proc. Delaware County Inst. Sci. v, 1910.

PALMER, T. C. & KEELEY, F. J. ('00). The Structure of the Diatom Girdle. Proc. Acad. Nat. Sci. Philadelphia, 1900.

PELLETAN, J. ('92) in Journ. de Micrographie, xvi, 1892.

PETIT, P. ('77) in Bull. de la Soc. Botan. de France, Paris, 1877.

PFITZER, E. ('71). Untersuchungen über Bau und Entwicklung der Bacillariaceen. Hanstein's Botan. Abh. Heft 2. Bonn, 1871.

PROVAZEK, S. ('00). Synedra hyalina, eine apochlorotische Bacillarie. Österr. botan. Zeitschr. l, 1900.

RICHTER, O. ('03). Reinkulturen von Diatomeen. Ber. Deutsch. Bot. Ges. xxi, 1903.

SCHILBERSZKY ('91) in Hedwigia, xxx, 1891.

SCHILLER, J. ('09) in Ber. Deutsch. Botan. Ges. xxviii, 1909.

SCHRÖDER, B. ('02). Untersuchungen über Gallertbildungen der Algen. Verhandl. d. Naturhist.-med. Vereins zu Heidelberg. vii, 1902.

SCHULTZE, MAX. ('65). Die Bewegung der Diatomeen. Archiv für mikr. Anat. i, 1865.

SCHÜTT, F. ('96). 'Bacillariales' in Engler und Prantl, Die Natürl. Pflanzenfamilien I, i b, 1896.

SELK, H. ('12). Coscinodiscus-Mikrosporen in der Elbe. Ber. Deutsch. Bot. Ges. xxx, 1912.

SEWARD, A. C. ('98). Fossil Plants. Vol. I. Camb. Univ. Press. 1898.

SMITH, W. ('53; '56). A Synopsis of the British Diatomaceæ. London. I, 1853; II, 1856.

VAN HEURCK, H. ('85). Synopsis des Diatomées de Belgique. 2 vols. Anvers. 1885.

VAN HEURCK, H. ('09). Diatomées in Résult. Voyage du s.y. "Belgica" en 1897—99. Anvers, 1909.

VOIGT, M. ('01). Über eine Gallerthaut bei Asterionella gracillima und Tabellaria fenestrata usw. Biol. Centralbl. xxi, 1901.

WEST, G. S. (G. S. W., '12). Algological Notes X—XIII. Journ. Bot. Nov. 1912.

WEST, W. & WEST, G. S. (W. & G. S. W., '09). The Phytoplankton of the English Lake District. The Naturalist, March—Sept. 1909.

WHIPPLE, G. C. & JACKSON, D. D. ('99). *Asterionella*, its Biography, its Chemistry, and its effect on water supplies. Journ. New Engl. Waterworks Assoc. xiv, 1899.

WILLE, N. ('03). Über einige von J. Menyhardt in Südafrika gesammelte Süsswasseralgen. Österr. botan. Zeitschr. 1903.

CHLOROPHYCEÆ

(Green Algæ)

THE Green Algæ attain their greatest development in fresh water, and the number of known species is exceedingly large (probably about 5000).

The group-name Chlorophyceæ is here used in its original and proper sense to include *all* Green Algæ. The separation instituted by Wille ('97) into 'Conjugatæ' and 'Chlorophyceæ' is scarcely justified by our knowledge of the structure, life-histories, and probable inter-relationships of the Green Algæ; and the adoption of the name 'Chlorophyceæ' by Wille to include merely those Green Algæ other than the Conjugatæ is a misuse of the original name. Species of *Spirogyra* or Desmids are just as much Green Algæ as species of *Œdogonium* or *Selenastrum*.

In no other group of plants are there such wide differences in form and cytological structure, or such varied life-histories as can be found even in one section of the Green Algæ. It is this great diversity which makes the group at first so difficult for the student, and it is also the reason why it is impossible to treat of the Green Algæ as a whole in the comprehensive way adopted for the three preceding groups. Many forms, including some of the more primitive, are unicellular; some are cœnocytic, the whole plant consisting of what may perhaps be regarded as an aggregate of protoplasts within a common cell-wall; some are incompletely septate, each segment being a cœnocyte containing a number of protoplasts, the septation of the plant being quite independent of nuclear divisions; and others are multicellular or completely septate, each segment, which is in this case a single cell, containing one protoplast.

Not only are large numbers of Green Algæ unicellular or colonial in habit, but every degree of simple or branched filament is met with in the various groups; and in some the thalli are flat, almost parenchymatous expansions, or even cushion-like masses of tissue. In some of the Siphonales and Siphonocladiales the interlaced branches of profusely branched cœnocytes give rise to structures resembling the shoots and leaves of higher plants (consult figs. 149, 152 and 165). Except for the attenuation of the branches into hairs in some species, there is practically no differentiation among the vegetative cells, but there may be a marked distinction between the vegetative and reproductive cells.

The CELL-WALL varies much in its structure in the different groups of the Chlorophyceæ. It is in most instances composed of *cellulose*, but sometimes it consists largely of *pectose*. In all cases it is a secretion of the protoplasm, and arises on the outer surface of the protoplast as the result of complex katabolic processes. Many of the thicker walls are lamellose, the lamination being most easily seen after treatment by strong acids or other hydrating reagents which cause the wall to swell. These lamellæ represent successive layers of growth in thickness, and in the case of thick walls they are often very numerous. They are not always parallel to the outer surface of the cell-wall, but may be upwardly and outwardly divergent. The pectose compounds are sometimes sharply demarcated from the cellulose parts of the wall, and at other times the two constituents are to some extent in alternating layers. The cellulose parts give a distinct violet colouration with chlor-zinc-iodine, whereas the pectose constituents do not. In many of the Green Algæ the pectose constituents of the wall are in the form of gelatinous layers on the outside of the much more compact layers of cellulose. There is a copious secretion of jelly by large numbers of the Green Algæ, but the chemical composition of these gelatinous masses cannot yet be said to have been satisfactorily determined. Virieux ('10) considers that the mucilage in most of the Chlorophyceæ is of a pectic composition, although in two cases he found in addition mucilage of the nature of callose. It is highly probable in many cases that much of the mucus arises by the conversion of the outer layers of cellulose into mucilaginous substances of various degrees of solubility in water, and increments are constantly added by the gelatinization of successive layers. The refractive index of the jelly differs very little from that of water, and very often the extent of the gelatinous investment can only be detected by staining, or by immersion of the Algæ in Indian ink in the living state. The mucus stains with various aniline dyes such as fuchsin, safranin, methylene-blue, and gentian-violet. It is also coloured more or less deeply by ruthenium red.

Sometimes the gelatinization occurs only on one side of the wall, with the result that the cells appear to be stalked, as in *Hormotila* (fig. 115 *C*).

It is in the majority of the Protococcales, and in the Conjugatæ, that the cells, colonies, or filaments are so conspicuously enveloped in a gelatinous covering. In the former group the jelly is very largely the result of the gelatinization of the outer layers of the wall, whereas in most of the Conjugatæ the mucilage is secreted directly by the protoplast and passes through definite pores in the cell-wall. In the Desmidiaceæ it exhibits a distinct radiating fibrillar structure.

In many of the Green Algæ there are various inorganic salts present in small quantities in the cell-wall, and not infrequently traces of silica.

Protoplasmic continuity between the cells of the thallus has not been shown to exist in the multicellular Green Algæ, notwithstanding the pores which are sometimes present in the cell-wall. In the Conjugatæ the cells of most of the filamentous forms are quite able to lead an independent existence.

There is a definite and well-differentiated NUCLEUS present in the cells of all Green Algæ; but, of course, in the cœnocytic forms, and in the segments of the incompletely septate forms, there are many protoplasts, and therefore many nuclei, within the confines of the cell-wall. In outward shape the nucleus may be spherical, ellipsoidal or lenticular. The details

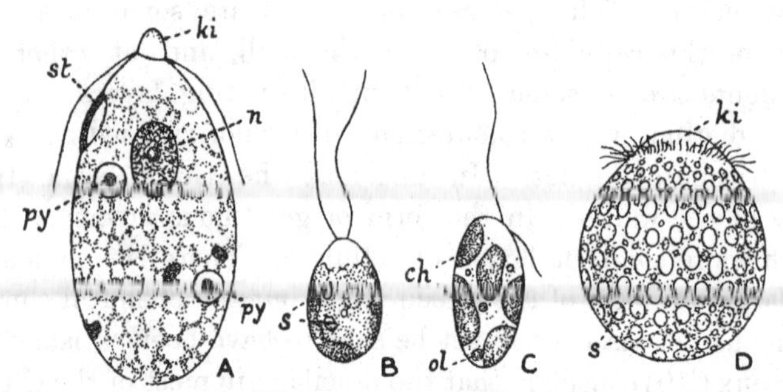

Fig. 88. Different types of zoogonidia. *A*, zoogonidium of *Cladophora*, ×about 2000 (after Strasburger); *B*, zoogonidium of *Microspora stagnorum* (Kütz.) Lagerh., ×1000; *C*, zoogonidium of *Tribonema bombycinum* (Ag.) Derb. & Sol., ×900; *D*, zoogonidium of a large species of *Œdogonium*, ×800. *ch*, chromatophores; *ki*, kinoplasm; *n*, nucleus; *ol*, oil globule; *py*, pyrenoid; *s*, starch; *st*, stigma (or pigment spot).

of the nuclear structure are as yet very meagre, but, in general, the resting nucleus presents a fine reticulum with a few thickenings at the angles of the meshes. There is a conspicuous nucleolus in many cases, but this structure may be somewhat irregular and compound, as shown by Lutman ('11) in *Closterium Ehrenbergii*. Nuclei with several nucleoli have also been described. In most of the Chlorophyceæ the nuclei are very small, a fact which materially increases the difficulties of their detailed study. The general structure of the nuclei of the Chlorophyceæ thus far investigated is essentially the same, and is not materially different from the nuclear structure exhibited by higher plants (*vide* McAllister, '13).

In all those Green Algæ in which the nuclear structure has been studied in detail the nuclear division has been shown to be mitotic. The genus *Spirogyra* has received most attention in this respect, and Pfeffer has shown that by etherizing the living cells the division could be changed from mitotic to amitotic. Gerassimoff also found that the cells of *Spirogyra* underwent amitotic division on loss of tone due to numbers of associated bacteria and other organisms. The published accounts of the mitosis in *Spirogyra* are not in agreement, but it is very probable that such discrepancies as occur are due to differences of interpretation of the same phenomena and perhaps also to defective technique. McAllister ('13) has shown that in the mitosis of *Tetraspora lubrica* 'the conduct of the chromatin in spireme formation, the origin and development of the spindle, and the mode of formation of the cell-plate are processes the same as in Angiosperms.'

In the multicellular forms division may occur in any, or all, the cells of the thallus, so that growth is to a great extent intercalary; or cell-division may be restricted to the apical cell of each branch. In the Zygnemaceæ and in the Siphonocladiales the transverse walls arise as a ring-shaped septum which gradually extends inwards. On the other hand, the formation of the cell-plate in *Tetraspora* and in *Œdogonium* begins between the daughter-nuclei, and the wall extends outwards as in higher plants.

In the Volvocineæ the vegetative cells are provided with cilia, and the zoogonidia and gametes of other groups are also ciliated. In the living cell the cilia can but rarely be observed, and only when their movements have become feeble or sluggish. A 2 per cent. solution of cocaine is often useful in causing the cilia to come to rest. They can generally be well observed by rapidly killing the cells by means of 1 per cent. osmic acid or by iodine solution. They can also be stained by methods essentially similar to those employed in staining the cilia of Bacteria, especially the tannin and carbol-fuchsin method, which gives good permanent preparations in which the cilia are stained bright red. In the small sub-family of the Tetrasporeæ the cells are peculiar in the possession of motionless 'pseudocilia,' which in *Apiocystis Brauniana* are sometimes very distinct and can be plainly observed without staining.

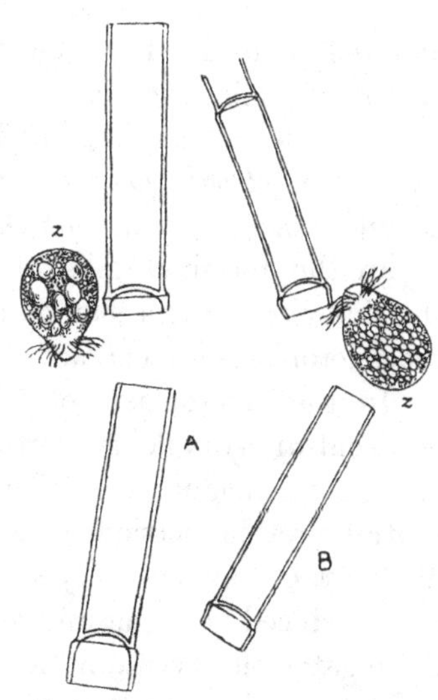

Fig. 89. Escape of the zoogonidium in *Œdogonium*; *A*, *Œ. Boscii* (Le. Cl.) Wittr.; *B*, *Œ. Hirnii* Gutw. × 460. Each zoogonidium possesses an anterior circlet of cilia.

In the large group of the Conjugatæ cilia do not occur.

The CHROMATOPHORES (which in all the Green Algæ are CHLOROPLASTS) are bright green or sometimes yellow-green in colour. The pigments are chlorophyll and xanthophyll, the former being as a rule greatly in excess of the latter. With the exception of certain of the Conjugatæ the Green Algæ do not apparently possess a protein pigment soluble in water such as the phycocyanin of the Myxophyceæ or the closely related red pigment of the Rhodophyceæ. The chloroplasts may be solitary or numerous, with entire margins, or with margins so lobed and incised as to present an almost infinite variety of form. In most families they are parietal, but the elaborate chloroplasts of the majority of Desmids, and those of *Zygnema, Prasiola,* etc., are axile. In some genera the parietal chloroplasts are ribbon-like or band-like, being wound spirally round the interior of the cell-wall (*Spirogyra* and some species of *Spirotænia*), and in others the axile chloroplast is furnished with spirally twisted ridges after the manner of screw threads (some species of *Spirotænia*). It is difficult to account for the extraordinary diversity in the form of chromatophores, especially as that diversity may be found even in the members of one family, as in the case of the Desmidiaceæ. They all perform the same function, and yet Algæ, with chromatophores entirely different both in form and disposition, live in association under precisely the same conditions. Oltmanns' comparison of the variability in the shape of chromatophores with the variability in the shape of the leaves of higher plants is a very apt one. In the lower types, and in a few of the higher types, the chloroplasts are sometimes very massive, in consequence of which it is difficult to ascertain their exact shape and limitations by direct observation; but very often they occupy only a relatively small part of the cytoplasm, in which case they are mostly clear and well-defined. In a few cases the chlorophyll is diffuse, no part of the cytoplasm being specially demarcated for the lodgment of the colouring matter. This has been definitely shown by Timberlake ('01) to be the case in *Hydrodictyon reticulatum,* and it seems probable that a similar condition may exist in other Green Algæ in which the chloroplasts are obscure.

In the terminal cells of piliferous branches, and also in the cells of rhizoids, chloroplasts may be reduced or entirely wanting. To some extent the chloroplasts are characteristic of the different families and genera.

In the chloroplasts of most Green Algæ are *pyrenoids,* which consist of a central crystalloidal portion of protein surrounded by a starchy envelope of variable magnitude. The most important part of the pyrenoid is the central protein portion, which is of the nature of an aleurone grain, and in many of the lower types the envelope of starch either does not exist or is reduced to a minimum. In the higher types and especially in the Conjugatæ, the accumulation of starch in the form of rounded or angular grains around the protein is often so great that the protein material is with difficulty detected in the living cell. The protein centre of the pyrenoid stains with practically all nuclear stains, and very deep staining may be obtained by acid-fuchsin after fixation with picro-sulphuric acid or an alcoholic solution of corrosive sublimate.

Pyrenoids sometimes divide equally or unequally on the division of the cell and they may multiply within the limits of any one chloroplast. They often disappear completely, and they can arise spontaneously in the chloroplasts of many of the filamentous and unicellular Green Algæ. They can be caused to disappear by starvation and they are often conspicuous in well nourished cells. Although there are many Green Algæ in which pyrenoids never occur, there is much evidence to show that in others their presence or absence is largely a question of nutrition. Under certain circumstances they may be entirely absent from the chloroplasts of Desmids or even those of *Spirogyra*.

Chmielewski's hypothesis that the pyrenoid was a permanent cell-organ always arising by the division of pre-existing pyrenoids is incorrect, as pyrenoids often arise *de novo*, especially in young cells. Also, Meyer's view that pyrenoids have no function other than that of a protein-reserve is not entirely correct, as Timberlake has shown very clearly in the case of *Hydrodictyon* that the pyrenoid is actually the seat of those processes resulting in starch-formation, and that a portion of it becomes converted into starch, presumably by other complex chemical processes. Timberlake takes the view that 'the

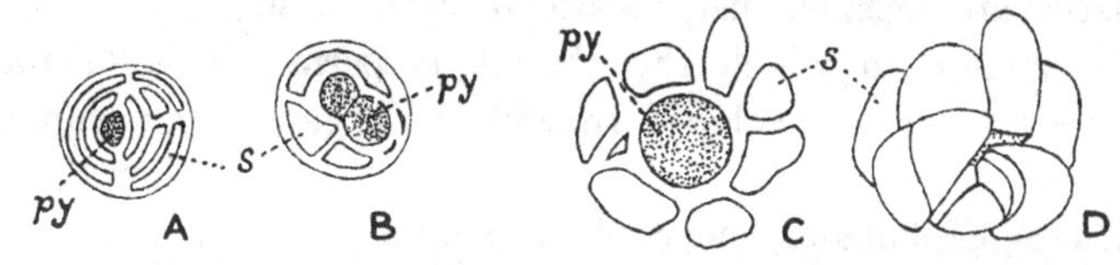

Fig. 90. *A* and *B*, optical sections of two pyrenoids (with associated starch grains) of *Hydrodictyon reticulatum* (Linn.) Lagerh. ×2250 (after Timberlake). *C* and *D*, pyrenoids (with associated starch grains) of *Closterium Ehrenbergii* Menegh. ×1250 (after Lutman); *C* is an optical section, and *D* the surface view; *py*, pyrenoid; *s*, starch grains.

pyrenoid is an active body, differentiated in the chlorophyll-bearing cytoplasm, which in co-operation with the latter acts as the base of the process of starch-formation.' In *Hydrodictyon* all the starch grains originate from the pyrenoids and gradually get pushed out into the cytoplasm. All the starch is therefore 'pyrenoid-starch.' This has been shown by Lutman ('10) to be the case in *Closterium Ehrenbergii.* But starch-formation does, of course, occur in the chloroplasts of Green Algæ which have no pyrenoids, and this starch might very well be regarded as 'stroma-starch,' since its method of formation is presumably different. Thus, although Klebs was wrong in attempting to discriminate between 'pyrenoid-starch' and 'stroma-starch' in *Hydrodictyon,* in which Alga Timberlake has shown that the difference does not exist, yet 'stroma-starch' occurs in those chloroplasts without pyrenoids, and there is no reason why it should not occur in some chloroplasts which possess pyrenoids.

In *Dicranochæte reniformis,* one of the setigerous members of the Protococcales, Hieronymus ('92) states that not only does the central portion of the pyrenoid consist of a protein crystalloid but that the envelope is also of a protein nature.

In the Chlorophyceæ the stored product of photosynthetic activity is for the most part starch. The exceptions to this are the Heterokontæ, in which the reserve is always a fatty oil, several of the larger species of *Mesotænium* in the Desmidiaceæ, certain species of *Conochæte, Protococcus, Asterococcus, Schizochlamys*, etc.; and the genus *Vaucheria*, in which oil is stored instead of starch. Oil globules also frequently appear in a number of other Green Algæ in the late autumn. In the genus *Trentepohlia* oil may be stored in considerable quantity, and there is frequently dissolved in it a red or orange-red pigment which is one of the carotins.

Some of the genera of the Chlorophyceæ, belonging to the Chætopeltidaceæ, Aphanochætaceæ, Coleochætaceæ, and Œdogoniaceæ, possess either hairs or bristles, which may be direct outgrowths of the cell-wall, as in the bristles of *Chætosphæridium*, *Conochæte*, or *Aphanochæte*, or finely attenuated, setigerous branches, as in various members of the Chætophoraceæ. In *Bulbochæte* the bristles originate as tubular outgrowths from the apical region of the cells, and possess a hollow, swollen base. In *Coleochæte* the delicate bristles are conspicuously sheathed at the base.

In those genera in which the thallus is permanently or temporarily attached there are often developed rhizoid-like organs of attachment, which are known as *hold-fasts* (or *haptera*). Except in the Cladophoraceæ, Ulvaceæ, Trentepohliaceæ, Siphonales and Siphonocladiales, these hold-fasts are, however, for the most part, only developed on young plants.

Multiplication occurs in the unicellular forms by cell-division (Desmidiaceæ and many Protococcales), and in some of the filamentous genera by the fragmentation of the filament (many of the Zygnemaceæ), or by the detachment of smaller or larger portions of the thallus. The last-mentioned method, which is really a proliferation of the thallus, is well shown in the genus *Caulerpa*, and to a less extent in *Monostroma* and *Prasiola*.

Vegetative propagation may also take place by *gemmæ* (which have also been termed 'cysts'). These are frequently unicellular, or 2—3-celled, and are mostly formed in the autumn as a means of surviving the winter in a vegetative condition. They occur only in a few Green Algæ, but are known in the Zygnemaceæ, Ulotrichaceæ, and Cladophoraceæ (in which they are *cœnocysts*); and in the Vaucheriaceæ, also, special short segments of the cœnocytes are often cut off for the same purpose. The walls of the gemma-cells are generally of considerable thickness.

Asexual reproduction by *zoogonidia* (or, as they are commonly termed, '*zoospores*') is general throughout the Green Algæ, although there is a notable exception in the Conjugatæ, in which motile reproductive cells are entirely wanting. They are also absent in the family Caulerpaceæ of the Siphonales and in the Autosporaceæ of the Protococcales. They arise in a zoogonidangium, which may be formed from an ordinary vegetative cell without change of form, or may be a considerably modified asexual reproductive organ.

The zoogonidia are small protoplasts formed singly by the rejuvenescence of the contents of a cell, or more frequently in numbers either by successive bipartition of the original protoplast or by free cell-formation[1]. The whole of the original protoplast, or in the case of cœnocytes the entire protoplasm, is not necessarily used up during the formation of the zoogonidia. In many cases a peripheral hyaline part of the protoplasm takes no part in this formation of reproductive cells, but becomes converted into a colourless mucilage which plays an important part in the escape of the zoogonidia. It is often largely owing to the pressure exerted by the swelling of this mucus that the weakest spot in the wall of the zoogonidangium is broken through and the zoogonidia (or 'swarm-spores') permitted to escape. These may swim away directly into the surrounding water or they may be held for a limited time within a delicate vesicle protruded from the orifice of the zoogonidangium.

It is possible to induce the formation of zoogonidia in some Chlorophyceæ by transferring them from a nutritive solution to distilled water or pure spring water, or to a solution of much less concentration. Placing in the dark sometimes assists the process, as in the case of *Vaucheria*. Change from a lower to a higher temperature (such as when Green Algæ are brought from their natural habitats into a laboratory) may also cause the production of zoogonidia.

The zoogonidia are naked protoplasts destitute of a cell-wall, and furnished with two[2], four, or many cilia attached to the narrower anterior end; that is, the end which is carried foremost in swimming. This anterior extremity is sometimes drawn out into a small beak-like or wart-like projection of a hyaline character, in which case the cilia are attached at or around its base, as in the zoogonidia of *Cladophora* (fig. 88 *A*) or of *Œdogonium* (fig. 88 *D*). This beak-like part of the protoplast is stated by Strasburger to consist of 'kinoplasm.' Each zoogonidium may possess one or more chloroplasts, and is frequently provided with two minute contractile vacuoles situated in front of the nucleus close to the anterior extremity. A red pigment-spot (or *stigma*) is often present, usually in an anterior and

[1] In the genus *Vaucheria* the single large zoogonidium is a compound structure which may in all probability be regarded as representing the fusion of a large number of small biciliated zoogonidia such as are typical of the Isokontæ.

[2] The zoogonidia of certain Algæ belonging to the Heterokontæ have been described as possessing only one cilium, but in those cases in which the zoogonidia have been carefully re-examined the second, much shorter, cilium has been discovered.

lateral position. The rapid vibratile action of the cilia causes the zoogonidium to swim quickly through the water, but the movements of some zoogonidia are much more rapid than others. After a time (of variable length, but rarely more than an hour or two) the zoogonidium comes to rest, the cilia disappear, the protoplast secretes a cellulose wall, and the quiescent zoogonidium develops into a new plant (consult figs. 185 *H—J*; 189 *G*). Many zoogonidia on coming to rest attach themselves by their anterior colourless extremity, which often develops into a hold-fast of greater or less complexity.

Pascher ('09) has found that in some of the Ulotrichales the swarm-spores may soon lose their cilia and then creep about in an amœboid manner for half to three-quarters of an hour, after which they settle down and germinate in the ordinary way.

All zoogonidia must be regarded as homologous structures, and as will be emphasized in subsequent paragraphs they must also be considered as the

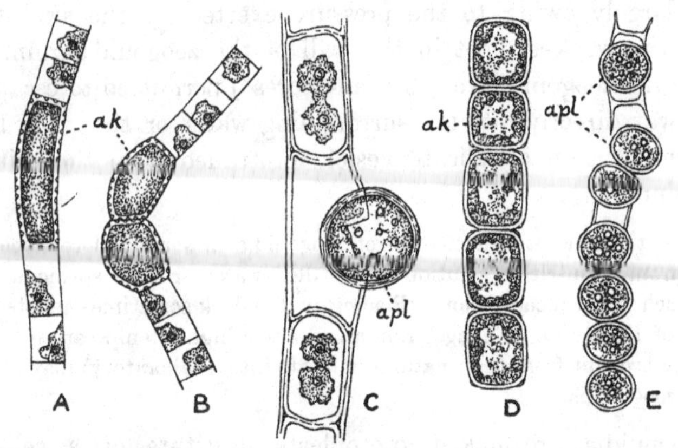

Fig. 91. *A* and *B*, *Ulothrix idiospora* G. S. West, showing scrobiculated akinetes (*ak*). *C*, a form of *Zygnema ericetorum* Kütz., showing aplanospore (*apl*). *D*, chain of akinetes (*ak*) of *Microspora floccosa* (Vauch.) Thur. *E*, aplanospores (*apl*) of *Microspora* sp. [probably *M. abbreviata* (Rabenh.) Lagerh.]. All × about 500.

direct representatives of the vegetative individuals of the primitive motile forms which were derived from the Flagellata.

In some groups (Protococcales, Ulotrichales, various Heterokontæ), single non-motile gonidia are frequently produced from the vegetative cells. If the gonidium is formed by the rejuvenescence of the cell and the original cell-wall is retained as part of the final thick wall, the gonidium is said to be an *akinete* (fig. 91 *A*, *B*, and *D*). If, on the other hand, the rejuvenescence results in the formation of an entirely new cell-wall around the protoplast, so that the gonidium is only set free by the rupture of the original cell-wall, the reproductive cell is known as an *aplanospore* (fig. 91 *C* and *E*). Some of these non-motile asexual gonidia germinate at once, but others undergo a variable period of rest before germination and are often termed 'hypnospores.'

SEXUAL REPRODUCTION. From the point of view of sexuality the Chlorophyceæ present a most interesting, and almost unique, series of forms. Within the group are to be found all conditions from a simple gamogenesis to the union of highly specialized gametes. Every degree of differentiation of the gametes is met with from those which are externally alike (*isogametes*) to those which are to a lesser or greater degree unlike (*heterogametes*). The union of the gametes results in a *zygote*, which may be either a *zygospore* or an *oospore*.

The term 'sexual' is often used to embrace all forms of gamogenesis, and is in this sense firmly established in botanical literature. It would, however, be more scientifically accurate to distinguish between gamogenesis (or the mere fusion of gametes) and sexual reproduction in the narrower sense (which should be restricted to those cases where there is a fusion of clearly differentiated ♂ and ♀ gametes). From this point of view, therefore, gamogenesis although including sexual reproduction is not identical with it. It must be remarked, however, that the gradation is so fine, especially in the Chlorophyceæ, that the distinction is scarcely worth making. Sometimes, as in many of the Zygnemaceæ, the gametes are morphologically indistinguishable but physiologically differentiated, and in these cases there is often a morphological differentiation of the gametangia (consult fig. 92 *A*).

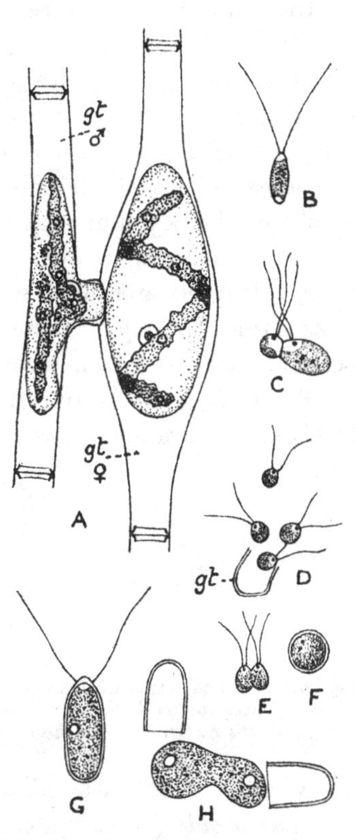

Fig. 92. Various forms of little-differentiated gametes. *A*, *Spirogyra tenuissima* (Hass.) Kütz. showing aplanogametes and slightly differentiated gametangia. *B—H*, Planogametes. *B*, Isogamete of *Trentepohlia Bleischii* (after Karsten); *C*, fusion of heterogametes of *Pandorina Morum* (Müll.) Bory; *D F*, escape of isogametes from gametangium, and fusion (*E*) to form zygote (*F*), of *Monostroma membranacea* W. & G. S. West; *G* and *H*, gamete (*G*) and gamogenesis (*H*) of *Chlamydomonas media* (after Klebs); *gt*, gametangium.

The zygospore is formed by the union of like (or nearly like) gametes, such as takes place in many of the Protococcales, Ulotrichales, Desmidiaceæ, etc., and is the result of a gamogenesis in which there are no external sexual differences between the gametes themselves. The gametes arise in *gametangia*. In the unicellular forms the cell itself becomes the gametangium (Desmidiaceæ, many Protococcales, etc.), and in the multicellular forms the ordinary vegetative cells become the gametangia with or without some external change of form (Zygnemaceæ, Ulotrichales, etc.). In the Conjugatæ the gametangium gives origin to only one large non-ciliated *aplanogamete*, but in other groups it is more usual for several ciliated *planogametes* to be formed in one gametangium.

The planogametes are usually very similar to the zoogonidia, although they are generally somewhat smaller and there may be slight differences in the ciliation.

In those Chlorophyceæ with dissimilar sexual cells the *female organ* consists of an *oogonium,* which is usually more or less spherical (Œdogoniaceæ), sometimes attenuated into a beak (*Vaucheria*), or more rarely produced at the apex into a long, narrow tube, the *trichogyne* (Coleochætaceæ). At the time of fertilization the oogonium contains only one female cell (the *egg-cell* or *oosphere*), except in the Sphæropleaceæ, in which there are numerous oospheres, the oogonium having in this case been originally a cœnocyte. The oosphere is generally a spherical cell containing chloroplasts, and often with a clear, colourless area at one side, known as the *receptive spot.* This spot is, as a rule, directly opposite the opening of the ripe oogonium and it is the point at which the antherozoid fuses with the oosphere.

The *male organ* is known as the *antheridium,* and is usually unicellular. In *Cylindrocapsa* and some species of *Œdogonium,* however, it may consist of

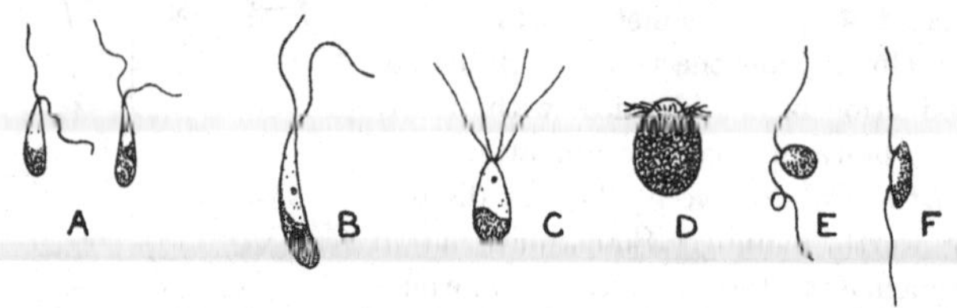

Fig. 93. Various forms of antherozoids of Green Algæ. *A, Sphæroplea annulina* (Roth) Ag.; *B, Volvox aureus* Ehrenb.; *C, Aphanochæte repens* A. Br.; *D, Œdogonium* sp.; *E, Coleochæte pulvinata* A. Br.; *F, Vaucheria sessilis* (Vauch.) D.C. All × about 900.

many cells. Each antheridial cell often gives rise to a number of male cells or *antherozoids* (= spermatozoids), but in the Œdogoniaceæ it gives rise to two and in the Coleochætaceæ to only one. The antherozoids may resemble the asexual zoogonidia, but are always much smaller. They are commonly pear-shaped, but may be elongate and almost rod-like. They possess two cilia, which are generally inserted at the pointed end, but in *Volvox globator, Vaucheria,* etc., the insertion is lateral (fig. 93 *F*), and in the Œdogoniaceæ there is a small circlet of cilia around the narrower, anterior end. They are as a rule faintly coloured, either yellow or green, and they often possess a red pigment-spot.

In all cases, with the exception of *Aphanochæte,* the oosphere is non-ciliated and remains *in situ* in the parent plant, being fertilized within the oogonium. The result of fertilization is an *oospore.*

It is in the Volvocaceæ that the evolution of sexuality is so remarkably displayed. There is in the different members of this group a complete series of transition-stages from the simplest form of gamogenesis to the highest type of oogamy, with antherozoids and passive oospheres. Even in the one genus *Chlamydomonas* the most primitive form of gamogenesis exists alongside a gamogenesis which has all the essential characteristics of true oogamy. In the Ulotrichales, also, there are all stages between the simplest gamogenesis, such as occurs in *Ulothrix*, to the highest type of oogamy such as is found in *Cylindrocapsa* and *Coleochæte*. In fact, in the genus *Ulothrix* the very beginnings of sexuality are observed, since the zoogonidia may be facultative gametes.

These statements, although very brief, are sufficient to show that in the Green Algæ the important upward step from asexual to sexual reproduction has been accomplished independently in more than one group.

ALTERNATION OF GENERATIONS. Much interest is necessarily attached to the life-histories of the Chlorophyceæ by reason of the slight indications in various families of what has been suspected to be an alternation of generations.

From the experimental work which has so far been done on the Green Algæ it is possible to conclude with some certainty that in the great majority of forms there is no obligatory succession of phases which can rightly be termed an 'alternation of generations.' The recurrent and alternating phases which are such a marked feature of Archegoniate Plants are at all times associated with distinct cytological differences. The nucleus of the cell of the sporophyte (or asexual generation) has twice as many chromosomes as the nucleus of the cell of the gametophyte (or sexual generation). On the sexual coalescence of the male and female gametes to form the zygote the number of chromosomes is doubled, and in the first division of the zygote, and in all subsequent divisions in the sporophyte generation developed from it, this number ($2n$) is maintained. The inauguration of the gametophyte generation is by a reduction division, the number of chromosomes being reduced to one half (n), and this number is retained in all subsequent nuclear divisions in the gametophyte. The gametophyte is thus known as the *haploid* generation in contrast to the *diploid* or sporophyte generation. The latter is regarded by some authorities as a post-sexual phase interpolated between sexual fusion and chromosome-reduction.

The question naturally arises as to how far this conception of the sporophyte can be applied to any phase in the life-histories of the various Green Algæ; and in view of the fact that the first land-plants very probably originated from forms not greatly different from some of the existing Chlorophyceæ, there is a further interest in the endeavour to obtain from these life-histories some clue to the origin of the sporophyte of the Archegoniatæ.

The researches of Klebs ('96) and others have shown that in certain of the Chlorophyceæ the various methods of propagation are largely dependent upon external conditions, and that to some extent the desired methods may be brought about at any time by employing the requisite combination of the operative factors. This may be partially or wholly true of many other Green Algæ; and since it is well known that the average Green Alga is exceedingly sensitive to external influences, it is not improbable that various combinations of the external factors really determine the successive phases of its life-history. There are, however, among the different groups of Algæ, a number of known cases in which certain successive phases are obligatory, and over which the external factors have little control. Among the Chlorophyceæ these examples are relatively very few.

Certain post-sexual phenomena of an obligatory character occur in the Ulotrichaceæ, the Conjugatæ, the Œdogoniales, and in some other groups of the Chlorophyceæ, notably in the Coleochætaceæ.

In *Ulothrix zonata* the quiescent zoogonidium grows into a typical *Ulothrix*-filament, but the zygospore develops into a dwarf plant quite different in appearance from the normal one. This dwarf individual then gives origin to a brood of zoogonidia, each of which ultimately grows into a typical filament. Thus there is in this Alga what may possibly be a first indication of an alternation of generations, but in the absence of all knowledge of the cytology of the phenomena such an interpretation must be entirely of the nature of a surmise.

In one or two of the Conjugatæ it has been ascertained that a twice-repeated division of the nucleus follows immediately on the sexual fusion in the zygote. Of these four nuclei, two are large and two are small, and one large and one small one go into each of the two cells resulting from the division of the zygote. The smaller nuclei gradually disappear, and although there is no positive evidence of reduction of chromosomes, yet as half the nuclei are discarded it is possible that the phenomenon is a reduction process.

The zygote of *Œdogonium* divides into four cells on germination, each of which becomes a ciliated zoospore. In some species it has been demonstrated that these zoospores produce only small asexual plants, and even several generations of asexual plants may be formed before the next sexual plant. This fact has given rise to the suggestion that there is in *Œdogonium* a rudimentary sporophyte generation, which is comparable to and homologous with that of the Archegoniatæ. That this is the case is very doubtful even though the so-called sporophyte occupies the same relative position in the life-cycle. It is quite possible that the division of the zygote into four may represent a chromosome reduction, although as yet there is no positive evidence of this.

A still more interesting case is that of *Coleochæte.* In this Alga a period of rest follows nuclear fusion in the zygote, and the latter on germination divides to form octants, and further divisions result in some cases in as many as 32 cells. Each of these cells becomes (or gives origin to) a zoospore of quite a different character from the ordinary zoogonidia of the gametophytic thallus. The zoospore on germination produces a small asexual plant, and there may be several generations of such asexual plants before a full-grown sexual plant is formed. This phase in the life-history of *Coleochæte* has generally been regarded in recent years as a rudimentary sporophyte generation, for the reasons that it is both post-sexual and asexual, and occupies the same relative place in the life-cycle of the Alga as the sporophyte of the Archegoniatæ. In a recent publication by Allen ('05), who followed out the nuclear changes on the germination of the zygote, it is stated that the first division is a reduction division. Some confirmation of this statement is certainly desirable since it means that the so-called sporophyte of *Coleochæte* is haploid and therefore part of the gametophyte generation. This development of *Coleochæte* is by analogy a simple sporophyte, and it seems probable that the precisely similar post-sexual development of *Œdogonium* is homologous with it; but in view of Allen's statements, especially if confirmed, it is scarcely possible to regard these post-sexual phenomena as phylogenetic fore-runners of the sporophyte of the Archegoniatæ.

The term "sporophyte" has in recent years acquired a definite interpretation as the non-sexual phase which occurs between sexual fusion and chromosome-reduction, and if it be true that in the Chlorophyceæ chromosome-reduction follows immediately upon sexual fusion, the sporophyte generation simply does not exist in these Algæ.

Since in the Chlorophyceæ the actual plant is the gametophyte, and the production of motile asexual reproductive cells does not involve any cytological change, they are referred to throughout this volume as 'zoogonidia.'

OCCURRENCE AND DISTRIBUTION. The Chlorophyceæ exist in the most varied habitats, occurring in every conceivable damp or wet situation to which light can penetrate. Some are epiphytes (*Coleochæte*, *Aphanochæte*, *Trentepohlia*, etc.), others endophytes (*Endoderma*, *Acrochæte*, *Blastophysa*, etc.), and a few are partial parasites (*Phyllosiphon* spp., *Cephaleuros* spp., *Coccomyxa Ophiuræ*). The great majority of the Green Algæ are exclusively confined to fresh water. The marine forms mostly belong to the relatively small groups of the Siphonales, Siphonocladiales, and Ulvales, all three of which also include a few freshwater representatives.

The Chlorophyceous vegetation of northern seas differs very much from that of tropical seas. In the more northern colder oceans the Green Algæ are of little importance except for the Ulvaceæ and some of the Cladophoraceæ, and such forms as occur are mostly confined to the uppermost part of the sublittoral region. In the tropics, on the other hand, not only are various members of Ulvaceæ and Cladophoraceæ abundant, but the three families Codiaceæ, Valoniaceæ and Caulerpaceæ, which are almost absent from northern seas, are developed luxuriantly, being represented by a large number of forms. The ability of these Algæ to grow in loose, muddy or sandy bottoms enables them to occur in great masses and extend to a depth of more than 20 fathoms. Many of them are also incrusted with lime and they contribute greatly to the various marine deposits, not merely in the bays and lagoons, but also in the more open sea. The gravelly sea-shores of these regions often consist very largely of the remains of calcareous Green Algæ.

There are some small marine genera of the Ulotrichales, mostly epiphytes, and a few marine species of *Ulothrix*. The latter appear to be confined to northern seas, being strangely absent from the tropics. There are also a few marine species of *Prasiola*, but of the large group of the Protococcales scarcely half-a-dozen live in marine habitats, and only one is of importance, viz.—*Halosphæra viridis*. There are no marine forms of the Heterokontæ, nor of the Conjugatæ.

The brackish-water Chlorophyceæ are very scanty, consisting of a few species of the Ulvales and of the Cladophoraceæ, together with some species of *Vaucheria* and *Ulothrix*, to which can be added a few members of the Protococcales, such as *Oocystis submarina* and *Brachiomonas submarina*.

It has been found that certain freshwater members of the Chlorophyceæ are able to adapt themselves to an existence in salt water. Richter ('92) stated that the lower the organization of the Alga the better its power of adaptation, but Comère ('03) found that only those Algæ with a robust structure could successfully withstand immersion in salt water. Comère succeeded in growing species of *Œdogonium* and a freshwater species of *Cladophora* in water containing 3·5 per cent. of sodium chloride, *Vaucheria sessilis* in water containing 2 per cent., and some of the large species of *Spirogyra* in water containing from 1·8—2 per cent. Richter affirmed that *Œdogonium, Spirogyra*, or *Vaucheria* have less power of adaptation to life in salt water than *Stichococcus* or *Tetraspora*, but this statement has not been verified. Notwithstanding the somewhat contradictory nature of Richter's and Comère's experiments, it appears that certain of the freshwater Chlorophyceæ can adapt themselves to an increasing salinity of the water in a manner comparable with the adaptation of a few forms of the Green Algæ to a life in hot water. Both these observers found that in all cases the salinity of the water caused the cells to increase in size, and when the concentration was high malformation of the cells invariably occurred. Starch at first disappeared from the cells, but reappeared when the adaptation was more complete.

Techet ('04) has found that many of the marine Chlorophyceæ exhibit great power of accommodation to changes in the salinity of the water. Experiments with *Cladophora trichotoma* showed that this Alga could withstand a salinity from 1·8 per cent. to 8·5 per cent., and that when the salinity reached 13·2 per cent. the plant produced quantities of zoogonidia and then perished. Techet also found ('08) in the genera *Halimeda, Udotea, Valonia,* and *Acetabularia,* that plants in water of weak salinity grew more rapidly and were less branched than in more strongly saline solutions.

The adaptation of certain freshwater Algæ to a life in salt water depends upon the power to absorb salts from outside and thus permanently acquire an increased osmotic pressure within the living cell[1]. Some Green Algæ are apparently able to bring about such osmotic changes in a very slow and gradual manner, others rather more quickly, but the great majority not at all. It is in this way that certain types of Green Algæ have been able to migrate from fresh water into brackish water and finally into the sea. The inability of most Green Algæ to adapt themselves to a considerable change in the osmotic strength of the cell-sap accounts for the fact that comparatively few of them are inhabitants of the sea.

A large number of the Green Algæ are minute forms occurring in bogs, pools and lakes among the leaves of submerged aquatic macrophytes, and often adherent to them by means of mucus. A considerable number are constituents of the freshwater plankton, not infrequently dominating it during the warm period of the year.

A few of the Chlorophyceæ have become constituents of the thalli of many Lichens, having entered into a symbiotic relationship with some fungus, although, as before explained (p. 37), this phase of symbiosis may in many cases be one of helotism. It is mostly one or two of the unicellular and colonial members of the Protococcaceæ which are thus found. These Algæ are for the most part indeterminable, the effect of their prolonged association with the fungus having so modified them that even cultures very often do not afford a clue to their exact identity. In the aquatic species of *Verrucaria* the algal constituent is also one of the Protococcaceæ.

Apart from members of the Protococcaceæ, which occur in large numbers of crustaceous, foliaceous and fruticose lichens, species of *Trentepohlia* occur in ARTHRONIA, CŒNOGONIUM, GRAPHIS, GYALECTA, and RACODIUM; and *Phycopeltis expansa* occurs in STRIGULA COMPLANATA.

Even species of *Cladophora* and *Vaucheria* are stated thus to associate themselves with a fungus; and the simplest, and probably the most primitive, of all lichens, BOTRYDINA VULGARIS, contains the Alga *Coccomyxa subellipsoidea* Acton ('09).

[1] Livingston ('00) showed that the stimulus which caused the alteration of form in the developmental stages of a species of *Stigeoclonium* (*Myxonema*) was the change in the osmotic pressure of the medium in which the Alga was growing. He showed that *a high osmotic pressure* (1) decreased vegetative activity, (2) inhibited the production of zoogonidia, (3) caused cylindrical cells to become spherical, and (4) freed the Alga from certain limitations as to the orientation of planes of cell-division. On the contrary *a low osmotic pressure* (1) increased vegetative activity, (2) accelerated the production of zoogonidia, (3) caused developing cells to become cylindrical, and (4) determined the orientation of the planes of cell-division. The inability of most Green Algæ to change more than very slightly the osmotic pressure within the living cell may therefore be in part responsible for the relatively small amount of polymorphism in the Chlorophyceæ (see p. 145).

Bialosuknia ('09) has carefully studied the Pleurococcaceous Alga in the thallus of LECANORA TARTAREA and named it *Diplosphæra Chodati*, but it might be better regarded only as a form of a *Protococcus* modified by its long association with a fungus.

Some of the Green Algæ have become symbiotically related to various animals. In the Protococcales 'zoochlorellæ' (of the genus *Chlorella*) occur in the cells of *Hydra viridis* and various Infusoria; a species of *Carteria* is intimately bound up with the nutrition and habits of the worm *Convoluta Roscoffensis*; a member of the Ulotrichales (probably a *Gongrosira*) is

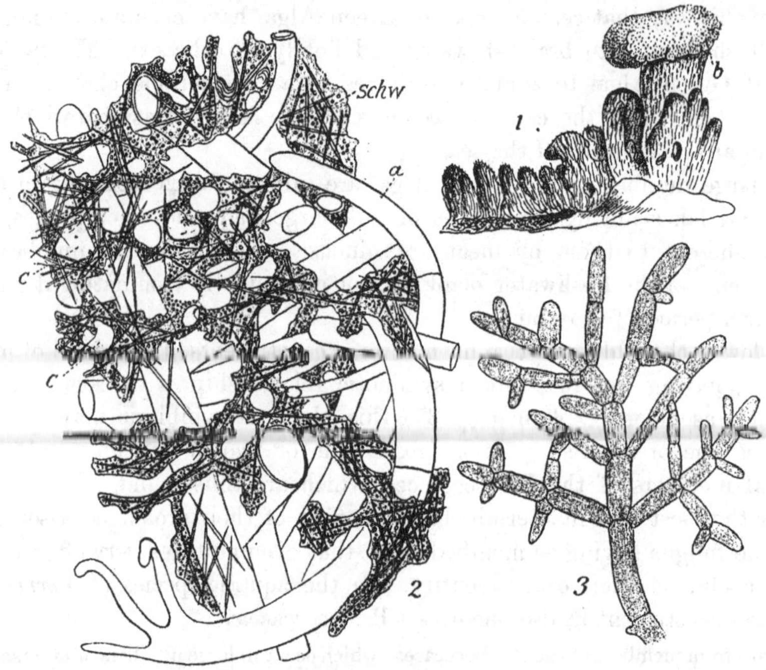

Fig. 94. The sponge *Halichondria* associated with the Alga *Struvea*. In *1* the *Struvea* is seen emerging from the sponge at *b*; *2*, a section through the sponge (*schw*) showing its canals (*c*) and the branches of the Alga (*a*) traversing it in all directions; *3*, some isolated branches of the *Struvea*.

apparently symbiotically associated with the common freshwater sponge, and the siphonocladiaceous genus *Struvea* is similarly related to the sponge *Halichondria* (fig. 94).

Quite a number of the Chlorophyceæ are inhabitants of the Arctic and Antarctic areas, and are capable of withstanding prolonged freezing. Among these are several species of *Chlamydomonas*, and another species of this genus (*Chl. nivalis*) is the Red Snow plant. In this connection the observations of Teodoresco ('09) on the zoogonidia (= vegetative cells) of *Dunaliella* are interesting, as he showed that their movements only ceased when the temperature of the salt water had been lowered to somewhere between – 17° C. and – 22·5° C.

CULTURES. It is often very desirable to be able to cultivate Green Algæ after they have been removed from their natural surroundings. In this way it is sometimes possible, by the exercise of due care, to obtain an insight into phases of their life-history which have previously been very obscure. The Algæ should be grown as nearly as possible under natural conditions, and filtered rain-water or tap-water may be used for some of the coarser kinds, which also require a considerable volume of water. Filtered water from *Sphagnum*-bogs is most useful as a medium in which to grow numerous Green Algæ normally occurring in stagnant situations.

In some cases, more particularly for experimental purposes, it is desirable to use a nutritive medium containing rather more of the available raw food-materials than is to be found in a natural medium. It is, however, necessary from the very outset to warn the investigator that many of the Green Algæ are very plastic organisms, and if the conditions of the culture are to a large extent abnormal (as compared with natural conditions), it is more than likely that strange phases and monstrous forms will be obtained, such as seldom if ever occur in nature. At the same time, it must be observed that a strange phase arising under abnormal conditions of culture may sometimes afford valuable evidence concerning phylogenetic relationships.

To obtain *successful cultures*, therefore, it is necessary that the conditions be as natural as possible, otherwise the development of monstrous forms will tend to lessen the value of the results obtained.

Most of the purely aquatic forms of the Chlorophyceæ can be cultivated in distilled water to which a small quantity of some nutritive solution has been added. For the smaller species the vessels used may be Petri dishes, test-tubes plugged with cotton-wool, or better still, circular glass boxes with glass lids; but large Algæ, such as *Spirogyra, Œdogonium*, etc., are best given a large volume of water, preferably one or two gallons. The culture-vessels should be placed in a north window or in such a situation that they are never exposed to direct sunlight. Temperature is also one of the most important factors. Ordinary room and laboratory temperatures are in most cases rapidly fatal, and the culture-vessels are best kept in chambers the lower parts of which are lined with zinc, and through which cold tap-water can, if necessary, be kept constantly running.

There are a number of well-known culture solutions, all of which require to be used in a very dilute state. Some of them are as follows:

Knop's solution: Water 1000 grammes; 0·25 gr. of $MgSO_4$; 1 gr. of $Ca(NO_3)_2$; 0·25 gr. of KH_2PO_4; 0·12 gr. of KCl; and a trace of Fe_2Cl_6. This makes a strength of 0·172 per cent., which is about the right strength for ordinary purposes. Other strengths can be made according to requirements.

Klebs' solution: 4 parts of $Ca(NO_3)_2$; 1 part KH_2PO_4; 1 part of KNO_3; 1 part $MgSO_4$; made up in strengths from 0·2 to 1 per cent.

Artari's solution: Water 100 grammes; 0·25 gr. of NH_4NO_3; 0·1 gr. of KH_2PO_4; 0·025 gr. of $MgSO_4$; and trace of Fe_2Cl_6.

Beijerinck's solution: Water 100 grammes; 0·05 gr. of NH_4NO_3; 0·02 gr. of KH_2PO_4; 0·02 gr. of $MgSO_4$; 0·01 gr. of $CaCl_2$; and trace of $FeSO_4$.

Knop's solution in varying strengths, but always weak, gives very satisfactory results; and if greater concentrations are desired, the strength of the solution should be *very gradually* increased.

Many of the subaërial Chlorophyceæ, such as *Stichococcus*, *Coccomyxa*, etc., can be cultivated on sterilized plates of porcelain or plaster of Paris which have been saturated with the weak culture solution. Sterilized porous pots of soil may also be used for the terrestrial species of *Vaucheria*.

Marshall Ward ('99) used a mixture of the prepared culture solution and silica jelly, both of which were sterilized and poured into glass dishes.

It is sometimes desirable, especially if pure cultures are required, to begin with one or two individuals. These can be isolated, and placed either in a hanging-drop or in a small cell on an ordinary slide, the cell being constantly supplied with the culture solution by means of blotting-paper strips. Cultures of this kind, which it is often necessary to keep for a long time, are very liable to become infested with Bacteria and Fungi. This can be prevented without any detriment to the Algæ, by the addition of 0·05 per cent. of neutral potassium chromate to the culture fluid, as recommended by Klebs, or by the addition of 0·01 per cent. of potassium bichromate, which Palla states to be quite as efficacious.

Many Algæ grow excellently in a fluid consisting only of filtered sterilized water obtained from the pools or bogs in which they were growing naturally.

Solid media, for very obvious reasons, do not give good results, and very often merely induce the formation of monstrosities and abnormal states. Agar-agar is rather better than gelatin, especially if it be dissolved in 0·2 per cent. Knop's solution. Some few of the Protococcales and some of the Green Flagellates can be cultivated on this medium, but such results as are obtained are not often of much value[1].

Some of the Protococcales of the plankton can be cultivated by using the filtered water from a boiled plankton-catch, or by using the sterilized filtrate from a plankton-catch a few days old.

The Conjugatæ are decidedly difficult Algæ to cultivate, and it is soon found that the conditions of culture suitable for one species are often a complete failure in the case of another. Desmids thrive best in filtered water from upland *Sphagnum*-bogs. In some of the Zygnemaceæ conjugation can sometimes be hastened by exposing to sunlight, and also by placing in a 2—4 per cent. solution of maltose or saccharose. The same result may be obtained by first putting the Alga into weak Knop's solution (not more than 0·2 per cent.) for about a week and then transferring to distilled water.

As a general rule, cultures of the Green Algæ thrive best when exposed to diffuse daylight. In cultures under coloured light, Dangeard ('09) has shown that *Chlorella* develops best in red and orange light, feebly in violet and not at all in green light. Changes in intensity of light, from diffuse light to darkness or *vice versa*, sometimes promote the formation of zoogonidia, as may also changes in the composition of the culture solution.

[1] The author has obtained normal cultures of *Ulothrix subtilis* and of two species of *Cosmarium* on agar-agar dissolved in 0·2 per cent. Knop's solution.

For the most recent work on the culture of certain forms of Green Algæ on solid media consult Kufferath ('13) and especially Chodat ('13). The last-named author has attempted to split up the widely distributed Alga *Scenedesmus quadricauda* into a number of 'species' on cultural characters. He has also recently described many cultural 'species' of *Chlorella*.

POLYMORPHISM. Concerning the so-called 'polymorphism' in the Chlorophyceæ there has been written quite a considerable algological literature, much of which consists of records of superficial and inaccurate observations, accompanied by expressions of opinion based upon defective evidence. Polymorphism, by which is meant the occurrence in nature of several different alternative vegetative forms of one and the same species, occurs to a very much less extent in the Algæ than has been so persistently advocated by some authors.

Hansgirg, Wolle, Borzi, and others, have stated at different times that there is a very wide polymorphism among Green Algæ, that most of the lower types are merely stages in the life-histories of the higher types, and that many of the 'genera' of the Protococcales are merely the polymorphic forms of one Alga. Chodat, also, in his earlier essays, was a believer in wide polymorphism among many Green Algæ. Most of these statements are assumptions which are not supported in any way by modern scientific evidence, and they have in a large measure been due to misjudgment and lack of precise methods of investigation. The evidence rested in nearly every case on the occurrence together in one habitat, perhaps even in one matrix, of many different Algæ in various stages of growth[1]. There is a great similarity between the various stages of closely allied species, and sometimes between certain stages of widely separated species, and since they may be all commingled in one stratum or mass, very exact methods are required to unravel the separate life-histories.

In the natural state, many of the unicells almost invariably live intermingled with others; but because the physiological conditions of this associate life appear to be necessary for their existence, it does not follow that they are merely forms of one another, even if fancied intermediate states have been recorded[2].

It is more especially the methods of pure culture followed out by Klebs and his pupils, by Beijerinck ('90; '93), Grintzesco ('02), and later by Chodat ('02; '09; '13), that have given the false hypothesis of wide polymorphism its death-blow; although it must not be overlooked that *exact* observations of the living Algæ under natural conditions were equally against the theory of wide polymorphism (Archer '62; G. S. W. '99; '04). The methods of pure culture simply clinched the arguments for the stability of algal species. Pure

[1] See also p. 30.

[2] The most recent advocate of the doctrine of wide polymorphism is Playfair, who in a paper entitled 'Polymorphism and life-history in the Desmidiaceæ' (*Proc. Linn. Soc. N. S. Wales*, xxxv, 1910) has put forward ideas regarding the wide Polymorphism of Desmids which are in entire disagreement with all biological principles. The value of his statements may be judged by his remarks on the Peridinieæ, Flagellata, etc., concerning which he makes assertions which are contrary to all the known biological facts relating to these organisms—assertions which have repeatedly been proved to be fallacious by the careful work of numerous investigators.

cultures conducted by exact scientific methods have shown that specific stability, even among the Protococcales, Volvocales, Flagellata, etc., is quite as constant as among higher plants.

If the cultures are carried out under conditions which are nearly natural, very valuable results often accrue, but cultures in sugar solutions and on gelatin or agar require the most careful interpretation. Some Green Algæ are much more plastic than others, but even the least susceptible members of the Protococcales are generally profoundly modified when grown under such abnormal conditions. It is quite justifiable to say that some of the statements made in consequence of the superficial examination of such cultures (which often abound in monstrosities) have resulted in positive harm.

In one of his later works, Chodat ('09) has admitted that the belief in wide polymorphism is not borne out by such facts as can be obtained from pure cultures, and it would seem evident that his former views, although based upon culture experiments, were the outcome of contaminated or impure cultures. He states ('09) that there are certain Algæ which by their extreme variability merit the name 'polymorphic,' if by that name it is understood that a plant may present itself under many aspects without change of nature; but their polymorphism is of the same order as that which is exhibited by most plants. He is also unable to support the views of Hansgirg, and states that his pure cultures have shown that alongside those Algæ which are polymorphic 'there are always as many, if not more, which present a remarkable stability.' This is a great admission from one who formerly believed in wide polymorphism[1]; and had Chodat's pure cultures been under more natural conditions, his recently expressed views on polymorphism would perhaps have been even more circumscribed.

Such polymorphism as occurs in the Chlorophyceæ is of a very limited character. The *Protoderma*-state of *Protococcus* (*Pleurococcus* of most authors), the *Dactylococcus*-state of *Scenedesmus obliquus*, the *Palmella*-state of *Chlamydomonas* and of certain species of *Ulothrix* and *Stichococcus*, the *Hormidium*- and *Schizogonium*-states of *Prasiola*, and the *Gongrosira*-states of *Cladophora* are among the few definite examples.

SOME ECONOMIC ASPECTS OF GREEN ALGÆ. The prolific growth of members of the Ulvaceæ, more especially of *Ulva Lactuca*, has sometimes proved a nuisance to sea-side communities owing to the obnoxious smell

[1] Another conversion, which resulted from the acquirement of greater experience, was seen in the case of Klebs. The views he expressed in one of his earlier works (*Ueber die Formen einiger Gattungen der Desmidiaceen Ostpreussens*, Königsberg, 1879), in which he lumped together all sorts of unrelated species, have received almost greater condemnation from his own subsequent researches than from the work of other investigators.

of the decaying plants. Cotton ('11) has shown that two conditions are requisite for this extensive growth of *Ulva*: (1) the absence of rough water, and (2) the presence of a substratum which affords a suitable anchorage for the plants. In muddy estuaries the 'byssus' of mussels is found to be a particularly secure attachment, and where sewage pollution occurs the growth of the *Ulva* may be very rank, the smell becoming most offensive on the decomposition of the plants. In some localities this *Ulva*-nuisance has become an acute problem, as for instance in Belfast Lough.

Freshwater Algæ of various kinds frequently prove a nuisance in reservoirs and in ornamental lakes. It is possible to get rid of an objectionable growth of such Algæ (Chlorophyceæ, Bacillarieæ and Myxophyceæ) by the use of minute quantities of copper sulphate. It has been shown, more especially by Moore & Kellermann ('05), that the amount of copper sulphate necessary to eradicate the Algæ is exceedingly small, in most cases one part in two millions[1] being quite sufficient. They have also shown that the actual quantity necessary to bring about the required result is dependent to a great extent upon the species to be dealt with.

A few genera of Green Algæ (along with many Brown and some Red Algæ) are used as food by the Japanese. The only important species are *Codium mucronatum*, *Ulva Lactuca*, *Enteromorpha linza* and *E. intestinalis*. In Hawaii, also, Green Algæ belonging to the Ulvaceæ serve as food, 'Limu eleele' consisting of *Enteromorpha intestinalis*, and 'Limu pahapaha' of *Ulva Lactuca* var. *laciniata* and *U. fasciata* (consult Setchell '05).

A large species of *Spirogyra* is dried and sold in bundles as a vegetable for food-purposes in the markets of Upper Burma (W. & G. S. W. '07).

In an air-dry state these Algæ contain about 3 per cent. of fat, and from 5 to 13 per cent. of protein, although nothing is known of the utility of the protein. The percentage of carbohydrates varies very much, and the simple sugars are of rare occurrence. Most of the carbohydrates in the above-mentioned Green Algæ are polysaccharides of the nature of pentosans which are with difficulty utilized by the human digestive tract (Alsberg '12).

* * * *

The foregoing summary will be found useful as a brief general account of the Algæ included in the Chlorophyceæ, but owing to the great diversity in form and structure, and the wonderfully varied life-histories exhibited by the Green Algæ, no advantage would be gained by an extension of such a general review. Hence, for the sake of greater clearness, a more sectional treatment is both necessary and desirable.

PHYLOGENY AND CLASSIFICATION. During recent years so many discoveries have been made concerning the life-histories and relationships of

[1] Approximately 8 oz. of crude copper sulphate to every 100,000 gallons of water.

Green Algæ that their classification now rests upon a much more certain basis than at any previous time. In fact, notwithstanding the large amount of work yet remaining to be done, more particularly among the lower types, it is unlikely that subsequent discoveries will involve any considerable change of principle. The most important result of this recent work is the recognition by the more experienced of algological investigators that the various groups of the Green Algæ have originated by a progressive evolution, either directly or indirectly, from flagellated ancestors, and that the cytological structure of the motile zoogonidia furnishes a reliable key to phylogenetic relationship.

Following up the suggestion of Bohlin ('97) that certain of the Green Algæ, by reason of their cytological structure, formed a very natural group which should be embraced in the 'Confervales' of Borzi ('89), Luther ('99) discovered that in all cases when these Algæ had been carefully examined the zoogonidia had been found to possess two unequal cilia. He also pointed out that this was the case in certain Flagellate forms in which the cytological structure was very similar. He therefore proposed to remove all these Algæ from the Chlorophyceæ and place them in a group of equal standing which he named the 'Heterokontæ.' He brought forward at the same time considerable evidence to show that this group possessed other definite cytological peculiarities, and most probably had a direct Flagellate ancestry.

Bohlin ('01), in a very suggestive paper on the phylogeny of Green Algæ, attempted to trace each of the principal groups back to an independent origin from the Flagellata. His suggested classification was as follows:

Heterokontæ.

A. Confervales.
Families: (Chloramœbaceæ); Chlorosaccaceæ; Chlorotheciaceæ; Confervaceæ; Botrydiaceæ.

B. Vaucheriales.
Families: (Vacuolariaceæ); Vaucheriaceæ.

Chlorophyceæ.

A. Conjugatæ.
Families: Desmidiaceæ; Zygnemaceæ; Mesocarpaceæ.

B. Siphoneæ.
Families: Codiaceæ; Caulerpaceæ; Bryopsidaceæ; Dasycladaceæ; Valoniaceæ; Cladophoraceæ; Sphæropleaceæ.

C. Protococcoideæ.
Families: Hydrodictyaceæ; Volvocineæ; Chlamydomonadineæ; Protococcaceæ; Tetrasporaceæ; Oocystaceæ.

D. Ulotrichales.
Families: Ulvaceæ; ?(Stichococcaceæ); Ulotrichaceæ (incl. Cylindrocapsa); Ctenocladaceæ; Coleochætaceæ; ? Chroolepidaceæ.

E. Microsporales.
Family : Microsporaceæ.

F. Stephanokontæ.
Family : Œdogoniaceæ.

Shortly afterwards, Blackman & Tansley ('02) pursued somewhat further the lines suggested by Bohlin and Luther, separating not only Luther's 'Heterokontæ,' but also the Œdogoniales as Bohlin's 'Stephanokontæ,' and the Conjugatæ as the 'Akontæ,' from the remainder of the Chlorophyceæ, which were placed under the 'Isokontæ.' They also accepted Bohlin's suggestion that all these four groups were phylogenetically independent and that each group had most probably had a direct Flagellate ancestry.

The bold effort of the above-mentioned authors to establish a reliable basis for the classification of the Green Algæ was in every way admirable in its broad outlines. Many matters of detail have required readjustment, and certain theoretical questions have had to be considerably discounted; but notwithstanding its defects, this combined effort, due in the first instance to the inspirations of Swedish algologists, will always remain as the one which paved the way for the establishment of a sound classification of the Chlorophyceæ.

The following is a brief synopsis of the recent proposals concerning the classification of Green Algæ, with critical remarks thereon. It will be found useful for purposes of reference only, as a strict comparison of the various propositions is not possible from such a synopsis, owing to the frequent use of the same group-names in widely different senses.

I. Blackman & Tansley ('02) advocated the separation of the three groups of the Heterokontæ, Stephanokontæ, and Akontæ from the remainder of the Chlorophyceæ, which were placed in the Isokontæ. With these four primary divisions the present author is in entire agreement, provided they are regarded as groups of Green Algæ of equal standing. The ancestry of the Heterokontæ and Isokontæ is in some respects fairly clear, but that of the Akontæ and Stephanokontæ is most obscure. The separation of the Stephanokontæ is based upon the great constancy of the characters of the zoogonidia in the Isokontan and Heterokontan series, and the absence of intermediate forms between any of these types of zoogonidia and the multiciliated zoogonidium of the Stephanokontæ. In this classification the authors followed Bohlin in placing the Vaucheriaceæ in the Heterokontæ, and Palla in the arrangement of the Conjugatæ. They did not, however, accept Bohlin's disposition of the Microsporales as a group parallel with the Ulotrichales. Briefly, the arrangement was as follows :

Class I. *Isokontæ.*
Series 1. Protococcales.
Group 1. Volvocineæ.
Families : Polyblepharideæ ; Chlamydomonadaceæ ; Polytomaceæ ; Phacotaceæ ; Volvocaceæ.

Group 2. Tetrasporineæ.
Families: Tetrasporaceæ; Glœocystaceæ; Selenastraceæ; Phytheliaceæ; Pleurococcaceæ.
Group 3. Chlorococcineæ.
Families: Chlorococcaceæ; Endosphæraceæ; Hydrodictyaceæ.

Series 2. Siphonales.
Group 1. Siphoneæ.
Families: Protosiphonaceæ; Bryopsidaceæ; Derbesiaceæ; Caulerpaceæ; Codiaceæ; Verticillatæ.
Group 2. Siphonocladeæ.
Families: Valoniaceæ; Gomontiaceæ; Cladophoraceæ; Sphæropleaceæ.

Series 3. Ulvales.
Family: Ulvaceæ.

Series 4. Ulotrichales.
Families: Ulotrichaceæ; Prasiolaceæ; Microsporaceæ; Cylindrocapsaceæ; Chætophoraceæ; Chætosiphonaceæ; Coleochætaceæ; Chroolepidaceæ.

Class II. *Stephanokontæ.*
Family: Œdogoniaceæ.

Class III. *Conjugatæ* (*Akontæ*).
Series 1. Desmidioideæ.
Families: Archidesmidiaceæ; Eudesmidiaceæ.
Series 2. Zygnemoideæ.
Families: Spirogyraceæ; Zygnemaceæ; Mougeotiaceæ.

Class IV. *Heterokontæ.*
Series 1. Chloromonadales.
Families: Chloramœbaceæ; Vacuolariaceæ; Chlorosaccaceæ.
Series 2. Confervales.
Families: Chlorotheciaceæ; Confervaceæ; Botrydiaceæ.
Series 3. Vaucheriales.
Families: Vaucheriaceæ; Phyllosiphonaceæ.

II. In the *Treatise on the British Freshwater Algae* (G. S. West, '04) the classification put forward was not essentially different from that suggested by Bohlin, except that the genus *Prasiola* was placed in a separate group, the Schizogoniales. It would have been better, however, if the Heterokontæ had not been given such undue prominence, and the Conjugatæ removed from the middle of those Algæ which are now regarded as constituting the Isokontæ. The detailed treatment of the Conjugatæ was entirely different from that followed by Blackman & Tansley, and the Vaucheriaceæ were retained in the Siphoneæ. Bohlin was followed in regarding the genus *Microspora* as the sole representative of the group 'Microsporales.'

III. Oltmanns ('04) put forward a classification of Green Algæ in many ways peculiar. His inclusion of the diatoms within the Akontæ is not only extraordinary, but constitutes an extension of the group 'Akontæ' which can hardly be countenanced and is most decidedly confusing. Similarly, his division of the Conjugatæ into the Mesotæniaceæ, Zygnemaceæ and Desmidiaceæ, with the inclusion of *Genicularia* and *Gonatozygon* in the Zygnemaceæ, is not altogether in accordance with the present-day knowledge of this group.

There is little, if any, justification for the inclusion of the Tetrasporaceæ (especially as constituted by Oltmanns) in the 'Volvocales,' and it is very doubtful if the Volvocineæ have sufficient claims to be removed from the Protococcales and elevated to a group of equal rank (the 'Volvocales'). The 'Halosphæraceæ,' established to include *Halosphæra* and *Eremosphæra*, is purely artificial, as everything known concerning these Algæ indicates that they are in no way nearly related. There seems little need for the 'Scenedesmaceæ,' which if *Dictyosphærium* be excluded, is equivalent to the Autosporaceæ (the 'Protococcaceæ Autosporeæ' of Chodat) and also with very slight differences is synonymous with Blackman & Tansley's 'Selenastraceæ.' Oltmanns' inclusion of the Œdogoniaceæ within the Ulotrichales is in striking contrast to his whole-hearted acceptance of the Heterokontæ as a primary group; and his use of the name 'Protococcaceæ' is not altogether in agreement with its original meaning, but is in part synonymous with the modern conception of the Chlorochytriaceæ (= Endosphæraceæ). The Vaucheriaceæ is retained within the Siphonales. In outline the arrangement is as follows:

Heterokontæ.

Families: Chloromonadaceæ; Confervaceæ; Botrydiaceæ; Chlorotheciaceæ.

Acontæ (Zygophyceæ).

a. Conjugatæ.

Families: Mesotæniaceæ; Zygnemaceæ; Desmidiaceæ.

b. Bacillariaceæ.

Chlorophyceæ.

a. Volvocales.

Families: Polyblepharidaceæ; Chlorodendraceæ; Chlamydomonadaceæ; Phacotaceæ; Volvocaceæ; Tetrasporaceæ.

b. Protococcales.

Families: Protococcaceæ; Protosiphonaceæ; Halosphæraceæ; Scenedesmaceæ; Hydrodictyaceæ.

c. Ulotrichales.

Families: Ulotrichaceæ; Ulvaceæ; Prasiolaceæ; Cylindrocapsaceæ; Œdogoniaceæ; Chætophoraceæ; Aphanochætaceæ; Coleochætaceæ; Chroolepidaceæ.

d. Siphonocladiales.

Families: Cladophoraceæ; Siphonocladiaceæ; Valoniaceæ; Dasycladiaceæ; Sphæropleaceæ.

e. Siphonales.

Families; Codiaceæ; Bryopsidaceæ; Derbesiaceæ; Caulerpaceæ; Vaucheriaceæ.

IV. Chodat has recently ('09) proposed a 'natural system' of classification for the Green Algæ, but in many respects the proposed system is very confusing. It is briefly as follows:

A. *Meiotrichales.*

Series I. Protococcales.

Families: Volvocaceæ; Palmellaceæ; Protococcaceæ (Prot. Zoosporées; Prot. autosporées; Prot. hemizoosporées).

Series II. Pleurococcales.

Families: Ulotrichaceæ; Ulvaceæ; Pleurococcaceæ; Prasiolaceæ; Chætophoraceæ; Coleochætaceæ.

Series III. Chroolepoidales.
Family : Chroolepidaceæ.

Series IV. Siphonales (*sensu* Oltmanns).

B. *Pleotrichales.*
Series I. Œdogoniales.
Family : Œdogoniaceæ.

C. *Atrichales.*
Series I. Conjugatæ.
Families : Desmidiaceæ ; Zygnemaceæ.

It will be noticed that the Heterokontæ are missing from the above scheme, Chodat having placed them [along with the Peridinieæ, Bacillarieæ, and the Euglenaceæ] within the Phæophyceæ !

The creation of the three new names Meiotrichales, Pleotrichales and Atrichales seems scarcely necessary, as they are merely equivalent to the older and better names of the Isokontæ, Stephanokontæ and Akontæ. Moreover, they have the great disadvantage that the termination '-trichales' (and also '-trichaceæ') in the group-names of Algæ has so far been exclusively used in reference to hair-like branches and filaments, and not to cilia.

It is doubtful if there is any real justification for the inclusion of the 'Pleurococcaceæ,' Ulotrichaceæ and Chætophoraceæ within the one series of the 'Pleurococcales,' apart from the extraordinary use of this group-name to include filamentous and branched Green Algæ. The Chætophoraceæ are merely the branched forms of the Ulotrichaceæ, and the evolutionary series of the latter through such forms as *Geminella*, *Stichococcus*, etc., is almost perfect, the most probable origin of the lower types being from some of the Tetrasporine Protococcales with thin cell walls and simple parietal chloroplasts. Chodat also includes the Ulvaceæ in his 'Pleurococcales,' without giving any reasons for such a step, and he places *Bulbochæte* (surely by an accident) in the Coleochætaceæ.

[The foregoing classification has been very recently modified by Chodat ('13), the principal alterations being in the 'Meiotrichales,' in which 'series I Protococcales' is changed to 'series I Cystosporeæ' and 'series II Pleurococcales' is changed to 'series II Parietales.']

V. WILLE ('09) recognizes neither the Heterokontæ nor the Stephanokontæ, and his proposed new classification is in some ways a retrogression. He still maintains the Conjugatæ as a group equivalent to the remainder of the Green Algæ ('Chlorophyceæ'), and the various members of the Heterokontæ are scattered with little regard to their affinities among the families of the Protococcales and the Chætophorales. It is difficult to see any reason for the inclusion of both *Tribonema* and *Microspora* in the Ulotrichaceæ. Following West ('04) and Oltmanns ('04), *Aphanochæte* is regarded as the type of a separate family. Wille's 'Pleurococcaceæ' is a very mixed assemblage, including some members of the Palmellaceæ and Heterokontæ, and one lichen (*Botrydina*). *Elakatothrix* is also included in the Pleurococcaceæ, although this genus is so closely related to *Ankistrodesmus* as to be hardly separable from it. His Chætophorales is enlarged in scope and somewhat unwieldy, and is certainly not equivalent to the 'Confervales.' His system is briefly as follows :

Chlorophyceæ.
Class Protococcales.
Families : Volvocaceæ ; Tetrasporaceæ ; Botryococcaceæ ; Pleurococcaceæ ; Protococcaceæ ; Ophiocytiaceæ ; Hydrogastraceæ ; Oocystaceæ ; Hydrodictyaceæ ; Cœlastraceæ.

Class Chætophorales (Confervales).

Families: Ulvaceæ; Ulotrichaceæ; Blastosporaceæ; Chætophoraceæ; Chroolepidaceæ; Wittrockiellaceæ; Chætopeltidaceæ; Aphanochætaceæ; Coleochætaceæ; Cylindrocapsaceæ; Œdogoniaceæ.

Class Siphonocladiales.

Families: Valoniaceæ; Cladophoraceæ; Dasycladaceæ; Sphæropleaceæ.

Class Siphonales.

Families: Bryopsidaceæ; Caulerpaceæ; Derbesiaceæ; Codiaceæ; Vaucheriaceæ; Phyllosiphonaceæ.

Conjugatæ.

Families: Desmidiaceæ; Zygnemataceæ; Mesocarpaceæ.

In the present volume the general principles originally expounded by the Swedish algologists are maintained, except for certain differences necessitated by recent discoveries. The classification here set forth differs considerably in detail from those just reviewed and is based, first, upon the immense amount of recent work on the Green Algæ, and, secondly, upon the author's wide experience of these plants from many parts of the world. Throughout the whole arrangement it has been necessary to keep constantly in mind the tendency of so many modern investigators to give rather too much emphasis to recently discovered characters, many of which are often but of secondary or even of trivial importance.

The details of the classification will be allowed to unfold themselves gradually, but four primary divisions of the Chlorophyceæ are recognized as fundamental.

CHLOROPHYCEÆ
- Division I. Isokontæ.
- „ II. Akontæ.
- „ III. Stephanokontæ.
- „ IV. Heterokontæ.

LITERATURE CITED

ACTON, E. ('09). Botrydina vulgaris, Brébisson, a primitive Lichen. Ann. Bot. xxiii, 1909.

ALLEN, C. E. ('05). Die Keimung der Zygote bei Coleochæte. Ber. Deutsch. Botan. Ges. xxiii, 1905.

ALSBERG, C. L. ('12). A Discussion of the Probable Food Value of Marine Algæ. pp. 263—270 in the 'Fertilizer Resources of the United States.' Washington, 1912.

ARCHER, W. ('62) in Quart. Journ. Micr. Sci. new ser. ii, 1862.

BEIJERINCK, M. W. ('90). Kulturversuche mit Zoochlorellen, Lichengonidien und anderen niederen Algen. Botan. Zeit. xlviii, 1890.

BEIJERINCK, M. W. ('93). Bericht über meine Kulturen niederer Algen auf Nährgelatine. Centralbl. für Bakteriol. xiii, 1893.

BIALOSUKNIA, W. ('09) in Bull. Soc. botan. Genève, 2e sér. i, 1909.

BLACKMAN, F. F. & TANSLEY, A. G. ('02). A Revision of the Classification of the Green Algæ. New Phytologist, 1902.

BOHLIN, K. ('97). Studier öfver några slägten af Alggruppen Confervales Borzi. Bihang till K. Sv. Vet.-Akad. Handl. Bd. xxiii, no. 3, 1897.

BOHLIN, K. ('01). Utkast till de Gröna Algernas och Arkegoniaternas Fylogeni. Akad. Afhandl. Upsala, 1901.

CHODAT, R. ('09). Étude critique et expérimentale sur le Polymorphisme des Algues. Genève, 1909.

CHODAT, R. ('13). Matériaux pour la Flore Cryptogamique Suisse, vol. iv, fasc. 2. Monographies d'Algues en culture pure. Berne, 1913.

COMÈRE, J. ('03). De l'action des eaux salées sur la végétation de quelques Algues d'eau douce. Nuova Notarisia, xiv, 1903.

COTTON, A. D. ('11). On the growth of *Ulva latissima* in excessive quantity, with special reference to the *Ulva*-nuisance in Belfast Lough. Seventh Report of the Royal Commission on Sewage Disposal. Appendix IV. 1911.

DANGEARD, P. A. ('09) in Bull. Soc. Bot. de France, lvi, 1909.

GRINTZESCO, J. ('02). Recherches expérimentales sur la morphologie et la physiologie du Scenedesmus acutus Meyen. Bull. de l'Herb. Boiss. 1902, sér. 2, i.

HANSGIRG, A. ('85). Ueber den Polymorphismus der Algen. Botan. Centralbl. 1885.

HIERONYMUS, G. ('92). Ueber *Dichranochæte reniformis* Hieron., eine neue Protococcacea des Süsswassers. Cohn's Beiträge zur Biol. der Pflanzen, v, 1892.

KLEBS, G. ('84). Beiträge zur Physiologie der Pflanzenzelle. Untersuch. a. d. bot. Institut zu Tübingen, Bd. ii, 1884.

KLEBS, G. ('96). Die Bedingungen der Fortpflanzung bei einigen Algen und Pilzen. Jena, 1896.

KUFFERATH, H. ('13). Contribution à la Physiologie d'une Protococcacée nouvelle. Rec. de l'Instit. botan. Léo Errera, ix, 1913.

LIVINGSTON, B. E. ('00). On the nature of the Stimulus which causes the Change of Form in Polymorphic Green Algæ. Botan. Gazette, xxx, Nov. 1900.

LUTHER, A. ('99). Ueber Chlorosaccus eine neue Gattung der Süsswasseralgen. Bihang till K. Sv. Vet.-Akad. Handl. Bd. xxiv, no. 13, 1899.

LUTMAN, B. F. ('10). The cell structure of Closterium Ehrenbergii and Closterium moniliferum. Bot. Gazette, vol. 49, no. 4, April, 1910.

LUTMAN, B. F. ('11). Cell and nuclear division in Closterium. Bot. Gazette, vol. 51, June, 1911.

MCALLISTER, F. ('13). Nuclear Division in Tetraspora lubrica. Ann. Bot. xxvii, Oct. 1913.

MOORE, G. T. & KELLERMANN, K. F. ('05). Copper as an Algicide and Disinfectant in Water Supplies. U.S. Dep. of Agriculture. Bureau of Plant Industry. Bull. no. 76, April, 1905.

OLTMANNS, F. ('04). Morphologie und Biologie der Algen. Jena, 1904.

PALLA, E. ('90). Beobachtungen über Zellhautbildung an des Zellkernes beraubten Protoplasten. Flora, 1890.

PASCHER, A. ('09). Über merkwürdige amoeboide Stadien bei einer höheren Grünalge. Ber. Deutsch. Bot. Ges. xxvii, 1909.

RICHTER, P. ('92) in Flora, lxxv, 1892.

SETCHELL, W. A. ('05). Limu. Univ. of California Publications. Bot. vol. ii, no. 3, April, 1905.

TECHET, K. ('04) in Oesterr. Bot. Zeitschr. liv, 1904.

TECHET, K. ('08) in Nuova Notarisia, xix, 1908.

TEODORESCO, E. C. ('09) in Ann. Sci. Nat. Bot. sér. ix, 1909.

TIMBERLAKE, H. G. ('01). Starch-Formation in Hydrodictyon utriculatum. Ann. Bot. xv, Dec. 1901.

VIRIEUX, J. ('10) in Comptes Rendus, cli, p. 334, 1910.

WARD, H. MARSHALL ('99). Some Methods for Use in the Culture of Algæ. Ann. Bot. xiii, 1899.

WEST, G. S. (G. S. W. '99). On Variation in the Desmidieæ, and its Bearings on their Classification. Journ. Linn. Soc. Bot. xxxiv, 1899.

WEST, G. S. (G. S. W. '04). A Treatise on the British Freshwater Algæ. Camb. Univ. Press, 1904.

WEST, W. & WEST, G. S. (W. & G. S. W. '07). Freshwater Algæ from Burma, including a few from Bengal and Madras. Ann. Roy. Bot. Gard. Calcutta, 1907.

WILLE, N. ('97). Chlorophyceæ und Conjugatæ in Engler & Prantl, Die Natürlichen Pflanzenfamilien. Leipzig, 1897.

WILLE, N. ('09). *Ibid.* Nachträge zu I. Teil, Abteilung 2. Leipzig, 1909.

WOYCICKI, Z. ('09) in Bull. Internat. Acad. Sci. Cracovia, no. 8, 1909.

Division I. ISOKONTÆ

This group of the Chlorophyceæ is on the whole a very natural one and is based largely upon the constancy of the zoogonidium-characters. The name was first given by Blackman & Tansley ('02) to the large residue of the Green Algae after the Stephanokontæ, Akontæ (Conjugatæ), and Heterokontæ had been removed. Since the group is undoubtedly homogeneous, it is unlikely that its general concept will be much changed in the future.

The plant-body is unicellular, cœnocytic, or multicellular; and sometimes (as in the Siphonocladiales and Hydrodictyaceæ) it is an aggregate of cœnocytes. In many forms the cells are associated to form colonies of variable limitations, sometimes loose and indefinite (*Tetraspora*, *Coccomyxa*, etc.), at other times compact and regular (*Cœlastrum*, *Scenedesmus*, etc.). In the higher types the plant is a simple or branched filament, or even a plate-like expansion (Ulvales). The branches may be free (Chætophoraceæ) or concrescent (some species of *Coleochæte* and *Cephaleuros*), or interwoven to form a more or less compact thallus (some forms of the Siphonales and Siphonocladiales). The cell-wall is often to a considerable extent mucilaginous. Each cell contains one or more chloroplasts, and chlorophyll is the principal pigment, although xanthophyll is present in variable amount. *The principal food-reserve is starch* (or rarely, a fatty oil) *and one or more pyrenoids are generally present in each chloroplast.*

Reproduction occurs in most forms by zoogonidia of an ovoid form, *to the narrower and usually colourless end of which two* (*or more rarely four*) *cilia of equal length are attached.* Many of these zoogonidia possess a laterally placed pigment spot (or stigma), and a single bell-shaped or basin-shaped chloroplast usually occupies the rounder or posterior end of the cell. After a variable period of activity the zoogonidium comes to rest and soon germinates to form a new individual. Sometimes the zoogonidia are facultative gametes, but in many forms obligate gametes also occur. Such gametes are frequently similar to the zoogonidia and are known as *planogametes.* If they are all alike they are termed *isogametes*, but often they are differentiated into megagametes (female) and microgametes (male), the latter being the more active. The greatest differentiation of gametes occurs in those forms in which there are large non-motile egg-cells and very small ciliated antherozoids (*Volvox*, *Côleochæte*, etc.). The germination of the zygote is very variable, and may be direct or indirect. In the cœnocytic and multicellular forms multiplication may take place by fragmentation of the thallus (*Monostroma*, *Prasiola*, *Caulerpa*, *Vaucheria*, etc.). Reproduction also occurs in many of the families by akinetes or by aplanospores.

The group of the Isokontæ is of particular interest owing to the fact that it contains the great majority of the most primitive known forms of the Green Algæ. It is only quite recently that an exact knowledge of some of these forms has been acquired, and there are yet very many about which the available information is of the scantiest nature. The details of the arrangement of some of the lower types may, therefore, require considerable readjustment before a really natural system of classification is finally built up.

Chodat ('97), from his observations on the lower forms of Green Algæ, both in a state of nature and in cultures, suggested that the principal groups of the Chlorophyceæ all had an origin in the Palmellaceæ, one of the lowly families of the Protococcales. He recognized three conditions exhibited by the lower Green Algæ (*i.e.* the Protococcales): (1) the zoospore-condition, the other two conditions being only transient; (2) the sporangium-condition or unicellular motionless stage, the other conditions being realized accidentally; (3) the Tetraspora-stage where non-motile cells are connected by regular cell-walls at right angles.

Blackman ('00), following the lines first indicated by Luther and Bohlin, considered that whatever tendencies were exhibited by the lower Green Algæ, they had their origin in the motile unicellular Chlamydomonads. He defined more precisely the three divergent tendencies of the lower Green Algæ, originally suggested by Chodat, as follows: (1) a Volvocine tendency towards the aggregation of motile vegetative cells into gradually larger and more specialized motile true cœnobia; (2) a Tetrasporine tendency towards the formation of aggregations by the juxtaposition of the products of septate vegetative cell-division to form non-motile organisms of increasing definiteness and solidarity; (3) an Endosphærine tendency towards the reduction of the vegetative division and septate cell-formation to a minimum. There is every reason to believe that these divergent tendencies had their origin in the unicellular genus *Chlamydomonas*, and the Chlamydomonads must therefore be considered as the real primitive forms of green plants. The genus *Chlamydomonas* has been extensively studied in recent years, and more than 30 species are known to occur, all of which possess more or less distinctive cytological characters. Unfavourable conditions produce in this genus the so-called 'palmella-state,' which is the beginning of the non-motile existence such as predominates in the Palmellaceæ. In the latter family the cells sometimes escape from their old walls, develop a pair of cilia, and become motile zoogonidia. Blackman remarks that the 'formation of zoospores is then nothing but reversion to an ancestral type of vegetative existence for a biological advantage, and all the vegetative existence of the higher Algæ is phylogenetically a new intercalation into the life-history of the motile Chlamydomonad which is permanently in the zoospore-condition, though walled, and in which zoospore-formation and vegetative cell-division are one and indistinguishably the same thing.'

There seems to be no doubt that *Chlamydomonas* has itself arisen from certain of the autotrophic Flagellata, through the family Polyblepharidaceæ. In this family intermediate forms are found which, although exhibiting preponderating algal characters, yet afford clear evidence that it is quite impossible to draw a sharp line of demarcation between the Flagellata and the primitive Green Algæ. Zumstein ('99) has shown that some Flagellates may be either green and holophytic or colourless and saprophytic according to the available food-substances. Just as green and brown organisms occur among the Flagellata, so do colourless forms occur among the lower Algæ. *Polytoma uvella* Ehrenb. is really a colourless member of the Chlamydomonadeæ; and *Prototheca moriformis* Krüger, *Mycotetraëdron cellare* Hansg., and *Chionaster nivalis* (Bohlin) Wille should be similarly regarded as colourless members of the Autosporaceæ.

The first of the three lines of descent along which the Protococcales appear to have evolved, in consequence of the *Volvocine tendency*, has resulted in the production of a series of organisms of gradually increasing complexity, which constitute the Volvocaceæ. This line of evolution has terminated blindly in the genus *Volvox*.

The second line of evolution, resulting from the *Tetrasporine tendency*, is the most important, having been the chief line of descent of the great majority of the Isokontæ, including nearly all the higher types of the Green Algæ. Indeed, it is along this line that in all probability all the higher green plants have arisen. In its initial stages this tendency resulted in the production of a series of forms in which vegetative cell-division gradually replaced the formation of zoogonidia as the chief method of propagation. In this way the Palmellaceæ came into existence, and from this family most of the other Isokontan families have arisen.

The third line of descent from the Chlamydomonad type is to be associated with the *Chlorococcine tendency*[1], which has resulted in various Algæ from which vegetative cell-division is absent, multiplication of individuals taking place only by the formation of zoogonidia or motile gametes. The plants are mostly cœnocytes and unseptate, with a gregarious habit. This

[1] Previously known as the Endosphærine tendency.

line of evolution has resulted in no Green Algæ of importance, and like the Volvocine series has not emerged from the confines of the Protococcales, unless Blackman's suggestion as to the origin of the Siphonales is a correct one (see p. 223). The tendency to retain the unicellular or unseptate form in the adult plant and to restrict multiplication entirely to small motile cells is found in the Chlorochytrieæ, Dicranochæteæ, Characieæ and Halosphærieæ. These small groups have therefore been placed alongside one another in the family Planosporaceæ. A similar multiplication also occurs in the cœnobic Hydrodictyaceæ.

The evolutionary lines within the Protococcales may be represented as follows:

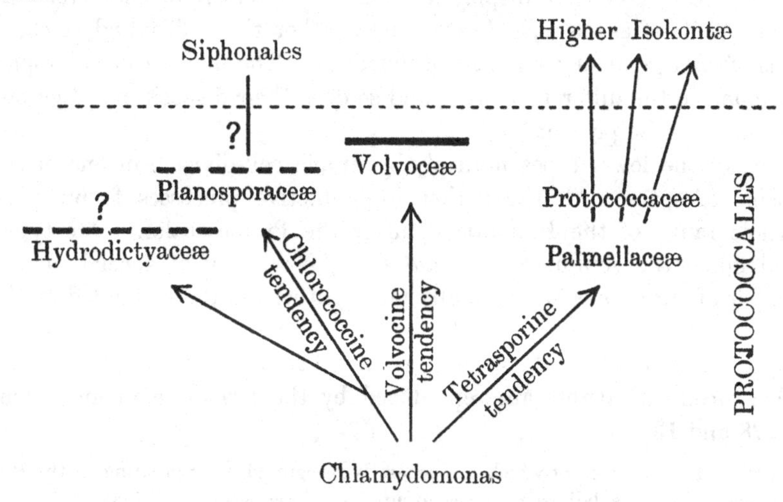

A careful consideration of our present knowledge of the Isokontæ leads one to the conclusion that there are only six natural groups into which it can be divided. These groups, both for convenience and for the sake of consistency, are here termed 'orders,' and are as follows:

Order 1. Protococcales.
2. Siphonales.
3. Siphonocladiales.
4. Ulvales.
5. Schizogoniales.
6. Ulotrichales.

Order 1. PROTOCOCCALES.

The Algæ of this order are *mostly unicellular*, motile or non-motile, or less frequently simple cœnocytes. *The cells are often aggregated to form loose irregular colonies and are frequently embedded in a copious mucilage. In a few forms there is a definite cœnobium*[1], which may be a cœnobium of cœnocytes, but a definite multicellular thallus does not occur. The order includes the lowest and the most primitive of the Green Algæ, forms through which all the higher groups of the Isokontæ have been evolved. Various members of the Protococcales are to be found in every conceivable situation in which Algæ will grow, and they display a wonderful diversity of character. The cell-wall is sometimes very delicate, but at other times firm and thick, and there is often a great development of mucilage. The number of chloroplasts varies much in the different genera, and so does their disposition. Pyrenoids may or may not be present.

Many of the lower types multiply by simple cell-division in one or more directions of space. All the various reproductive processes found within the whole range of the Isokontæ occur in the Protococcales. One type of asexual spore (or gonidium), the *autospore*, is worthy of special mention. A number of autospores are generally produced within the distended mother-cell, each spore usually developing the complete (or almost complete) characters of the parent-cell before liberation (figs. 121, 125, 126, etc.). In some of the cœnobic forms the plants are reproduced by the formation of autocolonies (figs. 128 and 133).

The three lines of descent which are generally recognized as originating in the Protococcales necessitate a subdivision of the group into three series as follows:

Sub-order *Volvocineæ*.
- Family Polyblepharidaceæ.
- Family Sphærellaceæ.
- Family Volvocaceæ.

Sub-order *Tetrasporineæ*.
- Family Palmellaceæ.
- Family Dictyosphæriaceæ.
- Family Protococcaceæ.
- Family Autosporaceæ.
- Family Chætopeltidaceæ.

Sub-order *Chlorococcineæ*.
- Family Planosporaceæ.
- Family Hydrodictyaceæ.

[1] A *cœnobium* is a compact colony of definite form, the number of cells in which remains constant so long as the colony exists. All the cells arise either as gonidia or zoogonidia within the original mother-cell, and therefore belong to the same generation. The adult cœnobium is formed by the regular grouping of the motile or non-motile gonidia, no further increase in the number of cells ever taking place.

Sub-order VOLVOCINEÆ.

The Algæ of this sub-order are unicellular or they consist of definite cœnobia of cells. *They are distinguished from all other members of the Protococcales by being ciliated and motile during their dominant or vegetative phase.* The cœnobia consist of cells which are regularly disposed within the swollen wall of a mother-cell, being either disconnected or joined together by protoplasmic processes. With the exception of a few marine or brackish-water forms among the unicells, all the members of the Volvocineæ are inhabitants of fresh water, occurring in ponds, pools, lakes and rivers. Many of them have a distinct liking for rain-water pools.

The cells are rounded or ovoid, more rarely compressed, *with two or rarely four cilia attached to the anterior end, which is generally narrowed.* Sometimes there is an expanded wing-like part of the cell-wall which gives the cell an angular or somewhat irregularly lobed contour (*Pteromonas, Lobomonas*). The protoplasm of the anterior region of the cells is hyaline in character, and (except in the Sphærellaceæ) usually contains two contractile vacuoles the pulsations of which are said to be alternate. The single nucleus generally occupies a central position in the cell. There is *one chloroplast in each cell,* very variable in form and disposition, but *often basin-shaped* or *bell-shaped* and filling up most of the broader posterior part of the cell. It contains one or more pyrenoids. A red pigment-spot (or stigma) occurs at the periphery of the cell in a lateral position, generally towards the anterior end but sometimes median or posterior in position. A distinct cell-wall is present in all the Volvocineæ with the exception of the few members of the Polyblepharidaceæ.

In the unicellular forms multiplication takes place by the division of the contents of the mother-cell into 2, 4 or 8 daughter-cells, and the latter are new motile individuals strictly homologous with the zoogonidia of other groups of the Isokontæ. In the more primitive types, such as the Polyblepharidaceæ, the plane of the first cell-division is longitudinal, but in all the higher forms this plane is transverse to the longitudinal axis of the cell. In the cœnobic forms, multiplication occurs by the formation of daughter-cœnobia, which are developed from some or all of the cells of the mother-cœnobium, and ultimately set free by the dissolution of the mucilage which bound together the cells of the original colony. The young colonies are really 'autocolonies' in every way comparable with those which are formed in the cœnobic genera of the Autosporaceæ.

Gamogenetic reproduction occurs in most of the forms by the union of isogamous planogametes, formed in a similar manner to the asexual daughter-cells (zoogonidia), but in greater numbers. In some forms, including all the higher types, there is marked heterogamy, sexual reproduction occurring by the union of heterogametes. Various grades of heterogamy can be observed in the Volvocine series, the highest types exhibiting true oogamy with highly specialized antherozoids and oospheres. The zygote invariably rests for a period before germination.

In passing from the lower to the higher members of the Volvocineæ there is a more striking progressive evolution of forms than is exhibited in any other family of Algæ; indeed, it is one of the most perfect evolutionary series known to the biologist. There is a progressive increase in the number of cells and in the size of the cœnobium, accompanied by a marked differentiation of reproductive from vegetative (or somatic) cells, and associated with these phenomena is a gradual replacement of isogamy by heterogamy, ultimately reaching the highest condition of oogamy.

The Volvocineæ connect the Green Algæ with the Flagellata, an extensive group of unicellular organisms of very varied character from which it is now generally recognized that the Chlorophyceæ have been phylogenetically derived. Klebs ('92) was the first investigator to study in a careful and comprehensive manner the varied organisms included in the Flagellata, which are a group of the Protista exhibiting a mixture of animal and vegetable characters. Since then our knowledge of the Flagellates has greatly increased, and it is well known that among the heterogeneous assemblage of forms embraced in the group the distinction between animal and vegetable organisms entirely breaks down. In the Flagellata are found the phylogenetic starting-points of many lines of descent, both animal and vegetable, and the group is one which should be studied by every serious student of biology. Klebs drew up the characters which distinguish the Flagellata from the Volvocineæ, and he showed that a great consensus of characters united the Volvocineæ with the Green Algæ and separated them from the organisms included in the Flagellata. All subsequent researches have consistently supported Klebs' statements, and the Volvocineæ must therefore be regarded as a group of motile Green Algæ[1].

Oltmanns ('04) and Pascher ('12) have both placed the Volvocineæ as a group—the 'Volvocales'—equivalent in rank to the Protococcales or Ulotrichales. It must be confessed, however, that a careful enquiry into the general organization of the Volvocineæ scarcely supports this view, which gives undue prominence to the unicellular and colonial motile forms of the Isokontæ. These certainly constitute a distinct group, not to be confused with any other; and indeed, the beautiful evolutionary series of the Volvocaceæ is

[1] The inclusion of the Volvocineæ in a modern text-book of Zoology is difficult to reconcile with the present state of our knowledge of the Green Algæ and the Flagellata, and can only be regarded as a relic of the chaos which existed a quarter of a century ago.

Dangeard's work ('98; '01) on the nuclei of the Chlamydomonadeæ has also demonstrated that in contrast to the primitive and divergent types of nuclear structure and nuclear division exhibited by the Protozoa, the mitosis, even of the lowest of the Volvocaceæ, does not differ essentially from that which occurs in higher plants.

almost unique: but even *Volvox*, which is the culminating form of the series, can hardly be considered as having emerged from the Protococcales. It is still but a fragile colony, the vegetative units of which are far more primitive than those constituting the cœnobia of *Cœlastrum*, *Sorastrum*, or *Pediastrum*. The high degree of differentiation of the gametes in the higher forms, resulting in passive oospheres of large size and active antherozoids of small size, is paralleled by the similar degree of sexual differentiation which has arisen in several families of the Ulotrichales, and in *Vaucheria* among the Siphonales.

Although there have been three attempts at the formation of cœnobia within the Volvocineæ, each along a different line, that through *Pandorina* and *Eudorina* has apparently been the only successful one. Nothing appears to have been developed beyond *Stephanosphæra* in the Sphærellaceæ and

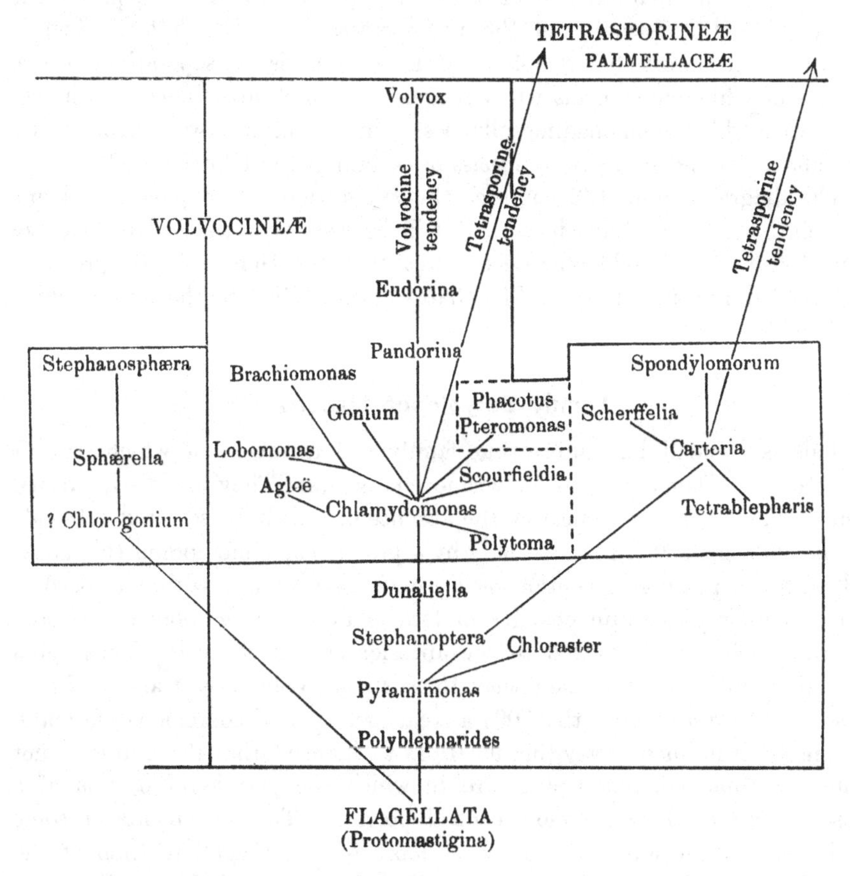

nothing beyond *Spondylomorum* in the Carterieæ, the cell-grouping in these types being biologically unsatisfactory.

One of the most important of recent suggestions concerning the Volvocineæ was made by Schmidle ('03), who showed that the affinities between *Sphærella* (= *Hæmatococcus*) and *Stephanosphæra* were such that they must be regarded as closely related. For these two genera he proposed a new family, the Sphærellaceæ. This proposition has been further supported by the researches of Wollenweber ('08). A new classification of the 'Volvocales' has recently been set forth by Pascher ('12), in which he accepts the family Sphærellaceæ and proposes another, the Carteriaceæ, to include those members of the Volvocineæ which are furnished with four cilia. The three families of the Carteriaceæ, Sphærellaceæ and Chlamydomonadaceæ, as outlined by Pascher, are open to criticism since they do not appear to be parallel groups. The Sphærellaceæ are primitive with very distinctive cytological characters, but the Carteriaceæ and Chlamydomonadaceæ agree absolutely in the main points of their cytological structure except for the four cilia of the former and the two possessed by the latter. Such a character can hardly be accepted as the sole basis of separation of two families of Chlamydomonads when zoogonidia (which are themselves homologous with Chlamydomonadine cells) possessing sometimes two and sometimes four cilia are found *in the same species* of certain of the Ulotrichales[1].

The diagram on p. 163 contains the suggestions for a possible scheme of evolution of the Volvocineæ. It will be seen that the Volvocaceæ are derived from the Polyblepharidaceæ, but that the Sphærellaceæ probably originated from some group of Flagellates closely allied to those from which the Polyblepharids were themselves evolved.

Family **Polyblepharidaceæ.**

This is a small but interesting family the members of which exhibit a mixture of Flagellate and Volvocine characters. They are distinguished from all the other Volvocineæ by the absence of a definite secreted cell-wall. The body of the cell is invested only by a protoplasmic membrane (fig. 95 *E*) and, with the possible exception of *Pyramimonas*, is able within prescribed limits to undergo certain changes of form. Two, four, or more cilia (often termed flagella) are attached to the anterior end of the cell. These cilia are rather thick (much thicker than those in the Volvocaceæ), and in *Pyramimonas delicatulus* Griffiths ('09) a weak solution of commercial formalin blisters them in such a way (fig. 95 *H*) as to indicate that the cilium is not a homogeneous rod, but a structure in which the peripheral part is of a denser character than the more central portion. The movements of some species have been stated to be rather more of the Flagellate than of the Volvocine type, but this is not so in *Pyramimonas* and *Dunaliella*. The chloroplast is of the typical Volvocine kind and contains a pyrenoid. A pigment-spot is present in some forms but absent in others.

[1] **Pascher has attempted to overcome this difficulty by splitting up the genera *Ulothrix* and *Stigeoclonium* (=*Myxonema*) on grounds which appear to be inadequate.**

Multiplication takes place by longitudinal fission, in some forms commencing at the anterior and in others at the posterior extremity. Resting-cells are formed by the direct encystment of the motile individuals. Gametes of a more or less isogamous character occur in *Dunaliella* (Teodoresco, '05), and they fuse in pairs to form zygotes.

The most primitive genus of the family is *Polyblepharides*, in which the body of the cell is conical with a broad anterior end to which are attached from six to eight cilia. In *Pyramimonas*, which possesses only four cilia, the cell-body is also more or less conical, and is four-lobed in the anterior region. *Chloraster*, which is furnished with five cilia, is

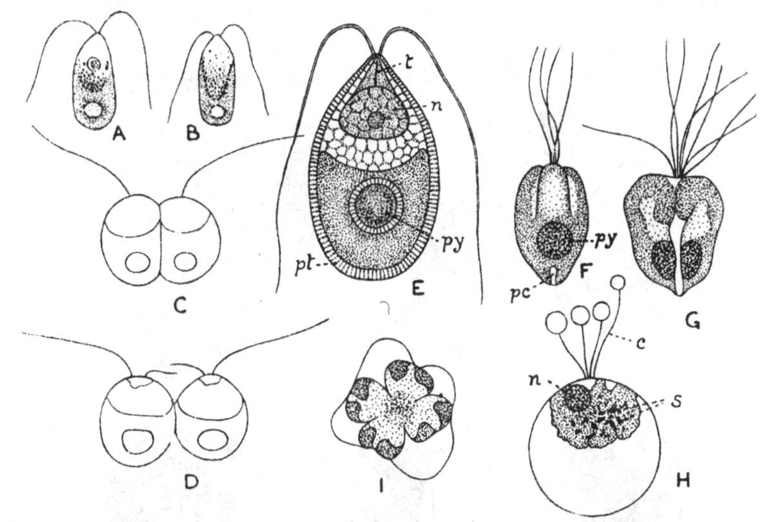

Fig. 95. *A—E*, *Dunaliella salina* (Dunal) Teodoresco; *A* and *B*, normal form of zoogonidia (= vegetative cells); *C* and *D*, stages in division, in *D* the second cilium making its appearance; *E*, single cell showing detailed structure. *F—I*, *Pyramimonas delicatulus* Griffiths; *F*, vegetative cell; *G*, beginning of longitudinal fission, the pyrenoid having divided and 8 cilia developed; *H*, peculiar blistering and disintegration of cell on treatment with 4 per cent. formalin, showing swelling of cilia at their extremities; *I*, anterior axial view of *H* at an earlier stage in disintegration showing lobed chloroplast. *c*, cilia; *n*, nucleus; *pc*, posterior excavation of chloroplast; *pt*, protoplasmic membrane; *py*, pyrenoid; *s*, minute grains of starch; *t*, plasmatic thread from nucleus to base of cilia. *A—D*, after Teodoresco, × 1000; *E*, after Hamburger, × 2250; *F—H*, after Griffiths, × 800.

much more prominently four lobed in the middle region of the cell-body; and the biciliated *Asteromonas* has the cell-body extended into six rounded ridges which run antero-posteriorly and have their greatest development in the median part of the cell.

Dunaliella possesses only two cilia and is in every way the nearest approach to the Chlamydomonads. *Stephanoptera*, recently described by Dangeard ('10) is very similar to *Pyramimonas*, but is only furnished with two cilia. It is thus a natural connecting-link between *Pyramimonas* and *Dunaliella*.

The genera included in this family are: *Polyblepharides* Dangeard, 1887; *Chloraster* Ehrenberg, 1848; *Pyramimonas* Schwarda, 1850; *Tetratoma* Bütschli, 1887; *Dunaliella* Teodoresco, 1905; *Stephanoptera* Dangeard, 1910; *Asteromonas* Artari, 1913; *Spermatozopsis* Korschikoff, 1913.

Family **Sphærellaceæ.**

Although but recently proposed (Schmidle, '03) and including only two genera, the unicellular *Sphærella* (= *Hæmatococcus*) and the colonial *Stephanosphæra*, this family appears to be established upon a sound basis. There are two leading characters which distinguish the Sphærellaceæ from the Volvocaceæ: (1) the form of the protoplast with its outstanding branched processes, and (2) the numerous contractile vacuoles. The presence of more than one pyrenoid in the chloroplast is an important although not a dis-

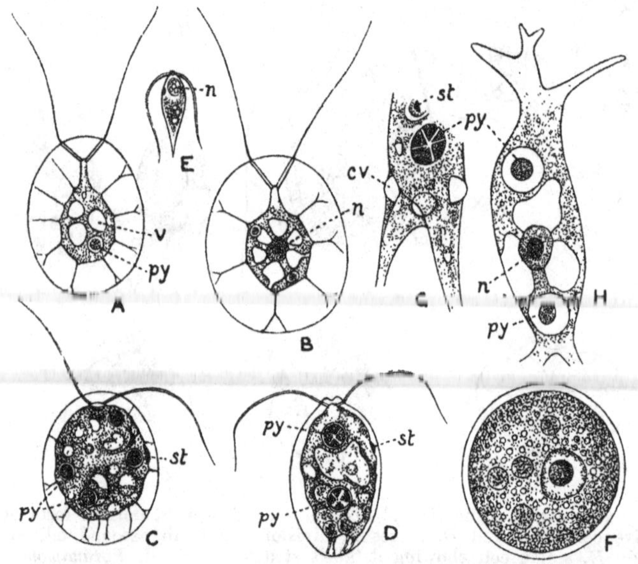

Fig. 96. *A—C*, vegetative individuals (=zoogonidia) of *Sphærella lacustris* (Girod.) Wittr. *A*, surface view; *B*, optical section (after Schmidle); *C*, zoogonidium from a 'snow-water-medium' (after Wollenweber). *D—F*, *Sphærella Drœbakensis* (Wollenw.) G. S. West. *D*, vegetative cell (zoogonidium) in optical section; *E*, gamete; *F*, resting cell (all after Wollenweber). *G* and *H*, portions of individual cells of *Stephanosphæra pluvialis* Cohn, to show general cytology; *G*, after Wollenweber; *H*, after Schmidle. All × about 1000. *n*, nucleus; *cv*, contractile vacuole; *py*, pyrenoid; *st*, stigma; *v*, vacuole.

tinctive character. The peculiar nature of the protoplast, with its branched processes, has resulted in a corresponding modification of the cell-wall, which consists of a firm outer membrane, extending around all the branched tips of the processes, and an inner mucilaginous portion filling up the intervening space between the protoplast and the outer wall.

In *Sphærella* the protoplasmic extensions are generally distributed all round the protoplast (fig. 96 *A—D*), but in *Stephanosphæra* they are mostly confined to the two extremities of the elongated cell (fig. 104 *K*). In both genera the chloroplast is more or less peripheral and appears to be reticulate in the mature cell, this condition being brought about by the formation of cell-sap vacuoles in the interior of the cell, such vacuoles gradually extending outwards and so breaking down the chloroplast at various points. In *Stephanosphæra* there are only two pyrenoids in the chloroplast (fig. 96 *G* and *H*), but in *Sphærella* there are from two to eight. There are numerous contractile vacuoles in the peripheral part of the protoplast, Wollenweber recording as many as 60 for *Sph. Drœbakensis.* In both genera the bright red pigment hæmatochrome is prominent, and in *Sphærella* both motile and resting-cells may be so completely tinged with red that the cytological characters of the living cell are entirely obscured. The hæmatochrome usually makes its first appearance in that part of the protoplast immediately surrounding the nucleus.

The motile vegetative cells (= zoogonidia) of *Sphærella* readily assume the condition of spherical resting-cells, which are completely red, 10—70 μ in diameter, and surrounded by a thick cellulose wall (fig. 96 *F*). All structural details are hidden by the hæmatochrome, which according to Zopf ('95), is partly in solution in oil-drops and partly in a microcrystalline form. Divisions within the resting-cell result in the formation of 4, 8 or 16 daughter-cells. These become, under favourable conditions, vegetative motile cells (or megazoogonidia) which grow much in size while swimming about. If the conditions are unfavourable they remain as further resting-cells, so that it is possible for the resting-cells to increase greatly in numbers and gradually form a stratum. The motile cells may divide repeatedly to form further megazoogonidia or they may return to the resting-state. They may also form microzoogonidia (identical with the megazoogonidia except in size), 32 or 64 being formed in each cell. Some confusion appears to have existed in the past between the microzoogonidia and gametes; the former frequently die, or they may come to rest, form a cell-wall, and grow into normal resting-cells. Both Wollenweber ('08) and Peebles ('09) have observed isogametes (fig. 96 *E*), as many as 100 being formed in one gametangium. Conjugation of the gametes to form zygotes has also been observed.

In *Stephanosphæra*, in which the cœnobium normally consists of eight cells arranged in an equatorial zone within a tough spherical or ellipsoidal investment, multiplication occurs by the division of each cell of the cœnobium, after having assumed a more or less globular form, into 4 or 8 daughter-cells, each group forming a new cœnobium. Gamogenesis occurs by the conjugation of isogametes, of which 8, 16, or 32 are formed within the mother-cell. As in *Sphærella*, the gametes are fusiform in shape and they fuse laterally to form spherical 'zygozoospores,' which soon become quiescent and turn yellow-brown in colour.

Wollenweber ('08) has shown that by appropriate methods of culture, *Sphærella* may be made to pass into any of its different states, yielding zoogonidia, zygotes, aplanospores, or palmella-state. He also showed ('07) that a pigment-spot (or stigma) occurs in all the three known species of *Sphærella*. The value of the hæmatochrome, which is found mainly in the resting-stage and is especially an accompaniment of diminished vitality, probably lies in its greater stability than chlorophyll, and perhaps also in its protective and heat-producing powers (Hazen, '99).

The resting-cells develop motile zoogonidia when transferred from old foul water into distilled water, and from long-continued darkness into light (Freund, '07).

The only genera are *Sphærella* Sommerfeld, 1824 (= *Hæmatococcus* Agardh, 1828)[1] and *Stephanosphæra* Cohn, 1852, both of which are perhaps more confined to rain-water pools than any other members of the Volvocineæ.

Note.—It would seem most natural to place the genus *Chlorogonium* Ehrenberg, 1830 (inclus. *Cercidium* Dang., 1888) with the Sphærellaceæ, since its chloroplast is often as indefinite in its limitations as that of *Sphærella*, and is furnished with from 2 to 5 pyrenoids (or more in *Chl. euchlorum*); moreover the protoplast contains a number (12—16) of contractile vacuoles.

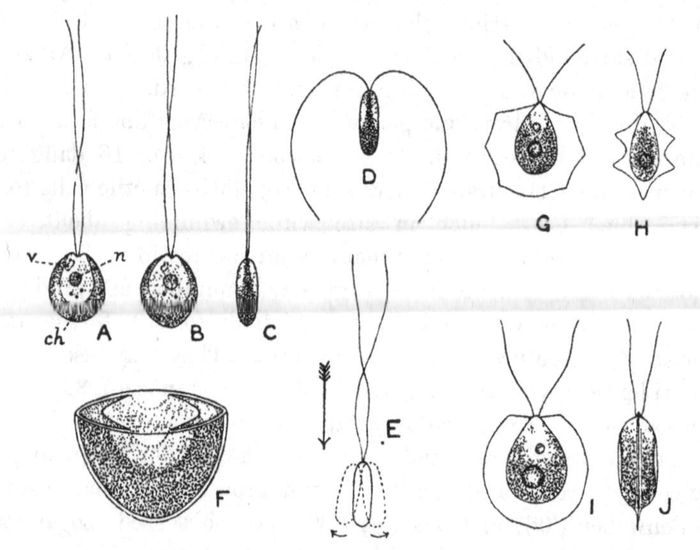

Fig. 97. *A—F*, *Scourfieldia complanata* G. S. West; *A—C*, three individuals after staining with iodine; *ch*, chloroplast; *n*, nucleus; *v*, vacuoles (2?); *A* and *B*, seen from the front; *C*, side view showing compression of cell; *D*, side view showing resting position of cilia in living cell; *E*, diagram to show direction and nature of movements; *F*, diagram to show cross-section of cell and nature of chloroplast. *G* and *H*, *Pteromonas Chodati* Lemm., front and side views. *I* and *J*, *Pteromonas angulosa* (Carter) Lemm., front and side views. *A—E*, ×2000; *G—J*, ×800.

Family **Volvocaceæ**.

This is much the largest family of the Volvocineæ, and includes both unicellular and colonial forms. The most important unicellular genus is *Chlamydomonas*. The colonies consist of motile cœnobia of Chlamydomonadine cells, generally embedded in a mucilaginous envelope through which the cilia project. The cells possess a distinct cell-wall, which in most cases is in

[1] Consult Hazen ('99); Wille ('03); and W. B. Grove, 'Sphærella v. Mycosphærella,' *Journ. Bot.* 1912.

close contact with the protoplast. This character, and the presence of only two contractile vacuoles[1] situated at the anterior end of the cell close to the base of the cilia, are the leading distinctions between the Volvocaceæ and the Sphærellaceæ. In the majority of forms the chloroplast contains only one pyrenoid, although several normally occur in *Chlamydomonas giganteus* Dill and *Pleodorina illinoisensis* Kofoid; and Bachmann ('05) has occasionally observed two or three pyrenoids in specimens of *Chlamydomonas inhærens.*

In the few forms included in the Carterieæ (the Carteriaceæ of Pascher) the cells are furnished with four cilia.

The Volvocaceæ may be subdivided as follows :

A. Cells with four cilia ; unicellular or cœnobic. Sub-fam. *Carterieæ.*

B. Cells with two cilia.

a. Unicellular.

*Cell-wall thin but distinct. Sub-fam. *Chlamydomonadeæ.*

**Cell-wall firm and often thick, sometimes consisting of two loosely connected halves. Sub-fam. *Phacoteæ.*

b. Cells aggregated to form cœnobia with very definite characters. Sub-fam. *Volvoceæ.*

Sub-family CARTERIEÆ. All the genera of this sub-family possess four equal cilia attached to the anterior end of the cell. The principal genus is *Carteria,* in which the cells are spherical, ovoid or ellipsoid, and provided with a thin cell-wall. Near the attachment of the cilia are two small contractile vacuoles; and a conspicuous pigment-spot is present, although its position is variable. The chloroplast is large and massive, hollowed out anteriorly into the form of a basin or cup, and is usually furnished with a single conspicuous pyrenoid in its posterior thicker part (fig. 98 *A* and *B*). The presence of four cilia is the only feature distinguishing this genus from *Chlamydomonas.* In the formation of zoogonidia the first division-plane is longitudinal or slightly oblique. In several species quadriciliated isogametes are also known, which

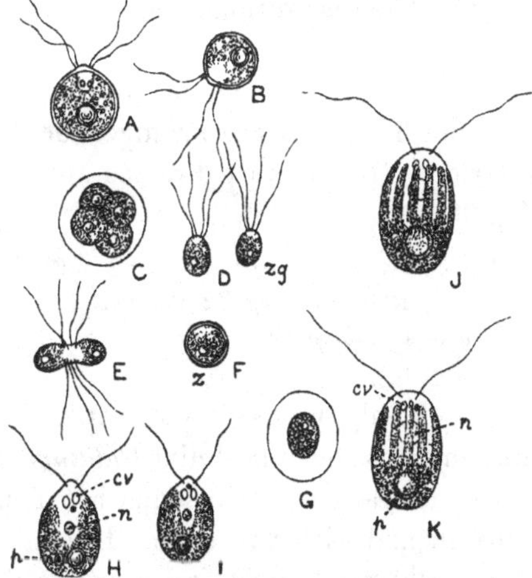

Fig. 98. *A—G, Carteria multifilis* (Fresen.) Dill; *A* and *B*, vegetative cells (= zoogonidia); *C*, four daughter-cells within old wall of mother-cell; *D*, gametes ; *E*, conjugating gametes ; *F*, zygote. *H* and *I*, *Chlamydomonas Debaryana* Gorosch. *J* and *K*, *Chl. grandis* Stein (= *Chl. Kleinii* Schmidle). All × 475. *cv*, contractile vacuoles ; *n*, nucleus ; *p*, pyrenoid ; *zg*, gamete ; *z*, zygote.

[1] In *Agloë* there is an additional group of three contractile vacuoles at the posterior end of the cell (*vide* Pascher, '12).

fuse in pairs to form zygotes (fig. 98 *D—F*). One species of *Carteria* occurs symbiotically in the marine Planarian worm *Convoluta roscoffensis* (Keeble & Gamble, '07); this species is also of interest on account of the very few marine species of the Volvocaceæ which are known to exist. An interesting genus is *Scherffelia* Pascher ('12), in which the cell is compressed, so that in cross-section it has somewhat the appearance of a thin biconvex lens. There are two plate-like chloroplasts, sometimes united posteriorly in *Scherffelia Phacus*, and no pyrenoid is present. In *Sch. Phacus* the cell-wall is laterally expanded into the rudiments of wings such as are more prominently developed in *Pteromonas*.

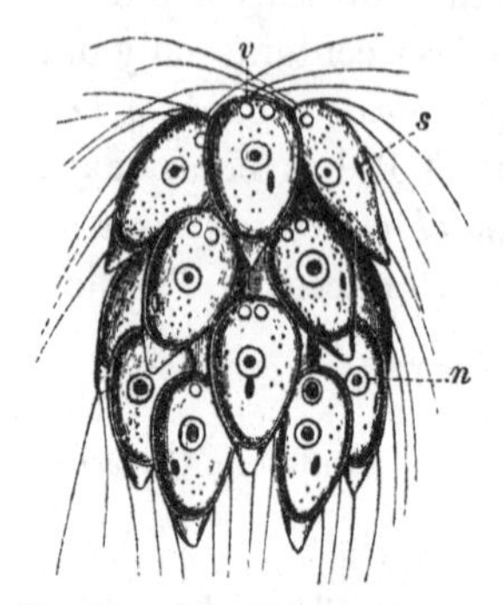

Fig. 99. *Spondylomorum quaternarium* Ehrenb., a colony of 16 cells. *n*, nucleus; *s*, stigma; *v*, contractile vacuole. (After Stein, from Wille.) ×650.

Tetrablepharis (fig. 102 *G*) is a colourless genus which is probably a degenerate derivative of *Carteria*, having assumed a saprophytic mode of existence.

The only colonial type is *Spondylomorum quaternarium* Ehrenb. which although known from both the New and the Old Worlds is much the rarest cœnobic form of the Volvocaceæ. The disposition of the cells forming the cœnobium is unique amongst the Volvocineæ. There are sixteen cells arranged in four tiers of four cells each, those of any one tier alternating with those immediately in front or behind. The cells are arranged on a central gelatinous axis so that their anterior extremities are directed approximately to the front end of the colony (fig. 99).

The genera are: *Carteria* Diesing, 1868 [inclus. *Pithiscus* Dangeard, 1888, and *Corbiera* Dangeard, 1888]; *Scherffelia* Pascher, 1912; *Spondylomorum* Ehrenberg, 1848; and the colourless *Tetrablepharis* Senn.

Sub-family CHLAMYDOMONADEÆ. The most important genus of this sub-family is the unicellular *Chlamydomonas*. The cells are spherical, ovoid, subcylindrical, or rarely somewhat fusiform in shape, provided with a thin cell-wall and with two cilia. The chloroplast is very variable in form, but it is typically cup-shaped and occupies the posterior region of the cell, more or less surrounding a central part of the protoplast in which the nucleus is lodged (fig. 98 *H* and *I*). The genus includes approximately 30 species, in most of which there is a single pyrenoid in the chloroplast; but in some there are several pyrenoids and in others none. There is usually a pigment-spot in a lateral position, generally anterior but sometimes posterior. Each cell possesses a distinct cell-wall which sometimes gives a cellulose reaction and sometimes does not; and there is evidence to show that in the older individuals the cellulose reaction is entirely obliterated owing to the impregnation of the wall with some other substance. The normal motile cells of *Chlamydomonas* are equivalent to the zoogonidia of other Isokontæ, and their characters are sufficiently constant to furnish a basis for the discrimination of the various species. The zoogonidia multiply rapidly by division which takes place usually after the cell has come to rest. Two, four, or eight daughter-cells are formed, the first division-plane being transverse, longitudinal, or oblique. The daughter-cells quickly acquire the characters of the parent and are set free by the conversion of the old mother-cell-wall into mucilage.

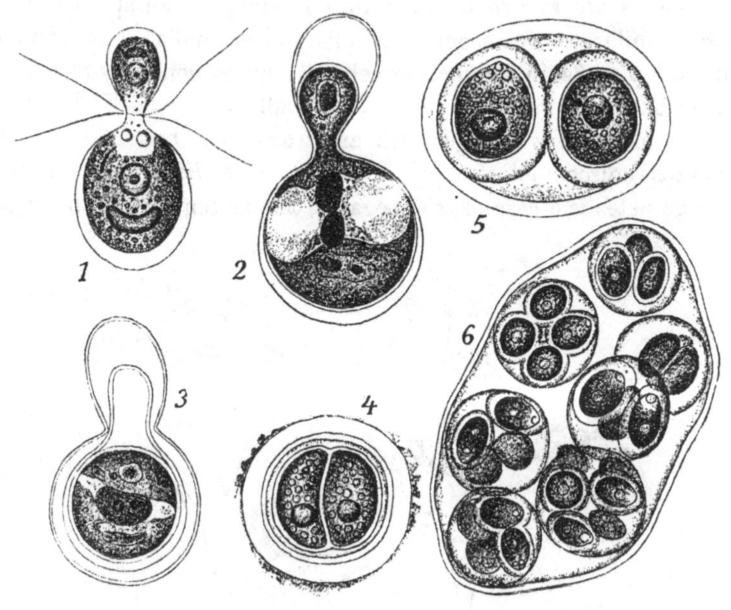

Fig. 100. *Chlamydomonas monadina* Stein (= *Chl. Braunii* Gor.). *1—3*, fusion of hetero-gametes; *4—6*, palmella-states. Very highly magnified. (After Goroschankin, from Oltmanns.)

All the species of *Chlamydomonas* can enter into a 'palmella-state,' in which the cells lose their cilia, become enveloped in a copious jelly, and undergo divisions of such a nature (often in oblique planes) that the mucilaginous colony extends in three directions of space (fig. 100 *4—6*). There is in this manner a great increase in the number of individuals, and when the requisite combination of external factors once more supervenes they all become motile and swim out of the jelly. This transitory palmella-state of *Chlamydomonas* is of great interest as it clearly indicates how the Palmellaceæ originated from the Chlamydomonadeæ by an extension of the period of quiescence and the gradual retention of a permanent gelatinous colony, the zoogonidium-state having become of much less importance and only being reverted to occasionally.

The palmella-state can be directly induced by cultivating *Chlamydomonas* in nutritive solutions or on more or less solid media. It is probable that some of the described species of *Glœocystis*, such as *Gl. vesiculosa* Näg., are merely the palmella-states of species of *Chlamydomonas*.

A vegetative resting-state has been found to occur in certain species (*Chl. alpina, Chl. gigantea, Chl. nivalis,* etc.) and Wille ('03) has suggested that several of the described species of *Trochiscia* (= *Acanthococcus*) are merely the resting-cells (akinetes) of Chlamydomonads. There is, however, as yet no definite proof of this.

Planogametes are known to occur in a number of the species; they are in all cases similar to the vegetative cells (= zoogonidia), but smaller and are produced up to 64 in the mother-cell. They are generally naked, but in some species they may be clothed with a cell-wall (*Chl. media* Klebs; fig. 92 *G* and *H*). In most cases the gametes are isogamous, but in some species heterogametes (anisogametes) occur, and in *Chl. monadina* Stein the conjugation of the gametes is a case of true oogamy, the contents of the megagamete

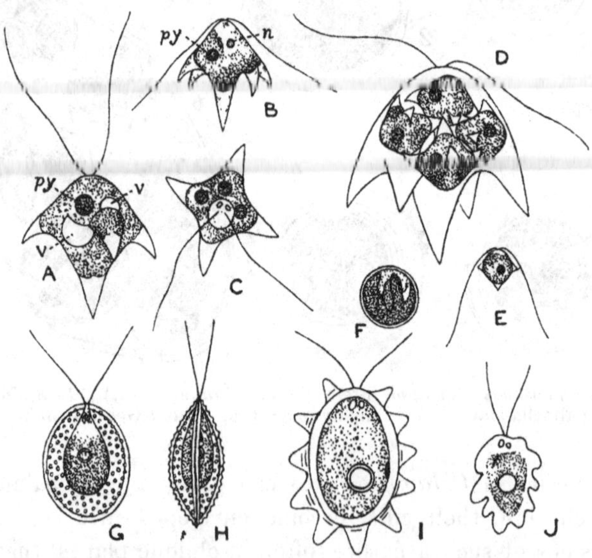

Fig. 101. *A—F, Brachiomonas submarina* Bohlin; *A* and *B*, side views of vegetative individuals (zoogonidia); *C*, anterior view; *D*, formation of four zoogonidia (daughter-cells) within old mother-cell-wall; *E*, gamete; *F*, zygote. *G* and *H*, *Phacotus lenticularis* Stein, two views showing compression of cell. *I*, *Lobomonas stellata* Chodat; *J*, *Lobomonas Francei* Dang. *n*, nucleus; *py*, pyrenoid; *v*, vacuole. *A—F*, × 1000; *G* and *H*, × 800; *I*, very highly magnified (after Chodat); *J*, × 1000 (after Dangeard).

acting as an egg-cell (fig. 100 *1—3*). The zygote often exhibits distinct specific characters, being adorned with short spines, warts, etc., or variously sculptured.

'This genus [*Chlamydomonas*] holds a unique position among the Green Algæ, and indeed among the whole of the Green Plants. It may be regarded as the phylogenetic starting-point of the various lines of Chlorophyceous descent. The history of these is a history of the intercalation of a vegetative phase between two successive motile (Chlamydomonadine) generations, these motile phases being retained for reproductive purposes as zoospores and gametes; in the oogamous types the male gamete alone remains motile, and constitutes in the Archegoniate series the last remaining representative of the Chlamydomonadine cell.

'The co-existence within the limits of an undoubtedly natural genus of the most primitive form of gamogenesis (the conjugation of equal clothed gametes) with a gamogenesis which has the essential characteristics of true oogamy is also a feature of unique interest.' (Blackman & Tansley, '02.)

As the genus is of such importance it has received considerable attention from a number of investigators, notably Goroschankin ('90—'91), Dill ('95) and Wille ('03).

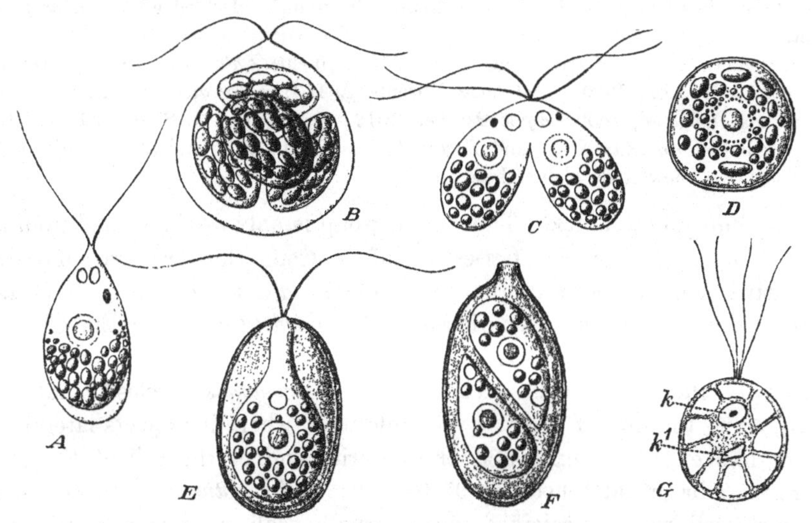

Fig. 102. Colourless members of the Volvocaceæ. *A—D*, *Polytoma uvella* Ehrenb.; *B*, formation of daughter-cells; *C*, fusion of motile cells; *D*, zygote. *E* and *F*, *Chlamydoblepharis brunnea* Francé; *F*, division of cell. *G*, *Tetrablepharis globulus* (Zach.) Senn. *A—F*, after Francé, ×660; *G*, after Zacharias (from Wille).

Two of the most interesting genera of the Chlamydomonadeæ are *Brachiomonas* and *Lobomonas*. The first-named occurs only in submarine habitats and possesses a cell-body furnished with five horns, all directed backwards: one straight posterior horn, and four curved horns regularly disposed around the antero-median part of the cell (fig. 101 *A—F*). These horns are hollow outgrowths of the cell, the protoplast extending to their extremities, although the chloroplast generally does not. *Lobomonas* has only been found in freshwater lakes and pools, and is characterized by the possession of wart-like lobes either at the posterior end of the cell only (*L. Francei* Dang.; fig. 101 *J*) or all round the periphery (*L. stellata* Chodat; fig. 101 *I*).

Scourfieldia is a recently described genus (G. S. W., '12) which bears exactly the same relationship to *Chlamydomonas* that *Scherffelia* does to *Carteria*. The cells are greatly compressed and contain a cup-shaped chloroplast without a pyrenoid; they are very minute and furnished with two very long cilia (fig. 97 *A—F*). This organism is also peculiar in moving backwards, the body of the cell being pushed in front by the movements of the cilia which are carried behind.

Polytoma uvella Ehrenb. (fig. 102 *A—D*) is a colourless saprophytic member of the Chlamydomonadeæ in which, although as a rule there is no chromatophore-stroma, starch-grains are stored in the cell. The cells divide while in the motile state, and the daughter-cells are facultative gametes.

The Chlamydomonadeæ occur most abundantly in stagnant water, especially in the spring months. They prefer rain-water, and are often found in great abundance in water-butts, tanks, small pools, and in canals, frequently giving a decided green colour to the water.

The genera are: *Chlamydomonas* Ehrenb. 1832 [inclus. *Chloromonas* Gobi, 1900[1] and *Dangeardia* Bougon, 1900]; *Polytoma* Ehrenberg, 1831; *Brachiomonas* Bohlin, 1897; *Lobomonas* Dangeard, 1899; *Agloë* Pascher, 1912; *Scourfieldia* G. S. West, 1912. It is possible that *Glœomonas* Klebs and *Kleiniella* Francé (1894) should also be added, but at present we know little about these genera.

Sub-family PHACOTEÆ. This small group is only distinguished from the Chlamydomonadeæ by the possession of a strong thick wall around each vegetative cell. In both *Pteromonas* and *Phacotus* the cells are compressed as in *Scherffelia* in the Carteriaceæ and *Scourfieldia* among the Chlamydomonadeæ. Reproduction is only known to occur by the longitudinal division of the motile cells. The wall of *Pteromonas* is closely adherent except along the line of greatest circumference, where it projects laterally as a wing-like expansion except at the anterior end of the cell at the point where the cilia are attached (fig. 97 *G—J*). In *Phacotus* (fig. 101 *G* and *H*) the thick wall, which is calcified and externally sculptured, is outstanding all round the cell, and consists of two loosely connected plates which separate on the escape of the daughter-cells.

It is probable that *Coccomonas* should be included in this sub-family, since it is a Chlamydomonad with a hard outstanding cell-wall, which is sometimes four-angled.

The genera are: *Phacotus* Perty, 1852; *Coccomonas* Stein, 1878; *Pteromonas* Seligo, 1886. A colourless form is seen in *Chlamydoblepharis* Francé, 1892 (fig. 102 *E* and *F*).

Sub-family VOLVOCEÆ. The Algæ of this sub-family consist of a motile cœnobium of Chlamydomonadine cells generally embedded in a copious and definite mucilaginous envelope of relatively firm consistency. The cells, in all except certain species of *Volvox*, are to a great extent independent units of the colony and not in any way protoplasmically connected. The two cilia of each cell project outside the colony, passing through minute canals in the mucous envelope, although in *Gonium*, in which the amount of mucus is

[1] *Vide* G. S. West, '12, p. 328.

relatively very small, the canals can scarcely be detected. In most of the genera the cells are all equivalent and capable of reproducing the plant by division, but in the higher types there is a differentiation, some cells being purely vegetative, having lost the power of division, whereas others are solely reproductive.

Gonium is the simplest of the Volvoceæ, either 4 (*G. sociale* and *G. lacustre*; fig. 103 *B—F*) or 16 (*G. pectorale*; fig. 103 *A*) cells forming a flat, plate-like cœnobium. The amount of mucus around the colony is relatively small, and the ovoid Chlamydomonadine cells are all turned one way, so that the cilia are all on one surface of the plate. Each cell is also enveloped in a special

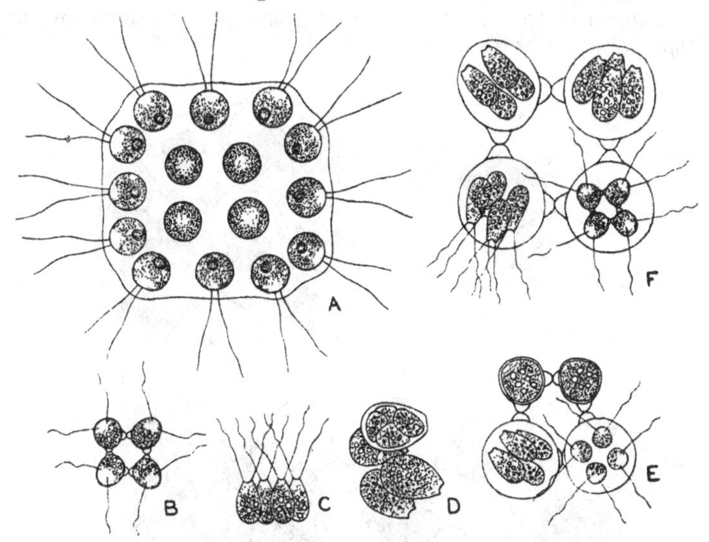

Fig. 103. *A*, *Gonium pectorale* Müll.; *B—F*, *Gonium lacustre* G. S. West. *E* and *F* show formation of daughter-colonies in mother-cells. All × 475.

rather denser mucilaginous coat, small projections from which unite the adjacent cells of the colony. In fact, *Gonium* is merely the aggregation side by side of either 4 or 16 Chlamydomonadine cells to form a permanent colony.

Harper ('12) has given reasons, based on careful study of the development of *Gonium* colonies, for believing that forms and space relationships of the cells are due rather to the mechanical interaction of the cells, regarded as colloid droplets, during the growth of the colony, than to any predetermined hereditary factor.

In *Pandorina* the aggregation of cells is entirely different, from 8 to 32 (usually 16) forming a compact, approximately spherical colony (fig. 104 *A*), which exhibits a distinct polarity. The cells are broadly ovoid in shape, often a little angular by pressure of contact, and with the narrower ends towards the centre of the colony. A tough and rather close mucous investment surrounds the cœnobium, and sometimes a wider, less dense mass of jelly is also evident.

In *Eudorina* the cœnobium consists, as a rule, of 32 cells, distantly arranged as a single layer within the periphery of an almost spherical, or rarely ellipsoid, mucous investment (fig. 105 *A*). The cells are globose, and the bell-shaped chloroplast may possess more than one pyrenoid. In *Pleodorina* (fig. 106) there is a differentiation into smaller, purely vegetative (somatic) cells, which occupy the anterior region of the cœnobium, and larger reproductive (or gonidial) cells which are alone capable of division. This differentiation is possibly constant in *Pleodorina californica* Shaw ('94), especially in view of the observations of Chatton ('11); but in *P. illinoisensis* Kofoid ('98) the differentiation is not at all constant, and every intermediate state can be observed between the cœnobium as originally described and that of *Eudorina elegans*[1]. In any case, the two forms are of great interest since they are links between *Eudorina* and *Volvox*.

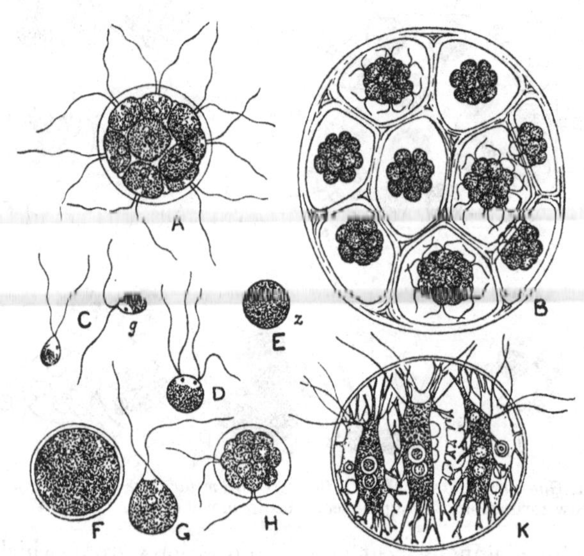

Fig. 104. *A—H*, *Pandorina Morum* (Müll.) Bory; *A*, normal colony; *B*, daughter-colonies within the swollen mother-cell-wall, ×475; *C—H*, gametes (*g*), formation of zygospore (*z*), and its development (after Pringsheim). *K*, *Stephanosphæra pluvialis* Cohn, ordinary vegetative colony (after Hieronymus, ×320).

Platydorina, described by Kofoid ('99) from the United States, is one of the most curious of the Volvoceæ, the cœnobium consisting of a twisted plate of 16 or 32 cells, the enveloping jelly being produced posteriorly into 3 or 5

[1] The author has examined large numbers of *Pleodorina illinoisensis* from Yorkshire, the west of Scotland, the west of Ireland, and Madras. The somatic cells may be entirely absent, or they may vary in number from 1 to 12; moreover, every gradation in size occurs between the reproductive and the somatic cells, and the latter are somatic or non-somatic according to the degree of differentiation in size. In many specimens somatic and reproductive cells are indiscriminately mixed. The author is inclined to agree with Powers ('05) that there are so many transition-types, even in the same collection, between *Eudorina elegans* and *Pleodorina illinoisensis* that the latter should be regarded merely as a state of the former.

symmetrically arranged, blunt processes (fig. 107 *A* and *B*). In contrast to the cell-disposition of *Gonium*, the cilia of adjacent cells project alternately upon either face of the plate. This flattened form cannot therefore have had an origin from *Gonium*, but has most likely originated from *Eudorina* by a compression of the colony. The presence of the 'caudal' lobes of the enveloping mucus also supports this view since these are sometimes developed (as posterior mamillate projections), but to a less extent, in the more ellipsoid colonies of *Eudorina*[1].

In *Volvox*, which is the most highly developed genus of the Volvoceæ, the cœnobium ranges from 200 to 2500 μ in diameter, consisting of a large number of Chlamydomonadine cells arranged in a single peripheral layer within a globose or ovoid mucous investment. The cœnobium is to a great extent a hollow sphere in its later life, although in the younger stages much of the

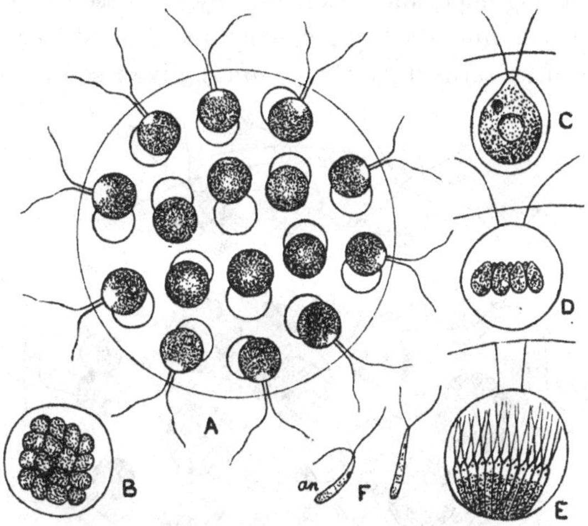

Fig. 105. *Eudorina elegans* Ehrenb. *A*, normal adult colony, ×475; *B*, young daughter-colony formed by division of contents of mother-cell, ×730. *C—E*, development of antherozoids from mother-cell; *F*, antherozoids (after Goebel).

interior is occupied by a watery mucus. The number of cells varies from about 2000 (rarely as few as 200) in *V. aureus* to rather more than 50,000 in *V. Rousseletii* (G. S. W., '10). In some species the cells are joined by distinct protoplasmic connections, which in *V. globator* are very broad (fig. 109 *C*), but in others so far as can be ascertained there are no such connections. The spacing of the cells on the periphery of the colony is also very variable, the extremes being in *V. perglobator* Powers and *V. Rousseletii* G. S. West. In

[1] Since both Chodat ('02) and Conrad ('13) have described the regular presence of posterior mamillations of the enveloping mucus, it would appear probable that there are two distinct races of *Eudorina*. The colonies which possess these posterior lobes are markedly ellipsoid or even ovate-ellipsoid, whereas the colonies without such lobes are almost globose. The author has examined thousands of colonies from the British Islands, N. America, India and Australia, and in all cases they were almost spherical with no trace of posterior mamillations.

the former the cells are so far removed from each other that the whole colony appears as a 'sponge-like reticulum,' whereas in the latter they are so compactly arranged that the whole cœnobium has a very firm and solid aspect. The cœnobium of *Volvox*, like that of *Pleodorina*, has a definite polarity, and Overton ('89) has pointed out that in *V. aureus* the stigma ('eyespot') of each cell lies on the side turned towards the anterior pole, probably in relation to the positive phototaxy of the cœnobium. In *Pandorina* and *Eudorina* the stigma is large in the anterior cells, smaller in the equatorial cells and invisible in the posterior cells (Conrad, '13).

In the lower forms of the Volvoceæ, *Gonium*, *Pandorina* and *Eudorina*, reproduction takes place by the formation of a daughter-cœnobium from every cell of the mother-colony (figs. 103 *F* and 104 *B*). The daughter-cœnobium is formed within the wall of the mother-cell, which swells up and becomes mucilaginous, and ultimately by its dissolution sets the young colony free. In the highest type, namely *Volvox*, only certain of the reproductive cells, often termed *parthenogonidia*, give rise to daughter-cœnobia.

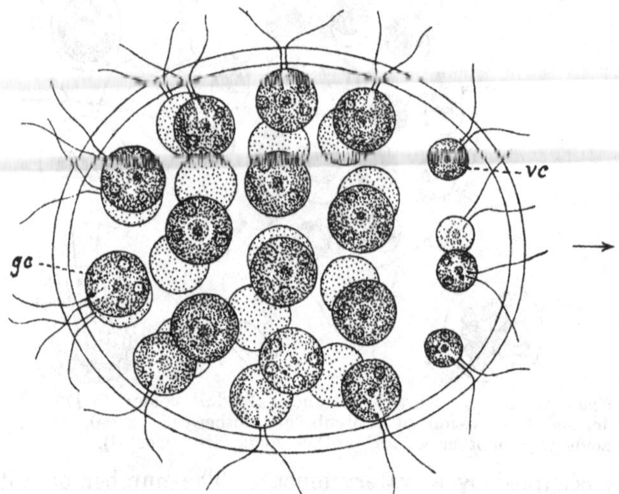

Fig. 106. *Pleodorina illinoisensis* Kofoid. A specimen from Madras, ×500. *gc*, gonidial or reproductive cell; *vc*, vegetative cell. The arrow marks the direction of translocation.

An intermediate condition is found in *Pleodorina californica*, in which the formation of daughter-colonies is restricted to certain purely reproductive cells located in one half of the colony; and in *Pl. illinoisensis* there is a similar but sometimes a much less obvious restriction of the reproductive cells. In *Volvox* the parthenogonidia vary from 1 to 24 (commonly 4 or 8) and are scattered irregularly among the somatic cells. In *Platydorina* the young daughter-colony is at first cup-shaped, but subsequently becomes flattened and twisted. The normal development in *Volvox* begins with the enlargement of the mother-cell, followed by successive divisions into

2, 4, 8, etc., until a large number of cells are produced forming a spherical young colony. All these cells are included within the periphery of the enlarged wall of the mother-cell, which rapidly becomes converted into a firm jelly. As gelatinization of the wall proceeds the young cells, although still dividing, become spaced further apart. The daughter-cœnobia project into the cavity of the mother-cœnobium and finally become quite free within this cavity ultimately escaping into the surrounding water by the rupture and death of the parent. In the various species of *Volvox* three generations of individuals can often be seen within one another; and in *V. africanus*, in which the daughter-cœnobia grow to a large size and become compressed before being set free, four generations can be distinctly observed (G. S. W., '10).

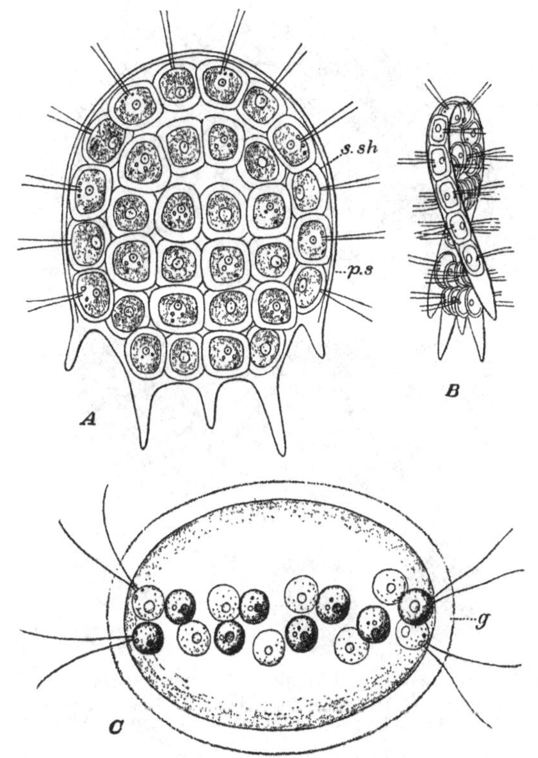

Fig. 107. *A* and *B*, *Platydorina caudata* Kofoid, ×420. *A*, colony seen from the flat side; *B*, seen from the edge. *C*, *Stephanoon Askenasii* Schewk., individual with 16 cells, ×600. (*A* and *B*, after Kofoid; *C*, after Schewiakoff, from Wille.) *p.s*, outer firm mucilaginous envelope; *s.sh*, secondary mucilaginous coat around each cell.

Powers ('05) states that in *V. Weismanniana* many of the young colonies are completely everted (*i.e.* turned inside out) during their development. 'The process begins at various periods before the closing of the young colony and is finally completed by a reclosure of the colony, the surfaces of which are now reversed, at a somewhat later period.'

Reproduction occurs in *Pandorina* by zoogonidia which are precisely similar to the mature cells of *Chlamydomonas* (Dangeard, '00). They arise by the longitudinal division of the contents of the mother-cell, and each one secretes in addition to its own membrane a mucous outer coat which ultimately forms the common investment of the colony. Schröder ('98) also observed zoogonidia in *Pandorina*, and states that they lose their cilia, vacuoles, and pigment-spot before dividing to form a new colony.

Gamogenetic reproduction has been observed in all the genera except *Platydorina* and *Stephanoon*. In *Gonium* it is by the fusion of isogametes which may arise from any cell of the cœnobium. In *Pandorina* the gametes are formed as in *Gonium*, but they exhibit considerable variability in size.

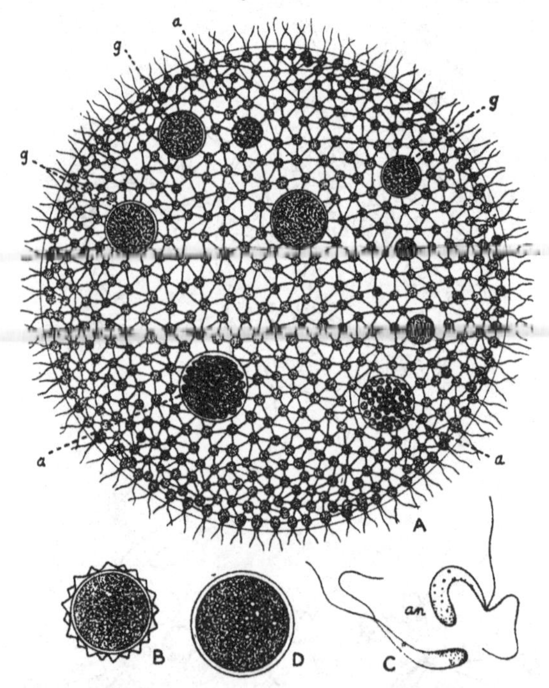

Fig. 108. *A*, *C*, and *D*, *Volvox aureus* Ehrenb. *A*, monœcious sexual colony, ×210; *C*, two antherozoids (after Klein); *D*, ripe oospore, ×475. *B*, ripe oospore of *Volvox globator* (L.) Ehrenb., ×475. *a*, androgonidia; *an*, antherozoid; *g*, gynogonidia.

They fuse in pairs, generally a smaller (male) gamete with a larger and more sluggish (female) gamete (fig. 92 *C*). In *Pleodorina*, *Eudorina*, and *Volvox* the differentiation of the sexual cells is altogether more complete, although it is in *Volvox* that the greatest differentiation is found. In this genus there are asexual (*i.e.* purely vegetative) colonies and sexual colonies, and the latter, which sometimes possess parthenogonidia also, may be monœcious or diœcious. There are in the sexual colonies certain reproductive cells which become either androgonidia or gynogonidia, the former giving rise to antherozoids and the latter developing into eggs. In *V. aureus*, which is the most abundant European species, 16 or 32 antherozoids are formed in a bundle by the division of the contents of an androgonidium, but in *V. globator* and other species as many as 128 are often formed. By the conversion of the mother-cell-wall into mucilage the antherozoids are at first set free in compact circular bundles, and are arranged with their long axes all parallel. Each antherozoid is an elongated cell, pointed at one end and rounded at the other, and the paired cilia may be inserted terminally (*V. aureus*) or laterally (*V. globator*). In the male colonies of *V. spermatosphæra*, in which all the cells form sperm-bundles, Powers has observed as many as 65,536 antherozoids (or sperms) all maturing at the same time in one cœnobium. The gynogonidia are usually few in number, averaging 6 in the colony in *V. aureus* and 30 in *V. globator*, but in *V. perglobator*

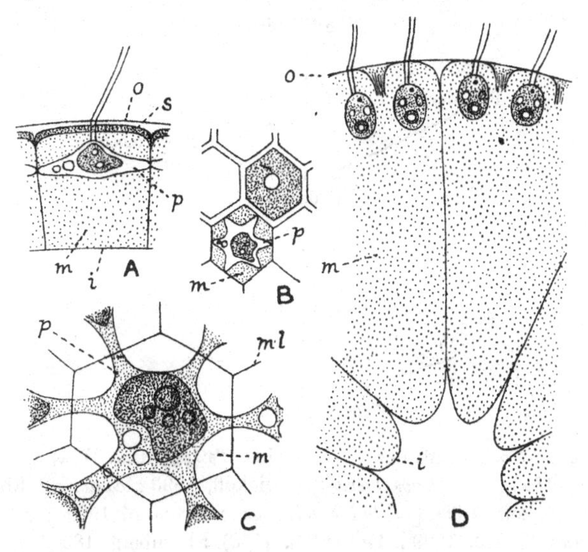

Fig. 109. *A—C*, *Volvox globator* (L.) Ehrenb. *A*, section of the peripheral region of colony showing a single cell; *B* and *C*, surface views, showing (in *C*) the broad protoplasmic connections between the cells. *D*, *Volvox aureus* Ehrenb., median section through part of colony. *i*, inner membrane of wall; *o*, outer membrane; *s*, special layer of wall; *m*, mucus; *ml*, middle lamella; *p*, protoplast. Very highly magnified (after A. Meyer).

from 300 to 400 have been seen in one cœnobium. They develop into globular oospheres, which while still unfertilized attain a diameter of over 90 μ in *V. Weismanniana*. Self-fertilization is said to take place in *V. globator*, but this is not the case in several of the other carefully investigated species. According to Overton ('89) each female colony of *V. aureus* has a 'polar plateau,' a slightly raised circular area of the firm limiting investment, situated at the posterior pole and free from vegetative cells. The polar plateau is about 42 μ in diameter, and is believed by Overton to be the point of entrance of antherozoids into the colony.

The fertilized egg-cells (oospores) are furnished with a thick wall (externally verrucose in *V. globator*) which becomes cutinized. They undergo a period of rest and then develop directly into new colonies.

Sycamina Dangeard (1880) appears to be a colourless saprophytic form of the Volvoceæ.

Volvox is the culminating genus in the evolution of motile cœnobic Algæ, and although its colony is but a delicate aggregation of primitive Chlamydomonadine cells, the organism as a whole has attained along its own line a stage of evolution which in some respects is comparable with the structural differentiation attained along other lines of algal descent.

The evolution of the Volvoceæ may be represented as follows:

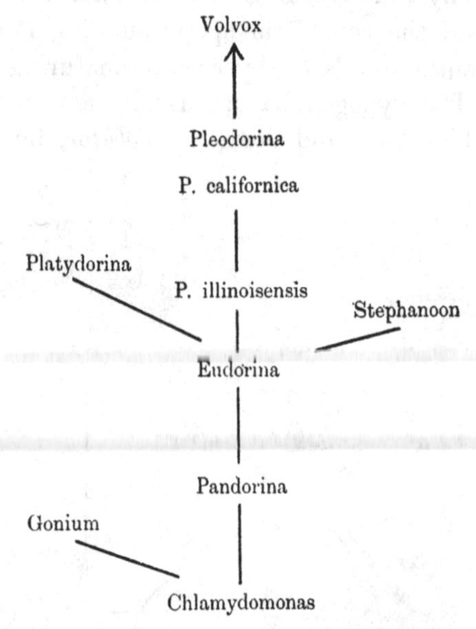

The genera are: *Gonium* Müller, 1773 [inclus. *Tetragonium* W. & G. S. West, 1896]; *Pandorina* Bory, 1824; ? *Mastigosphæra* Schewiakoff, 1893; *Eudorina* Ehrenberg, 1832; *Stephanoon* Schewiakoff, 1893 [= *Eudorinella* Lemmermann, 1900]; *Pleodorina* Shaw, 1894; *Platydorina* Kofoid, 1899; *Volvox* (L. 1758) Ehrenberg, 1830.

There is some evidence to show that the North American species of *Volvox*, though often called by the same names, constitute races distinct from the European forms, and the same may be true of some of the known African species.

The Algæ belonging to the Volvocaceæ have a wide distribution and occur plentifully in small ponds and pools. *Eudorina elegans* and *Volvox aureus* commonly occur in larger pools and in the plankton of lakes; and *V. africanus* is known only from the plankton of the Albert Nyanza. Species of *Volvox* are most abundant in small shallow ponds, especially if weedy.

Sub-order TETRASPORINEÆ.

The Algæ forming the second group of the Protococcales are either unicellular or they form colonies of indefinite extent. *The dominant vegetative phase is non-motile and the increase of the colony is very largely by vegetative cell-division.* The colonies are in most cases of an irregular character with no precise limitations, but in the Selenastreæ, Crucigenieæ and Cœlastreæ there are definite non-motile cœnobia. In the lower families the cells are of the Volvocine type, but in the higher families they are of various shapes, frequently greatly elongated, and sometimes furnished with spines or bristles. Each cell contains one or more chloroplasts. In the lower types (Palmellaceæ, etc.) the single chloroplast is usually of the Chlamydomonadine character, but in the higher types it is of various shapes, and often so massive as to render its exact form very obscure. The pigmented part of the protoplast is in some forms very indefinite. Both axile and parietal conditions are met with, but when a number of chloroplasts are present in a cell they are usually parietal. A stigma (or pigment-spot) is met with only in the lower forms, where it may occur both in the vegetative and reproductive phases.

Reproduction may occur by vegetative division of the mother-cell or by the formation of biciliated zoogonidia. In the Autosporaceæ zoogonidia are entirely absent, although it is possible that the autospores are really zoogonidia which have lost the power of movement. In some genera reproduction also occurs by isogamous planogametes.

In the Chætopeltidaceæ the cells are furnished with bristles of various kinds, sheathed or unsheathed at the base.

The Tetrasporineæ includes a number of important families of the Protococcales, one of which (the Palmellaceæ) is almost certainly on the direct line of descent of the higher Green Algæ.

Family **Palmellaceæ.**

This family is of considerable interest since it illustrates the first step in the Tetrasporine direction; that is, the formation of a colony which is non-motile in the adult state and which consists of the products of successive vegetative divisions held somewhat loosely in a mass of jelly. The colonies thus formed are of an irregular character, although the particular form assumed may be characteristic of certain species (*Tetraspora*) or even genera (*Palmodictyon, Palmophyllum,* etc.). The normal vegetative condition in the Palmellaceæ is thus quite distinctive, the cells as a rule being irregularly scattered through the mucilaginous mass, more especially near its periphery. In *Tetraspora* and *Palmophyllum* the grouping is somewhat more definite, in the former genus the products of vegetative division remaining more or less in aggregates of four, and in the latter the cells assuming a seriate disposition within a tough leaf-like expansion. The colonies are either microscopic (*Palmodactylon, Sphærocystis, Apiocystis*) or macroscopic (*Tetraspora, Collinsiella*), and may attain a length of several centimetres.

The cells are essentially Chlamydomonadine in their cytological details, exhibiting in most cases a single parietal chloroplast, often of a cup-shaped character, and furnished with one pyrenoid. In *Asterococcus* the chloroplast is central and star-shaped (Scherffel, '08 A), and the two contractile vacuoles and the pigment-spot of the Chlamydomonad are retained. There is a definite wall to each cell, but in most cases the mass of jelly is due to the conversion into mucilage of its outer layers. In many instances this is so complete that the jelly is structureless, but in others the cells retain to a variable degree their individual envelopes (*Palmodictyon*; fig. 115 *B*). In the dominant vegetative state of *Schizochlamys* the outer envelopes are more or less persistent and the ecdysis of the wall is by fragmentation (fig. 113 *B*). Division of the cells occurs transversely or obliquely in two or three directions, the daughter-cells rapidly attaining their full size and undergoing further divisions. In *Chlorangium, Prasinocladus* (fig. 110 *A*) and *Hormotila* (fig. 115 *C*) the mucilage is secreted in a unilateral manner and becomes tough, branched colonies of cells on gelatinous stalks being gradually built up.

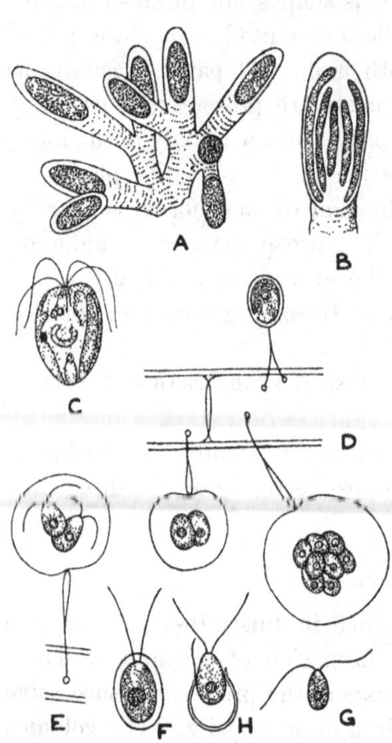

Fig. 110. *A—C, Prasinocladus lubricus* Kuck. *A*, part of a small colony, ×470; *B*, a single cell showing the chromatophores, ×960; *C*, zoogonidium, ×960 (after Kuckuck). *D—H, Physocytium confervicola* Borzi, ×600 (after Borzi). *D*, young colonies attached to a filament of *Spirogyra*; *E*, formation of zoogonidia; *F*, zoogonidium; *G*, gamete; *H*, germination of zygote.

The colonies frequently become dismembered into smaller portions by the disorganization of parts of the jelly, each portion increasing to form a new colony.

Asexual reproduction takes place by biciliated zoogonidia of the Chlamydomonad type. They arise either by the transformation of a vegetative cell into a zoogonidangium in which several (4 or 8) zoogonidia arise (*Apiocystis*; *Schizochlamys*, fig. 113 *C* and *D*), or by the assumption by the ordinary vegetative cell of the motile Chlamydomonadine condition (*Tetraspora*). The last-mentioned fact is a valuable piece of evidence in favour of the view that the Palmellaceæ have originated by the intercalation of a simple though well-marked vegetative condition between two formerly successive motile phases.

Gamogenesis has been observed in some of the genera by the fusion of isogametes, either similar in all respects to the zoogonidia and produced singly in a gametangium, or much smaller than the zoogonidia and produced in numbers from a gametangium (*Palmella*).

In *Tetraspora*, if the conditions are unfavourable, during the formation of zoogonidia the cells become invested in a strong resistant cell-wall and become hypnospores (Gay, '91; Chodat, '02).

The 'pseudocilia,' or motionless cilia, of the Tetrasporeæ are unique among the Protococcales.

Sub-family CHLORANGIEÆ. This is a small group of attached Algæ, fixed in all cases to a substratum by means of mucilaginous stalks. *Physocytium*, which is the simplest form, is unicellular; and the stalk of attachment is formed by the conversion into mucilage of the two cilia of the zoogonidium, after the latter has become quiescent with the tips of the cilia resting on the wall of some larger filamentous Alga (fig. 110 *D—H*). In *Prasinocladus*

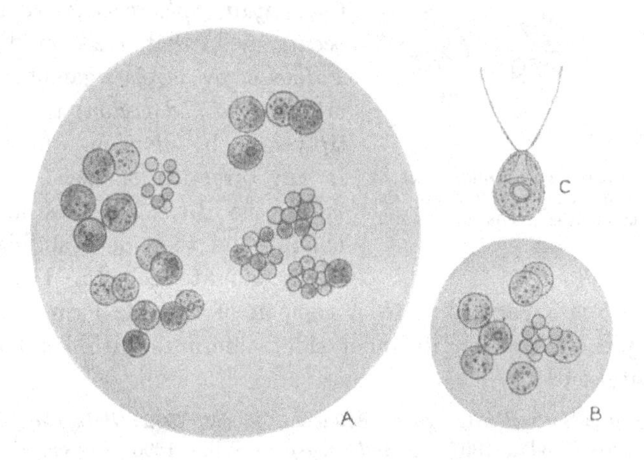

Fig. 111. ***Sphærocystis Schroeteri*** **Chodat.** ***A*** **and** ***B*****, normal colonies from the British freshwater plankton, × 450.** ***C*****. zoogonidium, × about 700 (after Chodat).**

(fig. 110 *A—C*), *Hauckia*, and *Chlorangium* the mucilaginous stalks are much thicker, and are formed by a unipolar secretion of mucilage. On the vegetative division of the cells, which is obliquely longitudinal, the mucilaginous stalks fork, so that small branched colonies are gradually built up.

In *Physocytium* the cells have the typical cup-shaped Chlamydomonadine chloroplast with one pyrenoid, but in *Prasinocladus* the chloroplast is reticulated or even divided into a number of parietal rod-shaped pieces, and has no pyrenoid. In *Prasinocladus* the zoogonidia possess four cilia, but in the other genera they have only two. There is little doubt that *Physocytium* represents one of the very first stages in the interpolation of a permanent resting-state into the Chlamydomonadine life-history. Some of these Algæ are freshwater, others brackish or marine.

The genera are: *Chlorangium* Stein, 1878; *Prasinocladus* Kuckuck, 1894 [inclus. *Chlorodendron* Senn, 1899]; *Physocytium* Borzi, 1883; *Hauckia* Borzi, 1883; *Ecballocystis* Bohlin, 1897.

It seems probable that the genus *Ecballocystis* is rightly placed in this group as the colonies are fixed by mucilaginous stalks which are mostly concrescent. The division of the ellipsoidal cells is also obliquely longitudinal (*vide* Bohlin, '97, t. 1, figs. 3 and 4).

Sub-family PALMELLEÆ. The Algæ of this small sub-family consist of aggregates of globose or ellipsoid cells irregularly grouped within a structureless mass of mucus. The latter is usually of indefinite extent, as in *Palmella*, but in *Palmodactylon* it is more or less cylindrical and often much branched, while in *Sphærocystis* it is mostly globose. In some of the genera (*Palmella, Coccomyxa, Sphærocystis*) reproduction occurs by biciliated zoogonidia, and in *Palmella* by isoplanogametes; but in others (*Palmodactylon*) the cells multiply only by division, which may be in any direction of space. In *Coccomyxa* the division-planes are mostly oblique and there is a sliding growth of the daughter-cells (fig. 112 *C* and *D*). The chloroplast is a parietal plate or cushion, of variable form, and with or without a pyrenoid, and on its inner side, sometimes within a hollow, the nucleus is situated.

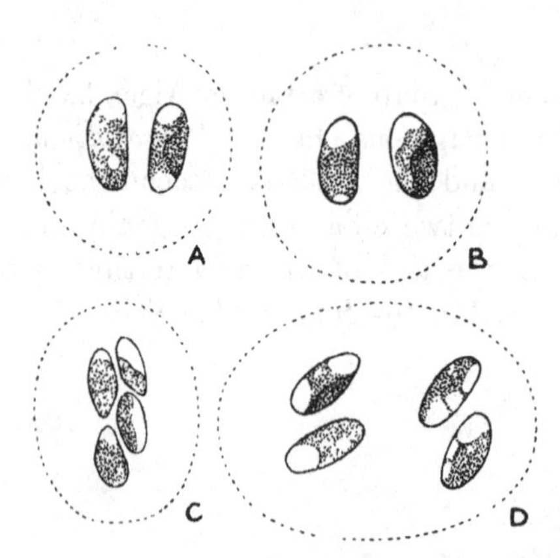

Fig. 112. *Coccomyxa subellipsoidea* Acton, ×1000. The dotted lines indicate the extent of the colourless mucus surrounding the cells.

The genera are: *Palmella* Lyngbye, 1819; em. Chodat, 1902; *Palmodactylon* Nägeli, 1849; *Sphærocystis* Chodat, 1897[1]; *Pseudotetraspora* Wille, 1896; *Coccomyxa* Schmidle, 1901.

All the genera are freshwater with the exception of *Pseudotetraspora*, which occurs in the sea off the Norwegian coast. *Coccomyxa* is generally subaërial, and *Sphærocystis* a frequent constituent of the freshwater plankton.

Evidence has been brought forward (G. S. W. '09) in support of the view that the plankton-Alga *Tetraspora lacustris* Lemm. is merely a stage in the life-history of *Sphærocystis Schrœteri* Chodat; and this view receives further support from the fact that pseudocilia have not been demonstrated in *Tetraspora lacustris.* On the other hand, both forms occurred simultaneously in equal abundance in the plankton of Loch Lomond (W. & G. S. W., '12), an observation which is rather confirmatory of Wesenberg-Lund's statements ('04) that the two are probably distinct.

[1] Wille ('03) has endeavoured to show that *Sphærocystis* (Chodat, 1902) should be regarded as synonymous with *Glœococcus* (A. Braun, 1851), but Chodat ('04) has rightly contested this suggestion. The microscopic limnetic colonies of *Sphærocystis* have nothing whatever in common with Braun's description of the macroscopic colonies of *Glœococcus*, which must be placed in the category of *genera dubia*. Wille's reference of the genus to the Chlamydomonadeæ (Wille, '03; '09) is also erroneous since the dominant vegetative phase is non-motile, and the experience of the present author is that a motile phase may be absent for very numerous generations.

Sub-family TETRASPOREÆ. The Algæ comprised in this group are distinguished from all other members of the Palmellaceæ by the possession of 'pseudocilia.' These are motionless, functionless cilia, each mature cell being provided with two or more. In *Tetraspora* the cells are largely grouped in fours at the periphery of a more or less extensive (macroscopic) mucilaginous

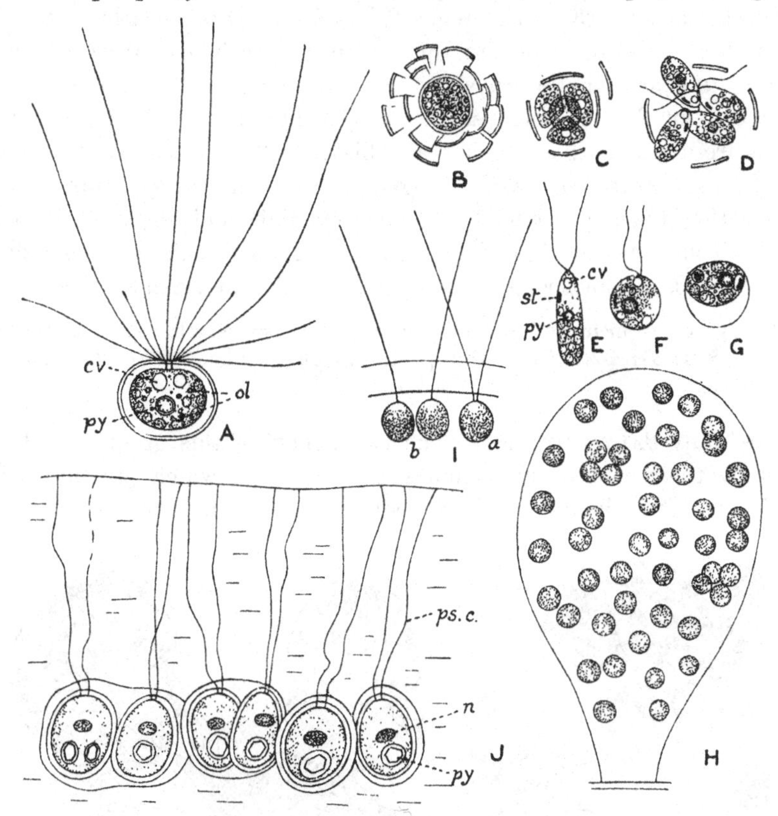

Fig. 113. *A—G*, *Schizochlamys gelatinosa* A. Br. *A*, vegetative cell showing pseudocilia, ×625; *B*, cell showing ecdysis of outer layers of wall, ×415; *C* and *D*, formation of zoogonidia, ×625; *E*, zoogonidium, ×830; *F* and *G*, zoogonidium changing to *Schizochlamys*-cell, ×830. *H* and *I*, *Apiocystis Brauniana* Näg. *H*, pear-shaped colony, ×430; *I*, three cells showing pseudocilia, *b*, two daughter-cells from a division, the second pseudocilium not yet developed, ×860. *J*, *Tetraspora gelatinosa* (Vauch.) Desv., periphery of colony showing a few of the cells with their pseudocilia, ×about 900. *cv*, contractile vacuole; *n*, nucleus; *ol*, oil globule; *ps.c.*, pseudocilia; *py*, pyrenoid; *st*, stigma (or pigment-spot). (*A—G*, after Scherffel; *J*, after Chodat.)

colony. The cells multiply by repeated division, chiefly in two directions in one plane, with the conversion of the walls of the mother-cells into mucilage. The pseudocilia are embedded in the mucilage of the colony (fig. 113 *J*), and each cell is of the *Palmella*-type, with a cup-shaped or a more or less indefinite Chlamydomonadine chloroplast. In *Apiocystis* (fig. 113 *H* and *I*) the colonies are relatively small, pyriform in shape, and usually attached by the base to other larger Algæ. The cells are disposed without any definite order near the periphery of the mucilaginous colony. The pseudocilia are very long and project through the mucus into the surrounding water. The cells divide in two or three directions in the peripheral plane. In this division one pseudocilium goes to each daughter-cell, a second one being subsequently developed (Correns, '93). The genus *Schizochlamys* has recently been shown by Scherffel ('08 B) to belong to this group, and each vegetative cell before it begins its characteristic ecdysis of the older parts of the wall (fig. 113 *B*) possesses a number of long pseudocilia (fig. 113 *A*). Both starch and oil occur as reserves in this genus. McAllister ('13) has observed the entire pyrenoid of *Tetraspora lubrica* segment to form several starch-bodies. Reproduction takes place by biciliated zoogonidia and isoplanogametes in *Tetraspora* and *Apiocystis*, and by biciliated or quadriciliated zoogonidia in *Schizochlamys*. Aplanospores (hypnospores) are also known in all these genera.

All the genera occur in fresh water; they are: *Apiocystis* Nägeli, 1849; *Schizochlamys* A. Braun, 1849; *Tetraspora* Link, 1809 [inclus. *Stapfia* Chodat, 1897]; ? *Tetrasporidium* Möbius, 1893.

Sub-family PALMOPHYLLEÆ. The two genera of this group are characterized by their macroscopic colonies, the mucus of which presents a firm outer layer. The colonies have in consequence somewhat definite shapes: they

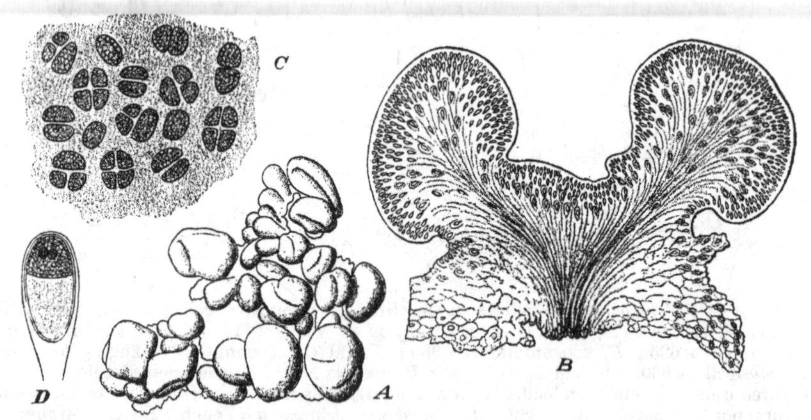

Fig. 114. *Collinsiella tuberculata* Setch. & Gardn. *A*, general appearance of colonies, ×35; *B*, section through colony, ×about 150; *C*, surface view of cells, ×300; *D*, single cell from the side, ×about 700. (After Setchell & Gardner, from Wille.)

are tuberculate in character in *Collinsiella*, and expanded and leaf-like in *Palmophyllum*. In the first-named genus the cells are chiefly aggregated towards the periphery, and there are obvious indications that the mucus is in reality a fusion of the mucilaginous stalks of a branched colony of cells. Division of the cells takes place in two directions at right angles to each other and to the surface of the colony (fig. 114). In *Palmophyllum* division of the cells is in two directions in a plane such that the colony becomes a flat expansion. The majority of the divisions are in one direction and the cells in consequence assume a seriate arrangement. The chloroplast is parietal and massive, with or without a pyrenoid.

The only genera are: *Palmophyllum* Kützing, 1845; *Collinsiella* Setchell & Gardner, 1903. Both are freshwater genera attached to submerged stones.

Sub-family GLŒOCYSTEÆ. In this sub-family the plants consist of small colonies of cells each of which is surrounded by a lamellose mucous investment. The concentric coats of mucus can be distinguished either round each individual cell or round a small group of daughter-cells. The greatest lamellation of the enveloping jelly is seen in *Asterococcus*, in which the envelope round a single cell (25—30 μ in diameter) may attain a diameter of 180 μ. In *Palmodictyon* (fig. 115 *B*) the groups of cells with their sur-

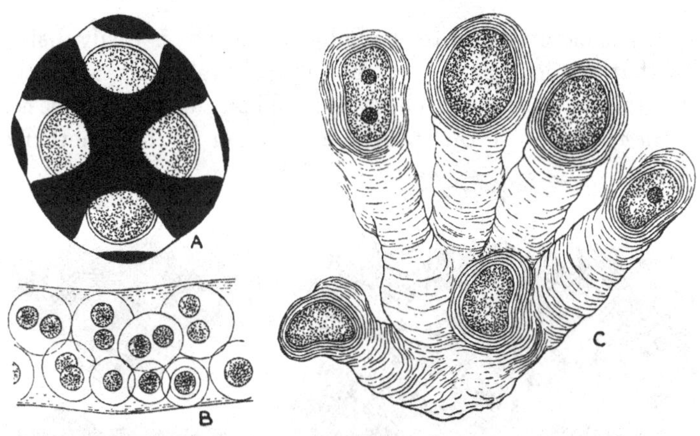

Fig. 115. *A*, four-celled colony of *Glœotænium Loitlesbergerianum* Hansg., ×430, showing black deposit in mucilaginous envelope. *B*, a small portion of a branched colony of *Palmodictyon viride* Kütz., ×430. *C*, colony of *Hormotila tropica* G. S. West, ×430.

rounding integuments are disposed in more or less cylindrical threads which branch and anastomose with each other; and in *Hormotila* the cells are usually aggregated to form a thin stratum, the secretion of mucus being mostly on one side so that each cell becomes possessed of a lamellate mucous stalk (fig. 115 *C*).

The chloroplast is in most cases parietal and massive, with or without a pyrenoid; but in *Asterococcus* (Scherffel, '08 A) there is an axile chloroplast with numerous radiating outgrowths, each of which becomes slightly expanded against the inner side of the cell-wall. There is a central pyrenoid, numerous starch-grains often fill the cells, oil is stored in small drops in the cytoplasm, and two contractile vacuoles and a conspicuous pigment-spot occur. Increase of the cells takes place by successive divisions of the mother-cell, or by the formation of a tetrad of four daughter-cells within the wall of the mother-cell, which then becomes mucilaginous. In *Palmodictyon* hypnospores with brown cell-walls are formed (G. S. W., '04), and after a period of rest these germinate directly into new elongated colonies.

The genera are: *Glœocystis* Nägeli, 1849 [in part]; *Palmodictyon* Kützing, 1845; *Hormotila* Borzi, 1883; *Asterococcus* Scherffel, 1908. [*Inoderma* Kützing is very doubtful.]

It has been asserted by various recent authors (Gerneck, '07; Wille '09; etc.) that Nägeli's genus *Glœocystis* merely includes developmental stages of members of the Chlamydomonadeæ and Ulotrichales, and should therefore be deleted. This is very likely true of a number of the so-called species which were at one time described, but there is as yet no proof that this is so in other cases. In *Glœocystis gigas*, for example, the colonies appear to be quite distinctive, and four daughter-cells arranged from the first as a tetrad are usually formed in each mother-cell. Much further investigation is required on some of these Algæ.

Family **Dictyosphæriaceæ.**

This family is closely allied to the Palmellaceæ, but is easily distinguished by the fact that portions of the old mother-cell-walls remain as thong-like attachments to the cells and cell-groups. This character is best seen in *Dictyosphærium*, in which the persistent portion of the mother-cell-wall

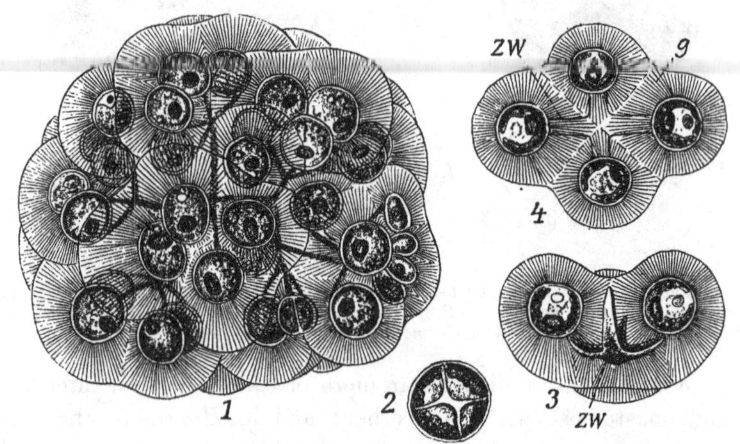

Fig. 116. *Dictyosphærium pulchellum* Wood. *1*, old colony; *2*, single cell dividing; *3* and *4*, two views of group of four cells. *g*, mucilaginous envelope; *zw*, cruciform remnants of old cell-wall connecting the daughter-cells. (After Senn, from Oltmanns.) × about 1000.

may become either a bifurcate or quadrifurcate thong (fig. 116) playing a considerable part in maintaining the definite globose or ellipsoid form of the colonies, which are in addition enveloped in a copious mucilage. In *Radiococcus* the relics of the mother-cell-walls are less important as binding-strands, possibly on account of the fact that the colonies are attached by one side of the large mucous envelope to the leaves of aquatic macrophytes, whereas those of *Dictyosphærium* are free-floating. In *Westella* and

Dimorphococcus (fig. 117) there is comparatively little mucus, notwithstanding the free-floating nature of the colonies, and the relics of the mother-cell-walls, which are to a great extent very irregular, are therefore of great importance as binding-threads. In all except *Dictyosphærium Hitchcockii* the chloroplast is parietal, often very massive, and with or without a pyrenoid.

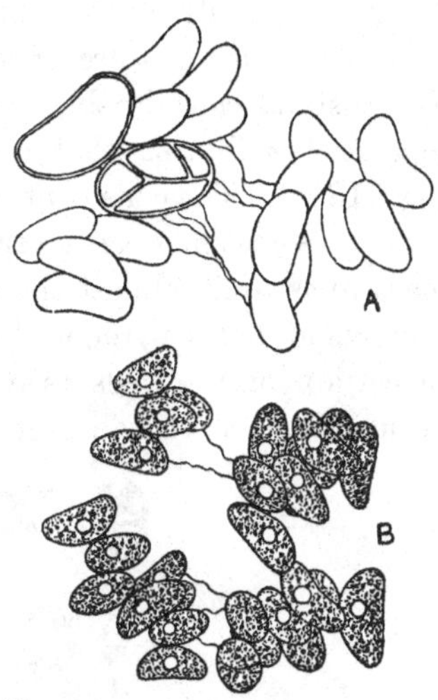

Fig. 117. *Dimorphococcus lunatus* A. Br. ×520. In *A* one cell shows division of contents of mother-cell into four daughter-cells.

Multiplication takes place by the formation of four daughter-cells (sometimes only two in *Dictyosphærium*) within the mother-cell. The division-planes are oblique; in *Westella* the four daughter-cells are always arranged in one plane, but in *Radiococcus* they are tetrahedrally disposed. The colonies frequently dissociate, each group of four cells being the commencement of a new colony. Reproduction by biciliated zoogonidia has been observed in *Dictyosphærium* by both Zopf and Massee, but does not occur under normal circumstances, nor has it been observed in any of the other genera.

The family may be conveniently divided as follows:

Sub-family Dictyosphærieæ: Colonies completely enveloped in mucus; cells in groups of two or four. *Dictyosphærium* Nägeli, 1849 [inclus. *Dictyocystis* Lagerheim, 1890]; *Radiococcus* Schmidle, 1902.

Sub-family Quaternatæ: Colonies with very little surrounding mucus; cells in definite groups of four more or less in one plane. *Dimorphococcus* A. Braun, 1849 [inclus. *Steiniella* Bernard, 1908]; *Westella* De Wildeman (in part) [*Tetracoccus* W. West, 1892 [1]].

Family **Protococcaceæ**.

This family as here constituted is the 'Pleurococcaceæ' of most authors. The cells are for the most part aggregated to form a definite stratum, but they may be scattered as unicells among other Algæ.

The most important member of the family is *Protococcus viridis* Ag. (= *Pleurococcus vulgaris* of nearly all authors)[2] which is found as a green

[1] In the Euphorbiaceæ there is a valid genus *Tetracoccus* Englem. ex Parry in *West-Amer. Scientist*, i, 1885, p. 13.

[2] Wille ('13) has examined the original specimens of *Protococcus viridis* from Agardh's herbarium and finds that they are identical with the common Alga usually known as '*Pleurococcus vulgaris*,' which was described by Chodat ('02) as *Pleurococcus Nägelii*. This note recently published by Wille is of the greatest general and taxonomic importance, for it not only finally decides that what is perhaps the commonest Green Alga in the world must in future be known as *Protococcus viridis* Ag., but it also determines the family Protococcaceæ, which cannot

incrustation in the temperate climates of the northern and southern hemispheres, generally on the windward side of tree-trunks, palings, stones, walls, etc. The cell-walls are strong and of some thickness, and the Alga is able to withstand considerable desiccation. The products of cell-division do not separate readily, so that compact groups of 2, 4, or 8 cells are frequent (fig. 118 *A*). The cells contain a single much-lobed parietal chloroplast, which appears sometimes as several parietal cushions and at other times is so massive as to fill almost the whole cell. Normally the chloroplast is without a pyrenoid, but a pyrenoid has often been described and figured as occurring in certain of the cells (see fig. 118 *A*, *py*), although it yet remains to be decided how far these records are correct[1]. Under certain conditions, usually

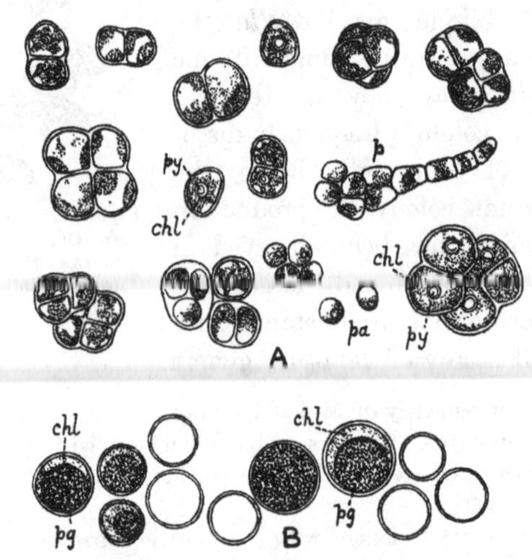

Fig. 118. *A*, *Protococcus viridis* Ag. (= '*Pleurococcus vulgaris* Menegh.' of most modern authors). *B*, *P. rufescens* Kütz. var. *sanguineus* (W. & G. S. West)—. All ×520. *chl*, chloroplast; *p*, protoderma-state; *pa*, palmella-state; *pg*, red pigment (hæmatochrome?) dissolved in globule of oil; *py*, pyrenoid.

of excessive moisture, short filaments of cells are formed which exhibit a simple type of branching. This form, which is almost entirely a culture-state and was described by Snow ('99) as '*Pseudopleurococcus*,' greatly resembles the genus *Protoderma* and is known as the 'protoderma-state.' A 'palmella-state' may also be developed in cultures. Reproduction occurs by the formation of one or many aplanospores. Zoogonidia and isogametes

in future be used in the sense of Oltmanns ('04), West ('04), Chodat ('09) or Wille ('09). This may seem undesirable, but the facts are so clear that the change must be made, and the sooner the better.

[1] It is possible that some of these records really refer to the initial stages of species of *Prasiola*, the cells of which are grouped in a manner very similar to their arrangement in the small colonies of *Protococcus viridis*. Each cell has, however, an axile stellate chloroplast with a central pyrenoid.

have been recorded, but there is much doubt about the observations, and under normal circumstances motile cells do not occur in *Protococcus viridis.*

Other species of *Protococcus* are not so well known and are less abundant. *P. rufescens* is of a brick-red colour owing to the presence of hæmatochrome which is dissolved in an oil. *P. dissectus* is characterized by the way in which the mother-cells divide, and the consequent disposition of the cells in the colonies. In the Antarctic there are several distinct and conspicuous species which are for the most part attached to papery sheets of the Blue-green Alga *Phormidium* (W. & G. S. W., '11; Fritsch, '12). These Antarctic forms are considerably larger than the species of more temperate regions, the cells of *P. antarcticus* forma *robusta* attaining a diameter of 100 μ. They possess a massive parietal chloroplast, with or without pyrenoids; they often store numerous small starch grains and not infrequently a fatty oil, the latter occurring sometimes in such quantity as to obscure all cytological details.

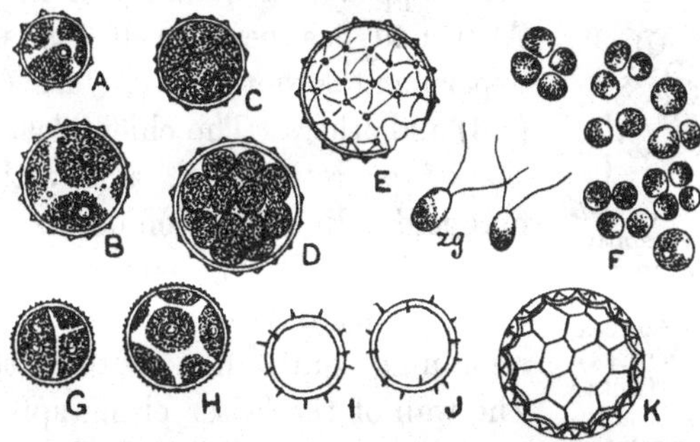

Fig. 119. *A—F*, *Trochiscia aspera* (Reinsch) Hansg.; *A* and *B*, vegetative cells; *C* and *D*, division of protoplast to form gonidia; *E*, empty cell from which gonidia have escaped; *F*, palmella state. *G* and *H*, *T. hirta* (Reinsch) Hansg. *I* and *J*, *T. paucispinosa* W. West; *K*, *T. reticularis* (Reinsch) Hansg. All $\times 520$. *zg*, zoogonidia found in a culture of *T. aspera*: there is no positive evidence that they belong to that species, or even to the genus *Trochiscia.*

A genus not far removed from *Protococcus* is *Trochiscia* (fig. 119). There are many species, some occurring in water and others on damp ground, but very little is known about most of them. They differ from the species of *Protococcus* in the external ornamentation of the cell-wall, which may be areolated or thickly clothed with denticulations or spines. Most of them are purely aquatic in habit. There are one or more parietal chloroplasts in each cell, and pyrenoids are frequently present. Reproduction is mostly by the formation of non-motile gonidia (aplanospores), 8 or 16 of which are formed in a mother-cell. There is no conclusive evidence that zoogonidia occur in this genus, although a palmella-state occurs in *T. aspera* (fig. 119 *F*). Wille ('01 A; '09) has stated that species of this genus are very probably the resting zygotes of members of the Chlamydomonadeæ, but available evidence does not support this view. His reference of the genus to the Volvocaceæ is not in accord with our present knowledge of the inter-relationships of the Protococcales.

The other important genus of the Protococcaceæ is *Chlorella* (fig. 120), which differs from *Protococcus* chiefly in the thinness of its cell-wall. Several distinct species are widely distributed, occurring on damp ground, on bark, and also in purely aquatic habitats. One of the most generally distributed is *Chlorella vulgaris* (fig. 120 *A*—*C*) which occurs in stagnant water, and is also found in symbiotic relationship with various Infusoria such as *Paramæcium*, *Ophrydium*, *Stentor*, etc.; and also in *Hydra viridis*. These symbiotic forms are known in a general way as 'zoochlorellæ.' *Chlorella vulgaris* often appears in quantity in impure cultures of Algæ and has been itself the subject of much experimental work by Artari ('92), Grintzesco ('03) and others. The chloroplast is parietal, bell-shaped or reticulated, and with or without a pyrenoid. Reproduction occurs by the successive bipartition of the original protoplast, the daughter-cells rapidly developing a delicate cell-wall and remaining quite free within the mother-cell. The wall of the latter either splits open or swells up and becomes diffluent, thus setting free the daughter-cells. In both *Trochiscia* and *Chlorella* the formation of the daughter-cells is almost exactly like the formation of zoogonidia in so many of the Protococcales, and it is not unlikely that they are reduced zoogonidia. Many of the recently described 'genera' of unicells found in cultures are referable to *Chlorella*.

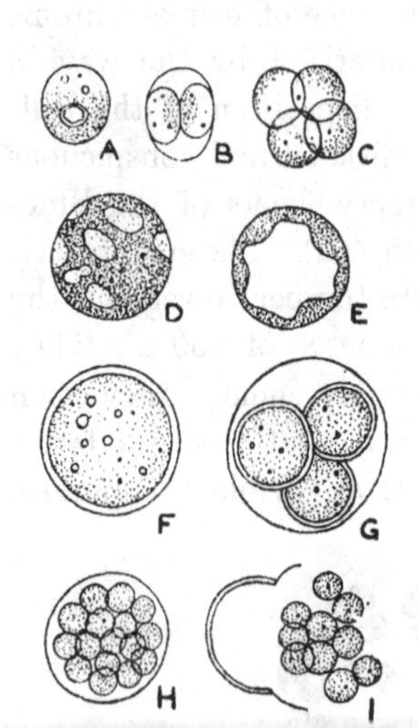

Fig. 120. *A*—*C*, *Chlorella vulgaris* Beijer.; *A*, vegetative cell; *B* and *C*, stages in division, ×about 800 (after Chodat). *D* and *E*, *Chl. faginea* (Gerneck) Wille, ×520 (after Gerneck). *F*—*I*, *Chl. miniata* (Kütz.) Wille (=*Palmellococcus miniatus* Chodat); *F*, vegetative cell; *G*, stage in division; *H* and *I*, formation and escape of gonidia. ×about 1000 (after Chodat).

The genera are: *Protococcus* Agardh, 1824 [inclus. *Pleurococcus* auct. but possibly not of Meneghini, 1842; ? *Diplosiphon* Bialosuknia, 1909]; *Trochiscia* Kützing, 1845 [=*Acanthococcus* Lagerheim, 1883]; *Entophysa* Möbius, 1889; *Chlorella* Beijerinck, 1890 [inclus. *Protococcus* auct.; *Palmellococcus* Chodat, 1894; *Chloroïdium* Nadson, 1906; *Krugeria* Heering, 1906; *Planophila* Gerneck, 1907; *Chlorotetras* Gerneck, 1907; *Acrosphæria* Gerneck, 1907; *Chlorosarcina* Gerneck, 1907].

Chlorella miniata (Kütz.) Wille is a frequent Alga in greenhouses, forming a soft brownish-green stratum, which often turns orange-red, on the outer surfaces of moist plant-pots, etc. Numerous 'species' of *Chlorella* have recently been described, mostly on cultural characters (*vide* Chodat, '09 and '13). An elaborate physiological investigation of *Chlorella luteo-viridis* var. *lutescens* Chodat has recently been made by Kufferath ('13).

It is quite possible that the Algæ described by Klebs ('83) and others as species of

'*Chlorosphæra*' should be referred to the genus *Protococcus*. The Antarctic species of *Protococcus* clearly show that there are no characters of importance by which they can be separated. On the other hand, '*Chlorosphæra*' appears to be related to *Chlorococcum* among the Chlorochytrieæ, since it produces zoogonidia (8 or more) from a single spherical mother-cell. It is not improbable that the Algæ described as species of '*Chlorosphæra*' are the relics of intermediate forms between *Protococcus* and the lowest form, viz. *Chlorococcum*, of the Chlorochytrieæ. They combine the zoogonidium-formation of the latter group with the formation of transverse walls in vegetative division such as occurs in *Protococcus viridis* and *P. dissectus*.

Family **Autosporaceæ**.

The Autosporaceæ are one of the most clearly defined families in the Protococcales. The Algæ included in it are free-floating, solitary or colonial, the cells being usually associated to form very small few-celled colonies. In some cases the colony is a cœnobium with a definite construction (Cœlastreæ, Crucigenieæ, *Scenedesmus*), but in most of the other forms the colonies readily

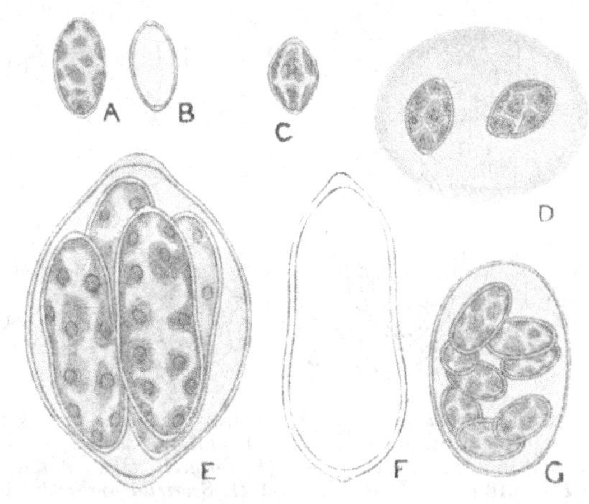

Fig. 121. *A* and *B*, *Oocystis solitaria* Wittr. *C* and *D*, *O. crassa* Wittr. *E* and *F*, *O. panduriformis* W. & G. S. West. *G*, *O. elliptica* W. West. All ×485.

dissociate into small groups of cells or single individuals. With few exceptions the amount of mucus surrounding the colonies is small, but in *Kirchneriella* and *Elakatothrix* there is a very copious mucous investment around each colony, and in a few other forms, such as *Crucigenia* and *Ankistrodesmus Pfitzeri*, there is a considerable mucous envelope. There is generally one parietal chloroplast in each cell, often very large, and not infrequently occupying most of the cell. It may or may not possess a pyrenoid. In *Eremosphæra* there are numerous parietal chloroplasts, and in most species of *Oocystis* there are several (*vide* fig. 121 *A*, *C*, *D* and *E*). The protoplast contains a single nucleus located in the central part of the cell.

Multiplication sometimes occurs by the division of the mother-cell along transverse (*Elakatothrix*) or oblique (*Ankistrodesmus falcatus* var. *acicularis*; fig. 129 *C*) planes into two or four daughter-cells. Reproduction takes place by the successive divisions of the protoplast to form 2, 4 or 8 spore-like bodies, which in most cases assume the characters of the mother-cell before being liberated. These are *autospores.* In the cœnobic forms each mother-cell gives origin to a new colony—an *autocolony* (figs. 128 *C*, and 133 *C* and *D*). The wall of the mother-cell is either ruptured or becomes converted into mucilage, and in *Oocystis* may become greatly distended to form a wide envelope enclosing the daughter-cells.

Neither zoogonidia nor gametes occur[1].

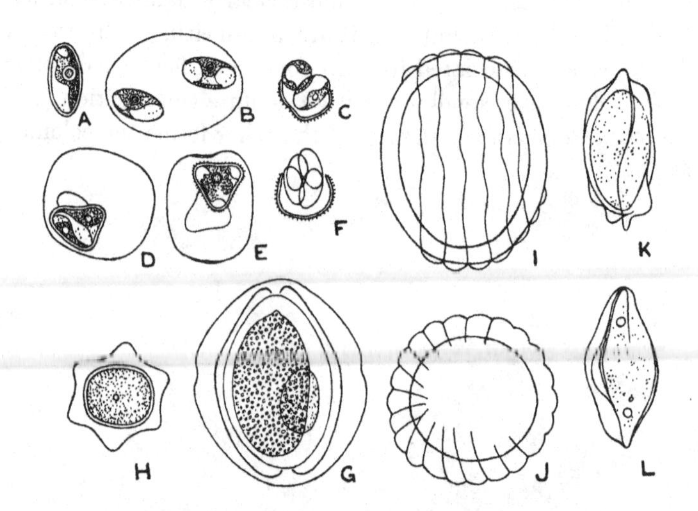

Fig. 122. *A—F*, *Oocystis submarina* Lagerh. *A*, vegetative cell; *B*, young colony; *D* and *E*, *Tetraëdron*-like cells formed within old mother-cell-wall of *Oocystis*; *C* and *F*, development of *Tetraëdron*-like cell to form two (*C*) or four (*F*) *Oocystis* cells. *A* and *B*, ×995; *D* and *E*, ×570; *C* and *F*, ×610 (after Wille). *G* and *H*, *Scotiella antarctica* Fritsch. *G*, single cell viewed with the principal ridges parallel to the substratum, ×830; *H*, optical section, ×430 (after Fritsch). *I* and *J*, *Scotiella polyptera* Fritsch. *I*, side view; *J*, oblique end view to show course of ridges, ×1100 (after Fritsch). *K* and *L*, two side views of *Scotiella nivalis* (Chod.) Fritsch, ×about 800 (after Chodat).

The Algæ of this family are little removed from the most primitive forms of the Chlorophyceæ, and it is probable that the characteristic autospores are merely arrested zoogonidia which at once develop either singly into a new cell or collectively into a new cœnobium. Under cultivation, and particularly in cultures on solid media, some of these Algæ are profoundly modified, but in their natural state they exhibit a truly remarkable constancy of character. Many of them are ubiquitous in all climates and nearly all are inhabitants of fresh water. The form of the cell and the nature of the colony are so varied that only a sectional treatment can give the student a clear idea of the family. There are six well-marked sub-families.

[1] Consult remarks on *Eremosphæra* (p. 198) and *Micractinium* (p. 199).

Sub-family OOCYSTEÆ. This sub-family is characterized by the globose or ellipsoid cells (curved or even sublunate in *Nephrocytium*) which are frequently retained within the distended wall of the old mother-cell. In nearly all cases the cells possess strong cellulose walls, which in the genus *Scotiella* (Fritsch, '12 A) of the polar and alpine snow-flora, are furnished with wing-like ridges extending from pole to pole (consult fig. 122 *G—L*). In *Oocystis* the cell-wall generally exhibits a slight thickening at each pole. Each cell contains from one (*Nephrocytium, Scotiella*) to many (*Eremosphæra, Excentrosphæra*) chloroplasts, parietally disposed except in *Oocystis natans* (Lemm.) Wille, and with or without a pyrenoid. In *Eremosphæra* the cell is exactly spherical, and the parietal chloroplasts are small and very numerous. The cells of this genus are much the largest of the unicellular Protococcales, attaining a diameter of 200 μ, and the centrally-placed nucleus is correspondingly big (fig. 123 *A*). The genus *Glœotænium*, which has recently

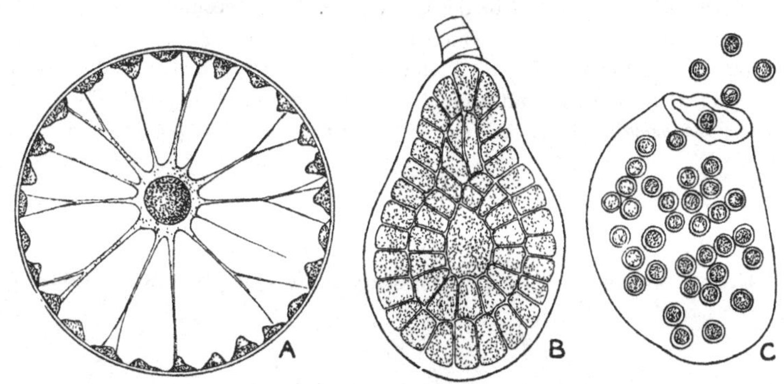

Fig. 123. *A*, optical section of *Eremosphæra viridis* De Bary, to show parietal disposition of numerous chloroplasts and centrally-placed nucleus, × 300. *B* and *C*, *Excentrosphæra viridis* Moore. *B*, surface view showing closely packed parietal chromatophores; *C*, escape of gonidia, × about 190 (after Moore).

been re-investigated by Transeau ('13), is remarkable for the presence of a black pigment in the old walls of the mother-cells so arranged as to form bands and caps (fig. 115 *A*).

Reproduction takes place by the division of the mother-cell into 2, 4 or 8 autospores, which grow and usually attain their full size while within the greatly distended wall of the mother-cell (figs. 121 *E* and *G*; 122 *B*). In *Oocystis* and *Nephrocytium* several generations may be contained within the remnants of an old mother-cell-wall. In *Nephrocytium ecdysicepanum* W. & G. S. West ('96) and in *Oocystis glœocystiformis* Borge ('06) several generations are aggregated in a fan-shaped manner owing to the ecdysis but incomplete dissolution of the old mother-cell-walls. Sometimes the wall of the mother-cell becomes converted into a structureless mass of jelly in which the daughter-cells are embedded (fig. 121 *D*). In *Ecdysichlamys* the outer layers of the cell-wall are thrown off in a manner comparable with the exuviation of the outer layers of the wall of *Nephrocytium ecdysicepanum.* The cells of *Ecdysichlamys* form a stratum, only one layer of cells in thickness, attached to wet sand-grains, a feature which at once distinguishes the genus from *Oocystis.*

Wille ('08) has found that *Oocystis submarina* may pass through a *Tetraëdron*-state, *i.e.* each cell may form a resting-spore (hypnospore) of a triangular shape which greatly resembles certain species of the genus *Tetraëdron.* These resting-spores form normal *Oocystis*-cells on germination (fig. 122 *C* and *F*). Chodat ('95) described the occurrence of zoogonidia in *Eremosphæra,* but this must be regarded as an error due to contaminated cultures[1] (*vide* Moore, '01 and G. S. W., '04). In *Excentrosphæra* Moore observed the formation of numerous gonidia (aplanospores) which were liberated by the bursting open of one end of the mother-cell (fig. 123 *C*).

The Oocysteæ are mostly found in bogs, but some forms occur in pools and others in the plankton of lakes. *Scotiella* is a constituent of the snow-flora.

The genera are: *Oocystis* Nägeli, 1845; *Nephrocytium* Nägeli, 1849; *Eremosphæra* De Bary, 1858; *Glœotænium* Hansgirg, 1890; *Excentrosphæra* Moore, 1901; *Scotiella* Fritsch, 1912; *Ecdysichlamys* G. S. West, 1912. *Prototheca* Krüger, 1894, is a colourless genus.

Sub-family MICRACTINIEÆ. The Algæ belonging to this small group (which has sometimes been called the Phytheliеæ) are at once distinguished from all other members of the Autosporaceæ by the presence of stiff bristles

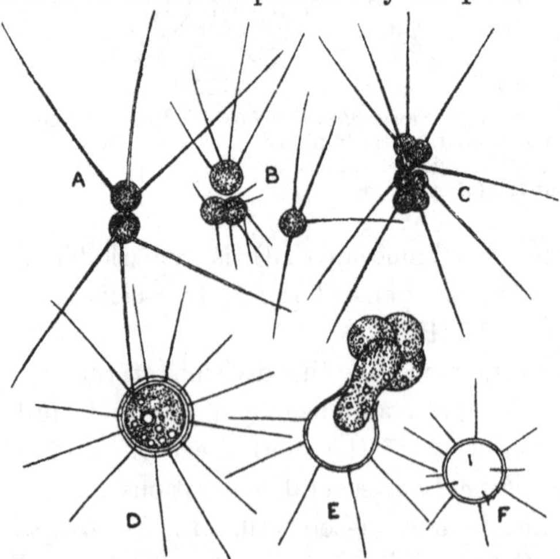

Fig. 124. *A*, *Micractinium pusillum* Fresen. [=*Richteriella botryoides* (Schm.) Lemm.], ×520 (after Lemmermann). *B* and *C*, *M. pusillum* forma *quadriseta* (Lemm.), ×450. *D* and *E*, *M. radiatum* (Chod.) Wille [=*Golenkinia radiata* Chodat], ×about 800 (after Chodat). *F*, *M. paucispinosum* (W. & G. S. West), ×450.

[1] Some of the supposed polymorphic states of *Eremosphæra viridis* described and figured by Chodat ('95) are merely vegetative cells of *Asterococcus superbus* (Cienk.) Scherffel.

variously disposed about the periphery of the cells. They are unicellular or they consist of small aggregates of cells, usually devoid of a mucous envelope. The bristles vary in number, length and attachment, but in most forms they are considerably longer than the diameter of the cells. They are in all instances rather delicate and easily overlooked. In addition to the more obvious bristles there are a number of others of a much more delicate character, similar to those which are known to occur on *Pediastrum* and *Scenedesmus.* (See pages 202 and 219, and figs. 130 *F* and 144 *H.*)

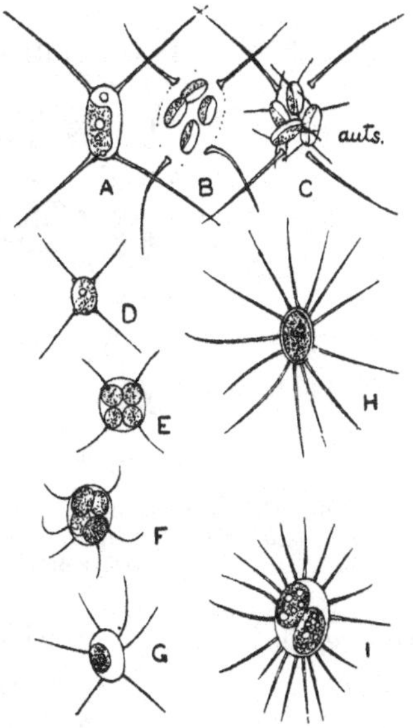

Fig. 125. *A—C*, *Lagerheimia genevensis* Chod.; *A*, vegetative cell; *B* and *C*, formation of autospores (*auts*), ×about 850 (after Chodat). *D* and *E*, *L. genevensis* var. *subglobosa* (Lemm.) Chod.; *D*, ×520 (after Lemmermann); *E*, ×450. *F* and *G*, *L. breviseta* (W. & G. S. West), ×450. *H* and *I*, *L. ciliata* (Lagerh.) Chod. var. *amphitricha* (Lagerh.) Chod., ×450.

Multiplication rarely occurs by simple division, and the usual method of reproduction is by autospores, formed 2, 4, 8 or 16 in a mother-cell, each autospore often acquiring the full characters of the adult cell before its liberation from the distended mother-cell-wall. It is incorrect to regard the small colonies of these Algæ as cœnobia, since the number of cells is not definite, not even in those forms of *Micractinium* described as '*Richteriella,*' and the aggregation of the cells is a loose one.

It has been generally supposed that most of the members of this characteristic group occur principally in the plankton of large lakes, but they certainly occur far more abundantly in small reservoirs, in ponds, and in canals than they ever do in lakes. Chodat ('94) described the occurrence of zoogonidia in *Micractinium* (= *Golenkinia* Chodat), but this is open to considerable doubt and may be an error of observation due to contaminated cultures, as in the case of *Eremosphæra.* No trace of zoogonidia has ever been observed by other investigators, and it seems unlikely that motile reproductive cells would occur in this sub-family.

The type genus, *Micractinium* Fresenius ('56—'58) had been for years entirely overlooked until Wille ('09 B) brought it to light again, although the original description and figures were quite good. The present author agrees with Wille's grouping of the recently described forms.

The genera are: *Micractinium* Fresenius, 1856—58 [inclus. *Archerina* Lankester, 1885; *Phythelios* Frenzel, 1891; *Golenkinia* Chodat, 1894; *Richteriella* Lemmermann, 1896]; *Acanthosphæra* Lemmermann, 1898; *Meringosphæra* Lohmann, 1908; ? *Echinosphæridium* Lemmermann, 1904; *Lagerheimia* (De Toni) Chodat, 1895 [inclus. *Tetraceras* Chodat,

1894[1]; *Pilidiocystis* Bohlin, 1897; *Chodatella* Lemmermann, 1898; *Bohlinia* Lemmermann, 1899]; *Franceia* Lemmermann, 1898.

Sub-family Tetraëdreæ. This small group is chiefly represented by

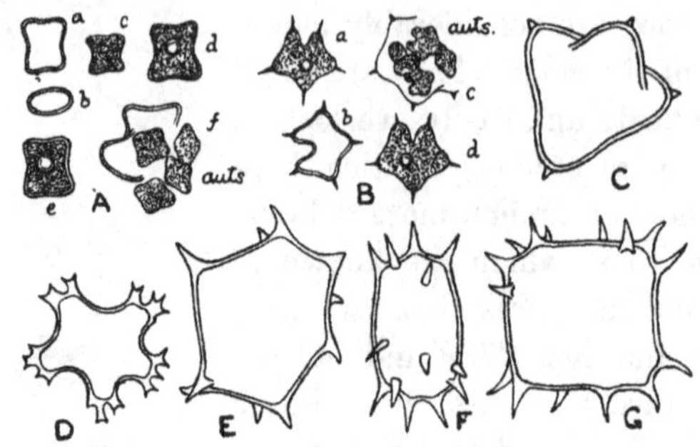

Fig. 126. *A*, *Tetraëdron minimum* (A. Br.) Hansg. *B*, *T. caudatum* (Corda) Hansg. *C*, *T. regulare* Kütz. *D*, *T. enorme* (Ralfs) Hansg. *E—G*, *T. horridum* W. & G. S. West. *auts*, autospores. All ×450.

the genus *Tetraëdron* in which the cells are free-floating without any mucous investment. They are compressed and angular, with a definite number of angles; or they are tetrahedral, octahedral, or polyhedral (fig. 126). The angles of the cells may be rounded, emarginate, or furnished with spines. Each cell contains a massive parietal chloroplast, usually with a distinct pyrenoid. Reproduction occurs by the formation of 4 or 8 autospores which are miniature adults when set free by the rupture of the mother-cell-wall (fig. 126 *A f* and *B c*).

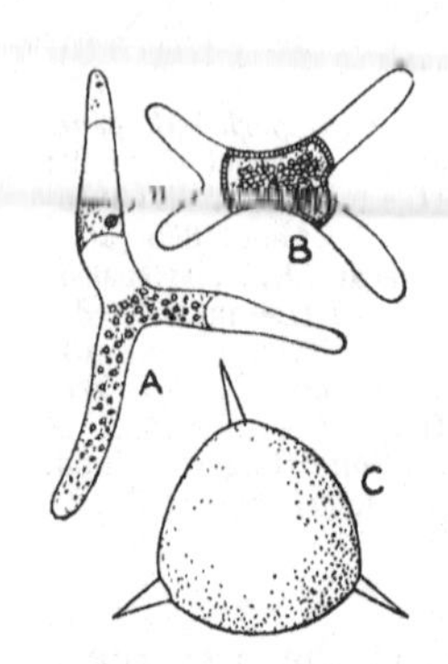

Fig. 127. *A* and *B*, *Chionaster nivalis* (Bohlin) Wille; *A*, vegetative cell showing nucleus (*n*); *B*, cell with aplanospore. ×450 (after Wille). *C*, *Mycotetraëdron cellare* Hansg. ×1500 (after Hansgirg).

The fact that *Tetraëdron*-like states of other members of the Protococcales are known (as in *Pediastrum*, *Hydrodictyon* and *Oocystis*) in no way interferes with the validity of this genus. Most of the forms occur in stagnant waters, and a few are frequent constituents of the freshwater plankton. In *Cerasterias* the angles of the cells are so greatly produced that the individuals are stellate in general appearance.

The genera are: *Tetraëdron* Kützing, 1845 [inclus. *Polyedrium* Nägeli, 1849; *Staurophanum* Turner, 1893; *Polyedropsis* Schmidle, 1898; *Dichotomum* W. & G. S. West, 1896 (in part)]; *Cerasterias* Reinsch, 1867 [inclus. *Astrocladium* Tschourina, 1909]; *Thamniastrum* Reinsch, 1888.

There are two interesting colourless members of this sub-family: *Mycotetraëdron* Hansgirg, 1890 (fig. 127 *C*) and *Chionaster* Wille, 1903 (fig. 127 *A* and *B*).

[1] Chodat (1895) states 'nomen ineptum ob affine Tetracera Dilleniacearum.'

Sub-family SELENASTREÆ. The Algæ of this group are characterized by the elongation of the cells, which are often very narrow with the extremities attenuated to fine points. They are frequently lunate or arcuate, and may be solitary or associated to form colonies often of a more or less fragile character (*Actinastrum*, fig. 130 *A* and *B*; *Selenastrum*, fig. 131), the cells in some instances being held in position only by an envelope of mucus (*Ankistrodesmus Pfitzeri, Kirchneriella, Elakatothrix*). In many species of *Ankistrodesmus*, in *Dactylococcus, Closteriopsis*, and most forms of *Scenedesmus* there is no enveloping mucus. The cell-wall is firm but delicate. Each cell contains a large elongated chloroplast which often fills almost the entire cell. In some forms pyrenoids do not occur, but in others one or more pyrenoids

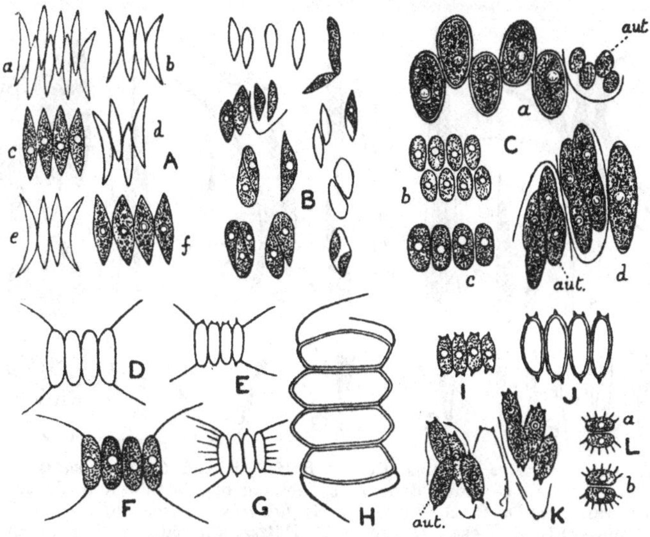

Fig. 128. *A*, *Scenedesmus obliquus* (Turp.) Kütz. *B*, the culture-state of *S. obliquus* formerly known as *Dactylococcus infusionum* Näg. *C*, *S. bijugatus* (Turp.) Kütz. *D—F*, *S. quadricauda* (Turp.) Bréb. *G*, *S. quadricauda* var. *horridus* Kirchn. *H*, *S. quadricauda* var. *maximus* W. & G. S. West. *I—K*, *S. denticulatus* Lagerh. var. *linearis* Hansg. *L*, *S. spicatus* W. & G. S. West. All ×520. *aut*, autocolonies.

are invariably present. In the ubiquitous species *Ankistrodesmus falcatus*, and especially in its var. *acicularis*, pyrenoids may be present in some individuals but not in others (fig. 129 *B*). *Scenedesmus* and *Tetradesmus* are the only genera with a definite cœnobium; in the former the cells are arranged side by side (with their long axes parallel) in one plane, and in the latter side by side in two planes (Smith, '13).

Reproduction takes place by the formation of autospores or, in *Scenedesmus* and *Tetradesmus*, of autocolonies (fig. 128 *C* and *K*). The division of the parent-protoplast is in one or more oblique planes, often nearly longitudinal, and results in the formation of 2, 4 or 8 daughter-cells. In those cases where the wall of the mother-cell does not become mucilaginous the oblique divisions have very much the appearance of causing a fragmentation of the cell (fig. 129 *C*). Sometimes the division-plane in *Ankistrodesmus* is almost transverse (fig. 129 *B c*), and in *Elakatothrix* it is generally quite transverse[1]. In *Actinastrum* the autospores diverge outwards but remain attached by their proximal ends, and as a result of the formation of successive generations of autospores the colonies may reach a comparatively large size. In *Selenastrum acuminatum*, a species connecting the two genera *Selenastrum* and *Ankistrodesmus*, and referred by Chodat to *Scenedesmus*, the young cells attained maturity in from 15 to 20 days, at the end of which period many of them again produced a new generation of autospores (G. S. W., '12).

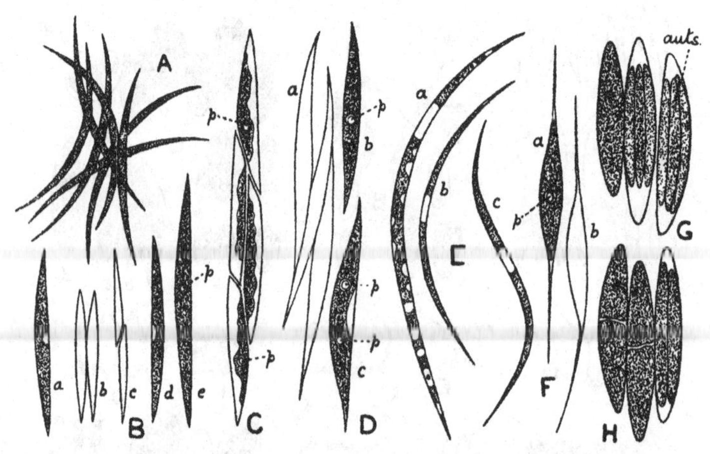

Fig. 129. *A*, *Ankistrodesmus falcatus* (Corda) Ralfs. *B* and *C*, *A. falcatus* var. *acicularis* (A. Br.) G. S. West; *C* shows division by oblique planes into four daughter-cells. *D*, *A. falcatus* var. *tumidus* G. S. West. *E*, *A. falcatus* var. *mirabilis* G. S. West. *F*, *A. setigerus* (Schröd.) G. S. West. *G* and *H*, *A. Pfitzeri* (Schröd.) G. S. West, the enveloping mucus not shown. All × 520. *auts*, autospores; *p*, pyrenoid.

The cœnobium of *Scenedesmus* is somewhat specialized, so that of all the genera of the group *Scenedesmus* is the least representative. *Scenedesmus quadricauda* possesses strong bristles attached to the end-cells of the cœnobium, and sometimes to the other cells also; in *S. denticulatus* the poles of all the cells are furnished with minute teeth; in *S. granulatus* the walls are granulated; and in *S. costatus* and *S. acutiformis* the cells are longitudinally ridged. It has been shown by Petersen ('11) that the cells of some species of this genus are furnished with very delicate bristles (fig. 130 *F*). Chodat

[1] Wille associates this Alga with *Coccomyxa*, but there is little doubt that it is most closely allied to *Ankistrodesmus*, especially to those forms which are normally enveloped in mucus; and although the actual division-plane of the cell is transverse, the daughter-cells as often as not grow by sliding obliquely past each other. *Elakatothrix gelatinosa* occurs in the lakes of Norway and the English Lake District (Wastwater).

& Malinesco ('93) and also Grintzesco ('02) have done considerable experimental work on *Scenedesmus obliquus* (= *S. acutus*), and have confirmed the observations of others that this Alga has a *Dactylococcus*-state (fig. 128 *B*)

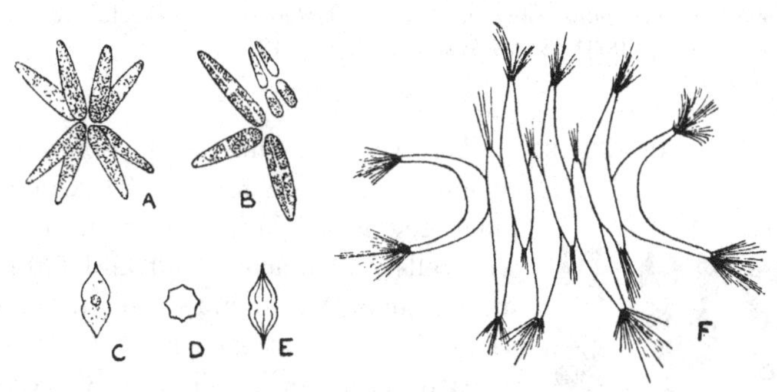

Fig. 130. *A* and *B*, *Actinastrum Hantzschii* Lagerh. × 400 (after Lagerheim); *B* shows the division of the mother-cells. *C—E*, *Desmatractum plicatum* W. & G. S. West, × 520. *F*, *Scenedesmus obliquus* (Turp.) Kütz. var. *dimorphus* (Turp.) Rabenh., showing the bristles at the extremities of the cells, × 730 (after Petersen).

which is identical with *Dactylococcus infusionum* Näg. Grintzesco also showed that remarkable malformations are produced by the culture of *Scenedesmus obliquus* on a nutritive medium of agar and glucose, and that this species can liquefy gelatin. He attributes the extensive geographical distribution of this plant to the ease with which it adapts itself to different media and different temperatures.

The members of the Selenastreæ usually occur in the smaller ponds and pools, where they are often found in great quantity; some of them are also constituents of the freshwater plankton, especially species of *Kirchneriella*, *Closteriopsis*, and *Ankistrodesmus*. *Actinastrum* (fig. 130 *A* and *B*) and *Lauterborniella* are found in the helioplankton of pools, canals, etc.; and *Desmatractum* (fig. 130 *C—E*) occurs in the paddyfields of Ceylon.

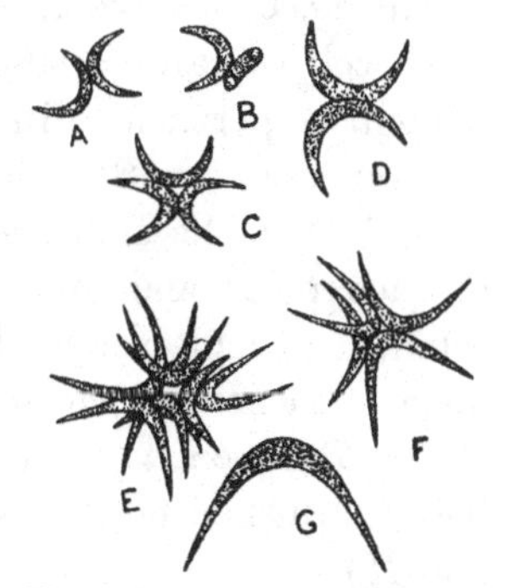

Fig. 131. *A—D*, *Selenastrum gracile* Reinsch. *E—G*, *S. acuminatum* Lagerh. All × 520.

One species of *Ankistrodesmus* (*A. nivalis*) is a constituent of the red snow flora, and other forms occur in the yellow snow. *Kirchneriella* is a genus which, although occurring in small pools and ponds, is a characteristic constituent of the summer plankton of certain lakes, its much-curved cells enveloped in wide mucous investments being very distinctive. *Closteriopsis longissima* is by far the most elongated member of the Selenastreæ, attaining a length of 530 μ; but there is some possibility that this Alga is a degenerate form of the Desmid *Closterium aciculare* var. *subpronum* (*vide* W. & G. S. W., '06), in which case its inclusion in the Protococcales would clearly be incorrect.

A number of imperfectly described species of *Reinschiella* are without doubt the resting-cysts of certain of the freshwater Peridinieæ.

The genera are: *Ankistrodesmus* Corda, 1838; em. Ralfs, 1848 [inclus. *Rhaphidium* Kützing, 1845; *Schröderia* Lemmermann, 1898]; *Selenastrum* Reinsch, 1867; *Actinastrum* Lagerheim, 1882; *Reinschiella* De Toni, 1889 [= *Clostridium* Reinsch, 1888]; *Kirchneriella* Schmidle, 1893 [inclus. *Selenoderma* Bohlin, 1897]; *Closteriopsis* Lemmermann, 1898; *Lauterborniella* Schmidle, 1900; *Desmatractum* W. & G. S. West, 1902; *Didymogenes* Schmidle, 1905; *Scenedesmus* Meyen, 1829; *Dactylococcus* Nägeli, 1849 (in part) [= *Ouracoccus* Grobéty, 1909]; *Tetradesmus* Smith, 1913.

Sub-family CRUCIGENIEÆ. The Algæ of this small group consist of colonies of few or many cells arranged with much regularity in the form of a flat plate. The cells vary much in outward form, but are mostly somewhat rounded and never elongate as in the Selenastreæ. Four autospores arise in each mother-cell. They are disposed in one plane, and in certain species of *Crucigenia* (*C. rectangularis*, fig. 132 *A*, and *C. Tetrapedia*, fig. 132 *F*) colonies of considerable size (128 cells) may be formed by the conversion into mucilage of the walls of the mother-cells of successive generations. In other species of *Crucigenia* (*C. quadrata*, *C. Lauterbornii*, etc.) the colonies never become very large, and consist mostly of from 4 to 16 cells. Even in the large colonies the cells remain in distinct groups of four, and there is always a quadrate or rhomboidal space in the centre of each group, small in *C. rectangularis*, *C. quadrata*, etc., but large in *C. Lauterbornii* and *C. fenestrata*. Each cell contains a single parietal chloroplast, often massive, and with or without a pyrenoid. In *C. irregularis*, which is known from the lakes of Norway and the Shetlands, the colonies are large and gradually become irregular; in the Madagascar species, *C. emarginata*, the cells are emarginate at each pole; and in *C. appendiculata* the wall of the old mother-cell is retained in four pieces which have the appearance of appendages, one piece being attached to the outer margin of each of the four daughter-cells.

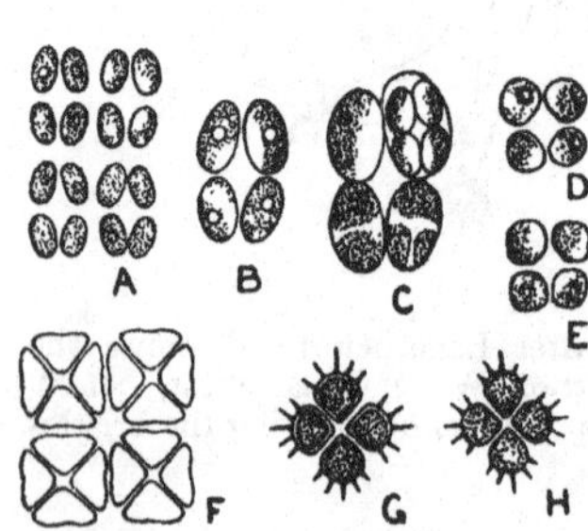

Fig. 132. *A—C*, *Crucigenia rectangularis* (Näg.) Gay. *D* and *E*, *C. quadrata* Morren. *F*, *C. Tetrapedia* (Kirchn.) W. & G. S. West. *G* and *H*, *Tetrastrum staurogeniæforme* (Schröd.) Chodat. All × 520.

In *Tetrastrum* the colony is always a four-celled cœnobium, each cell being furnished on its outer margin with from two to five spines (fig. 132 *G* and *H*).

The genera are: *Crucigenia* Morren, 1830 [inclus. *Staurogenia* Kützing, 1849; *Lemmermannia* Chodat, 1899; *Willea* Schmidle, 1900; *Crucigeniella* Lemmermann, 1900; *Hofmannia* Chodat, 1900]. *Tetrastrum* Chodat, 1895 [= *Cohniella* Schröder, 1897]. The various species occur in the quiet waters of ponds and boggy pools, and some are found in the benthos and plankton of lakes.

Sub-family Cœlastreæ[1]. This is the only group of the Autosporaceæ in which all the forms possess a definite and regular cœnobium. In this case, however, the cœnobium is not flat (as in *Scenedesmus* among the Selenastreæ) but spherical or polyhedral. The three known genera are essentially different in the grouping of the cells. In *Cœlastrum* the cells are rounded or polygonal, adhering closely by their margins to form a hollow sphere or cube (*C. cubicum*). In most species the outer surface of each cell is furnished with a short truncate process, which attains its maximum development in *C. cambricum* var. *nasutum*. In the African species, *C. compositum*, the normal single cell is replaced by a tetrad of four (*vide* G. S. W., '07). In *C. cambricum* the cells are joined by lateral truncate processes (fig. 133 *A*) and in *C. reticulatum*, which is mostly a plankton-species, similar processes occur, but of a more slender and elongate character. In *Burkillia*, a genus at present only known from Burma, the cells are more loosely coherent, being held in position

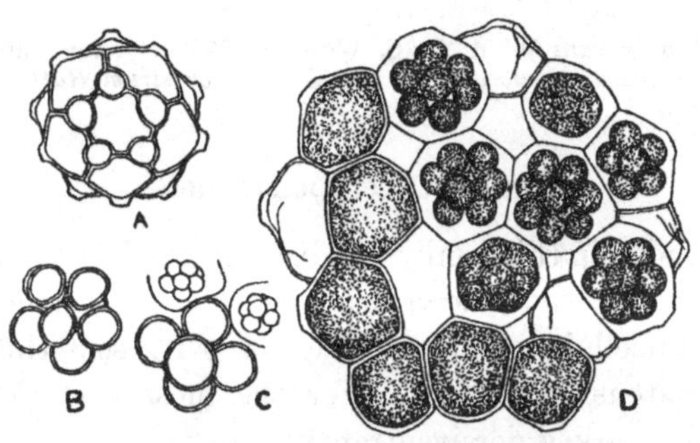

Fig. 133. *A*, *Cœlastrum cambricum* Archer. *B—D*, *Cœlastrum sphæricum* Näg. Both *C* and *D* show the formation of daughter-cœnobia (autocolonies). All ×475.

mainly by a small amount of mucus resulting from the conversion into mucilage of the wall of the original mother-cell. Each cell of the colony is provided with a solid conical horn of large size (fig. 134 *A*). In *Sorastrum* the cells are slightly compressed and lunate, each angle being furnished with a pair of spines (fig. 134 *B* and *C*). From the back of each cell there projects a colourless process, and all the processes meet in the central region of the colony where they form the facets of a small sphere.

The number of cells in a colony varies from 8 to 64, but 16 is the most frequent number. There is a single massive chloroplast in each cell, practically filling up the whole cell-cavity, and in it there is usually a centrally placed pyrenoid. Multiplication occurs by the formation of autocolonies which are miniatures of the adult form when set free, their liberation being

[1] G. S. West, '04. The 'Cœlastraceæ' of Wille ('09) appears to be an unnatural group in which are included a number of genera having little close affinity with each other. Genera such as *Ankistrodesmus*, *Selenastrum*, *Actinastrum*, etc., are much less advanced types than *Cœlastrum* or *Sorastrum*.

by a split in the wall of the mother-cell in *Cœlastrum* (fig. 133 *D*) and *Sorastrum*, but by the conversion of this wall into mucilage in *Burkillia*.

The Cœlastreæ mostly occur in bogs and at the margins of pools and lakes, although some species of *Cœlastrum* are more particularly confined to the freshwater plankton.

The genera are: *Cœlastrum* Nägeli, 1849 [inclus. *Hariotina* Dangeard, 1889]; *Sorastrum* Kützing, 1845 [inclus. *Selenosphærium* Cohn, 1879]; *Burkillia* W. & G. S. West, 1907.

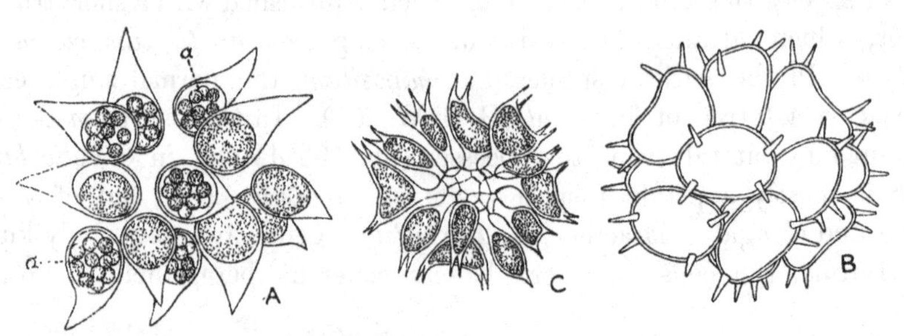

Fig. 134. *A*, *Burkillia cornuta* W. & G. S. West, ×455; *a*, young autocolonies. *B*, large cœnobium of *Sorastrum spinulosum* Näg. ×455. *C*, *Sorastrum Hathoris* (Cohn) Schmidle, ×530 (after Bohlin).

Family Chætopeltidaceæ.

The Algæ included in this family are distinguished from all other members of the Tetrasporineæ by the setæ or bristles which are attached to the cells. The plants are unicellular, or aggregates of cells, sometimes forming flat pseudo-parenchymatous expansions. For the most part they are attached, the cells showing a marked dorsiventrality.

In *Chætopeltis* (fig. 135), an Alga which lives attached to various aquatic macrophytes, the 'thallus' is flat and similar to that found in the Ulvaceæ except for its small size, and the fact that it is attached by the whole of one surface. A study of the development of this 'thallus' shows that it is not a concrescence of branches as usually described, but that cell-division may occur with considerable irregularity in any direction in one plane. The cells are embedded in a firm jelly, and simple elongated bristles usually grow out from the cell-walls of some of them. Reproduction generally takes place by quadriciliated zoogonidia, but biciliated isogametes are also known to occur.

Chætosphæridium (fig. 136 *A*—*C*) is much the most frequent genus and is found in stagnant water attached to various larger Algæ, or sometimes free-floating and unattached. The cells generally form irregular aggregates[1], sometimes enveloped in a copious mucus (*Ch. Nordstedtii*), at other times

[1] The statement that the cells of *Chætosphæridium* when mature are 'united by longer or shorter empty tubes to form a sympodial branch-system' requires some qualification. The author has examined hundreds of mature British and European specimens of this genus, and in no instance had such tubes been formed. The division of the cells was transverse, the lower cell simply sliding out during growth from under the upper one.

destitute of mucus (*Ch. Pringsheimii*). They are approximately spherical with a small conical sheath at the dorsal pole through which passes a long and delicate bristle. The fact that the conical projection is actually a sheath is very difficult to detect owing to the extreme fineness of the lumen. In *Dicoleon*, in which the cells are similar to those of *Chætosphæridium*, there is a double sheath at the base of the bristle, a longer inner sheath growing out of a shorter external sheath. In *Conochæte* (fig. 136 *D*) the cells possess several bristles, each of which has a thick, more or less gelatinous, basal sheath. In *Polychætophora* (W. & G. S. W., '03) the cells are loosely aggregated, or from 6 to 8 of them may form an irregular chain. Each cell has a very thick lamellate wall and is furnished with 8—12 long flexuose

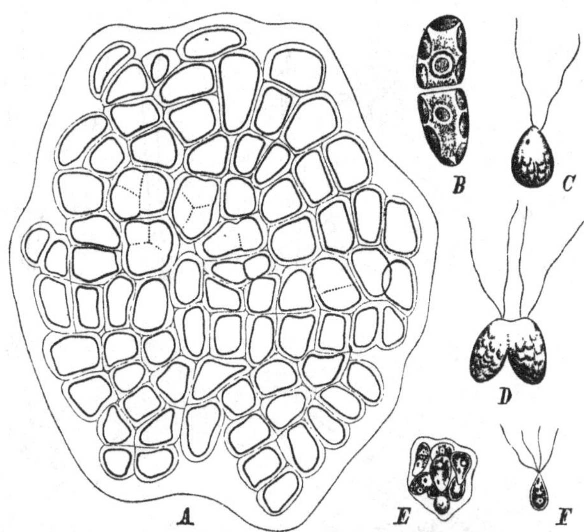

Fig. 135. *A—D*, *Chætopeltis minor* Möbius. *A* represents a complete colony, ×550; *B*, two cells showing the chromatophores and nucleus, ×700; *C*, gamete, ×950; *D*, fusion of gametes, ×950. *E* and *F*, zoogonidia of *Ch. orbicularis* Berth. ×540. *E*, zoogonidia not yet escaped from the surrounding vesicle; *F*, free zoogonidium. (*A—D*, after Möbius; *E* and *F*, after Berthold, from Wille.)

bristles entirely destitute of a sheath. In *Oligochætophora* the cell-wall is quite thin and there are 2—4 unsheathed bristles arising from the dorsal surface (fig. 136 *E—I*).

In all these Algæ the cells possess a parietal chloroplast (sometimes two in *Conochæte*) which in *Chætopeltis* is often much lobed and perforated. There is usually a single pyrenoid, and *Conochæte comosa* may store oil as a reserve (fig. 136 *D o*).

Multiplication of the cells occurs by simple division which in *Chætosphæridium* is transverse, the lower cell slipping out from under the upper one and at once developing a sheathed bristle.

The genera are: *Chætopeltis* Berthold, 1878 [inclus. *Myxochæte* Bohlin, 1894]; *Chætosphæridium* Klebahn, 1892; *Dicoleon* Klebahn, 1893; *Conochæte* Klebahn, 1893; *Diplochæte* Collins, 1901; *Polychætophora* W. & G. S. West, 1903; *Oligochætophora* G. S. West, 1911. *Nordstedtia* Borzi (1892) is a doubtful genus.

Some of the above genera have been regarded by various authors as reduced forms of the Ulotrichales. Their habit as colonial unicells and their methods of multiplication do not, however, lend much support to the view that they have originated from filamentous

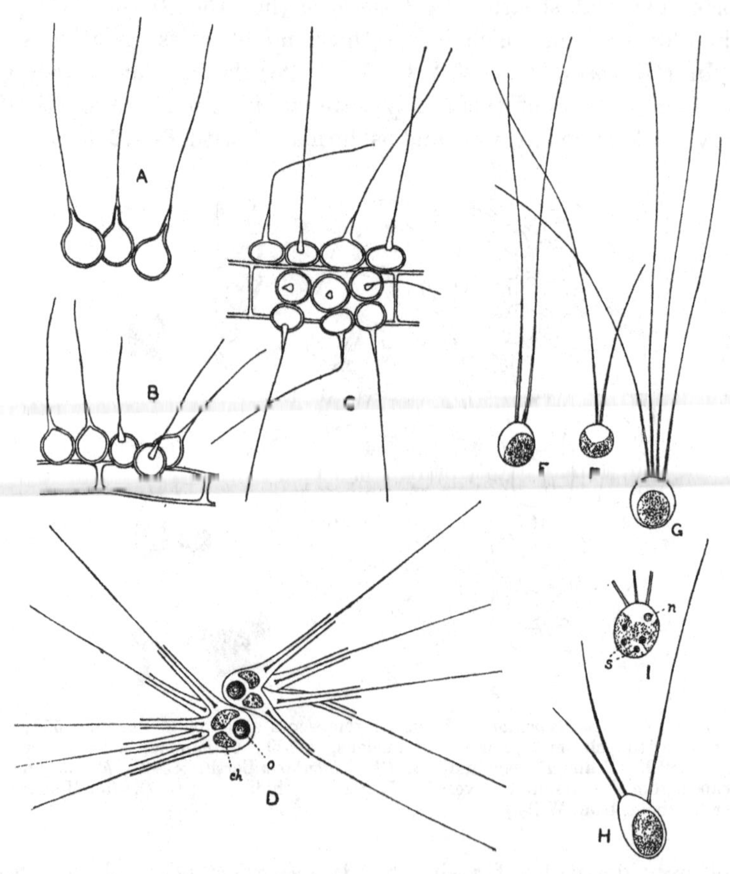

Fig. 136. *A* and *B*, *Chætosphæridium Pringsheimii* Klebahn. *C*, *Ch. Pringsheimii* var. *depressum* G. S. West. *D*, a form of *Conochæte comosa* Klebahn. *ch*, chloroplast; *o*, oil-globule. *E—I*, *Oligochætophora simplex* G. S. West; *I*, cell stained to show nucleus (*n*) and starch-grains (*s*). *A—D*, ×430; *E—I*, ×330.

forms. On the other hand, the sheathed bristle such as occurs in *Chætosphæridium*, *Conochæte* and *Dicoleon* is known elsewhere only in *Coleochæte*, the highest type of the Ulotrichales; and the curious unattached zoosporic forms of *Coleochæte* recently described by Lambert ('10 B), in which the thallus is reduced to a few irregularly grouped cells, might be said to throw light on the possible origin of the above-mentioned genera. There is, however, no evidence to show that the sheathed bristle is of any phylogenetic importance, and unsheathed bristles have certainly been developed independently in several groups of

Algæ since they occur in the Chætophoraceæ, in *Dicranochæte*, and in *Glœochæte*. It is probable that the Chætopeltidaceæ as here defined are a mixed assemblage, but our present knowledge of the included forms is insufficient to deal with them in a satisfactory manner.

Sub-order CHLOROCOCCINEÆ.

In this division of the Protococcales *vegetative cell-division does not occur. Reproduction takes place by zoogonidia and in some forms by isoplanogametes.* The zoogonidia on coming to rest immediately begin to grow into new individuals. The cells often grow to a considerable size and become simple cœnocytes by the division of the original nucleus. This non-motile vegetative phase is the dominant one, the cœnocytes assuming various forms. Most of the Chlorococcineæ are gregarious, but those belonging to the Hydrodictyaceæ consist of free-floating cœnobia of cœnocytes which arise by the apposition of the zoogonidia.

It has been suggested that the more complex cœnocytic types of the Siphonales and Siphonocladiales have been evolved from this group (Blackman & Tansley, '02), and although there is little direct evidence in support of this suggestion it is not improbable that the siphonaceous Algæ may have had such an origin. In the present volume the Chlorococcineæ is subdivided into two families as follows:—

Fam. *Planosporaceæ.* Cells or cœnocytes solitary or gregarious, for the most part fixed.

Fam. *Hydrodictyaceæ.* Cœnocytes united to form definite free-floating cœnobia.

Family **Planosporaceæ.**

This family is established to include all those non-cœnobic members of the Protococcales which are reproduced solely by zoogonidia or by isoplanogametes. It is quite a natural family, embracing a number of different genera which by virtue of their method of reproduction have in many cases established themselves as epiphytes and endophytes, the zoogonidia being able to swim direct to their destination and there grow.

Each individual is a single cell or cœnocyte, free-floating, epiphytic or endophytic. In *Chlorococcum* the aggregation of individuals may be so dense as to form a stratum. The outward form of the individual varies much, ranging from the globose cells of *Halosphæra* to the elongated stalked cells of *Characium*. In some of the endophytic forms, as in *Phyllobium*, long tubular outgrowths permeate the tissues of the host. *Dicranochæte* is unique in the possession of dichotomously branched bristles.

From 4 to 128 zoogonidia arise in a mother-cell, and are liberated either by the dissolution of the wall at some unlocalised point or by the separation of a distinct lid. In some of the Chlorochytrieæ isogametes occur, and the zygote germinates without any period of rest.

Sub-family CHLOROCHYTRIEÆ. As here constituted this group includes the genus *Chlorococcum* and those Algæ which have been placed for some time past in the 'Endosphæraceæ.' All of them are characterized by the rounded or ellipsoidal nature of their cells or cœnocytes, which are for the most part endophytic.

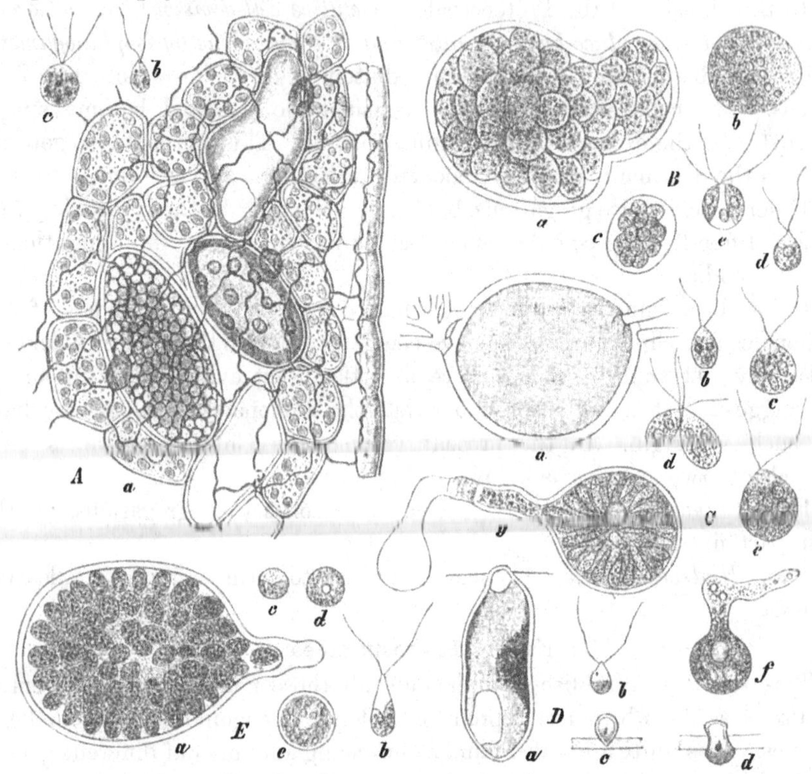

Fig. 137. *A*, *Chlorochytrium Lemnæ* Cohn; *a*, part of leaf of *Lemna trisulca*, with three *Chlorochytrium*-cells; a young cell showing pyrenoids, a fully-developed one, and an empty one; *b*, gamete; *c*, zygozoospore; *a*, ×360; *b* and *c*, ×720. *B*, *Chlorochytrium bienne* (Klebs) G. S. West; *a*, a large winter resting-cell which has divided into many cells; *b*, one of the cells; *c*, formation of gametes in this cell; *d*, gamete; *e*, zygozoospore; *a*—*d*, ×800; *e*, ×720. *C*, *Phyllobium dimorphum* Klebs; *a*, empty resting-cell; *b*, microgamete; *c*, megagamete; *d*, fusion of anisogametes; *e*, zygozoospore; *f*, germination of zygozoospore; *g*, young *Phyllobium*-cell; *a*, ×72; *b*—*g*, ×720. *D*, *Chlorochytrium Cohnii* (Wright) G. S. West; *a*, vegetative cell; *b*, gamete?; *c* and *d*, development of motile cells; all ×450. *E*, *Chlorochytrium paradoxum* (Klebs) G. S. West; *a*, resting-cell which has formed zoogonidia (*b*), ×360; *c*—*e*, germination-stages, ×720. (*D*, after Lagerheim; remainder after Klebs, from Wille.)

The best known genus is *Chlorochytrium* (fig. 137 *A* and *B*; fig. 138 *A*), the cells of which are ellipsoid, ovoid, or somewhat irregular, occurring either in the intercellular spaces or wedged in between the peripheral cells of aquatic macrophytes (*Ceratophyllum demersum, Elodea canadensis,* species of *Lemna*) and plants which grow in damp situations (*Mentha aquatica, Rumex obtusifolius, Lychnis Flos-cuculi,* etc.). Some species are marine, occurring in the peripheral portions of the thalli of various larger marine Algæ. Reproduction takes place by the formation of zoogonidia or isoplanogametes, or both. The contents of the cell become broken up by successive divisions, in a manner similar to that which occurs in *Characium*, into a large number of small zoogonidia or gametes. These are liberated either by a perforation in the wall of the mother-cell or by the gelatinization of the inner layer of the wall, which then becomes protruded as a large vesicle in which the motile cells 'swarm' for a short time. The biciliated zoogonidia or the quadriciliated 'zygozoospores' come to rest on the epidermis of the host-plant and germinate at once. Should the new plant be an endophyte, it penetrates the host either through a stoma or by forcing itself between two epidermal cells. Some of the vegetative cells become akinetes and pass the winter in that condition.

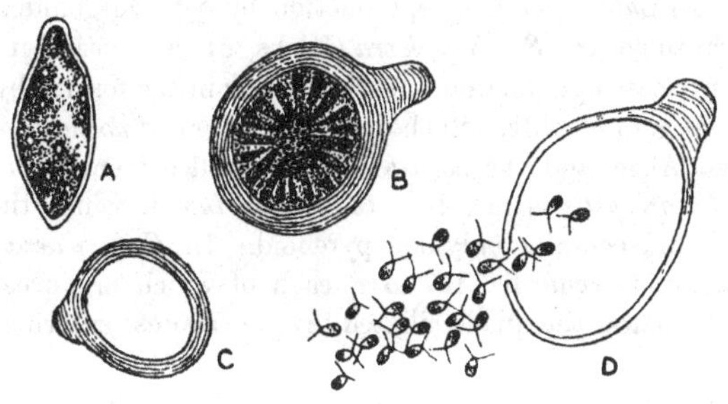

Fig. 138. *A*, young vegetative cell of *Chlorochytrium Lemnæ* Cohn, ×475. *B—D*, *Centrosphæra Facciolæ* Borzi; *B* and *C*, ×475; *D*, showing escape of zoogonidia (after Borzi), ×410.

The cell-wall is invariably thick and strong in the adult cells of all the members of this group. There is one extensive parietal chloroplast, usually lining the whole of the cell-wall and often with inwardly extending lobes, and containing one or many pyrenoids.

In *Chlorococcum* the cells, which live on damp ground or in water, are gregarious and often form a stratum. They grow much in size, become cœnocytic, and often angular by compression. From 8 to 32 zoogonidia arise in each mother-cell. The species of this genus are for the most part very imperfectly known. *Centrosphæra* is another gregarious form occurring as a rule unattached among various Algæ in the mud of pools and ponds. The cells are ovoid, with a thick lamellose cell-wall, provided with an asymmetrical button-like excrescence at one side (fig. 138 *B* and *C*). One species of this genus has recently been found as an epiphyte on the hairs of the leaves of *Callitriche*. Reproduction takes place by minute zoogonidia set free in large numbers from the mother-cells.

Phyllobium (fig. 137 *C*) is another endophytic genus in the leaves of moisture-loving plants such as *Ajuga* and *Lysimachia*, and one species (*P. sphagnicola*, G. S. W., '08) occurs on the leaves of *Sphagnum*. The plant-body is a branched cœnocyte, the branches traversing the intercellular spaces of the host-plant after the manner of *Phyllosiphon* among the Siphonales, and often becoming septate. Projecting from the surface of the leaf are numerous swellings of a bright green colour, which are resting akinetes. These are upwards of twenty times the diameter of the branched tubes from which they arise, and the chloroplast is a thin parietal layer from which numerous rod-shaped lobes (most of which possess a pyrenoid) radiate into the central cavity. Reproduction occurs in this genus by zoogonidia and also by anisogametes (fig. 137 *C b—e*), the zygote developing at once into a young *Phyllobium*-plant (fig. 137 *C f* and *g*).

The so-called 'genera' *Endosphæra*, *Scotinosphæra*, *Chlorocystis* and *Stomatochytrium*, should all be submerged in *Chlorochytrium*, as they are discriminated from each other and from that genus by the most trivial characters, not one of which can be regarded as of generic importance. In *Chlorochytrium* reproduction by both zoogonidia and gametes has been repeatedly shown to occur. *Scotinosphæra* (Klebs, '81) is identical in all respects with *Chlorochytrium*, except that reproduction is said to occur in the former by isogametes only and in the latter by zoogonidia only. Such distinctions *are of no taxonomic value* among lower forms of Green Algæ, and the separation of so-called 'genera' on such characters cannot be upheld. *Chlorocystis* is merely a *Chlorochytrium* in which the chloroplast is a little more restricted and contains only one pyrenoid. In *Endosphæra* the resting-cells form numerous gametangia (*vide* fig. 137 *B a*) each of which produces 8—10 biciliated gametes which fuse in pairs, the quadriciliated 'zygozoospores' entering the intercellular spaces of the host.

All the more recent work on this group of Algæ has shown that Cohn and other previous investigators were wrong in regarding certain forms as parasites, and that not merely can the endophyte live quite independently of its host-plant, but that in most cases the latter receives no injury beyond that which may be caused by the little mechanical pressure exerted by the growing endophyte. Freeman ('99) has suggested that the general biological conditions under which *Chlorochytrium* lives lend themselves to the development of parasitism, and that the allied genus *Phyllobium* is progressing in that direction[1]. *Chlorochytrium* (=*Chlorocystis*) *Sarcophyci* is the cause of deformities in the thallus of the seaweed *Sarcophycus* (Whitting, '93).

The genera are: *Chlorococcum* Fries, 1825 [=*Cystococcus* Nägeli, 1849]; *Chlorochytrium* Cohn, 1874 [inclus. *Endosphæra* Klebs, 1881; *Scotinosphæra* Klebs, 1881; *Chlorocystis* Reinhard, 1885; and *Stomatochytrium* Cunningham, 1888]; *Phyllobium* Klebs, 1881; *Centrosphæra* Borzi, 1883; ? *Dictyococcus* Gerneck, 1907.

Sub-family Dicranochæteæ. This sub-family includes only the single genus *Dicranochæte* Hieronymus (1892), which differs from all the other members of the Planosporaceæ in the possession of bristles. The latter are also of a unique character, being dichotomously branched; bristles of such a nature are entirely unknown in any other group of Algæ (consult fig. 139). The cells are solitary and attached to the leaves of submerged species of *Sphagnum* and *Hypnum*; they are uninucleate and possess a single massive chloroplast with or without pyrenoids. In *D. reniformis* Hieronymus ('92)

[1] The Archimycetes (Chytridieæ) show many striking resemblances to *Chlorochytrium*.

the bristle (or seta) arises from the base of the cell and passes upwards along a lateral groove, but in *D. britannica* G. S. West ('12) it is dorsal in its attachment. Reproduction occurs only by biciliated zoogonidia (fig. 139 *C*), formed 4 to 32 in a mother-cell, and liberated in *D. reniformis* by the detachment of a special lid.

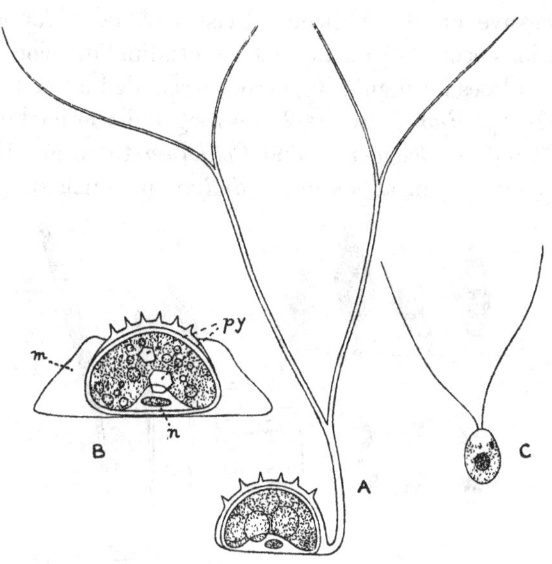

Fig. 139. *Dicranochæte reniformis* Hieronymus. ×1170 (after Hieronymus). *A*, cell showing dichotomously branched bristle. *B*, cell showing the cushion of mucilage (*m*) in which it is lodged and by which it is attached; *n*, nucleus; *py*, pyrenoids. *C*, zoogonidium.

Sub-family CHARACIEÆ. The Algæ included in this small group are unicellular and generally occur as epiphytes, either solitary or in clusters, on other larger Algæ. The vegetative cells are rounded or angular in *Sykidion* (fig. 141 *F* and *G*), but in the other genera they are distinctly stalked, in which case there is a differentiation into base and apex. The largest cells (up to 1 mm. in length when mature) are met with in *Codiolum*, in which the cell-body is ovoid and the base drawn out into a solid stalk of some length. In *Characium* (fig. 140) the cells are ovoid or fusiform, sometimes of considerable length, and in most cases asymmetrical. The basal stalk is in some species so short that the cell appears to be sessile, whereas in others it is long and slender. It is provided with a disc for attachment to the host, or more rarely with minute rhizoidal outgrowths. The apex of the cell is often acuminate and may be drawn out into a long apiculus.

Each cell contains a single parietal chloroplast, except in *Characiella,* in which the chloroplast is axile with short radiating processes and a central pyrenoid. In *Characium* the chloroplast is very massive, filling most of the cell, but in *Sykidion* it is more or less cup-shaped. A single pyrenoid is present in each of these genera. In *Codiolum* the chloroplast is reticulate with radial ingrowths, and contains several pyrenoids. In none of the genera is there more than one nucleus.

Vegetative cell-division is unknown. Reproduction occurs by the formation of zoogonidia through the division of the contents of the mother-cell, by either successive or simultaneous division-planes. As a rule several transverse divisions occur before the first longitudinal division, and in a short time each portion loses its angularity, becomes rounded off and forms an ovoid zoogonidium. In *Sykidion* there are 2 or 4 zoogonidia formed in each mother-cell (fig. 141 *F*), but in *Characium* and *Codiolum* there are 16 or 32. The zoogonidia are biciliated in all except *Codiolum* in which they are furnished

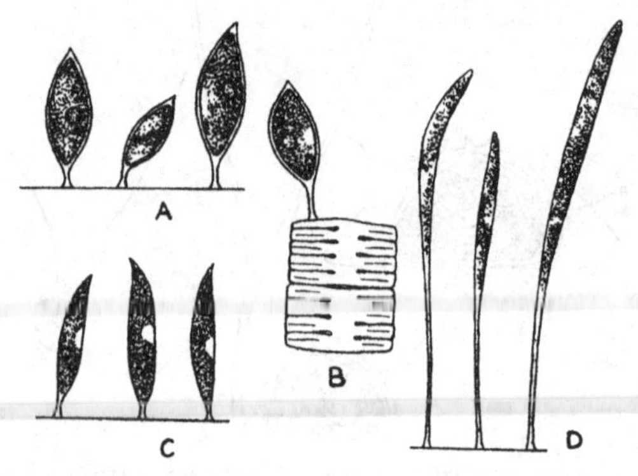

Fig. 140. *A* and *B*, *Characium Pringsheimii* A. Br. *B* is attached to a cell of *Tabellaria flocculosa*. *C*, *Ch. subulatum* A. Br. *D*, *Ch. ensiforme* Herm. All × 520.

with 4 cilia. The zoogonidia escape by either a terminal or a lateral (fig. 141 *D*) aperture, and on coming to rest at once germinate to form new plants. In some species the pyrenoid disappears during the formation of zoogonidia, pyrenoids being developed *de novo* in the young plants. In *Codiolum* biciliated gametes have been recorded, but they have not been observed to conjugate and must be regarded as doubtful. Aplanospores are known in *Sykidion* (fig. 141 *G*) and in *Codiolum*.

Characium is a common genus of freshwater Algæ, occurring as an epiphyte on various species of *Œdogonium*, *Vaucheria*, *Cladophora*, and other filamentous Algæ, often so thickly as completely to hide the host-plant. The gregarious habit is, of course, associated with the method of reproduction, large numbers of zoogonidia coming to rest in one place and germinating simultaneously. Much the largest species of the genus seems to be *Ch. graciliceps* Lambert ('10 A), which is epizootic on the Phyllopod *Branchipus vernalis*, the cells attaining a length of 480 μ (fig. 141 *A* and *B*).

Codiolum is marine or inhabits brackish water, occurring as a thin stratum on rocks, stones, the piles of harbours, and as a gregarious epiphyte on larger marine Algæ. *Sykidion* (with the possible exception of one species) is also a marine epiphyte.

The genera are: *Characium* A. Braun, 1849; *Codiolum* A. Braun, 1849; *Sykidion* Wright, 1879; *Characiella* Schmidle, 1903.

[It is necessary to mention here the genus *Characiopsis* founded by Borzi in 1895 to include a number of Algæ previously described as species of *Characium*. It is distinguished by the possession of two or more parietal chromatophores, the absence of pyrenoids, and the storage of oil as a reserve ; it thus belongs to the Heterokontæ.]

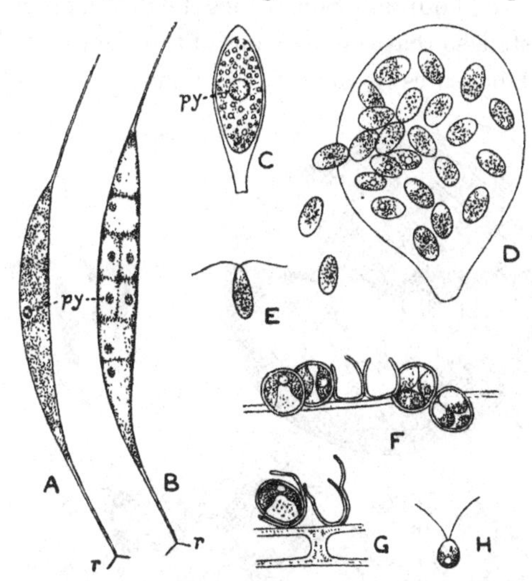

Fig. 141. *A* and *B*, *Characium graciliceps* Lambert, ×540 ; *A*, typical specimen ; *B*, showing one of the early stages in the division of the protoplast; *py*, pyrenoids; *r*, 'rhizoids.' *C—E*, *Characium Sieboldi* A. Br.; *C*, vegetative cell; *py*, pyrenoid; *D*, large cell from which zoogonidia are escaping laterally; *E*, zoogonidium; all ×about 540. *F—H*, *Sykidion Droebakense* Wille; *F*, formation of zoogonidia; *G*, showing formation of aplanospore; *H*, zoogonidium; all ×513. (*A* and *B*, after Lambert; *C—E*, after A. Braun; *F—H*, after Wille.)

Sub-family HALOSPHÆREÆ. This sub-family includes only the one genus *Halosphæra* Schmitz (1878), an Alga which is confined to the marine plankton. The cells are large and spherical (fig. 142 *1* and *2*), attaining a diameter of 600 μ, and float freely in the sea. The cell-wall is thin and the cytoplasm is mostly in a parietal layer. There is one nucleus, usually in the parietal cytoplasm but occasionally occupying a central position, in which case the small amount of cytoplasm surrounding it is connected by radiating strands with the parietal layer. In the young cell there are numerous parietal discoidal chloroplasts, but in older cells these apparently fuse to form a reticulum[1]. As the cells grow in size there is a repeated ecdysis of the cell-wall.

Reproduction takes place by biciliated zoogonidia of conical shape, the cilia being attached to the broad base of the cone, which is therefore anterior. This anterior end is also lobed at the periphery (fig. 142, *4*). The formation of zoogonidia commences by the repeated division of the nucleus, the numerous

[1] A similar condition exists in many genera of the Siphonocladiales, where the reticulated chloroplast is really formed by a fusion of a number of small parietal plates, which may in some cases be quite separate.

nuclei remaining in a peripheral position. The cytoplasm and chloroplasts then become aggregated around each nucleus, forming bodies of a plano-convex character which after division of the nucleus become constricted into two parts, each part forming a zoogonidium. Sometimes the development of motile reproductive cells is arrested, so that the divisions of the original protoplast result in the formation of aplanospores (generally 16 in the mother-cell; Cleve, '98).

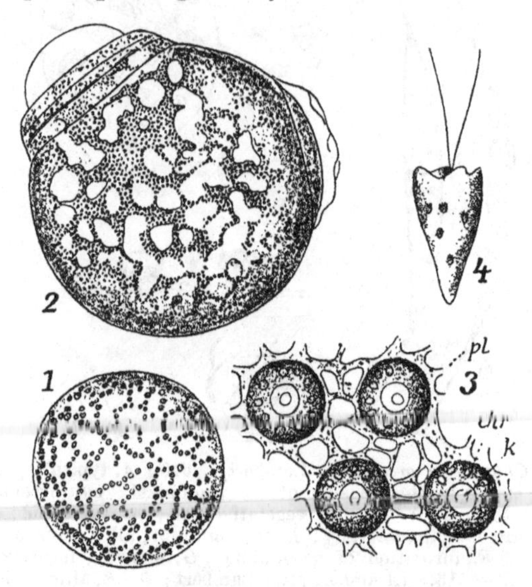

Fig. 142. *Halosphæra viridis* Schmitz. *1* and *2*, vegetative cells. *3*, part of the peripheral plasma during the formation of zoogonidia; *chr*, chloroplast; *pl*, cytoplasm; *k*, nucleus. *4*, zoogonidium. (After Gran and Schmitz, from Oltmanns.)

Halosphæra viridis Schmitz is really an inhabitant of the warmer temperate seas, although in the Atlantic Ocean it is carried northwards by the Gulf Stream. It is not unlikely that *Sphæra Kerguelensis* Karsten, described from the Antarctic Ocean, and *Pachysphæra pelagica* Ostenfeld, from the N. Atlantic, are developmental stages of *Halosphæra*.

Family **Hydrodictyaceæ.**

In this family of the Chlorococcineæ the plant-body consists of non-motile cœnobia of cœnocytes, floating freely in the water. The cœnocytes, which are of very varied external form, are disposed so as to form a flat plate in *Pediastrum* and arranged in the manner of a net in *Hydrodictyon*. In the disc-like cœnobium of *Pediastrum*, in which there may be more than 100 cœnocytes, those of the marginal series differ in external characters from those in the centre; but in the net-like cœnobium of *Hydrodictyon* they are all of the same cylindrical shape and many hundreds may be united to form a large irregular colony.

The normal method of reproduction is by the formation of new cœnobia by the close apposition of biciliated zoogonidia which have become quiescent. Every cœnocyte in the colony is capable of becoming a zoogonidangium, and since vegetative cell-division does not occur, the number of zoogonidia produced in it determines the number of cœnocytes in the daughter-cœnobium. In *Euastropsis,* however, several bi-cellular cœnobia arise in each mother-cell. This type of reproduction by the apposition of quiescent zoogonidia to form new cœnobia at once distinguishes the Hydrodictyaceæ from all other families

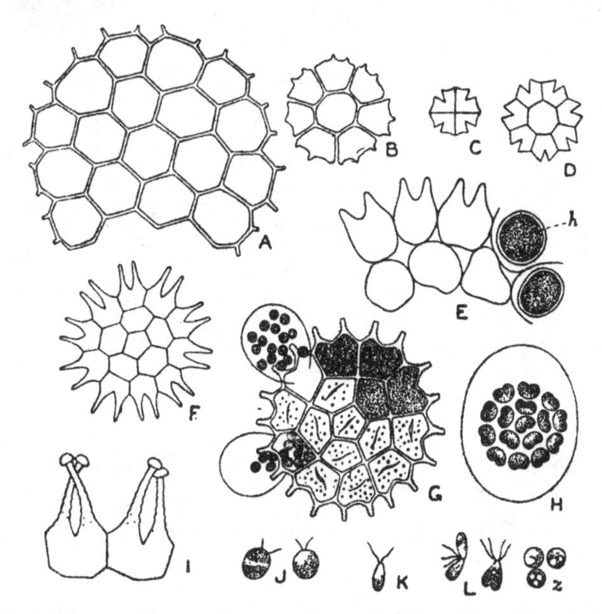

Fig. 143. *A*, *Pediastrum integrum* Näg. ×475. *B*, *P. tricornutum* Borge, ×475. *C* and *D*, *P. Tetras* (Ehrenb.) Ralfs, ×475. *E*, *P. duplex* Meyen, ×475; *h*, hypnospore. *F—H*, *P. Boryanum* (Turp.) Menegh.; *G*, showing escape of zoogonidia into vesicles; *H*, young cœnobium formed by apposition of quiescent zoogonidia; *F*, ×475; *G* and *H*, ×220 (after Kerner). *I*, two marginal cells of *P. glandulifera* Bennett, ×475. *J—L*, zoogonidia and gametes of *P. Boryanum* (after Askenasy); *J*, zoogonidia and *K*, gamete, ×500; *L*, conjugation of gametes to form zygospores (*z*); gametes ×730, zygospores ×220.

of the Protococcales. From the cœnobic forms of the Autosporaceæ they are distinguished by the cœnocytic character of the cœnobium, and by the production of zoogonidia, although it must not be forgotten that the autocolonies of the Autosporaceæ arise by the apposition of non-motile gonidia which may very likely be zoogonidia which have lost the power of swarming. This seems very probable since in *Pediastrum* itself the gonidia are sometimes non-motile and do not 'swarm,' but simply arrange themselves in position to form the daughter-cœnobium. Aplanospores, which are resting-spores (hypnospores), occur in the genus *Pediastrum.* Isoplanogametes are also known to occur, fusing in pairs to form zygotes the germination of which is indirect.

It is quite possible that the Pediastreæ and the Hydrodictyeæ have no very close affinity, but the phylogenetic relationships of these Algæ are at present very obscure.

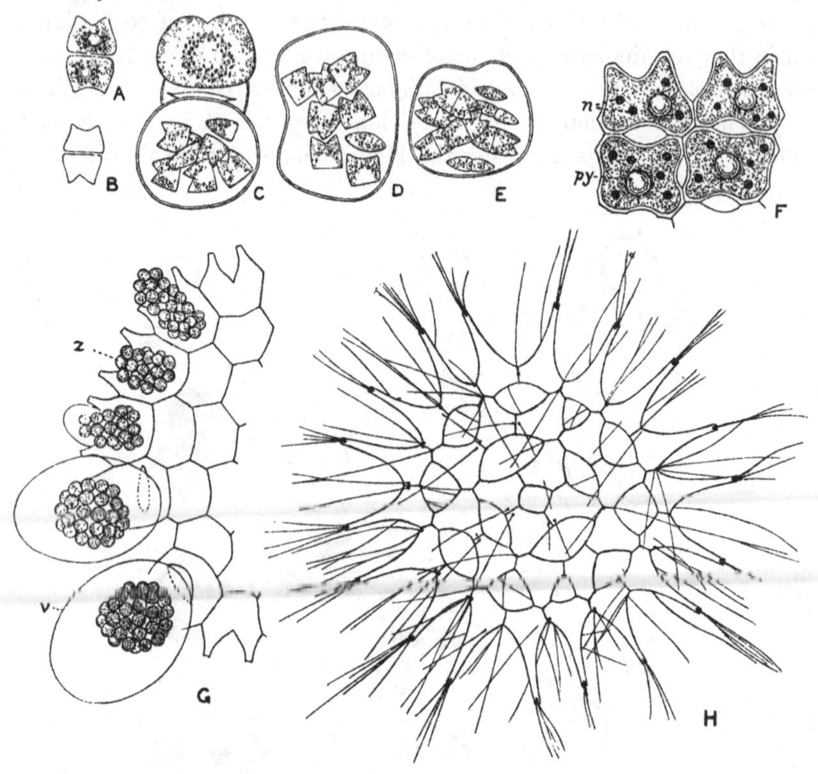

Fig. 144. *A—E*, *Euastropsis Richteri* (Schmidle) Lagerh. *A* and *B*, ×633; *C—E*, showing formation of young cœnobia, very highly magnified (after Lagerheim). *F*, four cells of a variety of *Pediastrum duplex* Meyen, stained to show the nuclei (*n*) and the pyrenoid (*py*), ×580. *G*, part of cœnobium of *P. Boryanum*, showing escape of zoogonidia (*z*) into the vesicles (*v*) just before swarming. *H*, *P. simplex* Meyen var. *clathratum* (Schroet.) Chod., showing the numerous fine bristles stained by Löffler's cilia-staining method, ×316 (after Petersen).

Sub-family PEDIASTREÆ. The Algæ of this sub-family are microscopic in size, consisting of a number of small cœnocytes firmly united to form a flat, disc-like cœnobium. The principal genus is *Pediastrum* (figs. 143 and 144 *F—H*), with a world-wide distribution in ponds, pools and lakes. The cœnobium is always free-floating, consisting of a single layer of small cœnocytes[1]. Either these are closely united or there are perforations of variable size between them which give the cœnobium a sieve-like aspect (fig. 144 *H*). In *Pediastrum simplex* var. *reticulatum* and *P. duplex* var. *reticulatum* the perforations of the plate-like colony are much larger than the cœnocytes

[1] In certain monstrous forms the cœnobium may consist of a double layer of cœnocytes in certain parts, or it may sometimes be very irregular. (Consult G. S. W., '07.)

themselves. In all known forms of the genus the cœnocytes forming the periphery of the disc differ in shape from the remainder, generally in the possession of marginal processes. The number of cœnocytes in a cœnobium varies from 4 in one of the common forms of *Pediastrum Tetras* (fig. 143 *C*) to 128 in some of the large forms of *P. duplex*. The few recorded instances of cœnobia of two cœnocytes probably refer to *Euastropsis*. The cœnocytes are often arranged in fairly distinct rings around a central one, 8, 16, 32, or more being the number constituting the cœnobium. Nägeli ('49) long ago pointed out that the cœnobia were usually constructed as follows :—colony of $8 = 1 + 7$; colony of $16 = 1 + 5 + 10$; colony of $32 = 1 + 5 + 10 + 16$; but this arrangement is not always observed and there are many irregularities due primarily to the death of one or more of the gonidia. In *P. Tetras* this sometimes results in a curious cœnobium of 3 cœnocytes (G. S. W., '07).

Each cœnocyte contains from 3 to 6 nuclei, and a massive parietal chloroplast with one pyrenoid (fig. 144 *F*).

Some species of *Pediastrum* are furnished with long bristles which play the part of a buoyancy apparatus and augment the floating capacity of the cœnobia (fig. 144 *H*). The bristles are rigid and elastic, and are mostly attached in tufts, largely at the ends of the marginal processes and therefore in the plane of the cœnobium. There are, however, many other tufts and also solitary bristles, attached to the inner cœnocytes, which project at right-angles to the plane of the cœnobium (Petersen, '11). The bristles are very delicate, but may sometimes be seen in plankton-specimens of *P. Boryanum* and *P. duplex* without any special treatment and with ordinary illumination. Dark-ground illumination usually shows them up very well. Petersen has shown that in *P. simplex* they are seasonal, disappearing during the winter.

The normal method of reproduction is by the successive division of the contents of a cœnocyte to form a number (4, 8, 16, 32, etc.) of zoogonidia, which are liberated into a delicate external vesicle through a slit in the wall of the mother-cœnocyte (fig. 143 *G*). The zoogonidia swarm in the vesicle for a time and then become quiescent, arranging themselves in one plane as a new cœnobium. In many cases, however, the gonidia do not appear to 'swarm'; they pass into the vesicle but are not ciliated and exhibit no movements except that they gradually arrange themselves as a plate. This fact is of considerable interest as it shows how the autocolonies of certain of the Autosporaceæ may have arisen. In fact, this method of reproduction in *Pediastrum* is really by autocolonies, and there is no doubt in this case that it is merely a modification of the normal method due to a suppression of the cilia.

Resting aplanospores (hypnospores) are not infrequently observed in certain species of *Pediastrum* (fig. 143 *E*). Their formation and subsequent germination were worked out by Chodat and Huber ('95).

Askenasy ('88) described the formation of biciliated gametes, much smaller than the zoogonidia, which conjugated in pairs (fig. 149 *K* and *L*) to form polyhedral zygotes. The latter then gave rise to new cœnobia by the segmentation of their contents. These observations have not been confirmed.

In the genus *Euastropsis*, originally established by Lagerheim ('94), the cœnobium consists of two flattened cells closely attached along their straight inner margins, the outer margins being widely notched (fig. 144 *A* and *B*). The entire cœnobium has a superficial resemblance to a minute species of the Desmidian genus *Euastrum*, and was, in fact, originally described as such. Each cell has one large parietal chloroplast with a single pyrenoid. Lagerheim states that there is one nucleus in each cell, but suggests that there may be more. Reproduction occurs by ovoid zoogonidia which swarm in a vesicle as in *Pediastrum*. From 2 to 32 of these are formed by successive divisions of the contents of the mother-cell. On becoming quiescent they arrange themselves in pairs, each pair gradually developing into an adult cœnobium (fig. 144 *C*—*E*). Thus, in this genus, as many as 16 daughter-cœnobia may be produced from one mother-cell.

The only genera are *Pediastrum* Meyen, 1829, and *Euastropsis* Lagerheim, 1894. The former genus occurs more particularly in small ponds and ditches, and not infrequently in quiet bog-pools. Certain forms are constant and regular constituents of the freshwater plankton. *Euastropsis* occurs only in bogs and is a very rare Alga.

Sub-family HYDRODICTYEÆ. Of this sub-family only the one genus *Hydrodictyon* Roth (1800) is known. The cœnobium is macroscopic, attaining a length of 20 centimetres, and consists of a net-like sack floating freely in the water. The meshes of the net vary much in size, and each one is bounded by five or six large cylindrical cœnocytes, the angles being formed by the junction of three cœnocytes (fig. 145 *A* and *B*). The protoplasm of each cœnocyte forms a fairly thick lining layer containing many nuclei, the central part of the segment being occupied by a large sap-vacuole. In *Hydrodictyon reticulatum* there are no definite and distinct chloroplasts. Both Artari ('90) and Klebs ('91) were wrong in describing a reticulated chloroplast, Timberlake ('01) having shown by means of sections that the chlorophyll is evenly distributed throughout the peripheral protoplasm of the cœnocyte. Numerous pyrenoids are present, and Timberlake has shown that they are directly the seat of the processes resulting in starch-formation.

Reproduction occurs normally by the formation of a very large number (7,000 to 20,000) of zoogonidia within the mother-cœnocyte, which swarm while still within the parent-wall and then become quiescent. They at once form a reticulated daughter-cœnobium by the apposition of their extremities (fig. 145 *C*). The old wall is then ruptured and the young cœnobium is set free. The zoogonidia are biciliated with one nucleus and a single pyrenoid; they are formed by the breaking-up of the contents of the mother-segment into large multi-nucleated masses, which in turn become subdivided into smaller masses, until each portion contains only one nucleus. Reproduction also occurs by isoplanogametes, which escape from the mother-cœnocyte by a lateral pore. The gametes, which are smaller than the zoogonidia, are formed in much greater numbers, and their escape is preceded by a swelling of the inner layer of the cœnocyte-wall. This layer ruptures the outer layers and protrudes as a large vesicle in which the gametes swarm, after which they conjugate in pairs, forming spherical zygotes. The latter rest for a short period and then produce two or four large biciliated zoogonidia, which on

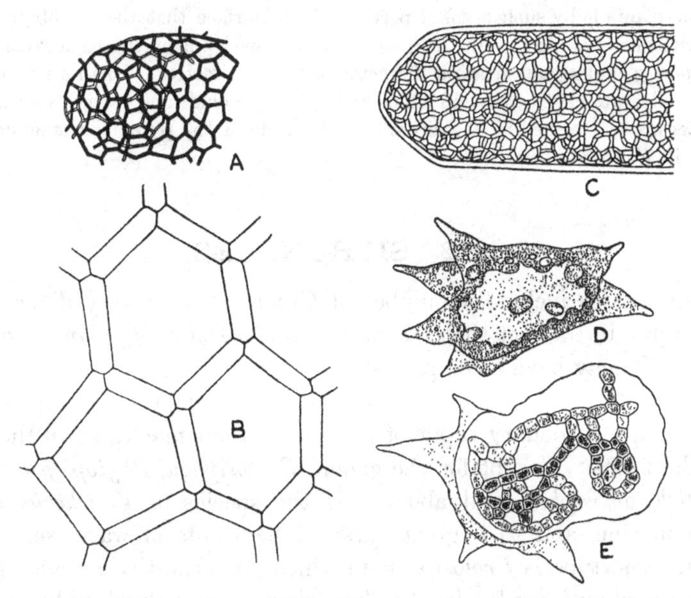

Fig. 145. *Hydrodictyon reticulatum* (L.) Lagerh. *A*, nat. size; *B*, small portion of a young colony, ×110; *C*, part of a large cœnocyte containing a very young colony, ×110; *D*, polyhedral form assumed by the biciliated zoogonidia which arise on the germination of the zygote; *E*, young colony formed by the apposition of quiescent zoogonidia which arise within the polyhedral cell. *D* and *E*, ×about 300 (after Pringsheim).

coming to rest assume a polyhedral form (fig. 145 *D*). The repeated division of the contents of this polyhedral body results in the formation of numerous zoogonidia which by apposition give rise to a new net-like cœnobium (fig. 145 *E*).

Hydrodictyon reticulatum has a wide distribution in both the eastern and western hemispheres, generally occurring in ponds and rivers. In the River Nile the young cœnobia float down the stream in myriads during the flood. The adult cœnocytes are 4—5 mm. in length just before they become zoogonidangia.

Klebs ('96) conducted experiments of much interest on *Hydrodictyon reticulatum.* He showed that the transference of cœnobia, which had been cultivated in a weak nutritive solution under bright illumination, to pure water resulted in the formation of zoogonidia, but in darkness the formation of zoogonidia could not be thus induced. In the case of cœnobia cultivated in sugar solutions (maltose) and then transferred to pure water, they formed zoogonidia if exposed to light, but gametes if kept in darkness. If, however, the cœnocytes had a strong disposition to form zoogonidia, then zoogonidia were formed in both light and darkness, and similarly cœnobia with a strong disposition to gamete-formation formed gametes whatever the nature of the illumination. The contest between the propensities toward zoogonidium-production and gamete-production appears to be perpetual, and the balance between the operative stimuli must be very delicate.

Quite recently another species, *H. africanum*, has been described by Yamanouchi ('13). This had developed in a culture of some soil obtained from near Cape Town, S. Africa. In this species the cœnocytes are for the most part ellipsoidal and about 60 of them form a very irregular net. They have a remarkable turgidity, and the connection between them in the larger plants is by such a small portion of the surface that the cœnobium cannot be moved without breaking it. In the final stages of development the cœnocytes become solitary, almost globular, and attain a diameter of 1·5 centimetres. In this species there are numerous definite chloroplasts, in which there are pyrenoids and reserve starch-grains, but the latter are not formed by the direct fragmentation of the pyrenoids as described by Timberlake in *H. reticulatum*.

Order 2. **SIPHONALES.**

The Siphonales include a number of Green Algæ of very diverse form, which all agree in the fact that *the entire plant consists of a single cœnocyte*, even though it may have a complex structure and attain a length of 40 cms. True septa only occur in connection with the formation of reproductive organs, although secondary septa of a peculiar kind may arise in the life of some of the forms. Except for the genera *Protosiphon*, *Phytophysa*, *Phyllosiphon*, *Dichotomosiphon*, and about half the species of *Vaucheria*, all the forms are marine, and to a great extent inhabitants of warm seas.

The simplest type is *Protosiphon*, in which the primitive rounded form of the cell is largely retained. In the Vaucheriaceæ and Phyllosiphonaceæ the thallus is filamentous and somewhat branched; in *Bryopsis* and *Caulerpa* the branching is much more extensive and exhibits a wonderful symmetry; in *Halimeda*, *Penicillus*, and others the thallus is a dense aggregation of branches bound together by a deposit of calcium carbonate; in *Codium* the branches are elongate hypha-like filaments which are densely interlaced to form a branched thallus of some solidity. It is this extraordinary complication of a single cœnocyte which is the distinguishing feature of most of the Siphonales; and of the known forms, those embraced in the genus *Caulerpa* may be regarded as amongst the most remarkable of all Green Algæ. The plants are in nearly all cases attached by strongly developed hold-fasts (*haptera*).

The protoplasm forms a parietal layer within the wall of the cœnocyte and contains numerous nuclei. Thus, the whole thallus consists of many protoplasts which are not individually separated by cell-walls but are all enclosed within a common wall. Numerous disc-shaped or lens-shaped chloroplasts occur in the lining layer of protoplasm, although in the complex forms they are of course mostly located in the more peripheral branches. In some of the Siphonales the nuclei are not infrequently larger than the chloroplasts.

Multiplication occurs in some forms by larger or smaller proliferous shoots which become detached, or by mere fragmentation of the thallus. Asexual reproduction takes place by non-motile gonidia (aplanospores), and by zoogonidia which arise in considerable numbers in zoogonidangia. In *Vaucheria* the zoogonidium is a large compound structure and is solitary, escaping from a terminal zoogonidangium. Gamogenesis occurs in some forms by the fusion of isogametes or anisogametes. In the family Vaucheriaceæ there are highly differentiated sexual organs.

It is not unlikely that the Siphonales have been evolved from the Protococcales by a further extension of the cœnocytic forms of the Chlorococcineæ. The genus *Protosiphon* is an example of an Alga which, after all, is not very far removed from *Phyllobium* among the Chlorochytrieæ, and it may possibly be a relic of those early forms in which originated the tendency which finally led to the Siphonales. It is exceedingly probable that *Halicystis* is the type from which *Valonia* has originated.

The view put forward by Oltmanns that the Siphonales are probably derived from the Siphonocladiales by the suppression of the septa, although deserving of consideration, suggests a line of evolution which appears very improbable.

There are a number of undoubted fossil siphonaceous Algæ, the calcified thallus of various forms having lent itself to preservation.

Family **Protosiphonaceæ.**

The principal genus of the family is *Protosiphon* Klebs (1896), which appears to be the simplest known form of the Siphonales. The cœnocytes, which, like those of *Botrydium*, live on damp earth, are bladder-like or subtubular (averaging about 100 μ in diameter) with a narrow, elongated, rhizoidal extension passing down into the soil. There is a single reticulate parietal chloroplast containing several pyrenoids. Starch is the principal food-reserve, and numerous nuclei occur in the lining layer of cytoplasm. The plants multiply by the budding off of new individuals from the green part of the older parent-cœnocytes, or by the division of the young cœnocytes into 4, 8, or 16 daughter-cœnocytes. Under certain conditions the protoplasmic contents become divided into a number of rounded 'cysts' (probably aplanospores) which turn red and rest for a time. They may germinate direct into new plants or they may form biciliated gametes. The ordinary vegetative cœnocytes may also become gametangia and produce ovoid isogametes like those formed from the aplanospores. The gametes fuse in pairs, producing a star-shaped zygote. After a period of rest the zygote grows direct into a new vegetative individual. Gametes which do not conjugate form parthenospores which soon grow into new plants. It would seem that the motile cells are probably of the nature of zoogonidia which are also facultative gametes.

Wille has placed *Protosiphon* alongside *Botrydium* in the 'Hydrogastraceæ'; and Oltmanns includes *Blastophysa* along with it in the Protosiphonaceæ, which he places as a family of the Protococcales. There is only one known species, *Protosiphon botryoides* (Kütz.) Klebs.

It seems natural to include the genus *Halicystis* (Areschoug, 1850) in the family Protosiphonaceæ. It is a small genus of marine Algæ in which the thallus consists of a single round or ovoid, bladder-like cœnocyte, about the colour and size of a green grape, having a short basal stalk terminating in a minute disc by which the plant is attached to its substratum. The substratum consists of shells or stones, or sometimes the disc bores its way into the calcified crusts of *Lithothamnion* (Kuckuck, '07). No rhizoids are formed. There is but one cavity, that of the bladder-like vesicle passing downwards into the cylindrical stalk. The wall is very finely stratified, and lining its interior there is a delicate layer of protoplasm containing numerous small nuclei and chloroplasts. The greater part of the volume of the lumen is filled with cell-sap. The chloroplasts are rounded or oval discs, wholly destitute of pyrenoids but containing starch. The nuclei are also flattened discs, scattered irregularly or sometimes in pairs through the entire lining layer of protoplasm (Murray, '93).

Zoogonidia are formed in the upper part of the cœnocyte, the protoplasm of the lower part remaining sterile, although there is no separating wall formed between the two portions. Both macro- and microzoogonidia have been observed by Kuckuck ('07). They are biciliated and escape through one or more apertures in the old wall. The microzoogonidia may perhaps also function as isogametes.

Halicystis has usually been placed alongside *Valonia* in the Valoniaceæ, but Schmitz, and also Murray, have given cogent reasons for its inclusion in the Siphonales. Although *Halicystis* bears a striking resemblance to certain species of *Valonia*, it differs in the more irregular disposition of the nuclei, in the rounded chloroplasts without pyrenoids, in the absence of rhizoids, in the feeble stratification of the wall, and in the fact that it never becomes at any time more than a single cœnocyte. *H. ovalis* (Ag.) Aresch. extends from subtropical regions into the N. Atlantic and North Sea. *H. parvula* Schmitz is confined to the Mediterranean.

Family **Chætosiphonaceæ.**

This family includes only the genus *Chætosiphon* Huber ('93) of which there is but one known species, *Ch. moniliformis*, an alga occurring as an endophyte in the dead and discoloured leaves of *Zostera marina*. The thallus consists of a richly branched cœnocyte, rather deeply constricted at frequent intervals, but without transverse walls except in connection with the zoogonidangia. The branches, which are from 10 to 15 μ in diameter, ramify through the tissues of the leaf, traversing both cells and intercellular spaces. Near the epidermis of the leaf the branches become swollen in a moniliform manner, attaining a diameter of 30 μ or more, and those which are in the epidermis often develop elongated tubular hairs (4—5 μ in diameter), devoid of any colour, which stand out more or less at right angles to the surface of the leaf. There is a parietal layer of protoplasm enclosing a central vacuole which is only interrupted at the constrictions of the cœnocyte. The chloroplasts are very numerous and discoidal, with polygonal outlines, and each one contains a small central pyrenoid. The nuclei are fairly numerous and large, each with a conspicuous nucleolus; they occur at scattered intervals in the lining layer of protoplasm and are always internal to the chloroplasts.

The zoogonidangia are separated from the rest of the cœnocyte by a wall, after which the nuclei divide and the disc-like chloroplasts become orientated at right-angles to the wall. Numerous zoogonidia (12—15 μ in length) are formed, which are at first clustered as a botryoidal mass, but they soon develop two cilia and finally escape through the tubular hairs. Each zoogonidium possesses several chloroplasts and a conspicuous pigment spot.

Huber's suggestion that *Chætosiphon* may be the extreme form of a chætophoraceous series passing through *Endoderma*, *Phæophila* and *Blastophysa* does not appear to be at all probable. From its siphonaceous thallus and the nature of its chloroplasts it is far more likely to be a member of the Siphonales allied to the Bryopsidaceæ, its special peculiarities being due to its endophytic habit.

Family **Bryopsidaceæ**.

In this family the thallus is a much branched cœnocyte, the lower branches becoming modified as root-like organs of attachment. The upper branches are arranged as shoots, each with a main ascending axis or 'stem' and branches of the first and second order which are often described as 'leaves.' The whole plant consists of graceful feather-like fronds of a deep green colour which have a superficial resemblance to certain species of *Caulerpa*. The 'leaves' are of a fusiform-cylindrical shape, having a marked constriction at their point of junction with the axis, and are arranged either in two rows or in a spiral. When they become fully mature in *Bryopsis* they are cut off from the axis by a wall and converted into gametangia (fig. 146 *1*). The latter eventually fall off leaving the separating walls as scars along the axis. In *Pseudobryopsis* the 'leaves' are from the first cut off by a transverse wall from the axis.

The cell-wall is thin and not encrusted. Wille ('97) stated that trabeculæ sometimes occur, such as those present in the cœnocytes of *Caulerpa*, but this observation is apparently doubtful. The protoplasm forms a parietal layer in which there are numerous nuclei and chloroplasts. The latter are small elliptic discs with a central pyrenoid. In the large cell-sap vacuole which occupies all the central space of the cœnocyte there are rounded and spindle-shaped albuminous bodies, first observed by Noll.

Vegetative propagation is known to occur by detached 'leaves' which grow into new plants, and by the formation of rhizome-like branches from the lower part of the thallus.

Zoogonidia are unknown.

Numerous biciliated anisogametes are produced in gametangia, which in *Bryopsis* are converted 'leaves' but in *Pseudobryopsis* are ovoid or pear-

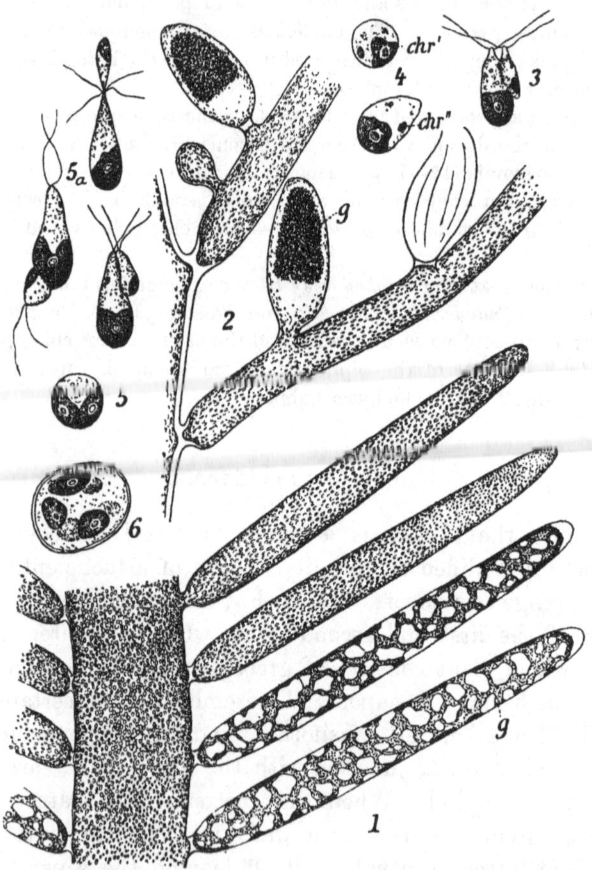

Fig. 146. *1*, portion of thallus of *Bryopsis cupressoides* Lamx. with almost ripe gametangia (*g*), × about 25. *2*, pinnæ of 'leaflet' of *Pseudobryopsis myura* (J. Ag.) Berth. with gametangia (*g*); *3* and *5 a*, copulation of gametes; *4*, zygotes soon after copulation showing chloroplasts of male (*chr''*) and female (*chr'*) gametes; *5*, zygote in which the chloroplast has divided; *6*, germination of zygote. (From Oltmanns; *2* and *3*, after Berthold.)

shaped outgrowths from the 'leaves' (fig. 146 *2*). Male and female gametangia may occur on the same axis or they may be on different plants. The female gametes are about three times the size of the male gametes, possessing a rather large posterior chloroplast with one pyrenoid and a pigment-spot. The male gametes, which are of a brownish-red colour, have only a very small chloroplast and no pigment-spot. The paired cilia are in each case attached to the anterior pointed end. The fusion of the gametes (*vide* fig. 146 *3* and *5 a*) results in a rounded zygote which is capable of immediate germination.

The Bryopsidaceæ are closely related to the Derbesiaceæ and the Codiaceæ; they occur in all seas, but more abundantly in the warmer oceans.

The genera are: *Bryopsis* Lamouroux, 1809; *Pseudobryopsis* Berthold, 1880.

Family **Derbesiaceæ.**

This family includes the single genus *Derbesia* first described by Solier in 1847. In general habit the thallus is tufted, numerous erect and more or less cylindrical branches arising from narrower, creeping rhizome-like branches, which are fastened to the substratum by branched holdfasts

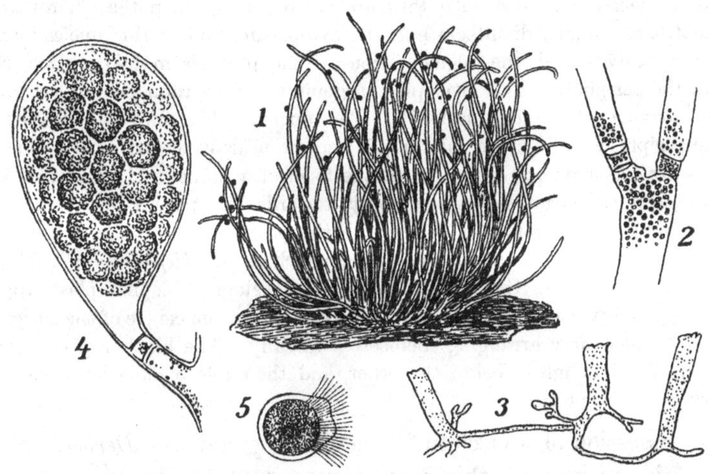

Fig. 147. *1* and *3*, *Derbesia Lamourouxii* (J. Ag.) Solier. *1*, general aspect, nat. size; *3*, 'rhizome' and bases of shoots, × 5. *2*, *4* and *5*, *D. marina* (Lyngb.) Kjellm. *2*, bases of branches; *4*, zoogonidangium; *5*, zoogonidium. (From Oltmanns; *2*, after Kjellman; *4*, after Kuckuck; *5*, after Solier.)

(fig. 147 *1*). The whole thallus is the development of a single cœnocyte, but the erect branches are often separated at the base by double cell-walls resulting in the formation of a small basal segment (fig. 147 *2*). There is no differentiation into 'stem' and 'leaves,' if one excludes the zoogonidangia which are homologous with the 'leaves' of *Bryopsis*, and the erect tubular branches have a great resemblance to the filaments of a thick *Vaucheria*. The cell-wall is thin and not encrusted. In the older parts of the thallus a few transverse walls may sometimes be formed, but they are seldom observed and have no definite position. The parietal protoplasm contains many nuclei and numerous discoidal chloroplasts. The latter are oval or elliptical and generally contain one or more pyrenoids, but in some species (*Derbesia neglecta*) pyrenoids are absent. In the large sap vacuole are found albuminous bodies similar to those which occur in *Bryopsis*, and also crystals of calcium oxalate.

Reproduction occurs only by zoogonidia which are produced in sub-spherical or club-shaped zoogonidangia borne laterally on the erect branches (fig. 147 *1* and *4*). The zoogonidangium, which is separated from the main axis by a thick, lamellate, basal wall, and often by a small basal cell, produces from 8 to 20 zoogonidia. These are comparatively large, ovoid in shape, and furnished with a circlet of cilia at the anterior end (fig. 147 *5*).

The protoplasm within the zoogonidangium contains at first several thousand nuclei of slightly larger size than the chloroplasts. Soon a process of nuclear differentiation sets in, certain of the nuclei increasing to about four or six times their original size. These rapidly acquire a definite cytoplasmic envelope, with radiating strands, and become distributed uniformly through the zoogonidangium. The smaller nuclei soon degenerate; they gradually decrease in size until they are much smaller than the chloroplasts, lose their chromatin, and finally disintegrate in the cytoplasm. When this nuclear degeneration has practically ended the segmentation of the protoplasm takes place, cleavage beginning at the periphery. The developing zoogonidia become rounded and the numerous chloroplasts arrange themselves radially. The nucleus then moves from the centre towards the periphery, and on the side of the nucleus nearest the periphery some of the protoplasmic strands arrange themselves in the form of a funnel and numerous granules which occur on these strands move outwards towards the plasma-membrane. These granules accumulate in a circle just underneath the plasma membrane and fuse with one another to form a deeply-staining, firm ring, which is the *blepharoplast*. The protoplasmic strands connecting the nucleus with the developing blepharoplast apparently disappear after its formation, and the nucleus passes back to the centre of the zoogonidium, the chloroplasts once more arranging themselves radially. The blepharoplast then splits to form two rings, one slightly below the other, and the circle of cilia is developed from the lower ring (Davis, '08).

In the possession of a circlet of cilia the zoogonidia of *Derbesia* resemble those of *Œdogonium*, but this resemblance does not in any way suggest affinity. The zoogonidia of *Derbesia* are formed from a multinucleate segment in which the nuclear changes are unusual, and their peculiarities may be entirely the result of the exceptional manner of their origin. The investigations of Davis ('08) indicate that it is in the highest degree probable that the ancestors of *Derbesia* produced numerous biciliated zoogonidia similar to those which are at present produced in many genera of the Siphonales, and that the peculiar zoogonidium, like that of *Vaucheria*, is only a late phylogenetic development. It is not unlikely that *Derbesia* may be distantly related to *Vaucheria*.

Family **Caulerpaceæ.**

This is another family of the Siphonales represented by a single genus—*Caulerpa* Lamouroux, 1809—of which about 60 species are known from tropical and subtropical seas. This genus is one of the most singular types in the vegetable kingdom, as although it consists of only a single branched cœnocyte, yet in external form the different species simulate the various types of habit found in higher plants. The thallus consists of a creeping

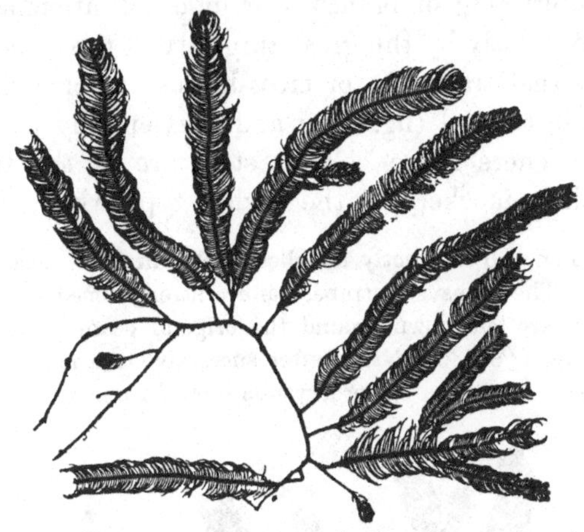

Fig. 148. *Caulerpa Holmesiana* G. Murray. ⅜ natural size.

rhizome-like axis, which gives off branched root-like rhizoids or 'holdfasts' from its under side and 'foliar shoots' from above. The latter are often of

Fig. 149. *Caulerpa verticillata* J. Ag. 1½ times the natural size (after Börgesen).

great beauty and may attain a length of 30 cms. The rhizome has an apical growing point and is often profusely branched, but the general character of both the rhizome and the root-like holdfasts varies much in the different species, depending upon the habit of the plant and the nature of its environment. The necessary mechanical support for the thallus is not obtained by the interlacing of branches or by calcification, as in so many of the Siphonales, but mostly by the great turgidity of the cœnocyte. A dense lattice-work of internal trabeculæ or cross-beams traverse the lumen of the cœnocyte from wall to wall (fig. 151) and prevent any over-distention or possible bursting. There is no cellular structure in the thallus and the plant is but a hollow mockery of the higher type which it simulates.

The cell-wall is thick and distinctly lamellose, with an outermost layer which gives rise to the trabeculæ. The transverse trabeculæ are strengthened by many of the later-formed lamellæ, which are laid down around the original beam as well as on the inner side of the wall. Correns ('94) found that after successive treatments with concentrated sulphuric acid and water numerous sphærocrystals were formed which were undoubtedly

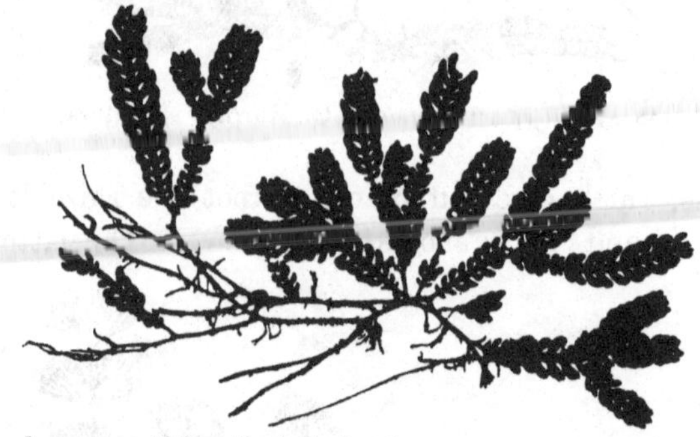

Fig. 150. *Caulerpa crassifolia* (Ag.) J. Ag. forma *mexicana* J. Ag. ½ natural size.

derived from the bulk of the membrane. These differed in several important respects from similar sphærocrystals of cellulose obtained by Gilson and others, and Correns concluded that the cell-wall of *Caulerpa* did not consist of true cellulose but of one of the allied substances. With iodine and sulphuric acid the wall colours only a golden yellow.

The building-up and the strengthening of the walls and trabeculæ of *Caulerpa* have furnished much evidence in favour of the growth of cell-walls by apposition.

The protoplasm is disposed as a parietal layer on the inner side of the wall, and it also covers all the trabeculæ, numerous anastomosing strands passing from the thin layer around each of the trabeculæ to the peripheral parietal layer. Most of the strands run more or less longitudinally, and it is not unlikely that one function of the trabeculæ is to give support to this delicate anastomosis of protoplasmic strands. In short, without this support it would scarcely be possible for the delicate protoplasmic threads to ramify through the large central vacuole of the cœnocyte. Numerous nuclei are present everywhere through the protoplasm, and there are also numerous disc-shaped chloroplasts, mostly aggregated in the peripheral layer. The chloroplasts, which are without pyrenoids, are according to Mme Weber van Bosse in some species relatively large. The peripheral layer of protoplasm is for the most part quiescent, but the more central anastomosis exhibits an upward and downward streaming movement which caused Janse ('90) to regard the streams as nutritive in character.

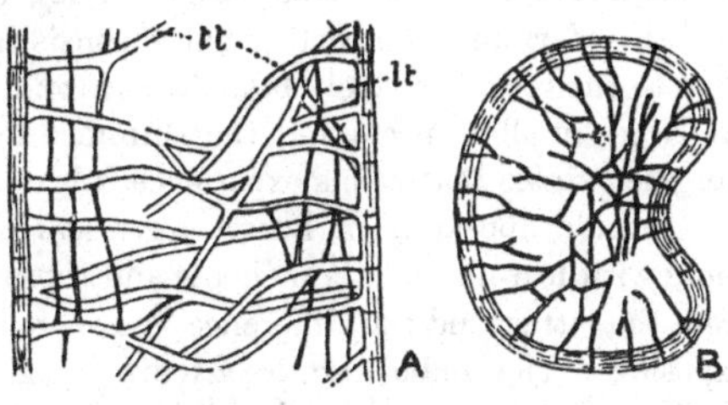

Fig. 151. *A*, longitudinal section of the 'leaf' of *Caulerpa prolifera* (Forsk.) Lamx., *tt*, transverse trabeculæ; *lt*, longitudinal trabeculæ; × about 80 (after Oltmanns). *B*, transverse section of 'stem' of *Caulerpa Holmesiana* G. Murray, × about 40 (after G. Murray).

There are no reproductive organs in *Caulerpa*, propagation taking place only by the separation of proliferous shoots. This is very well seen in *Caulerpa prolifera,* a species which is frequent in the Mediterranean, preferring deep water and sheltered places. It has been shown that in shallower, more turbulent water the proliferations are more abundantly developed.

The genus *Caulerpa* is divided into twelve sections based upon the type of habit, which resembles that of mosses, lycopods, ferns, *Zostera*, sedums, cacti, etc. Thus, there are the sections Charoideæ, Bryoideæ, Filicoideæ, Zosteroideæ, Sedoideæ, etc. Almost all the types of habit exhibited by higher plants are found to exist in this extraordinary genus. The genus was monographed by Mme Weber van Bosse ('98) and systematic and ecological studies have been published by Svedelius ('06; '07). It is to Börgesen ('00; '07), however, that we owe the splendid knowledge we now possess of the general biology of this genus. Börgesen has shown that Reinke's views ('00) that species of *Caulerpa* live under uniform external conditions and that they show no adaptations are not at all in agreement with facts.

Börgesen recognizes three ecological types in which the mode of growth as a whole depends upon environment.

(1) *Epiphytic or mud-collecting Caulerpas.* One of the best examples is *C. verticillata* (fig. 149), in which the rhizomes are almost thread-like and form a dense mat on the roots of Mangroves, accumulating mud and fine organic detritus.

(2) *Sand and mud Caulerpas.* Exemplified by *C. crassifolia* (fig. 150), *C. cupressoides*, *C. taxifolia*, etc., which are found in sheltered places in shallow water and in deeper water (down to more than 100 feet) where the influence of the surf is not felt. The rhizomes are strong and vigorous, growing extensively over a loose, sandy or muddy sea-bottom. In this way *C. prolifera* covers large areas of the soft bottom of parts of the Mediterranean. Börgesen likens the growth of *Caulerpa cupressoides* to that of *Carex arenaria*, the awl-shaped end of the rhizome boring its way through the sand in perfectly straight lines often more than a metre in length, and sending up at short distances the erect 'foliar shoots.' At about the same distance from one another as the erect 'shoots,' vigorous 'roots' grow down into the sand, undivided for about 2 or 3 cms., but then repeatedly branched to form numerous fine rhizoids which are firmly fastened to the sand and gravel. A similar attachment is found in *Penicillus*, *Halimeda* (consult fig. 154) and *Udotea* (three genera belonging to the Codiaceæ) when these Algæ are growing on a loose bottom, the loose material being knitted together so as to replace a fixed substratum.

(3) *Rock and coral-reef Caulerpas.* These are more especially forms of *C. racemosa*, although other species also occur. They are found in both exposed and sheltered localities, forming compact patches consisting of numerous entangled rhizomes firmly fixed to the rocks by richly branched rhizoids. In the more sheltered places the plants are larger in all respects and the rhizomes more vigorous. When attached to coral reefs Börgesen states that in his experience it is only to the dead parts of the corals.

It would appear from Börgesen's researches that there are two groups of *Caulerpas*, those with leaf-like bilateral shoots and those with radial shoots. The former are derived from the latter, and the difference is apparently caused by different degrees of light and exposure. The radial *Caulerpas* are essentially inhabitants of the littoral region in shallow water, being adapted to live in intense light, whereas the bilateral forms become more pronounced in deeper water which is often muddy and where the intensity of the light is much reduced. At depths of 20—30 metres the leaf-like segments of the 'foliar shoots' are quite distichous and even the branches may all be situated in the same plane.

In the absence of any form of reproductive organs the relationships of *Caulerpa* are not at all clear. It may possibly be related to *Bryopsis* since Correns finds that the cell-wall consists of a similar substance in each case, and Wille has declared that the characteristic trabeculæ of *Caulerpa* sometimes occur in *Bryopsis*.

Quite a number of fossil impressions have been attributed to the Caulerpaceæ, some being described under the name *Caulerpites* and others as species of *Caulerpa*. *Caulerpites cactoides* Göppert was described from Silurian and Cambrian strata, and *Caulerpa Carruthersi* Murray from the Kimmeridge Clay. These and many similar impressions are exceedingly doubtful, and it is more than probable that some are casts of animal origin rather than of plant origin (*vide* Seward, '00).

Family **Codiaceæ.**

The thallus in this family is of very varied form and in some genera is encrusted with calcium carbonate (*Halimeda, Penicillus,* etc.). The plant consists of a much-branched cœnocytic tube, the thread-like branches of which are so interwoven that they give rise to a thallus with a definite external form. The interweaving of the branches is sometimes rather loose (*Avrainvillea, Boodleopsis, Tydemania, Penicillus*) and at other times so compact as almost to form a tissue which has received the name of 'plectenchyma' (*Udotea, Halimeda, Cladocephalus,* etc.). In most cases the thallus is differentiated into a medullary region of interlacing branches and a cortical region in which short branches stand out at right-angles to the long axis. This cortical region often forms a superficial limiting layer which has been termed the 'palisade layer.'

The branches of the cœnocyte have only a limited apical growth and, after a certain length has been attained, one or two lateral outgrowths arise behind the now dormant apex of the branch, each outgrowth forming a new branch of about the same length as the parent-axis. The branching is usually dichotomous, either in one plane or in alternate planes, although trichotomous branching occurs in the basal part of *Chlorodesmis comosa*, in *Boodleopsis*, in *Tydemania*, in *Penicillus Sibogæ*, in the capitulum of *P. dumetosus*, and in the flabellum of *Udotea conglutinata* and *U. glaucescens*;

verticillate branching also occurs in *Tydemania expeditionis*, in *Boodleopsis* and in the capitulum of both species of *Rhipocephalus* (A. & E. S. Gepp, '11). In addition to the main branches there are often more or less prominent papillate outgrowths which may occur in great quantity, not infrequently cohering to form a continuous cortex covering the external surface of the thallus. This cortex may be uncalcified, as in *Flabellaria petiolata*, or calcified, as in many species of *Udotea*, etc. In *Cladocephalus* there are numerous 'pseudo-lateral' branches forming an uncalcified labyrinthine cortex (fig. 152 *B* and *C*). There are also 'pseudo-conjugating' filaments in

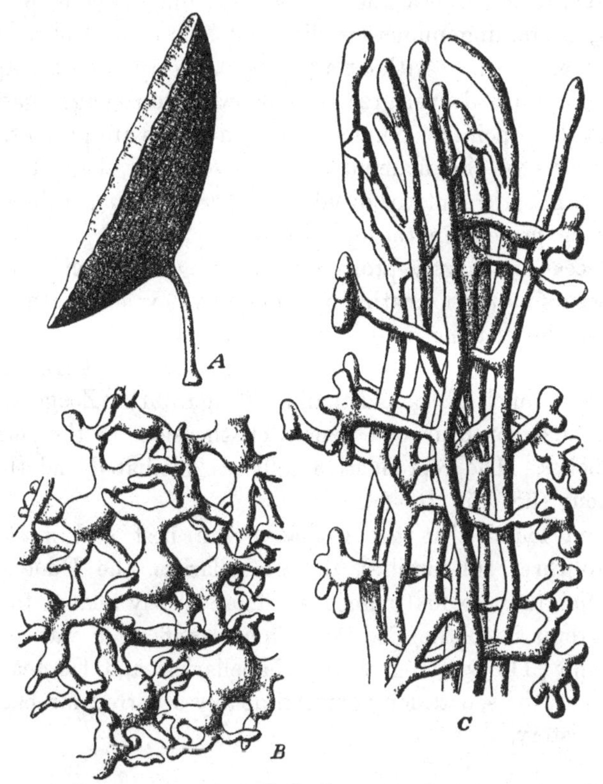

Fig. 152. *Cladocephalus excentricus* A. & E. S. Gepp. *A*, plant $\frac{2}{3}$ nat. size; *B*, section of the outer portion of the thallus showing the pseudocortex, ×245; *C*, longitudinal section near the apex showing the young branches of the cœnocyte, ×200. (From Wille, after A. & E. S. Gepp.)

Rhipiliopsis, consisting of lateral outgrowths which meet but never coalesce. In the Codieæ the cortex consists of contiguous branch-endings, which in the genus *Pseudocodium* are laterally coherent.

There is usually a constriction at the base of each branch of a dichotomy, and partial septa are often formed at these points due to the development of a ring-like ingrowth of the cell-wall. In *Callipsygma* this ingrowth is a septum with a central aperture, and in *Codium* and other genera the septa are frequently complete, forming 'stoppers' in various parts of the thallus.

The attachment of the Codiaceæ to the substratum is effected by numerous, repeatedly-branched rhizoids, which often penetrate deeply into a soft substratum, or fix themselves firmly to sand-grains, fragments of shells, remains of calcareous Algæ, etc. Some species, such as *Udotea Desfontainei* and *Halimeda Tuna*, only occur attached to stones or to the rhizomes of *Zostera*. The rhizoids possess fairly thick walls, but only a delicate protoplasmic lining and very few chloroplasts.

In the branches which form the 'shoot' the lining layer of protoplasm is much thicker, containing numerous discoidal chloroplasts of a rounded or polygonal form, and with or without a pyrenoid. Very often the apical parts of the branches are of a deeper green colour owing to the aggregation of the chloroplasts at these points. Numerous starch-grains are present, especially in the older parts of the filaments, and in outward shape they may be reniform, fusiform, or irregularly ovoid. In *Avrainvillea* a yellow or brown pigment may be present in the cytoplasm.

The Codiaceæ are largely propagated by fragmentation of the thallus, which often occurs by the formation of numerous new shoots, the older parts dying away and thus setting free the new shoots as young plants. At other times new plants are developed from the merest fragments, even from a single 'palisade' branch of the old thallus (Tobler, '11). Zoogonidangia and gametangia are developed in some genera on the sides of the peripheral (or palisade) branches. The reproductive cells are biciliated and the gametes are anisogamous (fig. 155 *D—G*).

It has been found that the Codiaceæ are best and most naturally separated into three sub-families, the Flabellarieæ, the Udoteæ, and the Codieæ, of which the two first-named are more nearly related to each other than either group is to the Codieæ. This classification was proposed by A. & E. S. Gepp ('11), who regard the Flabellarieæ and Udoteæ as fundamentally distinct, the separation having probably occurred far back in their phylogenetic history.

Sub-family FLABELLARIEÆ. This is the largest section of the Codiaceæ, including nine genera all of which are destitute of calcification. The thallus shows considerable diversity of form in the different genera, although the general tendency is towards the formation of a stalked and flattened expansion. *Chlorodesmis* and *Rhipidodesmis* are the most primitive genera, the former affording some clue to the ancestry of *Flabellaria* and probably of the flabelliform genera *Avrainvillea, Rhipiliopsis, Rhipilia,* and *Cladocephalus.* The largest genus is *Avrainvillea* in which the thallus consists of dichotomously branched, interwoven filaments forming subsessile or stalked, flabelliform 'fronds,' having a felt-like structure and attached below by rhizoidal branches (Murray & Boodle, '89). In this genus the branching is dichotomous, and the branches, which are invariably constricted at the base, are either cylindrical or moniliform, with thin walls. In *Avrainvillea* the filaments have no lateral branches, whereas in *Flabellaria* there are numerous lobulate lateral branchlets which usually interlock so as to form a close cortex over the expanded frond.

Callipsygma is but little more advanced, possessing a flabelliform frond with a complanate two-edged stalk. The filaments composing the stalk are approximately parallel, but those forming the expanded frond are not all in one plane as they frequently overlap and are partly superposed (A. & E. S. Gepp, '04; '11). In *Cladocephalus* (fig. 152) the branches of the cœnocyte are more irregular, forming an excentric infundibuliform thallus in which the subparallel medullary filaments are covered by a labyrinthine cortex composed of short, densely subdivided 'pseudo-lateral' branches.

Zoogonidangia occur in *Avrainvillea* as terminal expansions on filaments exserted from the flabellum.

The genera are: *Chlorodesmis* Bailey & Harvey, 1858 [inclus. *Rhytosiphon* Brand, 1911]; *Avrainvillea* Decaisne, 1842; *Rhipiliopsis* A. & E. S. Gepp, 1911; *Flabellaria* Lamouroux, 1813; *Rhipilia* Kützing, 1858, emend. A. & E. S. Gepp, 1911; *Cladocephalus* Howe, 1905; *Rhipidodesmis* A. & E. S. Gepp, 1911; *Callipsygma* J. G. Agardh, 1887; *Boodleopsis* A. & E. S. Gepp, 1911.

Of the above genera, *Boodleopsis* is unique and its affinities most obscure. *B. siphonacea*, the only known species, is a small Alga from the E. Indies forming dense flattened cushions growing on a muddy substratum. Its chief character lies in the abundant ramification of its branches and in the nature of the branching.

Sub-family UDOTEÆ. This sub-family contains five genera in all of which the thallus is calcified. The most primitive forms are certain species of *Penicillus*, although *Udotea javensis* is very little further advanced. The genus *Udotea* is of exceptional interest in view of the phylogenetic series exhibited by its various species, all of which may be traced back to *U. javensis*. Its stalked fronds are generally flabellate as in *Avrainvillea*, but are always encrusted with calcium carbonate. The calcareous matter is deposited in the outer gelatinous layers of the wall, but also penetrates into the inner cellulose wall. The calcium carbonate consists of either aragonite or calcite, but is never quite pure, being mixed with varying quantities of magnesium carbonate or calcium oxalate. The degree of calcification depends to some extent upon the species and also upon the insolation to which a given plant is exposed. There are two kinds of calcification: (1) the filaments composing the thallus are each enclosed in a porose calcareous sheath, and are either quite free (as in the capitulum of *Penicillus*), or laterally cemented side by side into monostromatic flabella (as in *Udotea javensis, U. glaucescens, Tydemania, Rhipocephalus*), or they are more or less completely conglutinated to form a pluriseriate flabelliform or cyathiform frond (as in *U. conglutinata, U. cyathiformis,* etc.); (2) the simple or branched lateral appendages of the filaments, which in this type cohere so as to form a cortical covering to the frond and its stalk, are thickened by a calcareous deposit along the laterally cohering walls. The extent of the calcification varies with the species, penetrating in some cases deeply into the thallus. This method of incrustation is found in the cortex of the stalks of *Udotea, Penicillus,* and *Rhipocephalus,* and in the frond of the corticated species of *Udotea.* It will be thus seen that the calcareous cement really fills the grooves between the filaments and binds them into a firm thallus.

Fig. 153. On the left, *Penicillus Lamourouxii* Decaisne, $\frac{2}{3}$ nat. size. On the right, *P. dumetosus* (Lamx.) Decaisne, $\frac{4}{5}$ nat. size.

In the first type of calcification 'pores' in the calcareous sheath are extremely numerous. Each pore is a bubble-like chamber in the thickness of the incrustation, being covered externally by a delicate calcareous pellicle in which is a minute ostiole. They mark the spots where bubbles of oxygen are evolved during photosynthesis (A. & E. S. Gepp, '11).

In the second type of calcification there are numerous 'windows' consisting of the apices of the lateral branches of the filaments which are free from calcification. These 'windows,' which are flush with the surface of the thallus, are for the entrance of light, and presumably escape calcification owing to the clinging of bubbles of oxygen evolved as the result of the photosynthetic work of the chromatophores lodged within (*vide* A. & E. S. Gepp, '11).

In most species the stalk consists of many parallel filaments, with numerous lateral branches forming a calcified cortex, but in *Udotea javensis*, a species in which the flabellate frond consists of a single layer of filaments, the stalk is monosiphonous.

A marked feature of the frond of *Udotea* is its beautiful zoning due to periodic growth. The thallus also shows marginal proliferation.

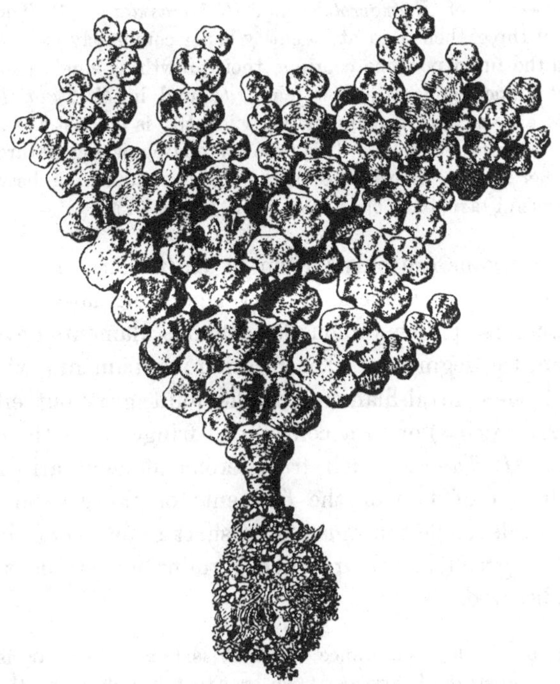

Fig. 154. *Halimeda incrassata* (Ellis & Sol.) Lamx. var. *simulans* (Howe) Börgesen. Nat. size (after Börgesen).

The highest type of the Udoteæ is *Halimeda* (fig. 154). This genus is certainly the most singular in the form of its fronds, which consist of branched chains of incrusted segments joined together in the manner of a cactus. The segments are compressed, more or less irregularly rounded, and in some species subreniform in shape. The structure of the segment is much like that of the frond of the more advanced types of *Udotea*, only between the axial strand of filaments and the peripheral cortical branches there is a subcortical layer, consisting of lateral branches of the axial filaments, which forms the main substance of the joint. The small cortical branches, which are really only the terminations of the filaments of the subcortical layer, are so compact that in surface view they have a more or less hexagonal appearance.

The axial filaments usually branch trichotomously, the middle branch continuing its course as a portion of the central strand (E. S. Gepp, '04). After a time the filaments cease growing and form the resting apex of the segment or joint. At this point they are in close contact and communication is established between them in one of three ways: (1) Free communication is established by means of large apertures in the walls of all the filaments, so that the whole central strand becomes welded into a connected mass. The individuality of the filaments is not lost, however, and on renewal of growth each filament continues again its individual course as it branches out to contribute to the tissues of a new joint (characteristic of *H. macroloba* and *H. incrassata*). (2) The filaments fuse in groups of two or three, their separate identity being completely lost, and at the end of the resting period the fused portions continue their growth as single filaments which by the usual type of branching form a new joint. (Found in *H. Tuna*, *H. gracilis*, etc.) (3) The filaments always fuse in pairs and their identity is not lost as they appear again almost at once as separate filaments. This type is a distinguishing feature of *H. Opuntia*, occurring in no other species. These three methods of communication between the central filaments furnish useful taxonomic characters (E. S. Gepp, '01).

Reproductive organs are only known in *Halimeda*. These are zoogonidangia producing large numbers of very small biciliated zoogonidia, and they arise on clusters of dichotomous fructiferous filaments developed on the dorsal margins of the segments. The fructiferous filaments, which are direct continuations of the central filaments of the joint, grow out either in small isolated tufts (*H. gracilis*) or as a continuous fringe along the dorsal margin of the segment (*H. Tuna*). Each fructiferous filament arises only as the result of the fusion of two of the filaments of the central strand. The zoogonidangia are developed in clusters on short branches of the fructiferous filaments, and are green in colour. The germination of the zoogonidia has not yet been observed.

An interesting point, the significance of which is not quite clear, is the fusion of filaments of the central strand before any fresh growth takes place, whether vegetative or reproductive. "Inasmuch as these fusions are found to precede the formation of new joints on the one hand, and of sporangiophores on the other, the obvious inference is that the fusions provide a powerful stimulus for further growth, whether vegetative or reproductive. They form new and vigorous growing-points. But what the factor may be which determines whether sporangia or whether a new joint shall result from the fusion is veiled in mystery" (E. S. Gepp, '04).

The interesting genus *Penicillus* (fig. 153) has probably arisen from ancestors common with those of *Tydemania* and *Halimeda*. It has a cylindrical or slightly flattened stalk consisting of closely interwoven filaments thickly incrusted with calcium carbonate. At the summit of the stalk the filaments diverge to form a mop-like mass of freely branching threads which, as a rule, are much less incrusted than the stalk. In some forms the apical mop of filaments is small and compact, with well calcified filaments, whereas in others it consists of long, straggling, and but slightly calcified filaments. In the former it is rarely more than 2 cms. in diameter, but in the latter it frequently attains a diameter of 10—15 cms.

The curious type *Rhipocephalus* has a general resemblance to *Penicillus* in habit and structure. Its stalk is round, and the capitulum, unlike that of *Penicillus,* consists of a cone of many small cuneate flabelliform fronds. These are 5—20 mm. in length, monostromatic, ascending and imbricating. They are normally arranged in subverticils, and in *R. phœnix* the flabelliform fronds of the same whorl are usually laterally connate into collars which more or less completely encircle the main axis.

The finely divided rhizoids with which all the Udoteæ are furnished, enable them to grow on a loose sandy or muddy bottom, the plants being fixed by the knitting together of the loose material in such a way that the latter replaces a fixed substratum. Upon a loose bottom of this kind in the West Indies Börgesen ('11) states that *Penicillus capitatus*, varieties of *Halimeda incrassata*, and *Udotea flabellata*, together with a few other species, form a very luxuriant vegetation. Sometimes they are mixed, but more often they occur in nearly pure unmixed societies. Thus, *Penicillus capitatus* grows so densely, head against head, that the ground is scarcely visible. The forms from the open sea, which occur down to 16 fathoms, are more firmly and compactly built than those from the lagoons. *Halimeda Opuntia* grows vigorously in the lagoons, where it often forms low banks or reefs, mud and organic detritus accumulating among its branches. In this way it may form mound-like banks which die in the centre and send out new shoots at the periphery.

Some species of *Udotea*, such as *U. cyathiformis*, *U. spinulosa*, etc., are only found in deep water down to a depth of about 40 metres (22 fathoms).

The Udoteæ although represented in more northerly seas (*Penicillus*, *Udotea* and *Halimeda* have each one species in the Mediterranean) are mostly tropical, and in the West Indies they are particularly abundant. *Udotea* and *Halimeda* are also abundant in the Eastern hemisphere.

The genera are: *Tydemania* Weber van Bosse, 1901; *Penicillus* Lamarck, 1813; *Rhipocephalus* Kützing, 1843; *Udotea* Lamouroux, 1812; *Halimeda* Lamouroux, 1812.

Owing to the calcification of the thallus the Udoteæ unquestionably lend themselves to preservation as fossils. Sollas has shown that the joints of species of *Halimeda* form no inconsiderable part of the material which goes to constitute some of the Pacific coral-reefs, and Börgesen points out that in the West Indies these joints often form an appreciable portion of the coarse particles of the calcareous sea-beach. There is every reason to believe that similar siphonaceous Algæ have played some part in the formation of the coral reefs and calcareous shore-deposits of previous geological periods; in fact, Fuchs has described a species of *Halimeda* (*H. Saportæ*) from the Eocene. A Tertiary Alga from the Paris basin very closely akin to *Penicillus* has been recognized in a fossil state by Munier-Chalmers, and the Eocene fossil described by Lamarck as *Ovulites* is very likely a siphonaceous Alga belonging to the Udoteæ. *Sphærocodium Bornemanni* described by Rothpletz as a fossil member of the Codiaceæ is exceedingly doubtful. For further information on these fossils consult Seward ('98).

Lorenz (1902) has given the name *Halimedites* to the fossil Alga he described as *Halimeda Fuggeri* in 1897. *Boueïna Hochstetteri* Toula (1884), from the Upper Neocomian of Servia, has a general habit similar to *Halimeda*. For a full account of the genus *Boueïna* consult Steinmann ('99).

Sub-family CODIEÆ. The plants of this small group have a spongy thallus with a variable external form. Some of them are globular cushions, some flat and lobed, and others are elongated cylindrical thalli which are more or less dichotomously branched; all are fixed to the substratum by means of rhizoids. The thallus consists of a central medullary mass of

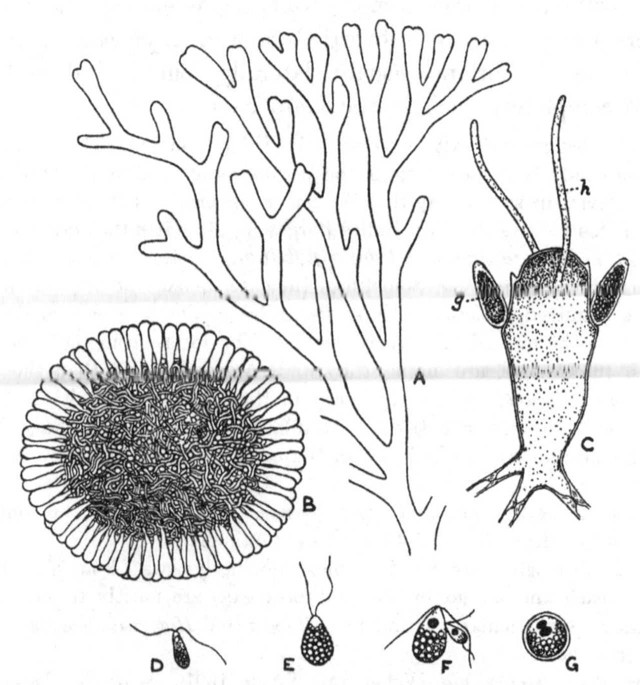

Fig. 155. *Codium tomentosum* (Huds.) Stackh. *A*, part of thallus shown only in outline, rather less than nat. size (after Murray); *B*, transverse section across thallus, × 10; *C*, one of the peripheral club-shaped branches of the cœnocyte, showing gametangia (*g*) and hairs (*h*), × about 60; *D*, male gamete; *E*, female gamete; *F*, fusion of gametes; *G*, zygote (very highly magnified). (*C—G*, after Oltmanns.)

densely interwoven filaments from which arise numerous, club-shaped, 'palisade' branches (the 'utricles' of Agardh) arranged perpendicularly to the surface. The latter are closely packed to form a pseudoparenchymatous tissue such as does not occur in the Udoteæ. The rhizoids are continuous with the medullary filaments.

Most of the medullary and rhizoidal filaments apparently become secondarily septate by the formation of annular thickenings which gradually increase until the lumen is completely occluded at those points. Thus, as the filaments get old they become fitted with 'stoppers' or 'plugs,' the beginnings of which appear quite early in the growth of the filament.

The parietal chloroplasts are very small and extremely numerous, especially near the apices of the branches: here they may be so densely aggregated that the green colour becomes of an intensity approaching black. In the lateral 'palisade' branches the chloroplasts are often grouped in clusters or arranged in strings.

The palisade branches, which are several times the diameter of the medullary filaments from which they arise, are developed acropetally, and they are not infrequently cut off at the constricted base by a stopper-like partition. The wall at the apex is thicker than elsewhere and is in some cases mucronate. It has been found that when *Codium* is exposed to strong light it tends to cover itself with hairs.

Reproduction occurs by the fusion of anisogametes. The gametangia are produced plentifully during the winter in *Codium*, and are borne on the upper lateral margins of the palisade branches (fig. 155 *C*). They are elongate-ovoid bodies, usually borne singly, and finally cut off by a plug-like wall from the palisade branch. Sometimes there is a succession of gametangia on the same branch, scars of attachment being left as the old gametangia fall away. Each gametangium has a distinctly two-layered wall, the outer layer being firm and thin, whereas the inner layer is thicker and able to undergo considerable swelling. The female gametangia were first described by Thuret ('50) and Derbes & Solier ('56), who showed that the green female gametes escaped *en masse* through a rupture in the apex of the gametangium. This rupture has been shown to take place as a result of the internal pressure exerted by the swelling of the inner layer of the gametangium-wall. The gametes soon separate and actively swim away. The male gametes, which are much smaller and of a golden-yellow colour, are produced in similar gametangia, but in larger numbers. The fusion of the gametes has been observed by Oltmanns ('04; consult fig. 155 *D—G*), although Berthold ('80) long ago showed that young plants could only be developed after the larger and smaller motile cells had been mixed. The zygote germinates at once.

Went ('89) found both kinds of gametangia on the same individual plants. His statement, that the megagametes can become parthenogenetic and develop directly into new plants requires confirmation.

Gibson & Auld ('00) have observed that the gametangia after reaching a certain stage of development, may become vegetative and be transformed into adventitious buds.

Only two genera are known, both of which may possibly have been derived from *Chlorodesmis*: *Codium* Stackhouse, 1795—1801; *Pseudocodium* Weber van Bosse, 1895.

In *Codium* the medullary filaments form an irregular plexus and the branches ('utricles') which form the cortical layer are contiguous but not adherent; whereas in *Pseudocodium* the medullary filaments are longitudinally arranged and the branches of the cortical layer are coherent, so that in a superficial view the cortex is hexagonally areolate. *Codium* occurs in both the Eastern and Western hemispheres, but *Pseudocodium* is known only in the temperate regions of the Eastern hemisphere. The plants are perennial, and in Western Europe reproduction takes place freely in the winter.

Family **Phyllosiphonaceæ.**

This family includes some interesting endophytic or endozootic Algæ which in most cases are partial parasites.

The thallus of *Phyllosiphon* consists of a richly branched cœnocyte (fig. 156 *B*), the rather irregular tube-like ramifications of which traverse in

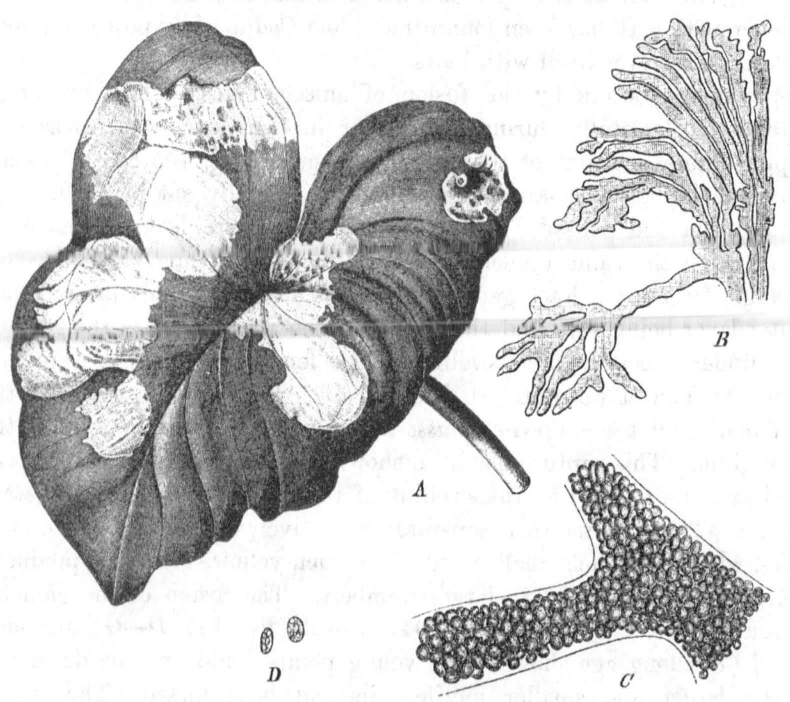

Fig. 156. *Phyllosiphon Arisari* Kühn. *A*, leaf of *Arisarum vulgare* showing the diseased areas caused by the *Phyllosiphon*; *B*, branches of the Alga in the leaf-stalk, ×68; *C*, portion of algal thallus with aplanospores, ×340; *D*, aplanospores, ×400. (From Wille, after Just.)

every direction the intercellular spaces of the leaves and leaf-stalks of various members of the Araceæ. Owing to this intrusion and the destructive action of the invader, the leaves of the host soon display diseased areas from 0·5—1 cm. in diameter, which often coalesce at their margins to form larger patches (fig. 156 *A*). These patches are at first yellow owing to the secretion of orange-yellow oleaginous droplets in the cells of the host due to the stimulation of the parasite. When the leaf wilts the oleaginous droplets disappear and the filaments of *Phyllosiphon*, which are now packed with green aplanospores, cause the affected spots to appear strikingly green on the withered leaf (Maire, '08). There are numerous nuclei in the branches of the cœnocyte and, except at the colourless tips of the branches, large numbers of parietal disc-shaped chloroplasts without pyrenoids. Both starch and oil occur as reserves. Motile cells are unknown, reproduction occurring only by the formation of numerous aplanospores, which are minute ovoid gonidia, set free from the ends of the branched tubes by a rupture caused by the swelling of the inner layer of the wall.

Phytophysa is a partial parasite in the cortical parenchyma of the stems of *Pilea*, an Urticaceous plant of Java, causing the formation of gall-like excrescences. The Alga consists of a vesicular cœnocyte of a deep green colour, furnished with a thick wall and attaining a diameter of 2·5 mm. Each vesicle is provided with a neck which projects towards the exterior of the pustule and through which the aplanospores ultimately escape. Oil globules and cellulose grains (?) are the reserves. Only the peripheral part of the large vesicular cœnocyte forms aplanospores, the inner part forming a mass of sterile cells between which and the outer wall the developing aplanospores become squeezed.

Ostreobium is another genus which consists of a richly branched system of cœnocytic tubes within the substance of the shells of marine bivalves or in the branches of corals. Numerous aplanospores are formed in club-shaped aplanosporangia developed at the extremities of the branches. These germinate directly to form new tubes.

The genera are: *Phyllosiphon* Kühn, 1878; *Phytophysa* Weber van Bosse, 1890; *Ostreobium* Bornet & Flahault, 1889.

Phyllosiphon Arisari is found in the leaves of *Arisarum vulgare* in Europe, and in the leaves of various Aroids in N. America and Java; also on *Arisarum simorrhinum* in Algeria, and on *Arum maculatum* near Lunéville in France. Lagerheim ('92) described *Ph. maximus*, *Ph. Philodendri* and *Ph. Alocasiæ* from S. America. *Phytophysa Treubii* is known only from Java.

Ostreobium is known in Europe, N. America, Samoa, and New Zealand. Nadson ('00) has stated that *Ostreobium Queketii* Born. & Flah. when growing in deep water develops the red pigment of the Rhodophyceæ, and in this state is identical with *Conocelis rosea* Batters.

Family **Vaucheriaceæ.**

In this family the thallus consists of a single and rather sparingly branched cœnocyte which often attains a length of 20 to 30 cms. As a general rule many of these filamentous thalli live as an intricate tangled mass, forming a dense mat of a very deep green colour. The filaments increase in length by apical growth and are in most cases attached to a substratum by means of 'hold-fasts' or rhizoid-like branches. In *Dichotomosiphon* (fig. 161) the filaments are dichotomously branched, each branch being constricted at the base and at various points along its length.

Considering the diameter of the filaments the cell-wall is as a rule relatively thin. This fact, combined with an absence of elasticity of the wall, causes the filaments to collapse very readily unless carefully handled.

The cytoplasm forms a thick lining layer in which numerous minute nuclei are embedded. The chloroplasts are very small and discoidal, with an oval or subcircular outline, and occur in very large numbers in the lining layer. They are without pyrenoids. The stored reserve material consists of globules of oil in *Vaucheria* and grains of starch in *Dichotomosiphon*. As in other cases in the Chlorophyceæ, and as pointed out by Fleissig ('00), these reserves are physiologically analogous.

Septa normally appear in the tubular thallus only in connection with the reproductive organs. Injury, however, results in the appearance of septa cutting off the injured parts, the uninjured portions developing into new plants (consult fig. 157 *A* and *B*).

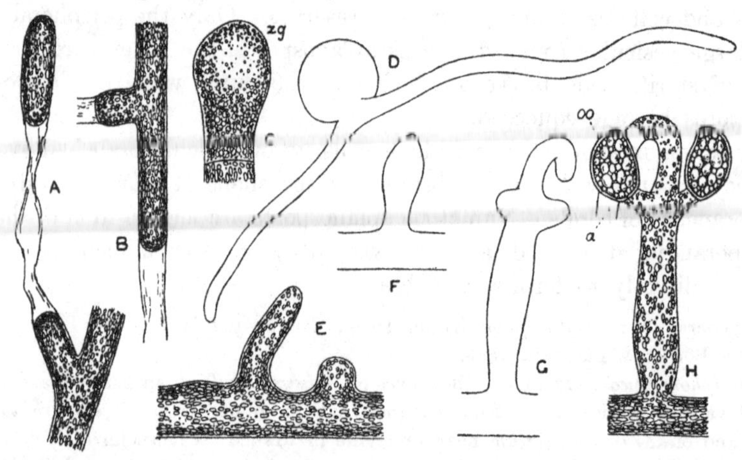

Fig. 157. *A* and *B*, portions of thallus of *Vaucheria* showing formation of septa on injury; *A*, *Vaucheria geminata* (Vauch.) DC.; *B*, *V. ornithocephala* Ag. *C*, apex of filament of *V. sessilis* (Vauch.) DC. showing the zoogonidangium. *D*, germination of the zoogonidium of *V. ornithocephala*. *E*, *V. sessilis* showing developing oogonium (on right) and antheridium (on left). *F—H*, *V. geminata* (Vauch.) DC. showing development of sexual organs. All ×75. *a*, antheridium; *oo*, oogonium; *zg*, zoogonidangium.

Asexual reproduction usually takes place by large solitary zoogonidia of a unique character, which have received the name of 'synzoospores' owing to their obviously compound nature. During the development of the zoogonidium the growing apex of a filament assumes a club-shaped form and becomes of an intense green colour, after which a transverse septum appears and the swollen end is cut off as a zoogonidangium. The contents of the gonidangium then become rounded off, forming an oval zoogonidium of large size, which ultimately escapes by an apical opening of much smaller diameter than itself through which it pushes its way (fig. 158 *B*). The whole surface of the zoogonidium is clothed with numerous short cilia, arranged in pairs, and embedded in the peripheral protoplasm under each pair of cilia is a small nucleus (fig. 158 *C*). The central part of the zoogonidium is occupied by a large sap vacuole traversed by delicate strands of protoplasm. In the outer zone of protoplasm, but within the layer of nuclei, are numerous small chloroplasts which give the zoogonidium a deep green colour. As first suggested by Schmitz ('79) this large structure can be regarded as compound in character, being constituted of an aggregate of numerous small biciliated zoogonidia. The disposition of the cilia in pairs, each pair being related to an underlying nucleus, certainly supports this suggestion, as does the fact that the whole of the contents of the gonidangium go to form a single motile gonidium with many paired cilia.

The zoogonidia generally escape in the morning, that is to say, after the plants have been in darkness for some time. They move but slowly and rarely continue active for more than twenty minutes. On coming to rest the cilia are at once withdrawn and a cell-wall is developed. Klebs ('96)

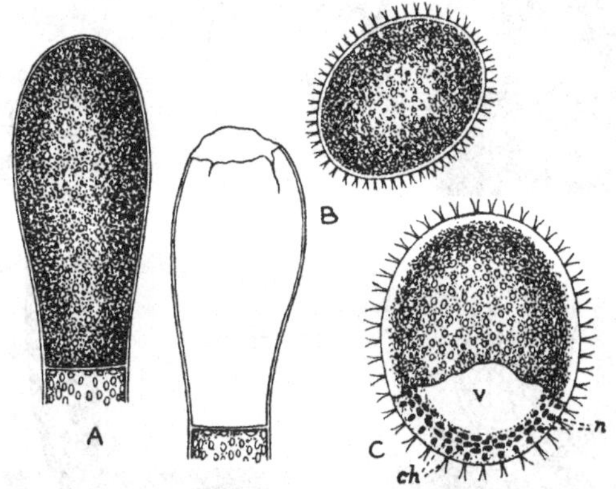

Fig. 158. *Vaucheria sessilis* (Vauch.) DC. *A*, zoogonidangium; *B*, escaped zoogonidium; *C*, somewhat diagrammatic representation of a zoogonidium, upper two-thirds in surface view, lower third in section; *ch*, chloroplasts; *n*, nuclei; *v*, central vacuole. *A* and *B*, ×160; *C*, ×220.

states that zoogonidia can always be produced when filaments which have been kept moist for some days are soaked with water, or when they are removed from running into still water, or when they are transferred from a dilute nutritive solution into pure water, or from light to darkness. After coming to rest the zoogonidia germinate almost immediately by the protrusion of one or more tube-like filaments (generally two at opposite poles), one at least of which attaches itself to the substratum by a colourless branched rhizoid.

Solitary aplanospores are sometimes produced in aplanosporangia, as in *Vaucheria geminata*, and their formation may be induced by cultivation in air which is only slightly damp. They are, however, only rarely observed under natural conditions.

Under some circumstances, such as when the plants become almost dried up, asexual bodies of the nature of 'cysts' or gemmæ may be produced. These cysts (sometimes rather misleadingly referred to as 'akinetes') were well described and figured by Stahl ('79), after they had been first noticed by Kützing[1]. They are shortly cylindrical bodies with thick cell-walls, and they may either germinate directly into a new filament or the protoplasmic contents may escape from the old wall before germination commences. Very often they break up into amœboid bodies, which soon round themselves off and become invested with a cell-wall, after which they either rest for a time or grow directly into new filaments. In some instances, due as a rule to unfavourable conditions of growth, septa appear at intervals, breaking up the filaments into a number of thick-walled segments which are also of the nature of rudimentary gemmæ or cysts.

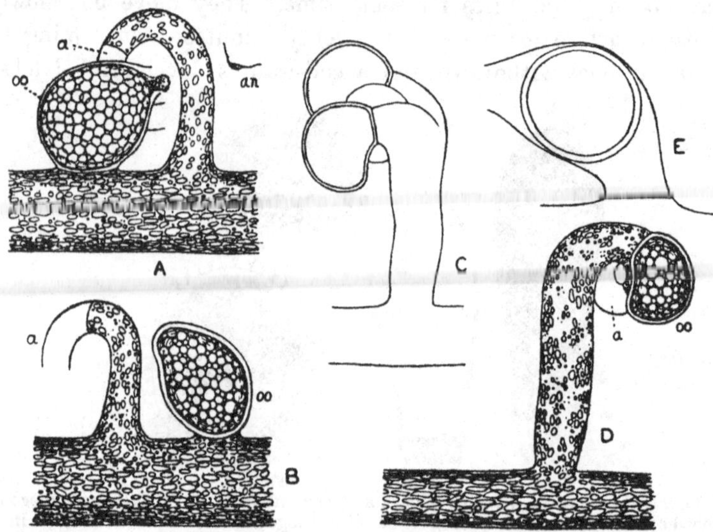

Fig. 159. The sexual organs of *Vaucheria*. *A* and *B*, *V. sessilis* (Vauch.) DC. ; *C* and *D*, *V. hamata* (Vauch.) Lyngb. *E*, oogonium and oospore of *V. ornithocephala* Ag. *A*—*D*, ×200 ; *E*, ×320. *a*, antheridium ; *an*, antherozoid ; *oo*, oogonium.

Dichotomosiphon produces curious tuber-like swellings at the extremities of the branches which may be compared with akinetes (fig. 161 *A* and *B t*). These always grow directly into new plants.

Sexual reproduction of a high type occurs in the Vaucheriaceæ, this family standing alone amongst the Siphonales in the possession of sharply differentiated oogonia and antheridia. The sexual organs are developed at scattered intervals along the cylindrical filaments, and, except in the

[1] Kützing (*Tab. Phyc.* IV, t. 98) figured this state of *Vaucheria* as '*Gongrosira dichotoma*,' an unfortunate error which has given rise to much confusion, since it has, of course, no relationship with the species of the true *Gongrosira* (a genus of the Ulotrichales). Consult G. S. West, '04, p. 111 ; and Herring, '07, p. 116. Kützing recorded the fact that his '*Gongrosira*' grew into filaments of *Vaucheria*.

diœcious species, the oogonia and antheridia usually arise side by side on the same filament (fig. 159 *A* and *B*) or on short lateral branches (figs. 157 *H*; 159 *C* and *D*). The oogonia, which are developed either as lateral outgrowths of the filaments or terminating short branches, swell out and assume a more or less rounded or ovoid form and are then cut off by a basal septum. The apex of the oogonium generally develops a rostrum or beak, which is in most

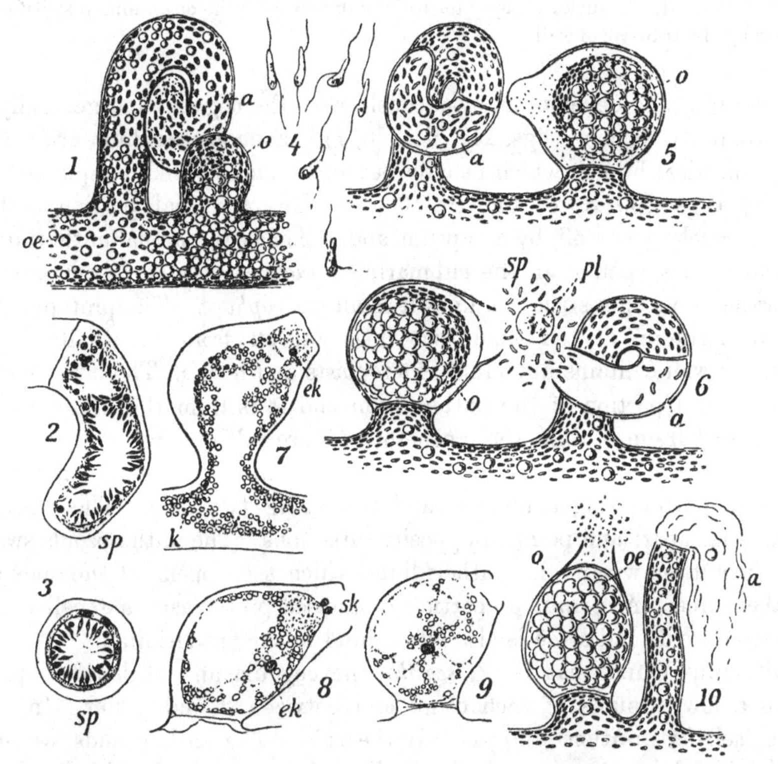

Fig. 160. *1—3, 5—10, Vaucheria sessilis* (Vauch.) DC. *1*, young sexual organs; *2*, partially ripe antheridium; *3*, transverse section of same; *5*, almost ripe sexual organs; *6*, the same at the moment of fertilization; *7*, longitudinal section through a young oogonium, in which there are many nuclei; *8*, older oogonium cut off from the filament by a wall, with only one nucleus (*ck*); *9*, fertilized egg; *10*, ripe zygote (oospore) and decayed antheridium. *4*, antherozoids of *V. synandra* Woron. *a*, antheridium; *ek*, nucleus of oosphere (egg-cell); *k*, nuclei; *o*, oogonium; *oe*, oil drops; *pl*, plasma-secretion from the oogonium; *sk*, nuclei of spermatozoids (antherozoids); *sp*, spermatozoids (antherozoids). (After Oltmanns.)

cases turned to one side, either towards the antheridium or away from it. The protoplasm of the oogonium contains much oil, numerous chloroplasts, and, after the appearance of its basal wall, only one nucleus. There is only a single oosphere or egg-cell which in most species of *Vaucheria* completely fills the oogonium[1]; its nucleus remains near the apex of the oogonium until just before fertilization when it takes up a central position.

[1] *V. ornithocephala* (fig. 159 *E*), *V. aversa* and *V. piloboloides* are notable exceptions.

Oltmanns ('95) stated that in *V. sessilis* just before the appearance of the basal wall all the nuclei except one were withdrawn from the young oogonium into the filament; but Davis ('04), from studies on *V. geminata* var. *racemosa*, declared that there were from 20 to 50 nuclei in the young oogonium, and that the oogonium was multinucleate at the time of the formation of the separating wall, after which all the nuclei degenerated except one, which grew to three or four times its original size. Heidinger ('08) studied the development of the oogonia in a number of species, and his observations support Oltmanns' statement that all the nuclei except one are withdrawn from the oogonium just before it is cut off by the transverse wall.

The antheridia develop simultaneously with the oogonia and generally in close proximity to them (figs. 157 *E*; 160, *1*). Each antheridium arises as a short cylindrical branch which usually becomes much curved on approaching maturity (figs. 159 *A*, *B*, and *D*; 160, *5*). The terminal portion of this curved branch is cut off by a septum and becomes the actual antheridium. In some species, such as in the submarine *Vaucheria synandra*, a number of antheridia occur on a structure known as an *androphore*. The protoplasm of each antheridium at the time that it is cut off from the thallus by a separating wall contains numerous chloroplasts and nuclei. The latter collect in the central portion of the antheridium and it is from the protoplasm of this central region that the antherozoids are ultimately formed. The antherozoids are very minute, each one consisting of a small amount of cytoplasm surrounding a nucleus and possessing two cilia. The cilia are attached far apart and point in opposite directions. The antherozoids swarm for a short time within the antheridium, which soon opens at the apex and sets them free. A certain proportion of unused protoplasm is expelled with the antherozoids and some is also left behind in the antheridium.

Oltmanns found that in *V. sessilis* the oogonia and antheridia opened within a few minutes of each other and between 2 and 4 a.m. In most species self-fertilization is apparently the rule. The antherozoids swarm in the vicinity of the oogonium, the beak of which has opened by the dissolution of the wall, a small quantity of mucilaginous protoplasm having been exuded from the open apex (figs. 159 *A*; 160, *6*). One antherozoid fuses with the oosphere and its nucleus passes towards the egg-nucleus, alongside which it remains for some time: according to Davis ('04) the male nucleus increases in size while in this position. At this stage the egg-cell becomes invested with a thick cell-wall. The fusion of the nuclei takes place slowly and the fusion-nucleus remains in a central position until the germination of the oospore, which takes place only after a rest of from one to three months, or more. Germination is direct.

In general it would appear that good nourishment and light of considerable intensity are necessary factors in the formation of the sexual organs. In *Dichotomosiphon* the sexual organs mostly terminate the branches.

Species of *Vaucheria* are widely distributed in temperate and tropical countries, occurring more especially in streams, cataracts, and boggy springs, where there is abundant aeration. The most ubiquitous species is undoubtedly *V. sessilis* (Vauch.) DC. Some species occur in brackish waters and some in the sea, but the majority live in fresh water or on damp ground. The terrestrial and some of the aquatic species show considerable variation, and there appears to be experimental evidence in support of the suggestion that *V. terrestris* and *V. geminata* are adaptational forms of the same Alga (Desroche, '10). The largest species of *Vaucheria* is *V. dichotoma*, in which the filaments attain a diameter of more than 200 μ. This species is also diœcious. *Dichotomosiphon* occurs in fresh water in Europe and North America.

Bohlin ('01), and following him, Blackman & Tansley ('02), transferred the family Vaucheriaceæ to the Heterokontæ, but the reasons for this change appear to be quite

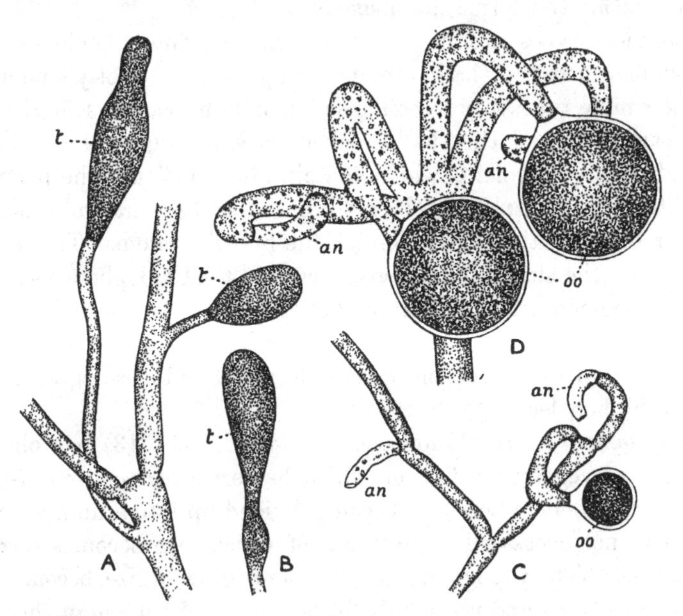

Fig. 161. *Dichotomosiphon tuberosus* (A. Br.) Ernst. *A* and *B*, part of thallus showing tubers (*t*); *A*, ×33; *B*, ×30. *C*, part of thallus with oogonium and antheridia, ×33. *D*, oogonia and antheridia, ×about 70. *oo*, oogonium; *an*, antheridium. (*A*, *C*, and *D*, after Ernst; *B*, after Virieux.)

inadequate. There is but one Heterokontan character in the Vaucheriaceæ, and that is the storage of oil instead of starch. Moreover, this occurs only in the genus *Vaucheria*, the closely allied genus *Dichotomosiphon* storing starch. The intense green colour of *Vaucheria* in the mass is never exhibited by any of the Heterokontæ, whereas such a depth of colour is quite common among the Siphonales; and the equal length of the paired cilia of the 'synzoospore' is also against the inclusion of the Vaucheriaceæ in the Heterokontæ. The nature of the thallus and of the chloroplasts points to a very close relationship with other families of the Siphonales, such as the Bryopsidaceæ and Codiaceæ.

The genera are: *Vaucheria* DC., 1803 [inclus. *Woronina* Solms, 1867]; *Dichotomosiphon* Ernst, 1902.

Order 3. SIPHONOCLADIALES.

In the Siphonocladiales, a group first established by Schmitz in 1878, *the thallus, in most cases, is septate from the earliest stages onwards. Septation is quite independent of nuclear division* and the segments contain from two to many nuclei. The transverse walls are more of the usual character and not of the nature of 'plugs' or 'stoppers' such as those which arise secondarily in the Siphonales. With the exception of *Sphæroplea* and some of the Cladophoraceæ all the forms are marine.

The simplest types are met with in the sub-family Valonieæ of the Valoniaceæ, in which the thallus in its young state is strictly siphonaceous. From these simple types a progressive series of forms can be traced in which a gradual partitioning of the thallus takes place, leading to the formation of smaller and smaller segments which contain fewer nuclei. The partitioning of the thallus is accompanied by branching, and as a general rule the greater the number of septa the more complex the branch-system. The maximum divergence from the siphonaceous type occurs in the Cladophoraceæ, in which the greatest segmentation of the thallus is found. The most symmetrical and elegant forms are members of the Dasycladaceæ.

The cytological details of the various forms are in close agreement with those of the Siphonales.

In many genera of the Valoniaceæ Börgesen ('05; '13) has observed a curious mode of vegetative division which he terms *segregative cell-division.* The entire contents of a segment become divided up into a number of small ball-like parts, not necessarily equal, each of which soon becomes surrounded by a membrane. The young daughter-segments grow in size, become angular by pressure of contact, and ultimately fill out the whole lumen of the mother-segment, which in this way becomes divided into a number of smaller parts. Each new segment contains many nuclei and chloroplasts, but the nuclei take no active part in the partitioning of the old segment. The daughter-segments subsequently grow into adult branches, and this process may be repeated several times giving rise to several orders of branches.

Reproduction occurs by zoogonidia in all the families except the Dasycladaceæ, in which the reproductive cells are aplanospores.

In *Pithophora* there are large barrel-shaped resting-spores of the nature of akinetes, which Wittrock originally termed 'agamo-hypnospores.'

Gamogenesis occurs in the Cladophoraceæ and Dasycladaceæ by the fusion of isoplanogametes.

Recent work on this group and on the Siphonales leads one to the inevitable conclusion that the Siphonocladiales have arisen from the Siphonales by the increasing septation of the thallus. The first steps of such a line of evolution are seen in the irregular walls which appear in the thallus of *Derbesia* and in the walls which sometimes cut off a few small outgrowths of the 'foliar shoots' of *Caulerpa*. The formation of the plug-like septa in the Codiaceæ is also a tendency in the same direction. The line of evolution suggested by Oltmanns ('04) in which the Siphonocladiales are derived from uninucleate groups such as the Ulotrichaceæ by repeated division of the nucleus, rather than from the Siphonales by the septation of the thallus, is an improbable hypothesis which would make the Siphonocladiales a more primitive group than the Siphonales. Such a view ignores the siphonaceous tendency exhibited by some of the Protococcales and also the significance which should possibly be attached to a comparative study of *Protosiphon* and a few genera of the Chlorochytrieæ (*vide* p. 223).

Family **Valoniaceæ.**

This is much the largest family of the Siphonocladiales, embracing a number of exclusively marine genera which attain their greatest development in tropical seas.

The thallus is at first siphonaceous and of a very simple character, consisting of a large simple or branched cœnocyte which is attached to the substratum by a branched system of rhizoids. In the more primitive forms of the Valonieæ the thallus develops very little beyond this stage, but in the Siphonocladeæ, Boodleæ and Anadyomeneæ there is a complex septate thallus which is frequently differentiated into 'stalk' and 'frond.' In *Dictyosphæria* the thallus is very compact and almost globular. The segments of the thallus are often firmly linked together by numerous holdfasts or haptera.

The chloroplasts are in the form of angular discs, disposed parietally in more or less anastomosing chains to form an open network.

Reproduction by zoogonidia is known to occur in several genera, and aplanospores are found in *Petrosiphon*.

Halicystis is removed to the Protosiphonaceæ (*vide* p. 224).

Börgesen ('13), as the result of his extensive West Indian investigations, discriminates between four sub-families of the Valoniaceæ.

Sub-family VALONIEÆ. In this, the most primitive division of the Valoniaceæ, the thallus consists at first, and in *Valonia ventricosa* always, of a single large sac-like cœnocyte which typically becomes branched. Septa soon appear, cutting off small lenticular, multinucleate, peripheral portions of the protoplasm (fig. 162 *a*, *a*). In *Valonia* these lenticular segments may remain small or they may grow out and form segments as large as the parent-segment from which they were derived (fig. 162 *A*). Successive formation of branch-segments sometimes results in a thallus with dichotomous or verticillate branches, although as a rule the branching is irregular. The main cœnocyte is attached to its substratum by numerous well-developed rhizoids or haptera (fig. 162 *C*), which grow out from the small lentiform segments that occur in great numbers at the basal end of the adult plant.

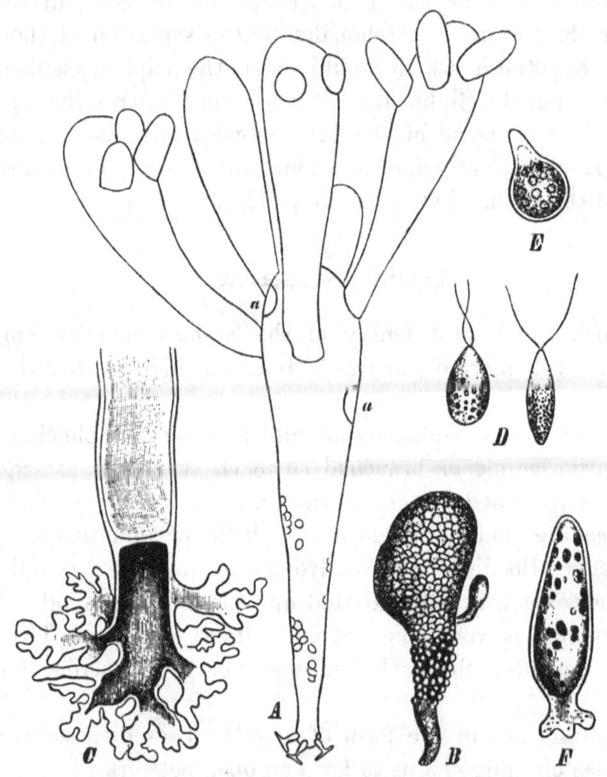

Fig. 162. ***Valonia utricularis*** **(Roth.) Ag.** ***A*****, complete plant showing a 'stem-cell' with five 'branch-cells' grown out from the apex and two 'rim-cells' (*a, a*) which have not yet grown out into 'branch-cells'; near the base of the 'stem-cell' are numerous smaller 'rim-cells' and three at the extreme base have grown into rivet-shaped rhizoids, ×4.** ***B*****, single cœnocyte developing zoogonidia.** ***C*****, older rhizoids.** ***D*****, zoogonidia.** ***E*** **and** ***F*****, developing zoogonidia. (*A*, after Schmitz; *B—E*, after Famintzin, from Wille.)**

In *Dictyosphæria* the contents of the very young cœnocyte, which may be ovoid, cylindrical or irregular, are divided by segregative cell-division, first into several cœnocytes, and afterwards by successive segregative divisions into numerous segments which are so compact as to be angular by compression. Numerous small lenticular cells (they have each only one nucleus) are formed in the peripheral parts of the segments. Some of those near the base develop into haptera for attaching the Alga to the substratum, but others, which are arranged with great regularity, form haptera which fasten the neighbouring segments together. *Dictyosphæria* is therefore not really a branched thallus, but rather a dense aggregate or colony of similar cœnocytes held together in a compact manner by haptera (called by Murray & Boodle 'tenacula'). The interior of this colony becomes hollow with its increasing growth, and eventually the colony bursts, forming an irregularly lobed thallus which may reach 12 cms. across. In this genus there are curious internal spines projecting from the walls of the cœnocytes into their central cavities.

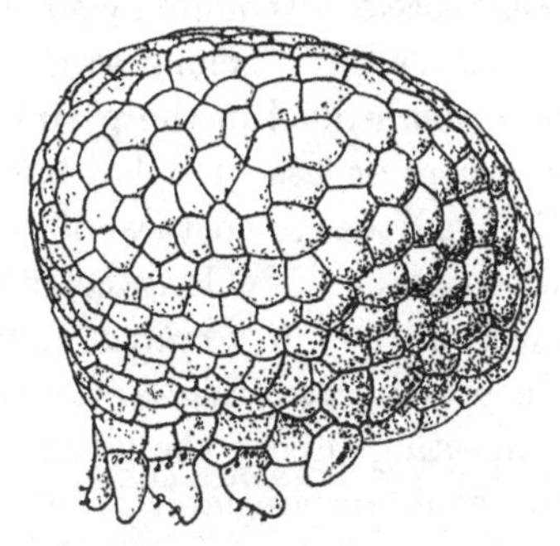

Fig. 163. A young plant of *Dictyosphæria favulosa* (Ag.) Decaisne. × 6 (after Börgesen).

The chloroplasts are rounded-polygonal and contain from 1 to 3 pyrenoids (Murray, '93; Kuckuck, '07; Börgesen, '13). In the protoplasmic layer there are starch-grains and amylum bodies. The walls of the cœnocytes are tough and elastic, and often striated on their outer surfaces.

The zoogonidia of *Valonia* are furnished with two (*V. utricularis*) or four (*V. macrophysa*) cilia and a red pigment-spot. They escape through numerous perforations in the wall of the cœnocyte, and germinate almost at once. Arnoldi ('13) has observed pyriform zoogonidia about 8—10 μ in length in *Dictyosphæria favulosa.* Börgesen ('13) did not find any zoospore-formation in large numbers of the W. Indian plants he examined, and he suggests that in some cases the young plants arise from aplanospores. He considers that most of the young plants arise from small lentiform cells which are loosened and set free from the thallus of other individuals.

The genera are: *Valonia* Ginnani, 1757, and *Dictyosphæria* Decaisne, 1842. Both are widely distributed in tropical and subtropical seas, and are especially abundant in the West Indies. They occur attached to rocks and stones, and also as epiphytes on other Algæ. They are almost equally abundant in shallow and in deep water, extending to a depth of from 30 to 40 metres. *Dictyosphæria favulosa* occurs frequently in shallow water on coral-reefs where it is constantly under the influence of the waves. In these situations it remains more or less globular and attains a diameter of about 5 cms.; on the other hand, in deep water (40 metres), where it is also often abundant, it becomes a flat expansion up to 12 cms. in diameter.

The swollen cœnocyte of *Valonia ventricosa* attains the size of a pigeon's egg or even of a small hen's egg. After death the plants become detached from the substratum, the green colour disappears, and they float away as translucent bladders. In this state they are often drifted ashore in the Bermudas and have received the name of 'sea-bottles' (Murray, '95).

NOTE: The genus *Blastophysa* Reinke ['88; '89 (inclus. *Phæophila* Hansgirg, 1890)], which is an epiphyte on the thallus of various larger marine

Algæ (*Urospora incrassata, Laminaria saccharina*, etc.) or an endophyte in the thallus of *Enteromorpha* spp. and *Nemalion* spp., is usually placed in the Valoniaceæ, although by some (Huber; Blackman & Tansley; Börgesen) it is considered to belong to the Chætophoraceæ. The cœnocytes are epiphytic on or embedded in the peripheral portion of the thallus of the host. They are more or less globular, but with lobed and sinuate outlines. *B. rhizopus* Reinke (the cœnocytes of which are 54—90 μ in diameter) is furnished with one or more long colourless hairs, which project into the surrounding medium, but *B. arrhiza* Wille has no such hairs. In *B. rhizopus* several cœnocytes are often arranged in a chain, being joined by intermediate connecting tubes; the latter at first contain protoplasm and chloroplasts, but the contents are later on withdrawn into the globular cœnocytes which are then cut off by septa. Sometimes new cœnocytes arise by 'budding,' in which case they are separated from the old ones only by a transverse wall. The chloroplasts are numerous, rounded-polygonal in shape (as in the Valonieæ), and some of them contain a single pyrenoid. There are many nuclei embedded in the peripheral protoplasm. Reproduction takes place by quadriciliated zoogonidia (15—23 μ in length) which issue from a tube-like extension of the cœnocyte (Huber, '93). The exact position of the genus *Blastophysa* is still to some extent doubtful, but the multinucleate character and the fact that one species has a septate thallus are good reasons for placing it in the Siphonocladiales and in proximity to the Valoniaceæ. This view is further strengthened by the nature of its chloroplasts and zoogonidia. Its peculiar characters may be largely the result of its epiphytic or endophytic habit.

Sub-family SIPHONOCLADEÆ. In all the genera of this group the primary cylindrical cœnocyte persists as a stalk or stipe, which is fastened to the substratum by unseptate or septate rhizoids. The basal part of the stipe (or, in *Chamædoris*, the whole stipe) has annular constrictions, the only exception being *Struvea anastomosans*, which is really a connecting link between *Struvea* and *Boodlea*. In this basal part the cell-wall is thick and stratified. Young erect 'shoots' sometimes grow out from the rhizoids. The latter may become creeping and rhizome-like, sending up new 'shoots' in great numbers, as in *Chamædoris*.

The septation of the thallus takes place by segregative cell-division, except in *Ernodesmis*. This type of division ultimately results in oblique walls in the upper parts of the thallus, each segment thus formed being able to grow out into a branch; and by a repetition of the process branches of the second and third order are formed. The lentiform cells so characteristic of *Valonia* are absent, although the verticillate branches of *Ernodesmis* originate in a manner not very different from the growth of branches in *Valonia*. In

Ernodesmis there are not only annular constrictions at the base of the main basal segment, but also a single annular constriction at the base of each branch. In *Apjohnia* there are several annular constrictions at the base of each branch and the branching is very regular. Moreover, in this genus the branches are not separated from their parent cœnocyte (at any rate in the younger parts of the thallus) by cross-walls. In fact, *Apjohnia* may be a transition-form between the Siphonales and the Siphonocladiales.

In *Struvea*, in which there is a flat frond on the end of a conspicuous stalk, there may be branches of the fourth order, and in the younger specimens the branching is very regular and the stalk consists of a single elongated cœnocyte with annular corrugations at the base. Later on in the life of the plant it becomes divided by transverse walls, more especially near its upper end. The branching is largely in one plane, and the upper part of each segment of a branch generally develops rhizoid-like organs of attachment ('tenacula') which fasten themselves to the wall of the branch nearest above. In this way there is formed an open net-work of branches bound together by 'tenacula' at the points of contact. As in other genera of the Siphonocladeæ, there are numerous decumbent creeping filaments attached to the substratum by rhizoids, and from them the erect 'leafy shoots' arise.

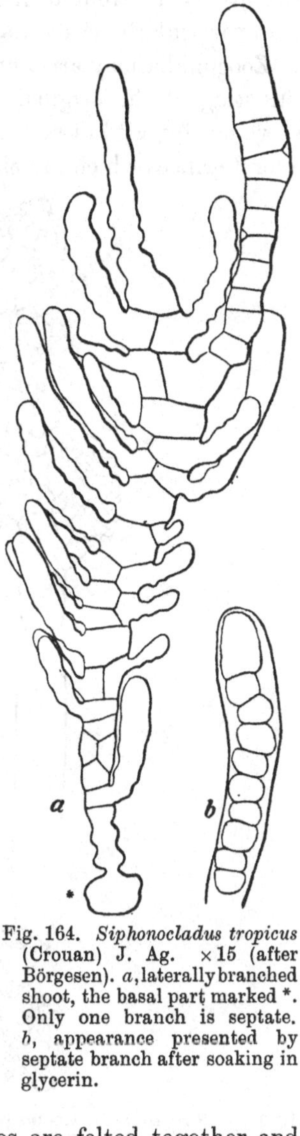

Fig. 164. *Siphonocladus tropicus* (Crouan) J. Ag. ×15 (after Börgesen). *a*, laterally branched shoot, the basal part marked *. Only one branch is septate. *b*, appearance presented by septate branch after soaking in glycerin.

In *Chamædoris* the stipe is 4 or 5 cms. in length, with annular constrictions from base to apex, where it terminates in a cup-shaped head, about 3 cms. in diameter, composed of large numbers of whorled branches of several orders, all arising by segregative cell-divisions. The branches are felted together and their coherence is increased by small rhizoids ('tenacula') growing out here and there and firmly attaching themselves to neighbouring filaments.

In all the Siphonocladeæ there are numerous small plate-like polygonal chloroplasts, many of which contain a central pyrenoid. Very often, but not always, the chloroplasts are connected by fine prolongations from their corners, so that in each cœnocyte there is apparently a net-like chloroplast with many pyrenoids. Internal to the chloroplasts there are numerous and very evenly distributed nuclei. Starch-grains are also present, but they are more particularly accumulated in the rhizoids.

Zoogonidia may arise in any of the branch-segments, the zoogonidangium consisting of the original cœnocyte of the mother-branch and the branch grown out from it but not yet cut off by a wall. The zoogonidia escape through several orifices which are slightly protuberant, and they germinate directly.

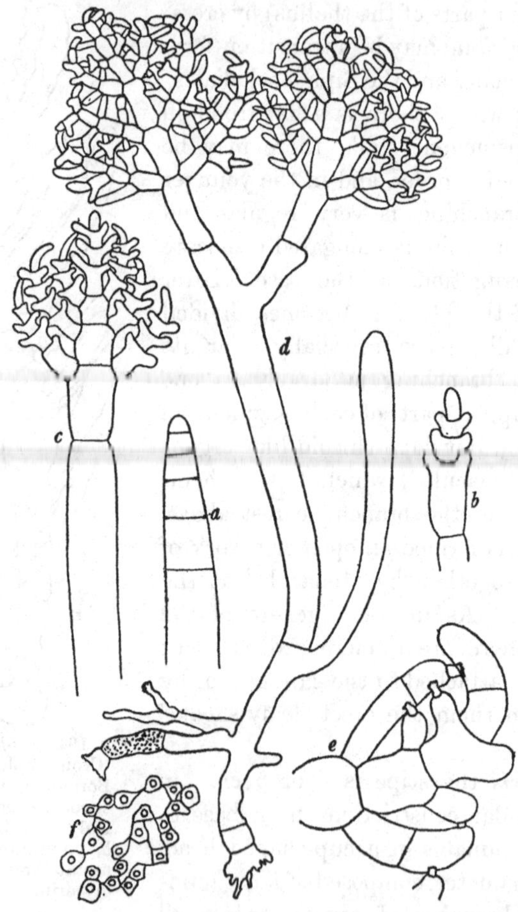

Fig. 165. *Struvea anastomosans* (Harv.) Piccone. *a* and *b*, extremities of young stalks showing development of the 'frond.' *c*, more advanced stage. *d*, single plant isolated from a tuft, showing the branched stalk with rhizoids. *e*, part of 'frond' showing how the branches are mutually attached by means of tenacula. *f*, chromatophores with pyrenoids. *a—d*, × about 6; *e*, × 20; *f*, × 250 (after Börgesen).

Most of the Siphonocladeæ are tropical and subtropical; they have a preference for sheltered localities and moderately deep water, although they do occur in less abundance in shallow water. *Siphonocladus tropicus* (Crouan) J. Ag. is often epiphytic on other algæ, or, as in *Ernodesmis verticillata* (Kütz.) Börges., it is not infrequently found in

Ægagropila-like clumps. *Struvea elegans* Börges. occurs in deep water down to 40 metres, and may become 10 cms. in length by 4 cms. in breadth; *St. anastomosans* (Harv.) Piccone frequently occurs in dense tufts in the fissures of rocks, the tufted habit being due to the numerous erect growths from the rhizome-like branches. *Chamædoris Peniculum* (Sol.) O. Kunze occurs from shallow water down to 50 metres, at which depth the stipe may reach a length of 15 cms. and the expanded head a diameter of 10 cms. *Apjohnia* is a southern type known from Australia and the Cape.

The genera are: *Chamædoris* Montagne, 1842; *Struvea* Sonder, 1845; *Apjohnia* Harvey, 1855; *Siphonocladus* Schmitz, 1878; *Ernodesmis* Börgesen, 1912.

Sub-family BOODLEÆ. This sub-family is characterized by the *irregularly branched filaments*, the branches as a rule growing out in all directions. In *Cladophoropsis* the branches are often twisted together and the plants form *Ægagropila*-like clumps on rocks, on other Algæ, or lying loose on the bottom. The attachment to the substratum is effected by branched and septate rhizoids. In *Boodlea* opposite branches may occur, and when this is repeated in the branches of the second and third order a *Struvea*-like appearance may result. In most cases, however, the ramification is very irregular and new adventitious branches contribute to it (Börgesen, '13). The branches are to some extent bound together by 'tenacula,' which may or may not be cut off by a wall, and which grow out from the apices of the cœnocytes in *Boodlea*, but from the sides of the filaments in *Cladophoropsis*.

There are no annular constrictions at the base of the segments, and the thin cell-wall is not stratified. The cœnocytes divide by segregative cell-division similar to that in *Struvea* and *Chamædoris*. Small lentiform cells occasionally occur on the inner walls of the cœnocytes, very like those found in *Valonia*, and from them branches may grow out. The thallus of *Petrosiphon* is encrusted with lime.

The parietal plate-like chloroplasts are very numerous and somewhat polygonal, being joined together in the young cœnocytes by very fine prolongations and thus forming a network. In older parts of the thallus the connections disappear and the chloroplasts are separate, each usually with a central pyrenoid.

The genera are: *Boodlea* Murray & De Toni, 1889; *Cladophoropsis* Börgesen, 1905; *Petrosiphon* Howe, 1905.

Sub-family ANADYOMENEÆ. In this small group the thallus is *irregularly frondose*, flat or basin-shaped, and consists of a leaf-like expansion of indefinite outline. It is fastened to the substratum by numerous thin-walled rhizoids which are mostly unbranched. *Microdictyon* is prostrate and sessile, but in *Anadyomene* the rhizoids grow alongside each other and form a short stalk, which broadens out at its base into a small disc composed of the irregular coralliform lobes of each rhizoid. The thallus recalls that of *Struvea*, but the branching has none of the regularity found in that genus. In *Microdictyon* the segments of the branches are more or less of the same size, and they form a close but irregular network. In *Anadyomene* the septate branches have long and short segments, and there are no interspaces between the branches. Moreover, in this genus the thallus may be clothed on both sides with an additional layer of closely united, isodiametric branch-segments. In *Microdictyon* the apices of neighbouring branches fix themselves to neighbouring segments and in order to strengthen their attachment a ring-like thickening of cellulose is formed at these points. In *Anadyomene* such a thickening is also found at the apex of each segment.

The chloroplasts are small polygonal plates, joined by prolongations of their angles in the young segments, but mostly isolated in older segments.

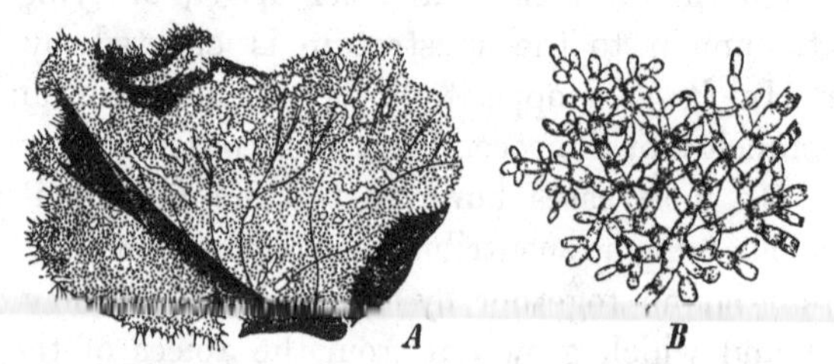

Fig. 166. *Microdictyon Montagneanum* Gray. *A*, nat. size. *B*, a piece of the periphery of the thallus showing the branches, × 20. (After Montagne, from Wille.)

Division of the segments takes place much as in the Cladophoraceæ, and lentiform cells do not occur.

The genera are: *Anadyomene* Lamouroux, 1812 [inclus. *Cystodictyon* Gray, 1866]; *Microdictyon* Decaisne, 1832; *Rhipidiphyllon* Heydrich, 1894. They are mostly tropical and subtropical, occurring both in shallow water and down to a depth of 50 metres.

Family **Cladophoraceæ.**

The septate thallus in this family consists of cylindrical or somewhat tumid cœnocytes forming *simple or branched filaments which are never bound together by tenacula.* The different genera of the family illustrate every condition intermediate between simple filaments, such as those of *Chætomorpha,* and the richly branched species of *Cladophora.* In *Urospora* there are short lateral outgrowths of the segments which are not separated by walls, and in the various species of *Rhizoclonium* there occur short unseptate or septate branches, generally of an irregular character. In the branched forms the thallus is tufted, or in the *Ægagropila*-section of *Cladophora* the branches are so arranged that they form a compact spherical or cushion-like mass.

There is always a basal attachment, generally by one or more holdfasts developed from the basal segment of the thallus, but the older plants of some genera break away from their attachment to the substratum and become free-floating.

Nearly all the forms are aquatic and most of them marine. A few are epiphytic or even epizootic, and more than one species of *Rhizoclonium* lives on damp soil.

The growth of the thallus is apical in *Cladophora* and *Pithophora*, but mostly intercalary in the other genera. Transverse walls are formed quite independently of nuclear divisions and by the gradual ingrowth of a thick ring-like septum (consult fig. 167 *C—F*).

In *Acrosiphonia*, in the *Ægagropila*-section of *Cladophora*, and especially in *Pithophora*, curious claw-like branches are sometimes developed.

The protoplasm lines the interior of the wall of each segment and usually contains many nuclei, although in the Rhizoclonieæ the number of nuclei may be reduced to two or even one (in *Spongomorpha*). There may be a single parietal, reticulated chloroplast, with many pyrenoids, or a number of isolated chloroplasts, each with one pyrenoid, and all intermediate states between these two extremes are of frequent occurrence.

In the *Ægagropila*-section of *Cladophora* the thallus often becomes fragmented by the death of the basal parts, each separate portion being able to grow into a complete new thallus. In *Acrosiphonia*, *Spongomorpha* and some species of *Chætomorpha* the rhizoids may form a pseudoparenchymatous tissue in which reserve materials are stored, and, should the ordinary vegetative filaments die away, this tissue remains alive and subsequently develops new filaments. Thick-walled 'cœnocysts' of the nature of akinetes are formed in *Rhizoclonium* (Wille; Gay; Brand) and *Urospora* from single segments of the thallus. On germination the cœnocyst forms a new filament in *Rhizoclonium*, but in *Urospora* it sometimes produces zoogonidia. Similar cœnocysts occur in *Pithophora*, but in this genus they have a very special origin.

Asexual reproduction by zoogonidia is known in all the genera except *Pithophora*. The zoogonidia are biciliated in *Chætomorpha*, *Rhizoclonium* and *Cladophora*, but quadriciliated in *Urospora*.

Isogametes occur in *Cladophora* and *Chætomorpha*, and anisogametes in *Urospora*. The zygotes germinate at once or after a period of rest.

The thallus is more or less destitute of a mucous covering, with the result that it is often loaded with smaller epiphytic Algæ. The latter may belong to the Bacillarieæ (fig. 84), Myxophyceæ or Chlorophyceæ, and are sometimes so thickly clustered as quite to obscure the host-plant.

The Cladophoraceæ have a world-wide distribution in both salt and fresh water. *Pithophora* is a freshwater genus mostly confined to tropical and subtropical areas.

It seems likely that the Cladophoraceæ have originated from the Valoniaceæ and not improbably from the Boodleæ through such a genus as *Cladophoropsis*. Wille's division of the family into the Cladophoreæ, Chætomorpheæ and Rhizoclonieæ is the best arrangement which has yet been suggested, but his inclusion of *Cladophoropsis* in the Cladophoraceæ is not quite in keeping with our present knowledge of the genus, Börgesen having recently shown that it is most nearly related to *Boodlea*. The view that the Cladophoreæ is the sub-family from which the other members of the Cladophoraceæ have evolved is supported by strong evidence; and, if it be a correct interpretation of the facts, then it follows that the small sub-families of the Chætomorpheæ and Rhizoclonieæ are rather specialized groups, the former having arisen by a suppression of branch-formation (the relics of which can be seen in *Urospora*), and the latter by a reduction of the branching and of the size of the cœnocytes, which has been accompanied by a corresponding reduction in the number of nuclei.

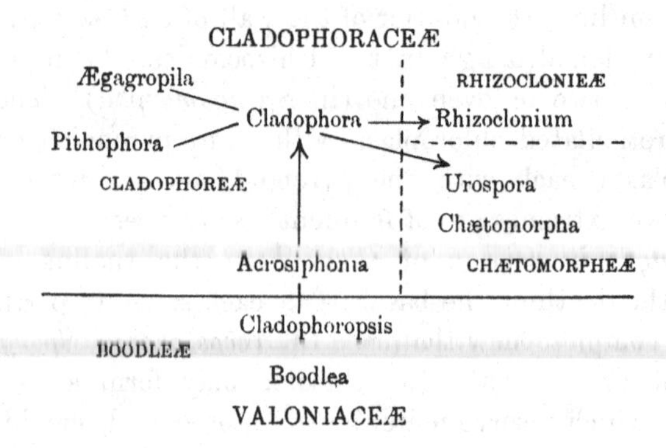

The above diagram illustrates the possible relationship of the Cladophoraceæ with the Valoniaceæ.

Oltmanns' extension of the scope of the Cladophoraceæ ('04), in which he included the genera *Anadyomene* and *Microdictyon*, forms, as pointed out by both Wille ('09 B) and Börgesen ('13), scarcely a natural group.

Sub-family CLADOPHOREÆ. The leading feature of this sub-family is the *richly branched thallus*, which may be the growth of a single individual or of several individuals in which the branches have become inextricably interwoven. The branches only rarely develop anything of the nature of 'tenacula' and therefore never form a connected net-like thallus such as occurs so frequently in the Valoniaceæ.

On the germination of the zoogonidium or of the zygote there is from the very first a differentiation of the thallus into a cauloid part and a rhizoid part. The latter usually consists of a few basal segments, but it may consist of a strong hold-fast resulting from the growing together of a number of rhizoidal filaments, as in *Cladophora rupestris*. Other rhizoids, which are developed from the cauloid part of the thallus, may be described as *accessory rhizoids*.

These are sometimes intracuticular, arising from segments high up on the axis and passing down between the lamellæ of the cell-wall, finally breaking through to the exterior in the lowermost part of the thallus (Brand, '09) This type of rhizoid is more particularly characteristic of *Spongomorpha* in the Rhizoclonieæ. In *Cladophora glomerata* var. *callicoma* there are often stoloniferous branches from which new erect shoots arise.

Fig. 167. *A* and *B*, *Cladophora incurvata* W. & G. S. West, a Cingalese species which shows polytomous branching very well; *A*, $\times 2\frac{7}{10}$; *B*, $\times 54$. *C—E*, successive stages in the formation of a transverse wall in *Cl. glomerata* (L.) Kütz., $\times 900$ (after Brand). *F*, older condition of same showing the joint (*j*), $\times 675$. *i*, inner layer of cell-wall; *o*, outer layer; *ou*, outermost layer.

The segments of the thallus are very variable in length, being usually from 6 to 12 (or even 20) times as long as their diameter. In *Cladophora fuliginosa* Kütz. very long segments occur, sometimes (as noticed by both Harvey and Börgesen) a whole branch being without a transverse wall. The cell-wall consists of an *inner* and an *outer* layer, both of which are lamellose, and an *outermost* layer which can be readily detected by treatment with acetic acid (Brand, '01; consult fig. 167 *C—F*). In some species, even in the living cells, the wall exhibits a cross-striation, and Brand's investigations ('06) show that this is caused by the presence of fibres and fibrils, which although not interwoven, pass near and over each other.

The branches arise as outgrowths of the upper part of the lateral wall of the parent-segment, and by growth on the part of the parent-segment are upwardly displaced so that eventually a dichotomous or polytomous appearance may result (Brand, '99; '01).

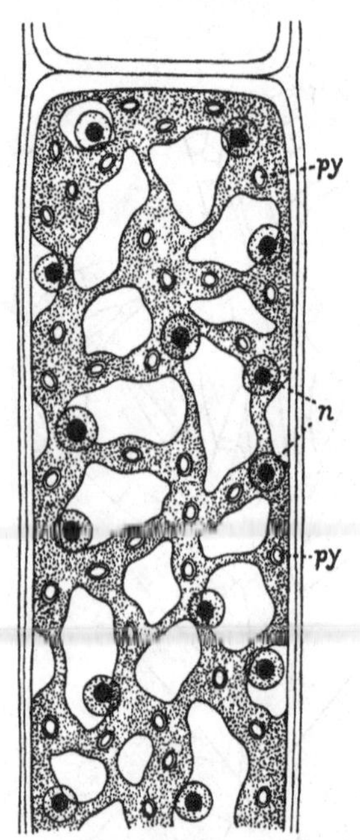

Fig. 168. Upper part of segment of *Cladophora glomerata* (L.) Kütz. stained to show nuclei (*n*) and pyrenoids (*py*). The reticulate character of the chloroplast is also well shown. ×610.

Each segment of the thallus contains many nuclei, although in some cases the number is much reduced. The chloroplast is normally a parietal reticulum (fig. 168), but under abnormal conditions and in segments which are degenerating all transitions occur between a true network and numerous isolated plates (Brand, '01).

In the division of the thallus into segments the transverse walls are formed by the ingrowth of a ring-like septum which originates from the inner layer of the cell-wall only (fig. 167 *C—F*). After the completion of the transverse wall a definite 'joint' is formed by the formation of a 'joint-space.' The joints of *Cladophora* have been carefully studied by Brand ('01; '08), who finds that the 'joint-space' between the inner and outer layers of the cell-wall is filled up by loosened lamellæ of the inner layer (fig. 167 *F*). This development of definite joints between the segments of the thallus adds much to the elasticity and flexibility of the filaments.

Vegetative propagation occurs by disjointed parts of the thallus (termed by Wittrock 'prolific cells' in *Pithophora*) and by large cœnocysts (which are really hypnocysts).

Asexual reproduction takes place by biciliated zoogonidia formed in large numbers in the segments of the thallus. They may arise in almost any segment of the thallus and they escape through an opening which arises by a complete dissolution of the cell-wall at some point near the apex of the segment. According to some observers the cilia are not always of equal length.

Gamogenesis occurs by the fusion of isoplanogametes, and the zygote germinates at once.

The *Ægagropila*-section of the genus *Cladophora* deserves special mention on account of the peculiar growth of the thallus. From the very earliest stages there is no differentiation into a cauloid part and a rhizoid part, and the plants grow in the form of compact cushions. The lower parts of the thallus (*i.e.* the oldest segments of the main axis) gradually die, so that the branches are set free from the base upwards. The plants are attached or free-floating, in the latter case forming compact globular masses from 2 to 14 cms. in diameter. These floating green balls are often found in great quantity in lakes, and also in sheltered seas, brackish lagoons, and mangrove swamps. They grow very slowly, live for a long time (Scourfield, '08)[1], and require very little light. An excellent account of these peculiar Algæ has been given by Brand ('02). Wesenberg-Lund's account ('03) of *Cladophora* (*Ægagropila*) *Sauteri* confirms many of Brand's observations. He found that much light was prejudicial to growth, and that the compact cushion-like masses resulted from the incessant destruction of the terminal segments of the branches. He regarded the globular shape of the thallus, which so many of these species present, as due to the beating of the waves and the constant friction against the bottom, but it is improbable that this is the full explanation of the ball-formation since these Algæ will live in perfectly still water for many years and yet maintain their spherical external form.

The genus *Pithophora* is of great interest. It was founded by Wittrock ('77) as the type of what he considered to be a new family of Green Algæ, the Pithophoraceæ. In the vegetative state the plants almost exactly resemble those of *Cladophora*, and as in that genus the growth of the thallus is apical. The only vegetative distinction between the two genera is in the invariable attachment of the branches of *Pithophora* some little distance below the apex of the supporting cells, but this insertion of branches may sometimes occur in *Cladophora* (Brand, '99). The branches are mostly solitary, but are sometimes opposite in pairs. The apical cœnocytes are the longest segments of the thallus, sometimes attaining a length 100 times as great as their diameter; they may become modified in some species by the development of terminal claw-like branches. Such apical cœnocytes were termed 'helicoid cells' by Wittrock, who found them commonly in *Pithophora Cleveana* but only occasionally in other species.

The chief character of the genus is the formation of large asexual resting-spores (the 'agamo-hypnospores' of Wittrock). These are either intercalary and cask-shaped or terminal and ovoid (or even fusiform), and a spore may be formed from almost every segment of the cauloid part of the thallus (fig. 169 *A*). The spores are usually developed at the upper end of a segment, which in most instances begins to swell, the main mass of the cytoplasm and chloroplasts moving upwards into the swollen part. Soon afterwards a transverse wall is formed, appearing first as a ring-shaped septum which gradually becomes a complete separating wall. The cell-wall of the whole spore now grows greatly in thickness. The spores are of an intense green colour and are completely filled by cytoplasm, nuclei, chloroplasts and food-reserves. In

[1] The present author has kept these balls alive for over seven years in a small dish into which tap-water was frequently allowed to drop.

general the formation of the spores is basipetal, the first-formed being at the tips of the branches. Twin-spores frequently arise from one mother-segment and even three in a row may be successively formed. In some instances spores are formed at the lower end instead of at the upper end of a segment. The spore germinates in opposite directions from the two poles and is very

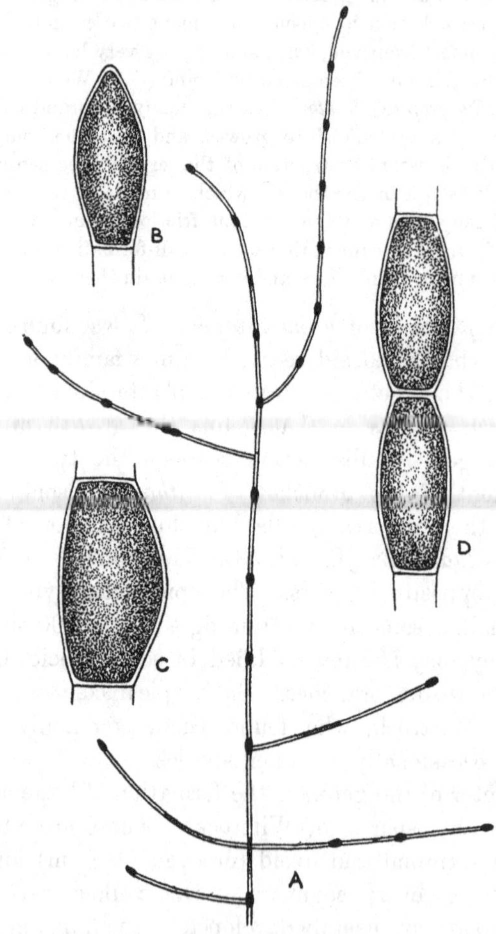

Fig. 169. *Pithophora Cleveana* Wittr. *A*, part of thallus with resting-spores, ×4; *B*, terminal spore; *C*, intercalary spore; *D*, paired intercalary spores. *B—D*, ×105.

early divided by a transverse wall, one half giving origin to the cauloid portion of the thallus and the other to the rhizoidal part. The latter consists generally of one segment and is rarely ramified.

Pithophora-plants are also able to propagate themselves in another manner by means of 'prolific cells' which are of the nature of hypnocysts. The latter arise from ordinary segments, usually of the main axis, without any change of external form, simply by an increase in the number of chloroplasts and in the food-reserves, more particularly starch.

That the genus *Pithophora* is closely allied to *Cladophora* there is no doubt; in fact, the relationship is so close that it is most undesirable to recognize the genus as the type of a distinct family—the Pithophoraceæ—as proposed by Wittrock. The curious resting-spores of *Pithophora*, somewhat erroneously termed 'akinetes' by Wille and others, are probably the outcome of a suppression of the formation of zoogonidia. The diameter of the filaments varies from 40 μ in the thinnest branches of *P. kewensis* to 190 μ in the thickest branches of *P. Roettleri.* The genus is a freshwater one and almost exclusively confined to the tropics. The plants are free-floating and at no period of their existence attached, except in the terrestrial forms of *P. Cleveana.*

The genera of the Cladophoreæ are: *Cladophora* Kützing, 1843 [inclus. *Ægagropila* Kützing, 1849]; *Acrosiphonia* (J. Agardh) Wille, 1909; *Pithophora* Wittrock, 1877. The freshwater species are very variable and present many growth-stages of a transitory character in which the specific characters are unrecognizable. In some species of the genus *Cladophora* the segments attain a diameter of 250 μ.

Sub-family CHÆTOMORPHEÆ. In this small group the *filamentous thallus is without branches,* except for the short unseptate protuberances which sometimes grow out from the segments of *Urospora.* The filaments are of a more or less uniform thickness except at the base and apex. The basal segment may possess annular constrictions and has a very thick wall; it is usually elongated and develops a strong hold-fast which firmly fixes the filament to the substratum (fig. 170 *1* and *2*). The hold-fast may consist of very irregular and much branched rhizoids, which sometimes contain a great deal of stored starch. In *Urospora* there may be a more or less large disc formed by numerous intertwined rhizoids which spring from quite a number of the segments in the basal part of the filament and grow downwards along the sides of the shoot; and in *U. mirabilis* Aresch. there are intracuticular as well as extracellular rhizoids.

In most forms there is a gradual attenuation near the apex of the filament, so that the apical segments are considerably narrower than those in the median part of the filament; they are also more elongated. The median segments are thick-walled and somewhat tumid, attaining a diameter of 500 μ in *Chætomorpha crassa* (Ag.) Kütz. The chloroplasts of the adult cœnocytes are reticulated parietal plates with many pyrenoids, and closely resemble those possessed by various species of *Cladophora.*

Reproduction takes place by comparatively small zoogonidia, which are biciliated in *Chætomorpha* and quadriciliated in *Urospora.* In the latter genus they are also more elongated and seen from the front are four-angled; they are also capable of slight changes in external form. Large numbers of zoogonidia are formed in each cœnocyte and are usually set free through a lateral aperture formed by the dissolution of the wall at that point (fig. 170 *4*).

Asexual reproduction also takes place by means of thick-walled cœnocysts, formed from single segments which become free. These either germinate directly or produce numerous zoogonidia after a period of rest. Gametes also occur in both genera; in *Chætomorpha* they are green and isogamous, but in *Urospora* they are anisogamous, the male gametes being smaller than the female and almost colourless. The zygote germinates only after a period of rest. Zoogonidangia and gametangia may occur simultaneously in the same filament (Börgesen, '02).

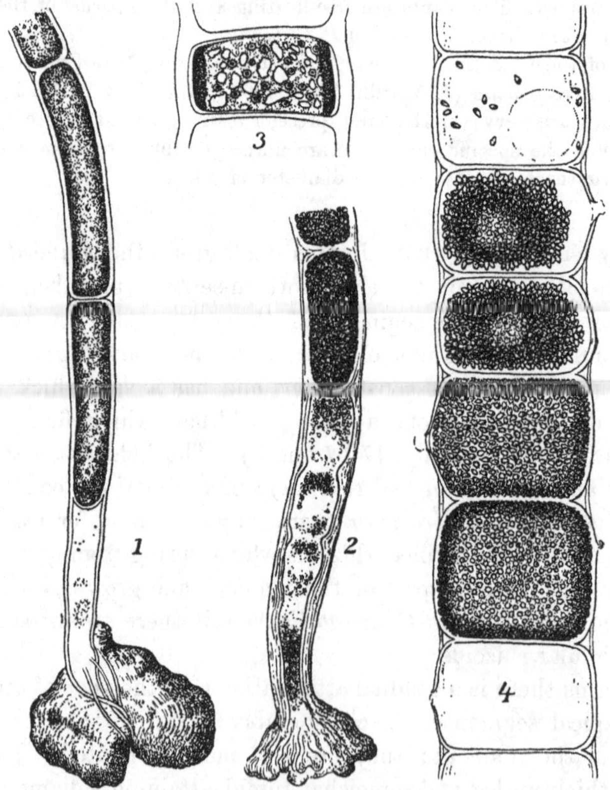

Fig. 170. *1* and *2*, *Chætomorpha aërea* (Dillw.) Kütz. *1*, basal part of young plant; *2*, base of an older plant. *3*, cell of *Urospora* showing reticulate chloroplast. *4*, part of a filament of *Chætomorpha aërea* during the formation of zoogonidia. (After Rosenvinge and Thuret, from Oltmanns.)

The genera are: *Chætomorpha* Kützing, 1845, and *Urospora*[1] Areschoug, 1866. Both genera are principally marine, but *Ch. Linum* (O. F. Müll.) Kütz. often occurs in fresh water. Most of the species are fixed to rocks on either exposed or sheltered coasts, although some (*Chætomorpha brachygona* Harv.) occur loose on a sandy or muddy bottom. Species of both genera may also be epiphytic on larger Algæ.

[1] From the remarks of Hazen ('02) it would appear that this genus should rightly be regarded as a synonym of *Hormiscia* Fries, 1835.

Sub-family Rhizoclonieæ. The filamentous thallus of these Algæ is *branched, or* less frequently *unbranched.* There is, as a rule, a basal rhizoidal attachment (fig. 171 *F* and *G*). In *Spongomorpha* the branching is considerable, but in *Rhizoclonium* the branches are generally short, often attenuated, and sometimes reduced to unseptate outgrowths (fig. 171 *B*). The filaments are usually slightly bent and twisted owing to the uneven growth of the individual segments, which are themselves in many cases asymmetrical, and in the vicinity of the branches not infrequently irregular (fig. 171 *C*). The cell-walls are strong and firm, sometimes conspicuously lamellose, and they may attain a great thickness, as in *Rhizoclonium profundum* and the subaërial species *R. crassipellitum* (fig. 171 *E*).

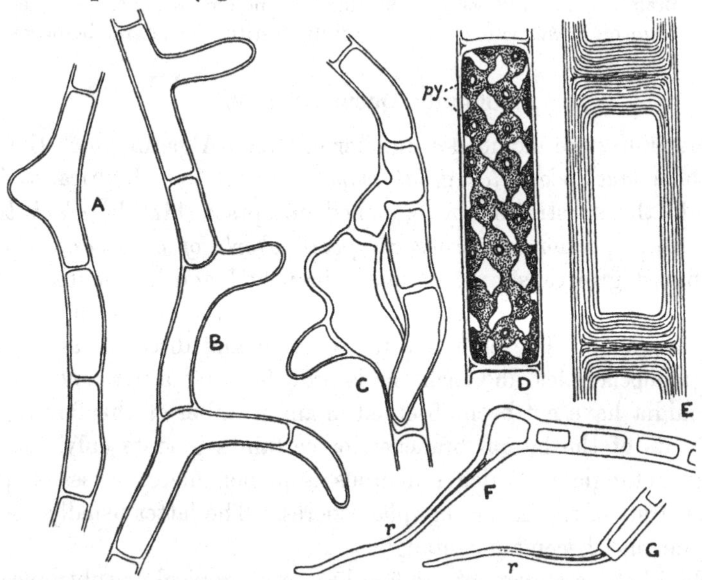

Fig. 171. *A–C*, parts of filaments of *Rhizoclonium Berggrenianum* Hauck var. *Dominicense* W. & G. S. West, with short lateral branches, × 450. *D*, single 'cell' of *R. hieroglyphicum* Kütz. to show reticulate chloroplast and pyrenoids (*py*). *E—G*, *R. crassipellitum* W. & G. S. West, a terrestrial species; *E*, showing great thickness of lamellose wall, × 450; *F* and *G*, bases of two filaments with rhizoidal outgrowth (*r*), × 105.

The *nuclei* of the segments *are greatly reduced in numbers* in this subfamily, only two being present in some species of *Rhizoclonium* and in *Spongomorpha* only one. The chloroplast forms a parietal network in the thicker parts of which pyrenoids are located (fig. 171 *D*). The reticulation is probably brought about as in *Cladophora* by the fusion of the produced angles of numerous small plates, each with a pyrenoid. It is sometimes not very obvious, and is best seen in young, actively growing filaments. In the autumn the segments of the thallus are often packed with starch grains, especially in the common freshwater species *R. hieroglyphicum.* Both Gay ('91) and Wille ('01 b) have described in detail the cytology of certain species of *Rhizoclonium*, and the systematic account of the genus given by Stockmeyer in 1890 is a thoroughly reliable one.

The genera are: *Rhizoclonium* Kützing, 1843; *Spongomorpha* (Kützing, 1843) Wille, 1909; *Chætonella* Schmidle, 1901.

Species of *Rhizoclonium* occur abundantly in marine and brackish situations, the two forms, f. *tortuosum* (Kütz.) Stockm. and f. *riparium* (Harv.) Stockm., of *R. hieroglyphicum* being characteristic of the mud of salt-marshes in north-west Europe; and several other forms of this species are also common in fresh water. A few species are subaërial in habit, occurring on damp soil, and some occur in the warm streams of active volcanic areas. *Spongomorpha* is exclusively a marine genus, and *Chætonella* is a small freshwater Alga which is sometimes epiphytic in the mucous coats of larger Algæ. Great care is required in the discrimination of species of *Rhizoclonium* owing to the variability of the vegetative characters (Brand, '98). In *R. profundum* there are no true rhizoidal branches, and the commonest form of *R. hieroglyphicum* is entirely destitute of lateral branches.

Family **Dasycladaceæ**.

In this family are included a number of Green Algæ in which the thallus consists of a *much elongated axial segment* (generally cylindrical and firmly attached to the substratum by branched unseptate rhizoids) *which bears* in its more apical portion *numerous acropetal whorls of branches.* The latter are of limited growth, simple or again branched, and segmented. All the segments are cœnocytes. In some forms the thallus is encrusted with calcium carbonate. The chloroplasts are small and discoidal, as in so many of the Siphonocladiales, and each one is provided with a pyrenoid.

Zoogonidia have not been observed in any member of this family, but all the segments of the lateral branches, or certain segments only, may either become gametangia in which numerous isoplanogametes arise, or produce small cœnocysts of the nature of aplanospores. The latter usually germinate directly, but may become gametangia.

Nearly all the members of the family live in tropical or subtropical seas.

There are quite a number of fossil forms of the Dasycladaceæ from Cretaceous, Eocene and Miocene deposits, and *Diplopora* and *Gyroporella* are known from the Trias. For information on some of these fossil Algæ consult Solms-Laubach ('91), Seward ('98), and Garwood ('13). Of the Cretaceous forms, *Neomeris cretacea* Steinmann (1899), *Munieria baconica* v. Hanklen (1883), *Diploporella Muhlbergii* Lorenz (1902), and *Triploporella Fraasi* Steinmann (1880) are the most important. The latter is intermediate between the Dasycladeæ and the Bornetelleæ, and might well form, as suggested by Oltmanns, the type of another sub-family, the Triploporelleæ. In this fossil form the lateral branches of the first order swell out and flatten themselves against each other to form a compact strobilus-like structure. The branches of higher orders were apparently rather delicate and deciduous.

Of the Eocene Dasycladeæ, *Decaisnella*, *Haploporella*, *Dactyloporella*, *Dactylopora*, *Uteria* and others have been described; and of the Acetabularieæ, *Acicularia Andrussowii* has been described from the Miocene rocks south of Sevastopol, and *A. miocenica* from the Miocene of the Vienna district. *A. pavantina* was described by D'Archiac from the Eocene sands of the Paris basin.

Sub-family Dasycladeæ. In the members of this sub-family the thallus is *not at all, or only slightly, encrusted with lime.* The main axis bears numerous lateral branches, dichotomously or polytomously arranged. All the terminal or lateral segments may be fertile.

In *Dasycladus* (fig. 172) the axial cœnocyte is clothed with whorls of branches, about twelve to each whorl. These branch-segments are again branched several times, the size of the segments diminishing with each successive branching. There is a slight calcification of the outer part of the wall of the axial cœnocyte (consult fig. 172 *3* and *4 k*). In a fertile plant

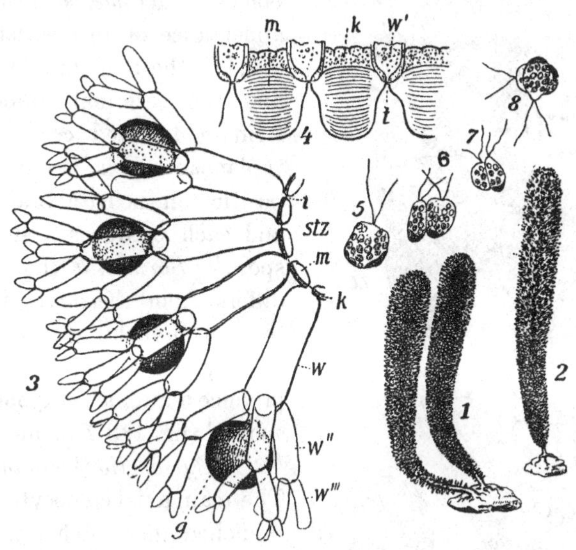

Fig. 172. *Dasycladus clavæformis* (Roth.) Ag. *1* and *2*, general habit of sterile and fertile plants, about nat. size. *3*, part of a whorl of branches, ×about 40; *stz*, 'stem-cell'; *w*, *w''*, *w'''*, whorl-branches of different orders; *g*, gametangium. *4*, transverse section through the wall of the 'stem-cell'; *m*, inner cellulose part of wall; *k*, outer calcareous part; *w'*, basal parts of branches of first order; *t*, minute canal. *5—8*, gametes and fusion of same. (*4*, after Nägeli, others after Oltmanns.)

a spherical, shortly stalked gametangium may be developed on the end of each of the basal branch-segments (fig. 172 *3 g*), especially on the upper part of the thallus (fig. 172 *2*). The gametes are isogamous and fuse in pairs (fig. 172 *5—8*) to form zygotes, but the germination of the latter has not yet been observed.

In *Chlorocladus,* a genus in which the terminal segments of the apical branches are piliferous, the fertile branches produce 'aplanosporangia' with numerous 'aplanospores,' the further development of which is unknown, although it is suspected that they ultimately become gametangia. In *Batophora* the general habit is of a simpler type and only 'aplanospores' are formed.

The genera are: *Dasycladus* Agardh, 1827; *Batophora* J. Agardh, 1854 [= *Botryophora* J. Agardh, 1887]; *Chlorocladus* Sonder, 1871. They are confined to tropical and subtropical seas, except for one species of *Dasycladus* which occurs in the Mediterranean.

Sub-family BORNETELLEÆ. In this small group the thallus is club-shaped and the branch-segments of either the first or second order are swollen out until they come into lateral contact, thus forming a continuous cortex, which is encrusted with calcium carbonate. *Neomeris* has much the appearance of an encrusted *Dasycladus*, but the branching is dichotomous and the terminal branch-segments form a facetted cortex. Spherical 'aplanosporangia' occur in a terminal position on the basal branch-segments, and each produces a single 'aplanospore.' *Bornetella* (fig. 173) mainly differs from *Neomeris* in its polytomous branching, and its laterally developed 'aplanosporangia' which produce a number of globular 'aplanospores' (fig. 173 *1 sp.* and *3*).

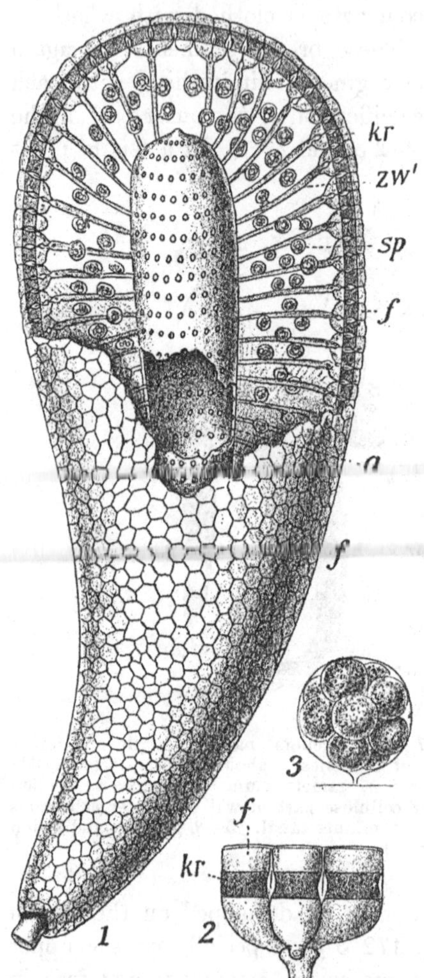

Fig. 173. *1* and *3*, *Bornetella oligospora* Solms. *1*, entire plant, slightly magnified, showing partly the external and partly the internal structure; *a*, axis; *zw'*, branches of the first order; *f*, peripheral facets; *kr*, calcareous zone of same; *sp*, 'aplanosporangia.' *3*, single 'aplanosporangium.' *2*, three peripheral facets of *B. nitida* (Harv.) Mun.-Chalm. (*1* and *3*, after Solms; *2*, after C. Cramer; from Oltmanns.)

In *Cymopolia* the habit is different. The main axial cœnocyte is repeatedly branched in a dichotomous manner, the branch-segments being all in one plane. Each branch-segment is calcified except at the constricted joints. The branches of the main axis bear closely set whorls of polytomous branches, the terminal segments of which are hair-like and deciduous. On the calcified parts of the thallus the cortex consists of the swollen branch-segments of the second order, but only the basal branch-segments persist at the uncalcified constrictions of the thallus. The 'aplanosporangia' are similar to those of *Neomeris* and are developed terminally on the basal branch-segments of the calcified parts of the thallus.

The genera, which are confined to the warmer seas, are: *Neomeris* Lamouroux, 1816; *Cymopolia* Lamouroux, 1816; *Bornetella* Munier-Chalmers, 1877.

Sub-family ACETABULARIEÆ. In this sub-family the main axis bears distinctly *differentiated sterile and fertile branches,* the former being polytomously branched. *All the genera are encrusted with calcium carbonate.*

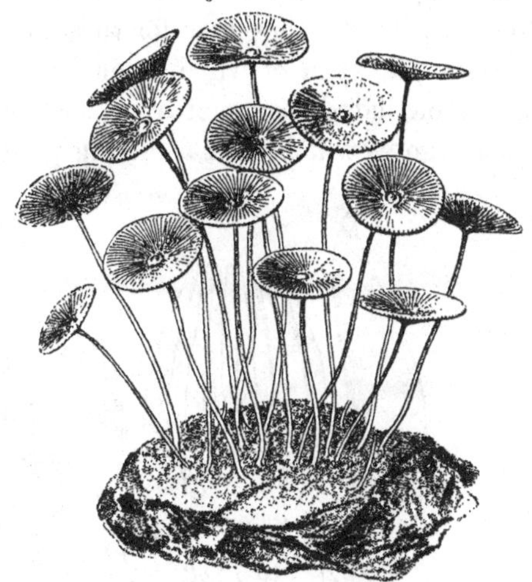

Fig. 174. *Acetabularia mediterranea* Lamx. Nat. size (after Oltmanns).

The best-known genus is *Acetabularia,* one species of which (*A. mediterranea*; fig. 174) has been very thoroughly investigated. The mature plant consists of a stem several centimetres in height terminated by a strongly calcified disc about 10—12 mm. in diameter. The latter is composed of a number of fertile branch-segments, radially arranged and laterally connate. The central area of the disc is occupied by a flat, circular membrane, which closes up the cavity of the large axial cœnocyte (or stalk), and between it and the bases of the radial branch-segments is a continuous circular cushion, the *superior* corona (fig. 175 *B sc*). This corona really consists of as many segments as there are radial branches forming the disc, and each segment bears the scars of deciduous branches. Below the disc there is another similar cushion, the *inferior* corona (fig. 175 *B ic*) but on this there are no hair-scars. The cavity of each radial branch of the disc is at its base in communication with the corresponding segments of the superior and inferior coronas, and the latter communicate with the large cavity of the axial cœnocyte by means of a small central opening in a separating fold of the cell-wall. For this perforated ingrowth of the cell-wall Howe ('01) suggested the name of *velum partiale.*

Acetabularia mediterranea grows for several years before a fertile disc is formed. In the first year the axis bears only a few irregular protuberances at its apex. The plant dies down at the end of the growing season, the lower part of the stalk (closed above by a wall) and the irregularly branched basal attachment remaining alive during the winter. The basal part increases in size with age and acts as a storehouse of reserve material for further growth. In the second year the axis bears several successive whorls of polytomously branched, deciduous, branch-segments ('leaves'), and terminally a disc which is not fertile. The deciduous branches are uncalcified, and, on being shed, rings of scars are left on the main axis. After several years of sterile growth the plant eventually develops a fertile disc. In the radiating branch-segments of the fertile disc numerous oblong-ellipsoid cœnocysts (often called 'aplanospores') are formed (fig. 175 *C apl*). These are ultimately set free after which they rest for three months, and then become converted into gametangia, each of which opens by a lid at one end, liberating numerous small, biciliated isogametes. These conjugate in pairs (fig. 175 *D—F*) to form zygotes which after a resting period germinate directly into new plants.

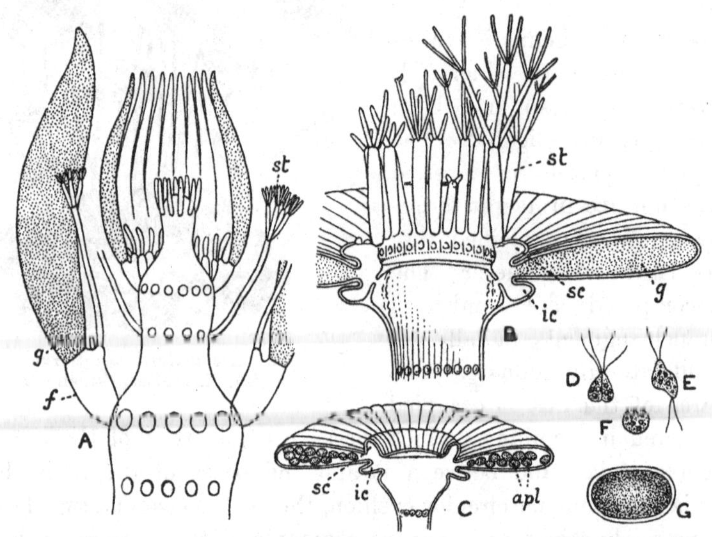

Fig. 175. *A*, *Halicoryne Wrightii* Harv., upper portion of thallus, ×58; *f*, whorl of fertile branches; *g*, branch-segment in which the cyst-like 'aplanospores' are formed; *st*, whorl of sterile branches. *B—G*, *Acetabularia mediterranea* Lamx. *B*, upper part of thallus in median section to show branches, ×5; *st*, sterile whorl; *sc*, superior corona; *ic*, inferior corona; *g*, fertile branch-segment. *C*, upper part of another individual showing formation of 'aplanospores' (*apl*), ×3. *D—F*, conjugation of gametes, ×450. *G*, 'aplanospore' (=gametangium) ×100. (*A*, after Cramer, from Wille; *B*, after Nägeli from Oltmanns; *C—G*, after Wille.)

In *Acetabularia crenulata* two or three fertile whorls of branch-segments are formed successively above each other. These form basin-shaped discs, and a fertile disc of a similar shape occurs in *A. Caliculus*. In the latter species the 'aplanospores' (gametangia) are spherical and there is no resting period, the gametes being set free while the gametangia are still within the radial branch-segments (Börgesen, '13).

The genus *Acicularia* differs from *Acetabularia* in the calcification of the walls of the 'aplanospores,' which adhere together as a calcareous mass. This genus, which was first described from fossil forms, has one living representative, *Acicularia Schenckii* (Möb.) Solms. Its interesting history has been well summarized by Seward ('98). In *Halicoryne* there are alternate whorls of sterile and fertile branch-segments (fig. 175 *A*), the former polytomously branched, the latter simple and completely free. Only a few 'aplanospores' are produced and they are calcified.

The genera are: *Acetabularia* Lamouroux, 1816 [inclus. *Polyphysa* Lamarck, 1816; and =*Acetabulum* Lamarck (*vide* Börgesen, '08)]; *Acicularia* D'Archiac, 1843; *Halicoryne* Harvey, 1859; *Chalmasia* Solms-Laubach, 1895.

Halicoryne is a link with the other sub-families of the Dasycladaceæ. Except for *Acetabularia mediterranea* all the representatives of the sub-family are tropical or subtropical, occurring mostly in shallow water, and often fastened to shells and stones. Species of *Acetabularia* are frequent in the mangrove-formation of the West Indies.

Family **Sphæropleaceæ.**

This family includes but one genus *Sphæroplea* which occupies a somewhat isolated position in the Siphonocladiales. The filaments of *Sphæroplea* are cylindrical and unbranched, consisting of series of greatly elongated cœnocytes. They are free-floating and never possess any organs of attachment. The transverse walls are sometimes of great thickness, whereas the side walls of the cœnocytes are comparatively thin. In rare instances the transverse walls are not fully closed owing to the imperfect development of the ingrowing ring-like septa. In the vegetative cœnocytes there is a thin lining layer of protoplasm in which there are apparently annular green chloroplasts separated by broad colourless bands (fig. 176 *A*). Each annular band is in reality a close aggregate of small parietal chloroplasts not essentially different from those present in other members of the Siphonocladiales; the aggregation is, however, so dense that the individual chloroplasts are scarcely discernible. Some of the larger chloroplasts (from four to six in each ring) contain a single pyrenoid, but the remainder are without pyrenoids. In the colourless zones between the annular aggregations of chloroplasts the lining layer of protoplasm is very thin, but in the coloured zones the protoplasm is much concentrated and often extends right across the lumen of the segment as a sort of diaphragm or plug. It is in these regions that the nuclei are situated (fig. 176 *A n*), generally just within the parietal chloroplasts. A few chloroplasts sometimes extend from ring to ring along strands of the protoplasm.

Asexual reproduction is unknown, but there is a sexual reproduction of a relatively high type. The sexual organs are antheridia and oogonia, formed without change of shape from any segment of the filament. Sometimes the oogonia and antheridia alternate, but more often they do not. In the formation of the antherozoids there is a repeated mitotic division of the nuclei, accompanied by a disappearance of the pyrenoids and a division of the chloroplasts, the latter ultimately becoming pale yellow in colour. Each antherozoid is a minute, elongated, spindle-shaped body, with a small posterior yellow chromatophore, a centrally placed nucleus and two anterior cilia. As many as 300 antherozoids may be formed from each 'ring' of the original cœnocyte during its conversion into an antheridium. They finally escape through small lateral apertures in the old wall.

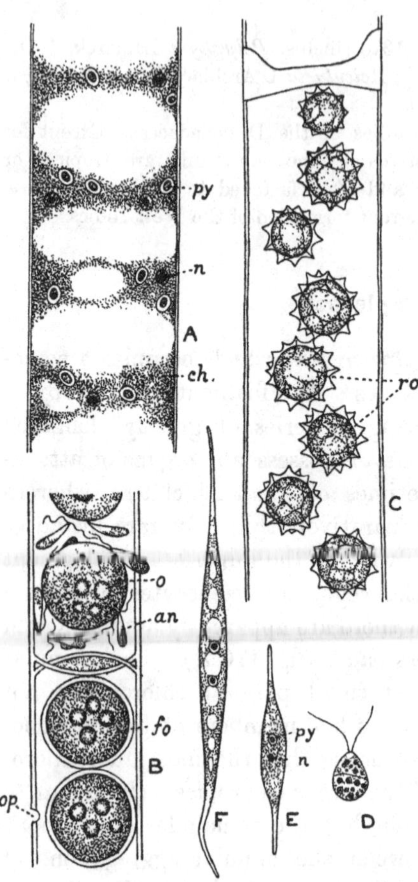

Fig. 176. *Sphæroplea annulina* (Roth.) Ag. *A*, part of cœnocyte stained to show annular aggregates of small chloroplasts (*ch*), nuclei (*n*) and pyrenoids (*py*), ×440. *B*, parts of two cœnocytes with gametes and fertilized eggs, ×440. *an*, antherozoid; *o*, oosphere; *fo*, fertilized egg; *op*, opening through wall of cœnocyte. *C*, part of cœnocyte with ripe oospores (*ro*) showing the verrucose walls, ×440. *D*, one of the zoospores formed on germination of the egg-cell, ×240. *E* and *F*, young plants formed from germinating zoospores, ×280. *n*, nucleus; *py*, pyrenoid. (*B* and *D*—*F*, after Cohn.)

In each oogonium there are numerous oospheres or egg-cells, dark green in colour but with a conspicuous receptive spot. According to Klebahn ('99) they may be either uninucleate or multinucleate. The antherozoids enter the oogonia through minute lateral orifices in the membrane (fig. 176 *B op*). After fertilization the wall of the egg is at first very thin, but a second verrucose wall arises underneath the first one, and the conical verrucæ of this membrane are connected by ridges. Subsequently the first membrane is shed. The egg-cells become filled with starch and oil, the latter having dissolved in it a bright red pigment. The ripe oospores rest for a long time and can survive dry conditions, even for several years.

On the germination of the oospore the chlorophyll becomes more obvious and the contents divide into 2, 4 or 8 (according to Meyer, '06, always 4) parts, which then issue from a split in the old wall as biciliated oval zoospores (fig. 176 *D*). These quickly undergo a transformation into spindle-shaped bodies, with much attenuated extremities, and soon an annular appearance of the contents is plainly visible (fig. 176 *F*). The young plants grow rapidly in size and become much elongated, acquiring numerous nuclei and pyrenoids, after which the first transverse wall appears.

Sphæroplea was first described by Agardh in 1824, but the first good account of the genus was by Cohn ('56). Since then others have added further and more detailed information, notably Rauwenhoff ('88), Golenkin ('99) and Klebahn ('99).

There is only one species, *Sphæroplea annulina*, with a rather local distribution in Europe, Asia, Africa and America. It occurs in pools, but more especially on plains which are liable to be flooded. Klebahn ('99) has suggested that *Sph. annulina* should be split up into *Sph. Braunii* and *Sph. crassisepta*, but K. Meyer ('06) rightly contests this view, and states that the variability is so great that only one species can be recognized, of which forma *Braunii* and forma *crassisepta* are the extreme forms.

Order 4. **ULVALES.**

The order Ulvales is characterized by an *expanded parenchymatous thallus, attached when young at one point to the substratum by rhizoids. The cells are uninucleate, with a single parietal chloroplast,* which contains one pyrenoid and is often sufficiently massive to fill most of the cell. In some cases the expanded thallus is flat and more or less irregularly lobed, but in others it is tubular and to some extent bent and contorted.

The thallus is sometimes propagated by proliferation or gemmation. Asexual reproduction takes place by zoogonidia, and there is a gamogenesis of isoplanogametes with two cilia, the zygote germinating at once.

The structure of the thallus indicates that the Ulvales are a rather isolated group which has probably had an origin in the Tetrasporine series of the Protococcales.

Family **Ulvaceæ.**

The Ulvaceæ are the only known family of the Ulvales, and the Algæ comprised in it are mostly marine and estuarine. The thallus may be a flat plate, ribbon-shaped or widely expanded, or it may be tubular and intestiniform. The simplest form is seen in *Monostroma* in which the flat thallus of the adult Alga is one layer of cells in thickness. In *Ulva* and *Letterstedtia* the thallus is composed of two layers of cells.

Monostroma has both freshwater and marine representatives, and in its early stages the thallus is bladder-like, splitting later to form an irregular membranous expansion which often floats quite freely. The cells may be so compactly arranged that their outlines are polygonal, or they may be distantly disposed and rounded in contour. In the latter case they are usually arranged in groups of four (*Monostroma bullosa*). In *Enteromorpha* and *Letterstedtia* the thallus is usually branched (consult figs. 177 *A* and 178), and in the latter genus lateral 'foliar' appendages are developed which subsequently fall off leaving the older parts of the thallus with irregularly toothed margins.

In *Enteromorpha* the adult thallus is tubular, having arisen by the splitting of a two-layered membranous expansion. The cells are very compact, with polygonal outlines (fig. 177 *B*), and in some species they are arranged in longitudinal rows. *Ilea* greatly resembles *Enteromorpha*, but is unbranched and the cells are arranged in regular groups of four.

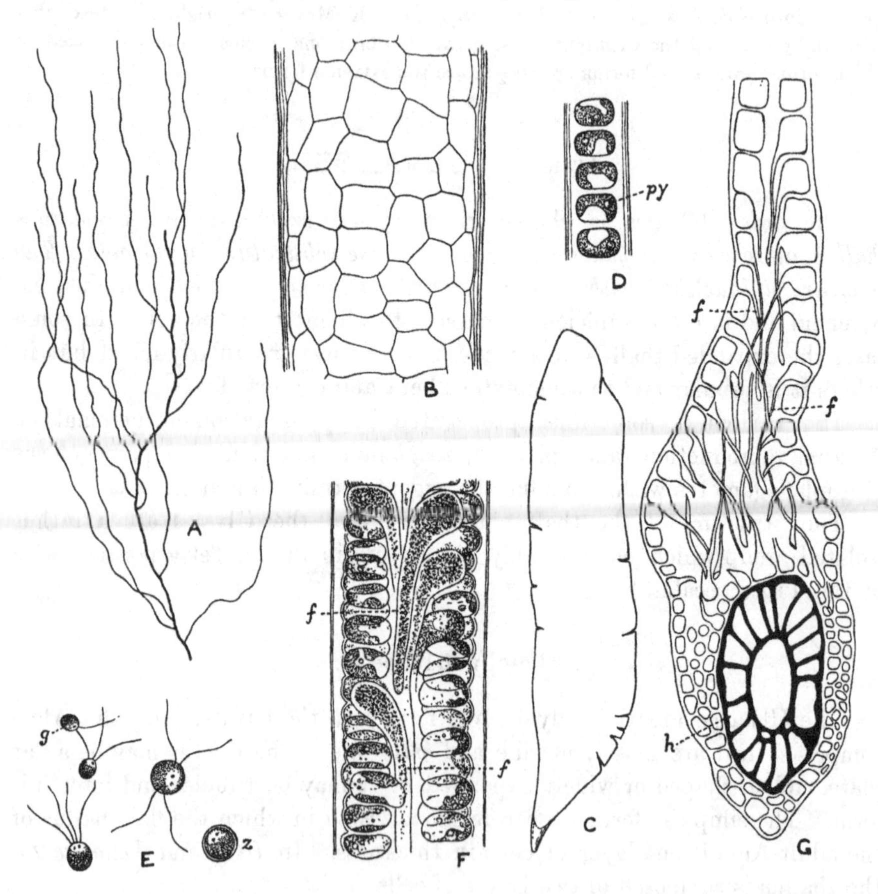

Fig. 177. *A* and *B*, *Enteromorpha gracillima* G. S. West. *A*, slightly less than natural size; *B*, surface view of small portion of branch showing outlines of cells, ×440. *C—E*, *Monostroma membranacea* W. & G. S. West. *C*, entire thallus rather less than natural size; *D*, section of thallus, ×495; *py*, pyrenoid; *E*, conjugation of gametes (*g*) to form zygote (*z*); gametes, ×495; 'zygozoospores' and zygote, ×690. *F*, section of thallus of *Ulva Lactuca* Linn. ×495; *f*, disc filaments. *G*, longitudinal section through the attaching disc of *Ulva*, somewhat simplified and diagrammatic; in this case the *Ulva* is attached to another seaweed which is acting as host (*h*), ×87; *f*, disc filaments. *F*, after Thuret; *G*, after Delf.

The cells are mostly disposed with their long axes at right angles to the plane of the thallus (fig. 177 *D* and *F*). Each cell contains a single nucleus and one large parietal chloroplast, often with lobed or deeply incised margins, and with one pyrenoid.

It is in the genus *Ulva* that the thallus is most securely attached. The attachment is by a basal disc, which is firmly adherent to stones, wood, or other Algæ, or as pointed out by Cotton ('11), it is held down in muddy estuaries by the byssus threads of mussels. The attaching disc was first described by Thuret ('78), and has been recently investigated by Delf ('12). The disc is composed of interwoven 'disc-filaments' which are tubular prolongations of the thallus-cells. The prolongations may be given off by any of the more basal cells of either of the layers composing the thallus; they pass down between the two layers of cells, pursuing a rather intricate and sinuous course in the lower part, and they sometimes branch. Thuret states that the filaments may attain a length of 6—10 mm., but Delf never found them to exceed 3 mm. At the periphery of the disc the extremities of the filaments swell out, become coherent and multinucleate, several transverse walls

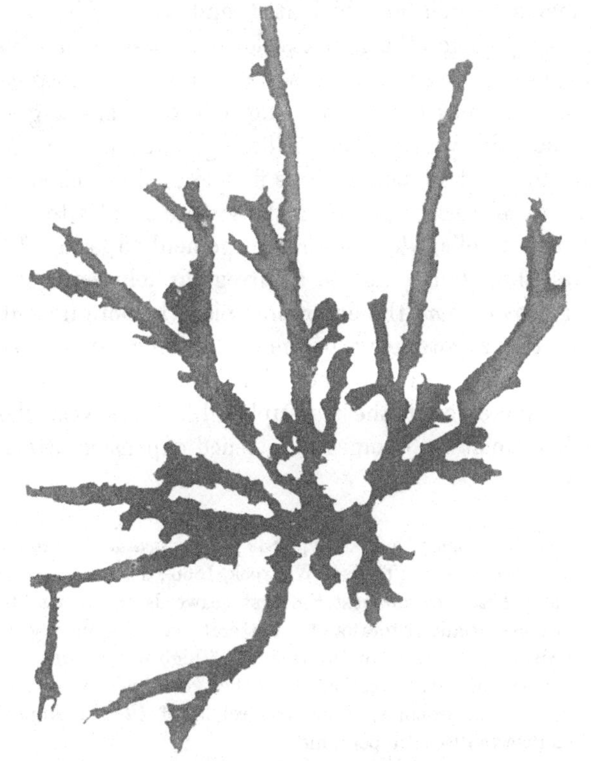

Fig. 178. *Letterstedtia insignis* Aresch. ½ nat. size.

often arising, so that a pseudoparenchyma is produced. The thallus-cells which give rise to the filaments appear to be always multinucleate, from three to five nuclei having been observed in their upper part; the minute nuclei also occur at intervals in the filaments and several are present in their swollen tips (Delf, '12).

In cases where *Ulva* is attached to certain of the smaller red seaweeds, Delf has found that the disc often completely encircles the host-plant (consult fig. 177 *G*) and that the filaments may penetrate in a parasitic manner into the cells of the host.

Propagation occurs in some species of *Monostroma* and *Enteromorpha* by gemmation or proliferation of the thallus, the small portions which are thus budded off growing into new plants. In *Ulva* new thalli can apparently be formed from 'multinucleate cells' which may be cut off from the swollen ends of the disc-filaments.

Asexual reproduction takes place by zoogonidia in *Ulva* and *Monostroma*. In the latter genus they may have either two or four cilia, but in *Ulva* they are always quadriciliated. On coming to rest they germinate at once.

Gamogenesis of isoplanogametes is known to occur in *Ulva, Monostroma* and *Enteromorpha*. Ordinary vegetative thallus-cells become gametangia without change of form and give rise to eight (sometimes four or sixteen) ovoid planogametes, which are biciliated and of smaller size than the zoogonidia. On conjugation the gametes slowly coalesce and a 'zygozoospore' with two pigment-spots and four cilia is first formed; this soon comes to rest, loses its cilia, and becomes a zygospore (fig. 177 *E*). The zygospore usually germinates at once, the young plants differing much in the various genera. In *Ulva* a short filament of four cells is first formed, which soon produces a flat expansion. In *Monostroma bullosa* the zygote divides to form a more or less globular mass of cells, which after enlargement to form a hollow sphere eventually splits, thus giving rise to an irregular foliaceous expansion. In other species of *Monostroma* the embryonic plant is more irregular. Reinke has found that the zygote may become a resting hypnocyst, which on germination divides into four and then eight cells arranged around a central cavity. By the increase of these peripheral cells a vesicular thallus is produced, which ultimately becomes a flattened expansion attached by a few rhizoids at its base.

The genera are: *Enteromorpha* Harvey, 1849; *Letterstedtia* Areschoug, 1850; *Ulva* (L.) Wittrock, 1866; *Monostroma* (Thuret) Wittrock, 1866; *Ilea* J. G. Agardh, 1883.

Enteromorpha and *Ulva* are amongst the first seaweeds to attract attention on the sea-shore. They are essentially littoral and sublittoral, occurring mostly from about half-tide level downwards, or sometimes in the vicinity of high-water mark. Various species of *Enteromorpha* are the principal constituents of the 'grass' which is so largely the cause of the fouling of the bottoms of ships. The flat 'fronds' of *Ulva* are annual, but it seems probable that the adhesive discs are perennial.

The Ulvaceæ are more especially marine, but there are several freshwater species of *Monostroma*, such as *M. bullosa* (Roth.) Wittr., *M. membranacea* W. & G. S. West ('03; fig. 177 *C—E*), and *M. expansa* G. S. West ('06), the latter being an Australian species with a thallus attaining 30 cms. in length. *Enteromorpha intestinalis* also extends from the sea-coast into brackish and fresh waters, being a frequent Alga in canals and ponds. *E. gracillima* G. S. West ('12 A; fig. 177 *A* and *B*) has been found in sulphurous springs in S.W. Africa. The widely distributed *Ulva* often extends into brackish waters, but *Letterstedtia* is confined to the coasts of Natal, Australia and Japan.

Ulva and *Enteromorpha* sometimes occur in great quantities in muddy estuaries and other coastal areas where there is some amount of sewage pollution (consult page 146). It has been suggested that both genera become partially saprophytic in the presence of the nitrogenous products of decay.

Order 5. SCHIZOGONIALES.

This order was instituted (G. S. W. '04) to include the single family of the Prasiolaceæ, in which the thallus is filamentous, or subparenchymatous, and often expanded into broad sheets. The expanded thalli sometimes arise by the concrescence of the filaments in one plane and sometimes by the regular division of the cells in two directions in one plane.

These Algæ are subaërial, and the expanded thalli are attached to the substratum by rhizoids.

The order is at once distinguished from the Ulotrichales by the *axile chloroplasts and by the division of the cells in two directions in one plane*, or even in three directions in young plants. From the Ulvales it is distinguished *by its chloroplasts and by the more regular arrangement of the thallus-cells consequent upon their division only in two directions at right angles.* It is not improbable that the Prasiolaceæ originated from the Protococcales along a line very different from that along which the Ulvales evolved.

Family **Prasiolaceæ**.

It is possible that all the forms of this family (= the Schizogoniaceæ of Chodat, '02, the Prasiolaceæ of West, '04, and the Blastosporaceæ of Wille, '09) are referable to the one genus *Prasiola* Agardh (1821), although there are some reasons for the retention of *Schizogonium* Kützing (1843) for the simpler types. The thallus is either a simple unbranched filament or it forms a flat expansion one layer of cells in thickness. The simple filament is the *Hormidium*-state and when several of these filaments fuse laterally they give rise to the *Schizogonium*-state. In the wider expansions the cells usually divide in two directions in one plane resulting in the *Prasiola*-state. In these flat expansions the cells are frequently arranged in regular groups of four or multiples of four, such groups being as a rule separated by rather thicker walls. In the *Hormidium*-state the cells are shortly cylindrical (fig. 179 *A* and *B*), but in the expanded *Prasiola* they are sometimes angular by compression, and so situated that their long axes are at right angles to the plane of the thallus. Each cell possesses a single nucleus and a central stellate chloroplast. The latter is much lobed and is furnished with one pyrenoid. The cell-wall is strong, hyaline and fairly thick. Most of the expanded thalli are fixed to the substratum by means of rhizoids, which may be developed from any cell of the thallus, although they are mostly found as outgrowths from the more basal cells (fig. 179 *G*).

Vegetative propagation often occurs by a proliferation (or gemmation) of the thallus, resulting in the detachment of small portions from the margin.

Asexual reproduction occurs by akinetes, formed either directly from the vegetative cells or by the division of the vegetative cells in two planes to form 'tetraspores' (Lagerheim, '92 B). The akinetes may grow directly into new plants on liberation or they may become aplanosporangia in which a number of rounded or ovoid aplanospores are formed. The latter escape by the bursting of the wall of the aplanosporangium and grow directly into new plants (Wille, '02; '06). Under unfavourable conditions both akinetes and aplanospores may become resting-spores.

Zoogonidia and gametes are entirely unknown.

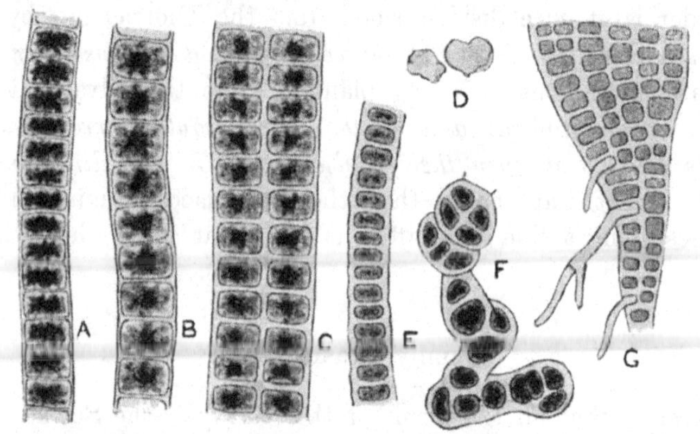

Fig. 179. *A—C, Prasiola crispa* forma *muralis* (=*Schizogonium murale* Kütz.), ×500. *D—G, Prasiola crispa* (Lightf.) Menegh.; *D*, two expanded thalli, nat. size; *E*; simple filament, ×500; *F*, part of irregular filament, ×500; *G*, basal part of an expanded thallus showing a few rhizoids, ×400.

The various species of *Prasiola* are subaërial Algæ, living on damp ground, damp walls, and on rocks, especially rocks of the sea-shore. In some forms the *Hormidium*- and *Schizogonium*-states appear to be permanently retained, whereas others almost invariably form expanded thalli. *Prasiola crispa* forma *muralis* (=*Schizogonium murale*) is invariably filamentous and is found almost exclusively on damp earth, especially in towns and villages, where it often forms a green stratum between paving stones (fig. 179 *A—C*). *P. crispa* (fig. 179 *D—G*) has an expanded thallus occurring on damp earth under walls, and also in quantity on rocky sea-shores and about the nesting-places of sea-birds. It has a world-wide distribution, extending far into the Arctic and Antarctic regions, always occurring where there is a plentiful nitrogenous food-supply. In one form of this Alga, which has been described under the name of *Gayella polyrhiza* Rosenvinge, a cell-mass is formed instead of a cell-plate.

Gay ('91), Chodat ('02) and Collins ('09) separate the genera *Prasiola* and *Schizogonium*, but other authors (notably Wille, '00; '09 B; Brand, '14) have united them under *Prasiola*.

Wille, and also Börgesen ('02), have each described some interesting marine forms of *Prasiola* in which the plants are more amply supplied with rhizoids. The expanded thallus in these forms does not attain so large a size as in the terrestrial forms. The thallus of *P. crispa* is normally from 4 mm. to 2 cms. in length, but in the vicinity of sewage works and on 'contact beds' the thalli may attain a length of 8 cms.

Letts ('13) has found a remarkable amount of nitrogen in specimens of *P. crispa*, amounting in some instances to 8·94 per cent. of the dried material. This percentage is almost double that which is found in ordinary vegetable substances, higher than that contained in dried cheese, and almost as high as that contained in dried meat.

Order 6. ULOTRICHALES.

This order, which is identical with the 'Chætophorales' of Wille, includes a relatively large number of Green Algæ which are mostly inhabitants of fresh waters. *The thallus is filamentous, sometimes simple, but more frequently branched.* The branches are generally attenuated and often piliferous. *The cells are uninucleate* (except in the peculiar genus *Wittrockiella*), *and in all the families of the order, with the exception of the Trentepohliaceæ, they possess a single parietal chloroplast with one or more pyrenoids.*

The only families in which simple, unbranched filaments occur are the Ulotrichaceæ, Microsporaceæ and Cylindrocapsaceæ; in fact, only one-fifth of the known genera are unbranched types. The majority of the unbranched forms are free-floating, whereas practically all the branched forms are attached, many of them being epiphytes. In the latter the branches are sometimes concrescent on the surface of the host, so that discoidal growths are formed, and the cells are frequently furnished with setæ or bristles.

Reproduction by biciliated, or more rarely quadriciliated, zoogonidia is general throughout the order, although it has not yet been observed in every genus. Aplanospores or akinetes, or both, are known in most genera. Isoplanogametes occur in very many of the genera, and in the Cylindrocapsaceæ, Aphanochætaceæ and Coleochætaceæ there are well-differentiated egg-cells and antherozoids, *Coleochæte* being, as regards its sexual reproduction, probably the highest type among the Green Algæ.

A few genera (*Arthrochæte, Endophyton*) are endophytic in larger red and brown sea-weeds, and others have epizootic (*Trichophilus*) and endozootic (*Endoderma*) representatives. *Tellamia* perforates the shells of marine and freshwater bivalves, and some species of *Cephaleuros* are destructive parasites on phanerogams.

The Ulotrichales are a natural and extensive group which appear to have been derived from certain of the Tetrasporine Protococcales by the development of the filamentous and ultimately of the branched habit. Most of them are aquatic, but the Trentepohliaceæ are a conspicuous family of subaërial Algæ. The Microsporaceæ and Trentepohliaceæ are rather anomalous groups, the peculiarities of the latter family being partly the result of complete adaptation to subaërial conditions.

The inter-relationships of some of the branched forms of the Ulotrichales are rather obscure, and Pascher ('07) has attempted to trace certain lines of descent by a survey of the characters of the zoogonidia and gametes.

The relationships of the unbranched genera are fairly clear, and the transition from *Geminella* to *Ulothrix*, through *Stichococcus*, is obvious and complete. *Binuclearia* and *Uronema* are both specialized forms closely akin to *Ulothrix*. *Rhaphidonema* may be a specialization from the subaërial form of *Stichococcus* or a reduction from the subaërial forms of *Ulothrix*.

It is highly probable that there have been several divergent lines of descent from members of the Ulotrichaceæ, along three of which oogamy has been independently attained.

The accompanying diagram may help the student to form an approximate idea of the phylogenetic relationships of the chief types of the Ulotrichales.

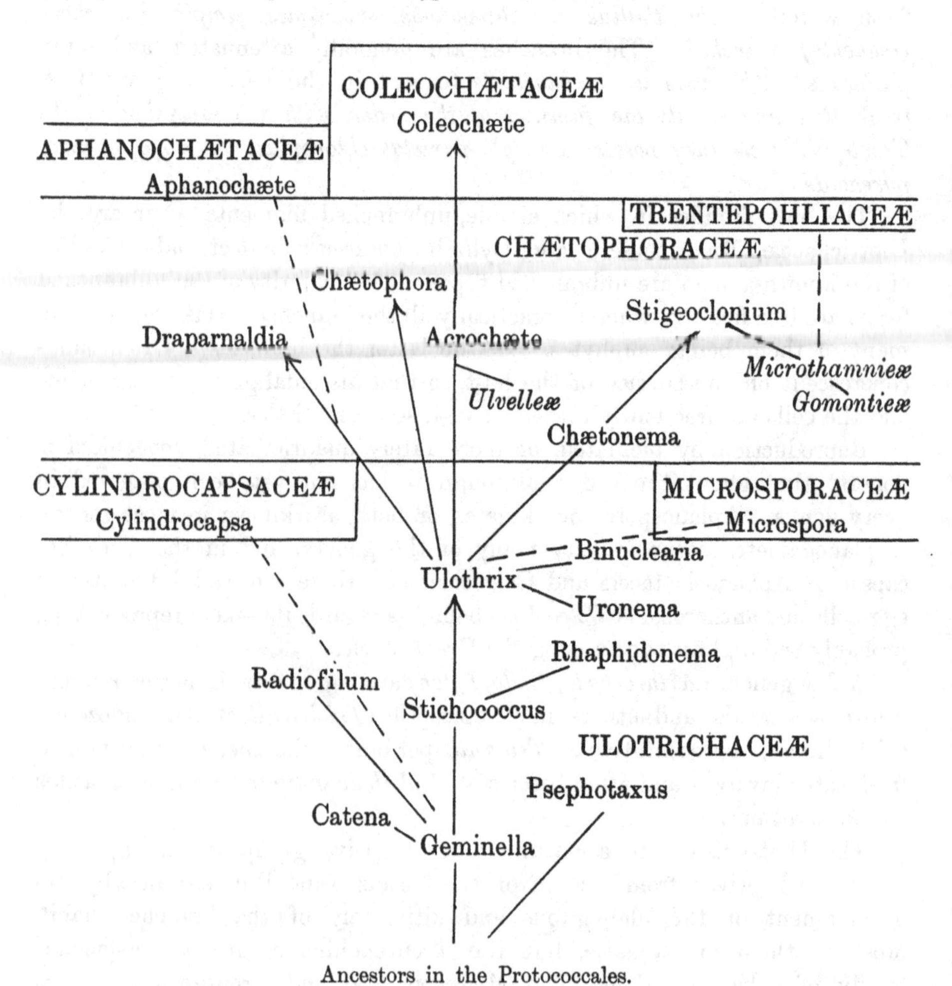

Family **Ulotrichaceæ**.

The Algæ of this family are unbranched and readily recognizable by their cytological characters. The lowest types are those in which the cells of the filaments are largely held in place by a conspicuous gelatinous envelope (*Geminella, Radiofilum*). More advanced are those genera in which there are short filaments of loosely connected cells without any mucous investment (*Stichococcus, Catena*). The highest types are those forms with cylindrical filaments of firmly united cells (*Ulothrix, Binuclearia*).

The cell-wall is always hyaline and colourless, but varies much in thickness. It is sometimes delicate (*Catena, Stichococcus, Rhaphidonema,* some species of *Ulothrix*), and sometimes thick and lamellose (*Psephotaxus, Binuclearia, Ulothrix zonata*). In the lower types (*Geminella, Radiofilum*) the outer layers become converted into a wide mucilaginous envelope. In *Binuclearia tatrana* and *Ulothrix zonata* the degree of lamellation of the cell-wall depends upon the state of growth.

There is a single parietal plate-like chloroplast in each cell, with an entire or variously lobed margin, usually occupying only a portion of the cell-wall. It generally possesses one or more pyrenoids, but may be entirely destitute of them. There is a small inconspicuous nucleus, usually placed internally to the chloroplast.

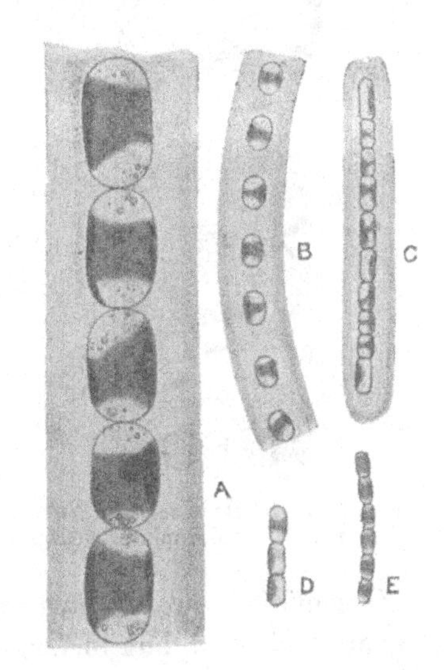

Fig. 180. *A*, *Geminella mutabilis* (Bréb.) Wille. *B*, *G. ordinata* G. S. West. *C—E*, *G. protogenita* (Kütz.) G. S. West. All × 440.

In the genus *Geminella* the end-walls of the cells become hemispherical soon after cell-division and the cells, although remaining as a filament, are only held in position by the mucilaginous sheath (fig. 180). In *Stichococcus* the external characters are intermediate between those of *Geminella* and *Ulothrix.* The cells divide to form cylindrical filaments which under normal conditions undergo fragmentation into short pieces consisting of only a few cells. This is caused by the apposed ends of certain pairs of adjacent cells becoming hemispherical, with the result that the filament becomes dislocated at these points. In some species the fragmentation is so complete that the plants are almost unicellular (fig. 183 *D—F*).

Dismemberment of the filaments of *Ulothrix* often occurs by the death of some of the cells, each dismembered piece growing into a new filament.

Asexual reproduction takes place by aplanospores, akinetes and zoogonidia. Aplanospores occur in a number of the smaller species of *Ulothrix*, in *Uronema* and in *Binuclearia* (G. S. W., '04). They are more rarely found in the larger *Ulothrix zonata* (fig. 183 *B* and *C*). Wille ('12) regards the aplanospores of *Stichococcus flaccidus*, which are formed singly within the vegetative cells, as reduced zoogonidia; they germinate at once, dividing either cruciately or tetrahedrally, and ultimately form a palmella-state. A similar palmella-state may sometimes be observed in *Ulothrix subtilis* (G. S. W., '04). In some instances they increase in size and become globular hypnospores with slightly asperulate walls. Akinetes are frequently formed in various species of *Ulothrix*, *Stichococcus* and *Geminella*. In *Ulothrix idiospora* G. S. West ('09) they are thick-walled and scrobiculate (fig. 91 *A* and *B*).

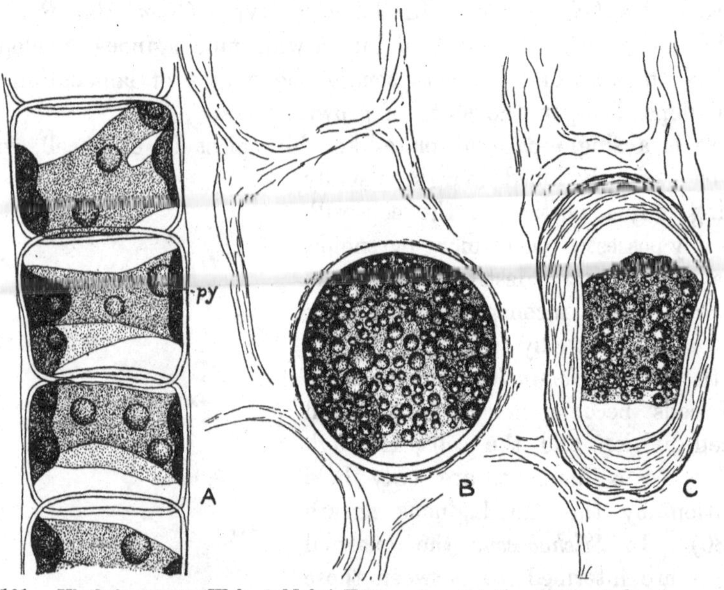

Fig. 181. *Ulothrix zonata* **(Web. & Mohr) Kütz.** *A*, **vegetative filament showing parietal ring-like chloroplasts with pyrenoids (*py*).** *B* **and** *C*, **aplanospores. All × 800.**

Zoogonidia occur in *Ulothrix*, *Stichococcus* and *Uronema*. In *Stichococcus* they are biciliated, but in *Uronema* they are quadriciliated; in both genera only one zoogonidium arises from a mother-cell. In the genus *Ulothrix*, in which the different species range from 4 μ in diameter (in the narrowest form of *U. subtilis*) to 70 μ (in the larger montane forms of *U. zonata*), the zoogonidia vary very much. From 1 to 32, or even more, may arise in a mother-cell. In the narrower species they arise singly or less often in pairs, and are furnished with either two or four cilia. In the large species *U. zonata*, in which zoogonidium-formation was very carefully studied by Dodel ('76), every gradation in size is met with between macrozoogonidia with four cilia and microzoogonidia with two cilia. Moreover, all the gradations of zoogonidia may be produced from the cells of the same filament. In *U. zonata* 16 or 32 microzoogonidia are produced from each mother-cell, but only 2 or 4 in *U. subtilis*; similarly, 2, 4 or 8 macrozoogonidia (often complanate) are formed in *U. zonata*, but only one in *U. æqualis* and *U. subtilis*. The zoogonidia germinate directly on coming to rest, and sometimes while still within the mother-cell (fig. 183 *B* and *C*). Owing to the great variability in the size of the zoogonidia of *U. zonata*, the filaments of this species vary greatly in thickness, but in other species of the genus there is a much greater uniformity in the thickness of the filaments.

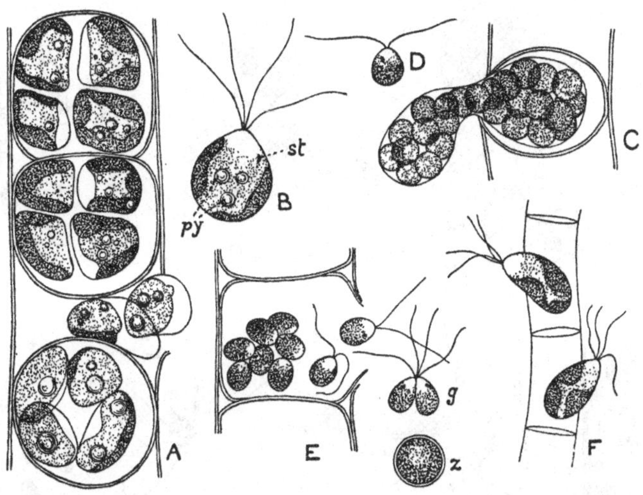

Fig. 182. *A—E*, *Ulothrix zonata* (Web. & Mohr) Kütz. *A*, formation of macrozoogonidia; *B*, macrozoogonidium showing 4 cilia, pigment-spot (*st*) and pyrenoids (*py*); *C*, formation of microzoogonidia; *D*, microzoogonidium; *E*, escape of gametes, their conjugation (*g*), and the zygote some days after its formation (*z*). *F*, the zoogonidia of a form of *U. tenerrima* Kütz. All × 800.

Gamogenesis takes place by the fusion of iso- or anisogametes. The gametes are always biciliated, and in some species of *Ulothrix* and *Stichococcus* they are sufficiently differentiated in size to be termed macro- and microgametes. They may conjugate isogamously or anisogamously, or germinate without conjugation (parthenogenetically). The gametes are frequently quite indistinguishable from the microzoogonidia, a fact which causes one to suspect that the latter are facultative gametes. The zygote becomes invested with a thick wall and undergoes a more or less extended period of rest. On germination it becomes converted into a unicellular germ-plant, which gives rise to 8 or 16 swarm-spores (zoogonidia).

The conditions necessary for the production of zoogonidia in *Ulothrix zonata* were investigated by Klebs ('96), but most of his conclusions have not been verified, even for

U. zonata, and they do not apply to several of the smaller species of *Ulothrix*[1]. Four of Klebs' main conclusions were: (1) that if, after cessation of zoospore-formation, the filaments were transferred from nutritive solutions to water, then zoospore-formation recommences; (2) that zoospores were formed by transferring filaments from a nutritive solution to water and then putting the cultures in the dark for three days; (3) that transference of the filaments from running water to quiet water, or bringing from a natural habitat into a room, resulted in zoospore-formation, the cause being attributed to the diminution of the amount of air in the water; (4) that observations were against a rise in temperature being a special cause in inducing the formation of zoospores.

Klebs' first statement does not apply to the smaller species of *Ulothrix* and repeated experiment has failed to confirm it in *U. zonata*. Experiments extending over two years

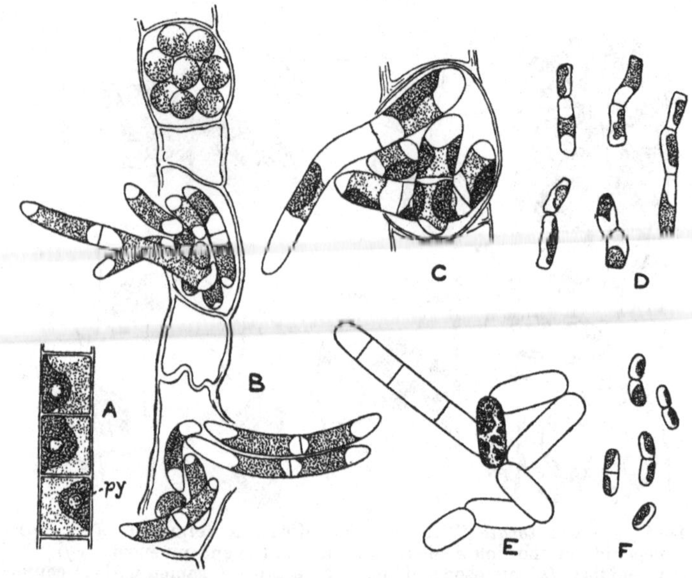

Fig. 183. *A*, part of filament of *Ulothrix æqualis* Kütz. to show single chloroplast with pyrenoid (*py*). *B* and *C*, *U. zonata* (Web. & Mohr) Kütz. *B*, part of filament showing germination of microgonidia (which have never been motile) to form new plants; *C*, similar germination of macrogonidia. *D*, *Stichococcus variabilis* W. & G. S. West. *E*, aquatic form of *S. flaccidus* (Kütz.) Gay, only in outline except for one cell which shows partial fragmentation of chloroplast. *F*, *S. bacillaris* Näg. All ×500.

do not bear out his second statement in the case of either *U. tenuissima* or *U. subtilis*. Concerning his third statement this was not found to be true in the case of *U. zonata* and *U. subtilis* unless the transference was accompanied by a decided change in temperature. In contrast to his fourth statement experiments conducted over a long period with *U. zonata*, *U. æqualis*, *U. tenuissima* and *U. subtilis* lead to the conclusion that a rapid alteration in temperature (whether increase or decrease) is a direct incentive to zoospore-formation, and is really a factor of great importance in the production of the zoogonidia of *Ulothrix*.

[1] Various species of *Ulothrix* have been under experimental investigation by the author and Miss C. B. Starkey in the botanical laboratory of the University of Birmingham for upwards of two years.

The genus *Uronema* differs little from *Ulothrix*, but the filaments are mostly rather short and fixed at the base; moreover, the apical cell is always conical. In *Geminella* the cells develop convex, and often hemispherical, poles, and are scarcely adherent, being held in position very largely, and in some cases entirely, by the copious mucous investment (fig. 180). *Binuclearia* is a montane and subalpine genus characterized by the great thickness of some of the transverse walls of the filaments, the cells having the general appearance of being arranged in pairs. In *Rhaphidonema* the filaments, which are attenuated at both ends, are the shortest of any genus of the family, consisting of only a few (4 to 7) cells. This genus was first described by Lagerheim ('92 C) from the snow-flora of the Andes, but both the original species (*Rh. nivale*) and another one (*Rh. brevirostre* Scherffel, '10) are known to occur in Europe.

The genera are: *Geminella* Turpin, 1828; em. Lagerheim, 1883 [inclus. *Hormospora* Brébisson, 1840; *Glœotila* Kützing, 1843; *Planctonema* Schmidle, 1903]; *Ulothrix* Kützing, 1833; *Stichococcus* Nägeli, 1849 [inclus. *Hormidium* Klebs, 1896 (non Kützing, 1843) and *Hormococcus* Chodat, 1902]; *Binuclearia* Wittrock, 1886; *Uronema* Lagerheim, 1887; *Rhaphidonema* Lagerheim, 1892; *Radiofilum* Schmidle, 1894; *Psephotaxus* W. & G. S. West, 1897; *Catena* Chodat, 1900.

With the exception of a few subaërial species of *Ulothrix* and *Stichococcus* all the members of the Ulotrichaceæ are aquatic.

Most of the species of *Ulothrix* pass the winter in the vegetative condition in the form of short filaments in which both the transverse and lateral walls have much increased in thickness. In the British Islands active cell-division generally begins by the middle of January (in the absence of prolonged frosts), the old thick walls being burst apart as the cells divide. In the winter state of *U. æqualis* and *U. subtilis* oil-drops are of frequent occurrence.

Quite recently Brunnthaler ('13) has attempted to show that *Radiofilum* should be transferred to the Desmidiaceæ, but there is little doubt that such a position would not be in accordance with the affinities of the genus. In describing some Australian forms of *R. conjunctivum* the present author (G. S. W., '09) instituted a comparison between the structure of the cell-wall in this Alga and that which occurs in the Placoderm Desmids. 'The wall of each individual cell is composed of two halves, and cell-division appears to take place much as it does in some of the simpler types of Desmids, such as *Penium*, by the interpolation of two new half-cells between the old ones. The line of junction of the old and new halves of the wall is distinctly visible in most specimens, and is particularly obvious at the marginal apiculations, the latter owing their prominence to the projecting suture at this region. Each fully-grown half is helmet-shaped, but in its earliest stages the young half is much flattened. The chloroplasts are parietal and cup-shaped, occupying about two-thirds of the interior of the cell-wall. They are disposed very largely back to back in pairs of adjacent cells.' This statement apparently induced Brunnthaler to transfer the entire genus (with its 3 species) to the Desmidiaceæ, but such a view suggests lack of direct knowledge of the Algæ in question. Neither *R. flavescens* G. S. West ('99) nor *R. irregulare* has any near affinity with the Conjugatæ, and the similarity in structure of the wall between *R. conjunctivum* and many Desmids can only be regarded as a parallelism of modification.

It is necessary to make some special mention of the genus *Schizomeris* Kützing (1843). Hazen ('02) states that he has no doubt concerning the validity of the genus *Schizomeris*, but that its affinity seems to be with the Ulvaceæ rather more than the Ulotrichaceæ. The filaments are attached to rocks and stones in streams; they are attenuated at both base and apex, and longitudinal division of the cells occurs. The manner of the dispersal of the zoogonidia is very striking and unlike anything in the Ulotrichaceæ. Hazen remarks that 'all the dissepiments in the upper part of the thallus appeared to be softened or broken down, and the masses of zoospores escaped through the open funnel formed by the outer cell-wall.'

Family **Microsporaceæ.**

This family includes only the one genus *Microspora* (Thuret, 1850; em. Lagerheim, 1888)[1] the species of which are amongst the most abundant and widely distributed of freshwater Algæ. The filaments are cylindrical and unbranched, and from the very earliest stages of development are free-floating. The cell-wall varies in character in the different species of the genus. In some it is apparently homogeneous, but in others it is distinctly lamellate, attaining its greatest thickness and irregularity in *Microspora amœna* var. *irregularis* W. & G. S. West ('06). The lamellæ consist of cellulose. The statement so frequently made that the lamellæ of the wall are so laid down that the filaments become disarticulated into H-pieces is only true of certain species of the genus. This occurs in *M. tumidula*, in the typical forms of *M. amœna* and *M. Löfgrenii*, and to some extent in *M. floccosa* and *M. quadrata*. In the two last-mentioned species the disarticulation is only partial and often irregular, parts of the lateral walls usually

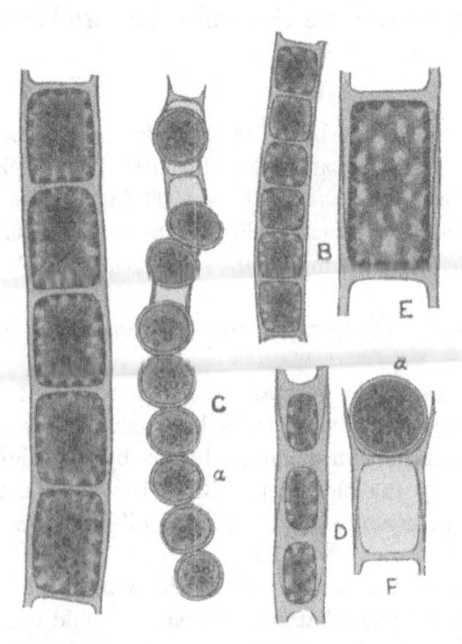

Fig. 184. *A*, *Microspora amœna* (Kütz.) Lagerh. *B* and *C*, ? *M. abbreviata* (Rabenh.) Lagerh.; *B*, vegetative filament; *C*, filament with aplanospores (*a*). *D*, *M. pachyderma* (Wille) Lagerh. *E*, single vegetative cell of *M. amœna* var. *crassior* Hansg., showing the reticulated chloroplast. The indistinct blur in the centre of the cell indicates the position of the nucleus. *F*, fragment of filament of *M. amœna* with aplanospore (*a*). All × 520.

[1] The main conclusions of the paper on "The Structure, Life-History and Systematic Position of the genus *Microspora*," which was communicated by the present author to the British Association at the Birmingham Meeting in September, 1913, are embodied in this account of the Microsporaceæ.

The recent paper by Meyer on *Microspora amœna* (December, 1913) adds nothing of importance to the present account. Meyer's figure of the formation of 'akinetes' (fig. 15) represents the formation of aplanospores.

becoming mucilaginous. In the common species *M. stagnorum*[1] the filaments almost always break at or near the transverse walls.

The cells are uninucleate, the nucleus in many cases occurring in a fairly broad band of cytoplasm stretching across the median part of the cell.

There is but one chloroplast in each cell, and the nature of this chloroplast is the chief distinguishing feature of *Microspora.* It consists of a number of small parietal green cushions which are united in a most irregular way to form a reticulum. The very irregularity of this reticulum, the form of which may be quite different even in adjacent cells

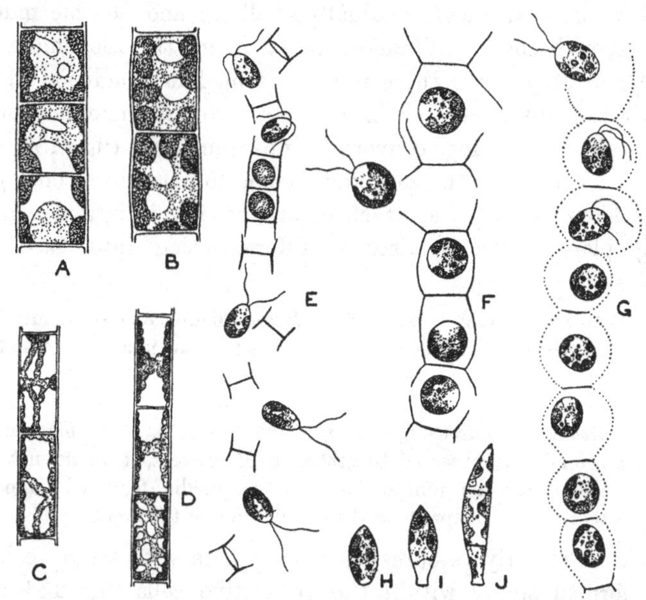

Fig. 185. *A*, ? *Microspora abbreviata* (Rabenh.) Lagerh., three cells to show the chloroplast, ×1000. *B*, two cells of *M. tumidula* Hazen, showing form of chloroplast, ×1000. *C* and *D*, portions of two filaments of *M. stagnorum* (Kütz.) Lagerh. showing characteristic variations in the chloroplast, ×500. *E*, escape of zoogonidia in *M. tumidula* Hazen, in which the cell-walls break up into H-pieces, ×500. *F*, escape of zoogonidia in *M. floccosa* (Vauch.) Thur., in which the H-pieces are more irregular and there is slight gelatinization of the wall, ×500. *G*, escape of zoogonidia of *M. stagnorum*, in which H-pieces are never formed but the wall is completely converted into mucilage, ×500. *H—J*, germination of zoogonidium of *M. stagnorum*, ×500.

(consult fig. 185 *D*), is one of the leading characters of the genus. The cushions of the chloroplast are usually spread over both lateral and terminal walls, and in some instances (frequently in *M. stagnorum* and occasionally in *M. floccosa*) they are disposed as irregular constricted bands united to form a loose network (fig. 185 *C*). There are no pyrenoids, but starch is stored in the form of small granules.

[1] The figure given by Hazen ('02, t. 24, f. 12) of the disarticulation of a filament of *M. stagnorum* during the liberation of zoogonidia does not agree with any state of this alga observed by the present author.

Reproduction occurs by zoogonidia developed singly (or more rarely in twos, threes or fours) in the vegetative cells (fig. 185 *E*—*G*). They are ovoid or subellipsoid, *biciliated*, and possess an irregularly lobed chloroplast the limitations of which are usually indefinite. As a rule most of the cells of a filament produce zoogonidia simultaneously, their formation and ultimate escape only occupying about an hour. The escape of the zoogonidia in the different species brings out clearly the differences in the structure of the cell-wall. In *M. tumidula* the entire filament becomes disarticulated into H-pieces (fig. 185 *E*) and the zoogonidia swim quickly away; in *M. stagnorum* the lateral walls of the cells gradually swell up and become mucilaginous, the mucus soon becoming diffluent and allowing the zoogonidia to escape (fig. 185 *G*); in *M. floccosa* there is an intermediate condition in which the filament becomes disarticulated in rather an irregular manner, some portion of the lateral walls being converted into mucilage (fig. 185 *F*). After swarming for a brief time the zoogonidia come to rest and at once germinate. The young filament may be attached, but usually it is not. In any case the young plant shows a distinct differentiation into base and apex (fig. 185 *H*—*J*).

Thuret ('50), in his original account of *M. floccosa*, figures 8 or 16 minute 'zoogonidia' arising from each cell of the filament. Such is not the case in any species of *Microspora* examined by the present author—not even in *M. floccosa*. In all observed cases, with the exception of *M. amœna*, only one biciliated zoogonidium was formed in each cell. From a careful comparison of Thuret's figures 5, 6 and 7 (on Pl. 17 of his work) with the formation of zoogonidia as observed by others in *M. floccosa*, it would not appear unreasonable to suggest that the minute elongated 'zoogonidia' figured by him were really isogametes and that his fig. 7 represents the germination of the zygote.

Spherical or slightly compressed aplanospores with strong cell-walls are sometimes formed singly within the vegetative cells (fig. 184 *C* and *F*). Akinetes are formed abundantly in some species, such as *M. floccosa*. They are subquadrate or rectangular, slightly tumid, and provided with thick walls (fig. 91 *D*). Both aplanospores and akinetes generally occur in long chains, and both are resting-spores.

Gametes have not been observed, unless the above suggestion concerning Thuret's 'zoogonidia' of *M. floccosa* is correct.

The systematic position of the genus *Microspora* has been a controversial question for some time. Bohlin ('01) regarded it as the type of a separate order, the Microsporales, and in this he was followed by West ('04). Blackman & Tansley ('02) placed the genus in the family Microsporaceæ of the Ulotrichales, and both Oltmanns ('04) and Wille ('09 B) referred it directly to the Ulotrichaceæ.

A careful investigation of the British species upholds the position assigned to this genus by Blackman & Tansley. It cannot with justice be placed in the Ulotrichaceæ, and yet in view of the occurrence of a reticulated chloroplast in the axial cells of *Draparnaldia platyzonata* (consult fig. 190 *B*) and the usual absence of pyrenoids from certain species of *Ulothrix* (such as *U. subtilis* and *U. æqualis*), it is probably best to regard the genus as somewhat peculiar member of the Ulotrichales for which it is necessary to provide the special family Microsporaceæ.

The germination of the akinetes has been studied by the present author. When kept in water it was found that they remained alive and apparently unchanged for a period of about two years, after which a small percentage germinated. The germination proceeded in one of two ways: (1) the protoplast divided into four as it escaped from the old wall of the akinete, four rounded cells being set free, each of which grew into a new filament (fig. 186 *D* and *E*); (2) the protoplast escaped without any division from the old wall, and divided first into two and then into four cells tetrahedrally arranged, each cell on becoming free developing into a new filament (fig. 186 *F*—*I*).

If the akinetes were allowed to become dry for a period of a week or ten days, soon after their formation, and then placed in water they germinated very readily. Each one elongated, the thick wall became largely mucilaginous, and transverse divisions soon resulted in a new filament (fig. 186 *A*—*C*). This quick germination is entirely different from the belated type which takes place when the akinetes are never allowed to dry, and indicates that a period of drought is favourable for their proper development.

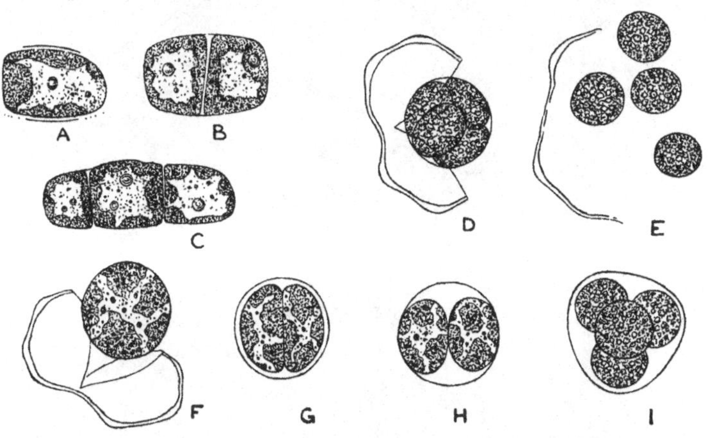

Fig. 186. Germination of the akinetes of *Microspora floccosa* (Vauch.) Thuret. *A*—*C*, direct germination of akinete in water, after having been dried for ten days. *D*—*I*, germination of akinetes after a resting period of two years during which they were never allowed to become dry; *D* and *E*, first method of germination; *F*—*I*, second method of germination. All ×850.

Family **Cylindrocapsaceæ.**

This family includes the single genus *Cylindrocapsa* Reinsch ('67), concerning which our knowledge is as yet in many ways defective. The thallus is filamentous and unbranched, with the cells arranged in a single series. The filaments are short and in some respects they resemble those of the more mucilaginous types of the Ulotrichaceæ. The cells are ellipsoid, subrectangular or less often subtriangular in shape, and they are frequently grouped in pairs at short intervals along the filament. They are provided with a firm cell-wall outside which are several mucilaginous lamellæ, and the row of cells comprising the filament is enclosed within a cylindrical, close-fitting sheath of tough mucus (fig. 187 *E* and *F*). The general arrangement of the cells is not unlike that in the genus *Geminella*, their frequent disposition in pairs resulting from the simultaneous division of many cells. Each cell possesses a massive parietal chloroplast, often with most obscure limitations, and generally furnished with one pyrenoid. Numerous small grains of starch can as a rule be detected in the healthy cell.

Asexual reproduction takes place by zoogonidia which arise singly, or in twos or fours, from any cell of the filament. The zoogonidium is ovoid or ellipsoid in form, biciliated, and provided with a pigment spot and two contractile vacuoles.

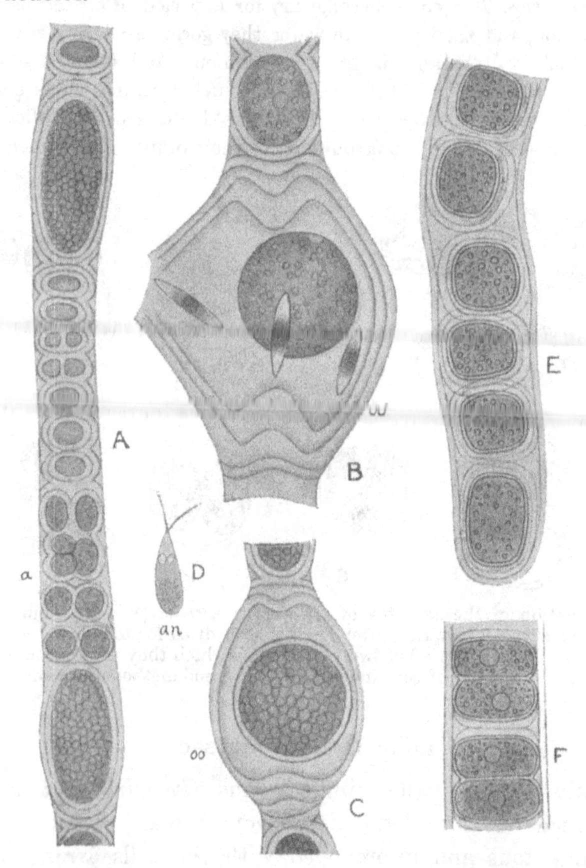

Fig. 187. *A—D*, *Cylindrocapsa involuta* Reinsch, × 480 (after Cienkowski). *E* and *F*, *C. conferta* W. West, × 520. *a*, antheridium; *an*, antherozoid; *oo*, oogonium.

Sexual reproduction of a comparatively high type was observed in *Cylindrocapsa involuta* by Cienkowski ('76). The gametes are well differentiated antherozoids and eggs. The antheridia result from the active division of some of the vegetative cells, and they finally consist of one, two or four longitudinal series of small rounded cells (fig. 187 *A a*). Each antheridial cell produces two antherozoids, which are similar in form to the zoogonidia, but slightly more elongated, with shorter cilia, and of a brownish-red colour (fig. 187 *D*). The oogonia arise from ordinary vegetative cells by an increase in size. They become ovoidal in shape and develop a thick lamellose wall. Only one oosphere is formed within the oogonium, which opens by a lateral pore to admit the antherozoids. After fertilization the oospore develops a thick wall and a brick-red colour, but it does not fill the oogonium (fig. 187 *C*).

Species of *Cylindrocapsa* are distinctly uncommon, although *C. geminella* Wolle is perhaps the most widely distributed. They usually occur in pools and lakes among aquatic macrophytes. *C. involuta*, the species in which Cienkowski observed the sexual reproduction, is an Alga of great rarity.

Family **Chætophoraceæ.**

The principal members of the Chætophoraceæ have doubtless originated directly from the Ulotrichaceæ by the branching of the thallus. In all the genera except *Microthamnion* the branches are attenuated, but in some cases the attenuation is much more pronounced than in others, as for instance, in *Draparnaldia*, *Stigeoclonium* (= *Myxonema*), etc., in which the branches are produced into long multicellular hairs. The thallus is recumbent or creeping in the Ulvelleæ and in some of the Microthamnieæ, but in the Chætophoreæ only the basal portion of the thallus is recumbent and often but feebly developed, the main portion consisting of erect tapering branches. In the creeping part of the thallus the cells are frequently so inflated that the branches are moniliform or torulose. In the erect part of the thallus the cells, although often more or less tumid, are elongated, especially towards the ends of the branches, the terminal cells of which are sometimes attenuated to form long hyaline hairs. Setæ or hairs occur, however, in a number of the genera of both the Chætophoreæ and Ulvelleæ, and they are of various kinds, usually with a lumen, but sometimes without. In the Gomontieæ, in which the thallus bores into the calcareous shells of Molluscs, the cells in the more superficial parts are often most irregular.

There is a single parietal chloroplast in each cell, usually plate-like and more or less irregular, and containing one or more pyrenoids. In the attenuated cells towards the ends of the branches of many of the Chætophoreæ the chloroplast becomes reduced, and the long hyaline terminal cells and setæ have no chlorophyll.

Asexual reproduction generally occurs by zoogonidia which exhibit a considerable range in size, the smaller ones being termed microzoogonidia and the larger macrozoogonidia. They possess either two or four cilia and a pigment-spot, and they may be produced in any cell of the thallus except those forming the rhizoids or the terminations of the branches. In most of the Chætophoreæ the cell which produces zoogonidia does not as a rule change its external form, and therefore the zoogonidangium may differ in no way from the vegetative cell in appearance (*Trichodiscus* is rather exceptional in this respect); but in the Microthamnieæ the zoogonidangia are usually inflated and of larger size than the vegetative cells. The zoogonidia germinate directly into new plants, and in the Chætophoreæ there is often a simultaneous germination of large numbers of zoogonidia which have congregated on becoming quiescent (fig. 189 *G*).

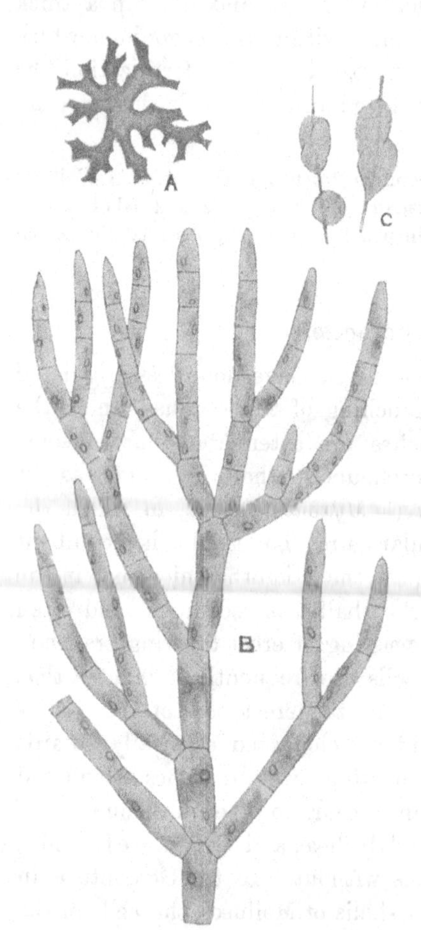

Fig. 188. *A* and *B*, *Chætophora incrassata* (Huds.) Hazen. *A*, entire thallus, nat. size; *B*, portion of branch-system, ×500. *C*, three thalli of *Ch. elegans* (Roth) Ag., showing attachment to aquatic macrophytes, nat. size.

Aplanospores and akinetes (both of which are hypnospores) are known in several genera. They are especially frequent in species of *Draparnaldia,* in which almost all the cells of a tuft of branches take part in spore-formation, one globular resting-spore being formed in each cell. They have a thick cell-wall, generally brown in colour, and often asperulate. The walls of the mother-cells rapidly undergo dissolution and the branches thus exhibit a moniliform appearance. In other genera, especially of the Microthamnieæ, several aplanospores are sometimes formed in each mother-cell.

A gamogenesis of biciliated isogametes occurs in *Stigeoclonium* (fig. 189 *F*), *Draparnaldia*, *Chætophora* and *Trichodiscus* among the Chætophoreæ, and in several genera of the Microthamnieæ. The gametes are scarcely to be distinguished from the biciliated microzoogonidia and it is not unlikely in some genera of the Chætophoreæ that the microzoogonidia are facultative gametes. In one instance both micro- and macrogametes have been observed. The gametangia may not be different in outward aspect from the vegetative cells or they may be much inflated, as in *Trichodiscus*.

It seems highly probable that the Algæ included in the Ulvelleæ and

Microthamnieæ are mostly reduced forms resulting from the adoption of an epiphytic or endophytic mode of life. The irregular character of the thallus in the Gomontieæ is also the result of specialization.

Sub-family CHÆTOPHOREÆ. In this sub-family are included all the best developed types of the Chætophoraceæ, in which, as a rule, the main portion of the thallus consists of free erect branches, much attenuated and often piliferous (*Chætophora, Draparnaldia,* etc.). The basal attachment, which consists of recumbent and more or less concrescent branches, is usually small and rather insignificant, but in *Bulbocoleon, Pseudochæte* and *Endoclonium* it is of more importance, and in *Trichodiscus* it is the most conspicuous part of the thallus.

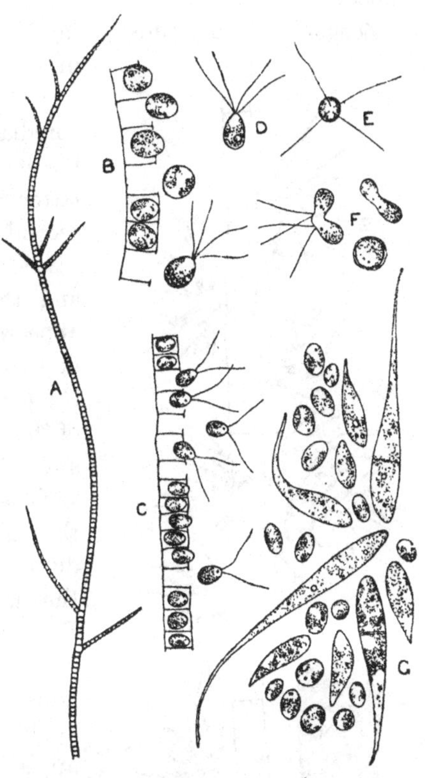

Fig. 189. *Stigeoclonium tenue* Ag. *A*, part of thallus, ×100; *B*, escape of macrozoogonidia; *C*, escape of microzoogonidia (or gametes?); *D* and *E*, macrozoogonidia; *F*, conjugation of gametes to form zygospore; *G*, germination of a cluster of zoogonidia. *B—G*, ×500.

In some of the genera, more especially *Chætophora* and *Draparnaldia*, the thallus is enveloped in copious mucilage, which in the first-named genus is so firm and tough that the thallus as a whole possesses a definite external form (consult fig. 188 *A* and *C*). A few of the Chætophoreæ are encrusted with lime, notably *Fridæa torrenticola* Schmidle and several forms of *Chætophora* (*Ch. incrassata* var. *crystallophora* Kütz., *Ch. calcarea* Tilden, '97, etc.). *Draparnaldia* is unique among the Chætophoreæ in the possession of a main filament of large cells forming a little-branched axis which bears numerous tufts of small branchlets (fig. 190 *A*). In the axial cells the chloroplast is frequently perforated, and in *D. platyzonata* is truly reticulated (fig. 190 *B*). Perforated chloroplasts also occur in *Bulbocoleon* and sometimes in *Chætonema.*

In the epiphytic genus *Thamniochæte* (fig. 191 *C*) the thallus has been reduced to very few cells, one or two of which bear long, tubular or solid hairs. This greatly reduced thallus represents the erect portion and there is no attempt at the formation of recumbent branches. Two other reduced forms, also epiphytic, are *Bulbocoleon* (fig. 191 *A*) and *Acrochæte* (fig. 191 *B*), but in these genera the main part of the thallus consists of basal recumbent branches.

Zoogonidia are formed either singly or rarely in twos, fours, or up to 32 in a mother-cell, and they may arise from almost any cell of a branch. The larger zoogonidia are quadriciliated, with a prominent pigment-spot and two contractile vacuoles, the latter contracting alternately about every fifteen seconds in *Draparnaldia* (Johnson, '93). Such a zoogonidium arises singly in the mother-cell and the first sign of its production is the appearance of the pigment-spot, which is formed at least twenty-four hours before the escape of the mature swarm-cell. The orifice in the wall of the zoogonidangium is often of much smaller size than the zoogonidium, which assumes various extraordinary shapes as it is squeezed through. The colourless end of the zoogonidium always comes through first and when free the swarm-spore darts away with great rapidity. Even when a number of zoogonidia are formed in the mother-cell they do not on their escape issue into a vesicle, but are expelled by the swelling of a gelatinous substance which disappears very rapidly in the surrounding water (Johnson, '93; Pascher, '06 B). The smaller zoogonidia are biciliated in certain genera and there is every probability that in some cases they are facultative gametes.

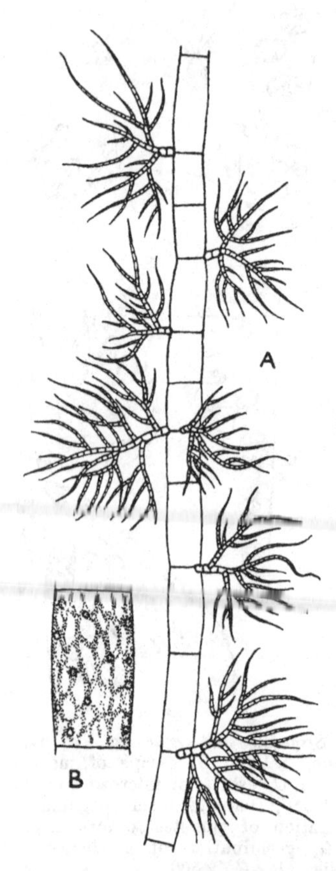

Fig. 190. *Draparnaldia platyzonata* Hazen. *A*, small portion of plant, × 70. *B*, single cell of main filament showing the reticulated chloroplast, × 180. (From a specimen collected in British Columbia by Mr F. L. McKeever.)

The gametes are always biciliated and the zygote rests for a short time before germination.

The zoogonidia as a rule only 'swarm' for about ten minutes, and they germinate at once on coming to rest. The cilia are lost and the cell rapidly increases in length, one pole generally elongating to form a hair-like multicellular outgrowth. The young plant very soon begins to branch and in some genera a basal recumbent portion is first developed, whereas in others the attachment is by a specially modified basal cell, as in *Thamniochæte aculeata* (*vide* G. S. W., '04) and in some forms of *Stigeoclonium* (Fritsch, '03)[1]. A variable development of rhizoids occurs

[1] Fritsch records the occurrence of a brownish-red salt of iron deposited about the attachment-surface of the basal cell in *St. variabile*, which probably serves the purpose of

in the young plants of the larger genera. The zoogonidia of *Stigeoclonium* and *Chætophora* often develop in masses, and in many instances the adult thallus is really a colony of plants which have grown intermingled.

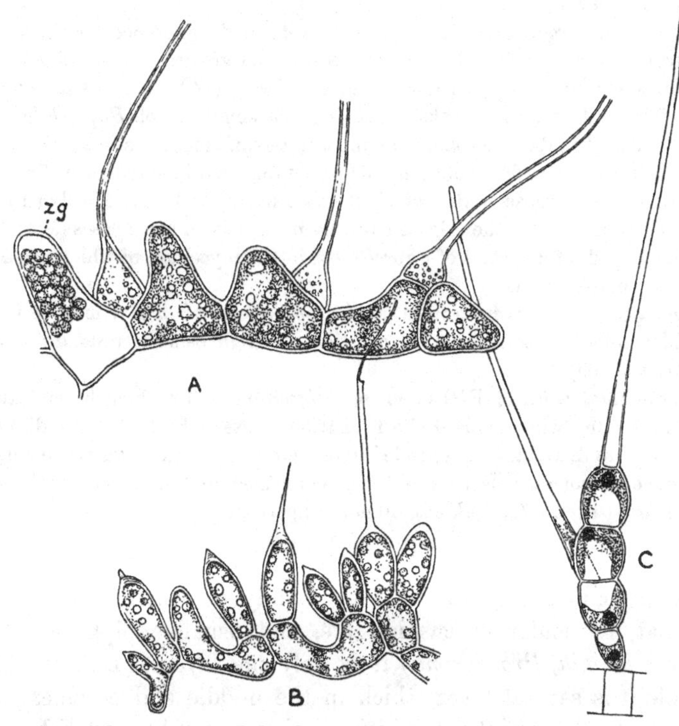

Fig. 191. *A*, *Bulbocoleon piliferum* Pringsh. ×240. Only the bases of the bulbous bristles are shown; on the left is a zoogonidangium (*zg*). *B*, small portion of plant of *Acrochæte repens* Pringsh. ×420. *C*, single plant of *Thamniochæte Huberi* Gay attached to filament of *Oscillatoria* sp. ×700. (*A* and *B*, after Pringsheim; *C*, after Gay.)

The genera are: *Chætophora* Schrank, 1789; *Draparnaldia* Bory, 1808; *Stigeoclonium* Kützing, 1843[1] [=*Myxonema*, Fries, 1825; and inclus. *Iwanoffia* Pascher, 1906]; *Acrochæte* Pringsheim, 1862; *Bulbocoleon* Pringsheim, 1862; *Chætonema* Nowakowski, 1876; *Phæophila* Hauck, 1876; *Endoclonium* Szymanski, 1878; *Thamniochæte* Gay, 1893; *Pseudochæte* W. & G. S. West, 1902; *Fridæa* Schmidle, 1905; *Ectochæte* (Huber) Wille, 1909; *Trichodiscus* Welsford, 1912; *Didymosporangium* Lambert, 1912.

As pointed out by Huber ('93), *Endoclonium* is so closely related to *Stigeoclonium* as

a cement. This he has also observed in young plants of *Œdogonium* and in *Characium Sieboldi*. The present author has found this substance at the point of attachment of the young plants of many Green Algæ, notably in the genera *Microspora*, *Chætophora*, *Thamniochæte*, *Characium*, *Tribonema*, *Characiopsis* and others; also in attached Flagellates such as *Colacium*.

[1] *Stigeoclonium* was adopted as one of the 'genera conservanda' at the Brussels meeting of the International Botanical Congress, 1910. Consult also Nordstedt in *Botaniska Notiser*, 1906, p. 123.

scarcely to be separated from it. *Chætonema* is a reduced type with a more or less endophytic habit. *Pseudochæte* is either a reduced form or one which has permanently remained in the juvenile state, and the two known species of this genus greatly resemble the young states of species of *Stigeoclonium*. One species of *Acrochæte* (*A. parasitica*) is a partial parasite in *Fucus*.

Stigeoclonium is largely a vernal type with a decided preference for the well-aërated water of streams and springs. In this connection the growth of *Stigeoclonium tenue* on living fish in stagnant waters, as mentioned by Hardy ('07) on *Carassius auratus* in a Melbourne fish-pond and by Minakata ('08) on the small fry of *Haplochilus latipes* in a tiny bog-pool at Tanabe in Japan, is of much interest, since by securing a foot-hold on a moving substratum the Alga has been able to obtain, even in stagnant water, conditions which approximate to those under which it more normally lives. Hardy ('10) has also shown that the presence of the Alga on the living fish is in some cases dependent upon the preceding growth of a species of *Saprolegnia*, in the mycelium of which the zoogonidia of the Alga become enmeshed.

A 'palmella-state' occurs in certain species of *Stigeoclonium* and may be induced by cultures under subaërial conditions, in solutions of high osmotic pressure, and in sea-water (Livingston, '05).

The experimental work of Klebs ('96) on *Stigeoclonium*, by which he endeavoured to show that the stimuli which caused the formation of asexual cells (zoogonidia) or sexual cells (gametes) were distinct, and depended upon definite changes of environment, requires repetition, since his conclusions have not been confirmed, and other observations, such as those of Welsford ('12) on *Trichodiscus*, do not support them.

Sub-family ULVELLEÆ. In this sub-family the thallus is more or less discoidal, flat, lenticular or cushion-like, and usually epiphytic. In some forms (*Pringsheimia*, *Protoderma*) it is only one layer of cells in thickness, but as a rule it is several layers thick in the middle and becomes gradually thinner towards the periphery, which is always one-layered. In the two genera *Ochlochæte* and *Chætobolus* (fig. 193) many of the cells are provided with long tubular setæ or hairs. In *Arthrochæte* there are also long hairs, which are several times septate near the base, the basal wall always occurring at the junction of the seta and the supporting cell. The other genera of the Ulvelleæ are without setæ.

The discs really consist of recumbent branches which have become more or less completely concrescent, and in *Ulvella* they may attain a diameter of 1·5 mm. In *Chætobolus* (fig. 193 *D* and *E*) all evidences of a branch-system have disappeared and the cells divide in all directions of space, giving rise to a hemispherical or sometimes almost a globular thallus. The thallus of *Pseudopringsheimia* is also very thick and cushion-like; it increases in diameter by the vertical and tangential division of its peripheral cells, but the increase in thickness is due to numerous transverse divisions parallel to the plane of attachment, a section of the median part of the disc having the appearance of a concrescent mass of vertical filaments.

There is one parietal chloroplast in each cell, disc-shaped or somewhat irregular, and often massive. Except in *Ulvella* it contains a single pyrenoid. The genus *Ulvella* is also peculiar in the fact that the 'cells' have several nuclei.

Reproduction takes place by ovoid zoogonidia, which are biciliated in *Protoderma* and *Ulvella,* but quadriciliated in *Pringsheimia, Pseudulvella* (fig. 192 *E*) and *Ochlochæte* (fig. 193 *C*). As many as 30 zoogonidia may arise in a zoogonidangium, and the latter, which are generally larger than the vegetative cells, are developed only in the more central parts of the thallus. The zoogonidia escape by an apical orifice or by the dissolution of a large part of the wall of the zoogonidangium.

Quadriciliated gametes are known in *Pringsheimia.*

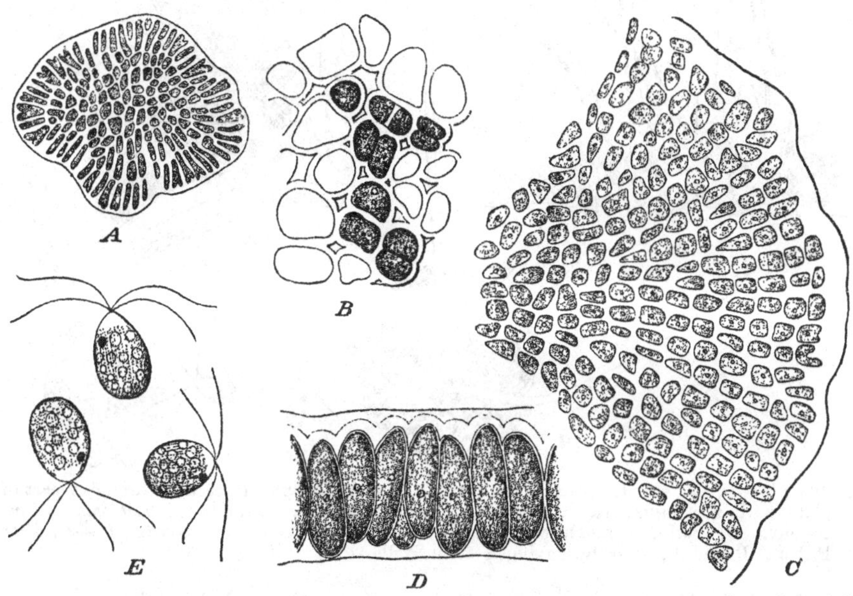

Fig. 192. *A* and *B*, *Ulvella Lens* Crouan. *A*, young plant, ×280; *B*, central portion of the thallus, ×750. *C—E*, *Pseudulvella americana* (Snow) Wille. *C*, portion of thallus from above, ×280; *D*, middle part of same in transverse section, ×820; *E*, zoogonidia, ×820. (*A* and *B*, after Huber; *C—E*, after Snow; from Wille.)

The Ulvelleæ are mostly marine Algæ; the genera are: *Ulvella* Crouan, 1859 [inclus. *Dermatophyton* Peter, 1886 (Sept.) and *Epiclemidia* Potter, 1886 (Nov.)]; *Pseudulvella* Wille, 1909; *Protoderma* Kützing, 1843, em. Borzi, 1895 [inclus. *Entocladia* Hansgirg, 1888]; *Pringsheimia* Reinke, 1889; *Pseudopringsheimia* Wille, 1909; *Ochlochæte* Thwaites, 1849; *Chætobolus* Rosenvinge, 1898; *Arthrochæte* Rosenvinge, 1898.

All the genera are epiphytic except *Arthrochæte*, which is sometimes endophytic in the thallus of *Turnerella Pennyi* (a red seaweed of the Rhodophyllidaceæ). One freshwater species of *Chætobolus*—*Ch. lapidicola* Lagerh.—is found on stones in streams in the north of Norway. *Protoderma viride* Kütz. is a common freshwater Alga occurring as an epiphyte on species of *Lemna, Callitriche,* etc. *Pseudulvella americana* (Snow) Wille is a freshwater epiphyte of N. America. *Ulvella involvens* (Savi) Schmidle (=*Dermatophyton radians* Peter) is epizootic on the carapace of the European water-tortoise (*Clemmys caspica*). *Ulvella fucicola* Rosenv. is partially parasitic on plants of *Fucus*, causing a destruction of the peripheral cells.

It is probable that all the Ulvelleæ have originated from the Chætophoreæ by a reduction and modification of the branch-system consequent upon the adoption of a completely epiphytic mode of life.

Arthrochæte should be compared with *Pseudochæte* since both these genera are really intermediate between the Chætophoreæ and the Ulvelleæ.

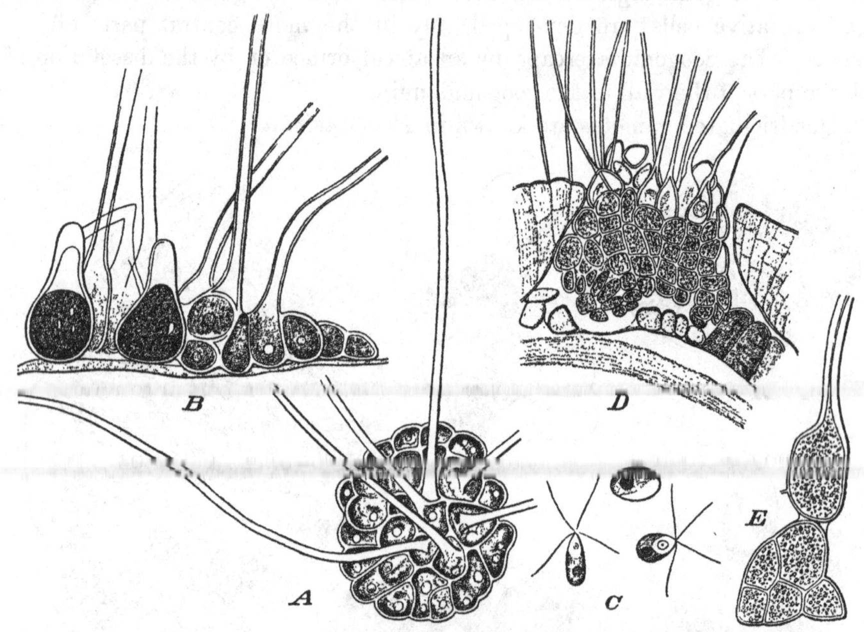

Fig. 193. *A—C*, *Ochlochæte ferox* Huber. *A*, young plant, ×383; *B*, section through a part of plant with zoogonidangia, ×383; *C*, zoogonidia, ×1050. *D* and *E*, *Chætobolus gibbus* Rosenv. *D*, section through plant, ×316; *E*, small part of same, ×526. (*A—C*, after Huber; *D* and *E*, after Rosenvinge; from Wille.)

Sub-family MICROTHAMNIEÆ. This group as here defined is equivalent to the Microthamniaceæ (West, '04) and almost the same as the Leptosireæ (Wille, '09). The Microthamniaceæ was originally separated from the Chætophoraceæ on two characters: first, the absence of multicellular hairs, and secondly, the restricted origin of the zoogonidia. Recent investigations have shown, however, that neither of these characters is sufficiently pronounced or constant to be utilized as a basis upon which a family 'Microthamniaceæ' can be rightly established.

The thallus is branched, with or without a mucous investment, sometimes procumbent, but more often branched so as to form a pulvinate mass. The branches are usually attenuated, but never piliferous, and hairs (or setæ) are never developed. The cells vary much in external form, and in some species of *Endoderma* and *Gongrosira* are most irregular (consult fig. 194 *E*). In

Pleurothamnion, which is a very close ally of *Gongrosira,* the thallus is encrusted with lime. In *Microthamnion* the branches are blunt without the slightest trace of attenuation, and all the species of this genus appear to be invariably free-floating and unattached. The greatest reduction of branches is seen in *Glœoplax,* which is epiphytic on the leaves of *Sphagnum.*

Each cell contains one parietal lobed chloroplast (rarely subdivided?), with or without a pyrenoid. In *Gongrosira* and *Endoderma* the chloroplast may sometimes contain from one to three pyrenoids.

The zoogonidangia may be developed from almost any cell of the thallus, and in most cases are clearly differentiated by their larger size (fig. 194 *C*;

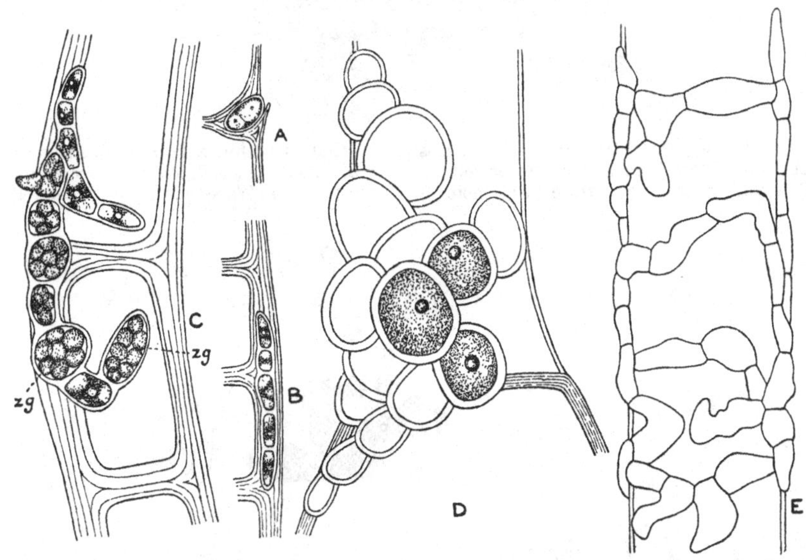

Fig. 194. *A—C, Endoderma Wittrockii* (Wille) Lagerh. *A*, young plant consisting of germinating zoogonidium which has penetrated the cell-wall of a species of *Ectocarpus*; *B*, later stage showing growth of *Endoderma* within the wall of the host; *C*, older plant with zoogonidangia (*zg*). All × 425 (after Wille). *D, E. Pithophoræ* G. S. West on the upper part of a spore and adjacent vegetative cell of *Pithophora Clevei* Wittr.; three cells show the massive chloroplast with a single pyrenoid, the rest in outline only. *E, E. polymorpha* G. S. West on the vegetative cells of *Pithophora Clevei* Wittr., outline of cells only. *D* and *E*, × 460.

fig. 196 *D* and *E*). The zoogonidia are ovoid or ellipsoid, biciliated in some genera (*Microthamnion, Gongrosira, Chloroclonium, Leptosira,* etc.), but quadriciliated in others (*Sporocladus, Trichophilus, Endoderma* and *Pseudendoclonium*). Only one zoogonidium arises in the mother-cell in *Glœoplax,* 2—8 in *Endoderma,* 4—8 in *Microthamnion* and *Pleurothamnion,* and many in *Gongrosira.* In *Microthamnion* the zoogonidia have no pigment-spot. The germination of the zoogonidium is generally direct, but in *Leptosira* a small *Characium*-like plant is first formed the contents of which divide to form 4 aplanospores, which are set free by the dissolution of the wall of the mother-cell, each growing into a new plant.

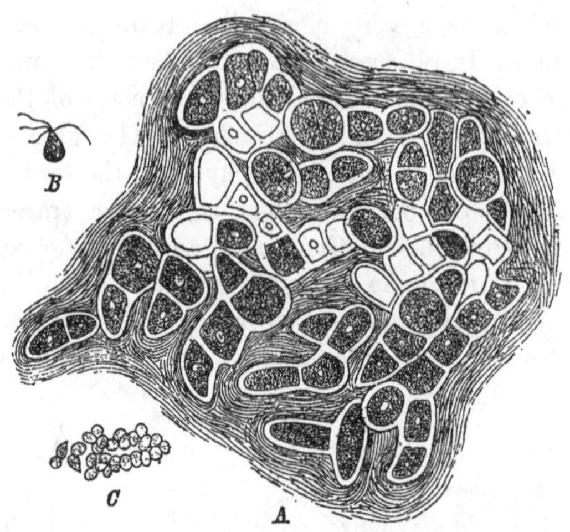

Fig. 195. *Trichophilus Welckeri* Weber v. Bosse. *A*, thallus within a hair of *Bradypus* (the Three-toed Sloth), with some of the cells forming zoogonidangia; *B*, a single zoogonidium; *C*, a group of smaller motile cells (gametes?). All × 470 (after Weber van Bosse, from Wille).

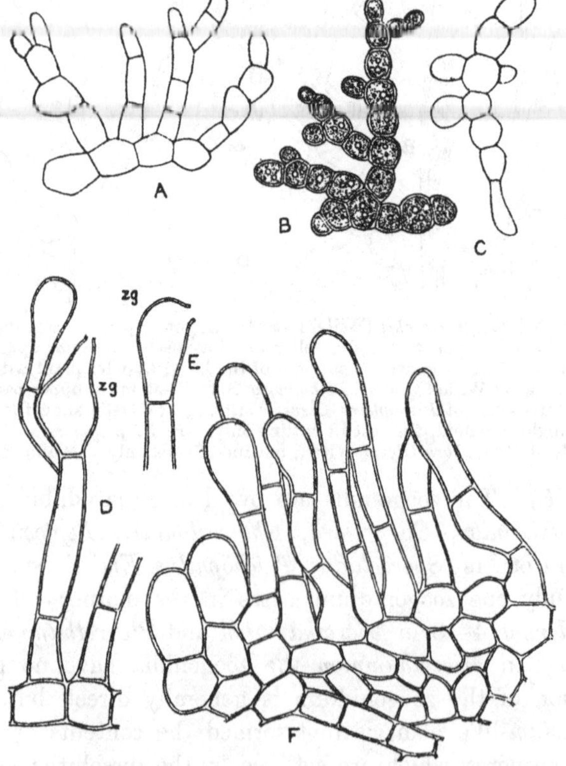

Fig. 196. *A—C*, *Gongrosira viridis* Kütz., × 500. *D—F*, *G. stagnalis* (G. S. West) Schmidle, × 200. *zg*, zoogonidangium.

Isogametes with two cilia are known to occur in several genera, and the gametangia may be formed from any cell, except in *Gongrosira*, in which they are formed only from the lower cells of the thallus. From 6 to 16 gametes arise in a gametangium in *Endoderma*, but they are much more numerous in *Gongrosira* and *Leptosira*. In the latter they fuse by their posterior extremities, forming first a spindle-shaped 'zygozoospore,' which then becomes a resting zygote.

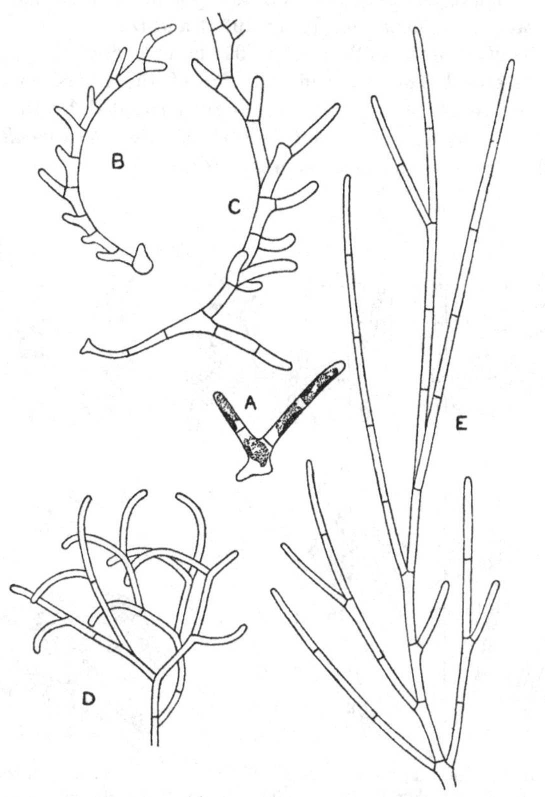

Fig. 197. *A—C*, *Microthamnion Kützingianum* Näg., young developing plants, × 500. *D*, small portion of thallus of *M. curvatum* W. & G. S. West, × 430. *E*, part of thallus of *M. strictissimum* Rabenh., × 500.

The genera are: *Microthamnion* Nägeli, 1849; *Chlorotylium* Kützing, 1843; *Gongrosira* Kützing, 1845 [=*Stereococcus*[1] as described by Wille, 1909; *Pilinia* Kützing, 1843; *Ctenocladus* Borzi, 1885]; ? *Acroblaste* Reinsch, 1879; *Endoderma* Lagerheim, 1883 [=*Periphlegmatium* Kützing, 1843 (in part); *Entocladia* Reinke, 1879; *Epicladia* Reinke, 1879]; *Leptosira* Borzi, 1885; *Trichophilus* Weber van Bosse, 1887; *Chloroclonium* Borzi, 1895; *Pleurothamnion* Borzi, 1895; *Sporocladus* Kuckuck, 1897; *Glœoplax* Schmidle,

[1] The generic name '*Stereococcus*' Kützing (in *Linnaea*, viii, 1833, p. 379) was discarded by him (Kützing, *Phycol. gener.* 1843) when he found that it had been given merely to an aggregation of crystals of lime. *Vide* Nordstedt in *Botaniska Notiser*, 1911, p. 263.

1899; *Pseudendoclonium* Wille, 1901; *Zoddæa* Borzi, 1906: *Endophyton* Gardner, 1909; *Pseudodictyon* Gardner, 1909.

With the exception of *Pseudodictyon* and *Endophyton*, which are marine, and the brackish genus *Endoclonium*, all the above-mentioned genera are inhabitants of fresh water. Some are epiphytic (*Endoderma* spp., *Glæoplax*, *Pseudodictyon*, *Chloroclonium*), some endophytic (*Endoderma* spp. and *Endophyton*), and others are attached to stones or shells. The latter usually develop a procumbent stratum of branches, often to some extent concrescent, from which arise numerous, erect or divergent, branches (consult fig. 196 *A*, *D*, and *F*). *Microthamnion* does not appear to be attached at any time, even though the young plants have a decided basal development (consult fig. 197 *A* and *C*).

Trichophilus Welckeri Web. v. Bosse (fig. 195) is an endozootic Alga in the hairs of *Bradypus* (the Three-toed Sloth), causing the fur of the animal to assume a peculiar greenish tint. The growth of the Alga is greatly facilitated by the dampness of the atmosphere in the gloomy forests in which the sloth lives. *Trichophilus Neniæ* Lagerh. is epizootic on the shells of the mollusc *Nenia* (*Clausilia*).

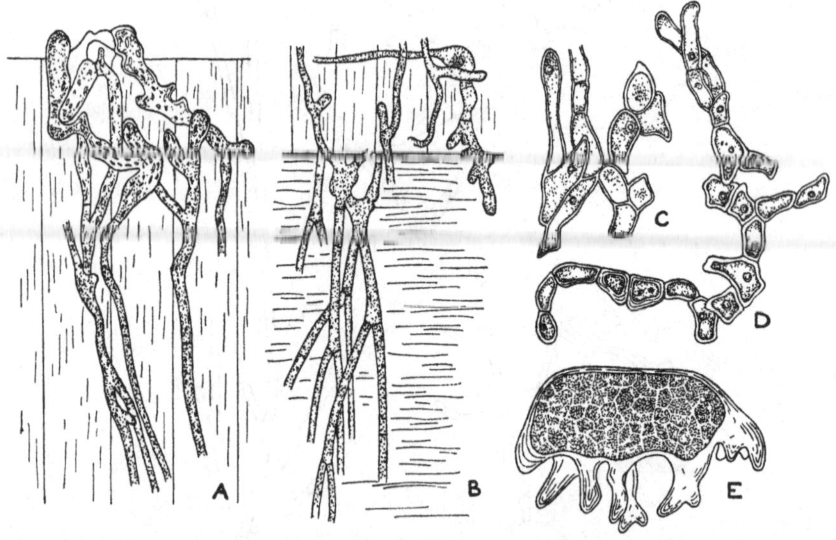

Fig. 198. *A* and *B*, *Tellamia perforans* (Chod.) Wille. *A*, the Alga penetrating the prismatic layer of the shell of *Anodonta*; *B*, the algal filaments passing from the prismatic to the nacreous layer, × about 750 (after Chodat). *C—E*, *Gomontia codiolifera* (Chod.) Wille. *C*, part of thallus with codioliform cells; *D*, typical filament; *E*, detached 'codiolum-state,' × about 500 (after Chodat).

Sub-family GOMONTIEÆ. The Algæ of this sub-family bore into the calcareous shells of various molluscs, especially bivalves. There is a peripheral dorsiventral thallus, with many branches more or less radiating from a central point, the cells of which are both irregular in outline and irregularly branched. From this part of the thallus, which in *Tellamia* is often pseudoparenchymatous, numerous branches descend almost vertically and penetrate deeply into the calcareous layers of the molluscan shell (fig. 198 *A* and *B*). The cells of the more deeply penetrating branches are narrow and cylindrical, forming quite a contrast to the irregular outlines of those situated in the peripheral region.

The cells possess parietal chloroplasts, discoidal or band-like, with one or more pyrenoids. In some cells of *Gomontia* up to five nuclei have been detected, but this may have been due to the incipient formation of resting akinetes. Starch is the usual food-reserve, with sometimes a fatty oil in the rhizoids.

Zoogonidangia are formed in the peripheral parts of the thallus by a swelling of the vegetative cells, 2—4 zoogonidia arising from each mother-cell in *Gomontia*, but many in *Tellamia*. The zoogonidia are ovoid, with four cilia and a stigma.

Resting akinetes occur in all the known species. They are formed from ordinary vegetative cells in the peripheral parts of the thallus, and they grow to a diameter of over 200 μ. They often become irregular in shape, their walls increase in thickness, and independent rhizoids are developed (fig. 198 *E*). On germination they produce numbers of either aplanospores or biciliated 'swarmers' (gametes ?).

The genera are: *Gomontia* Bornet & Flahault, 1888; *Tellamia* Batters, 1895 [=*Foreliella* Chodat, 1898]. Both genera include freshwater and marine species. *Tellamia perforans* (Chodat) Wille penetrates into the shells of the common freshwater mussel (*Anodonta*).

Family **Trentepohliaceæ.**

The Algæ of this family are the most *completely subaërial* of all the Ulotrichales, many of them occurring in situations in which they have to withstand much desiccation. They occur on rocks and stones, and as epiphytes on the leaves and bark of trees. Some of the epiphytic forms have also become parasites.

The plants are *filamentous and branched*, forming in *Trentepohlia* felt-like masses and tufts, but in *Phycopeltis* and *Cephaleuros* much more compact discoidal thalli. In *Trentepohlia* there is a basal creeping portion of the thallus, much less developed in some species than in others, from which numerous erect, branched filaments arise. In *Phycopeltis* and *Cephaleuros*, which are chiefly epiphytes on leaves, the basal part of the thallus is the principal part of the Alga, erect branches being either entirely absent (*Phycopeltis* spp.) or of the nature of multicellular hairs (*Phycopeltis nigra* and certain species of *Cephaleuros*). In these genera the thallus is for the most part discoidal, the disc consisting of concrescent branches with apical growth. In *Phycopeltis* the disc is one layer of cells in thickness and entirely epiphytic. In *Cephaleuros* it usually consists of more than one layer of cells, and some species are parasites, numerous rhizoid-like branches penetrating and destroying the tissue of the leaf.

The branches of *Trentepohlia* are usually bluntly rounded and growth is entirely apical. The cell-walls are lamellate and the lamellæ consist of cellulose. In some species the lamellæ are approximately parallel and the growth of the apical cell takes place by the proportionate distention and permanent increase in area of all the layers. In other species the lamellæ diverge upwardly and outwardly, and the growth of the apical cell takes place

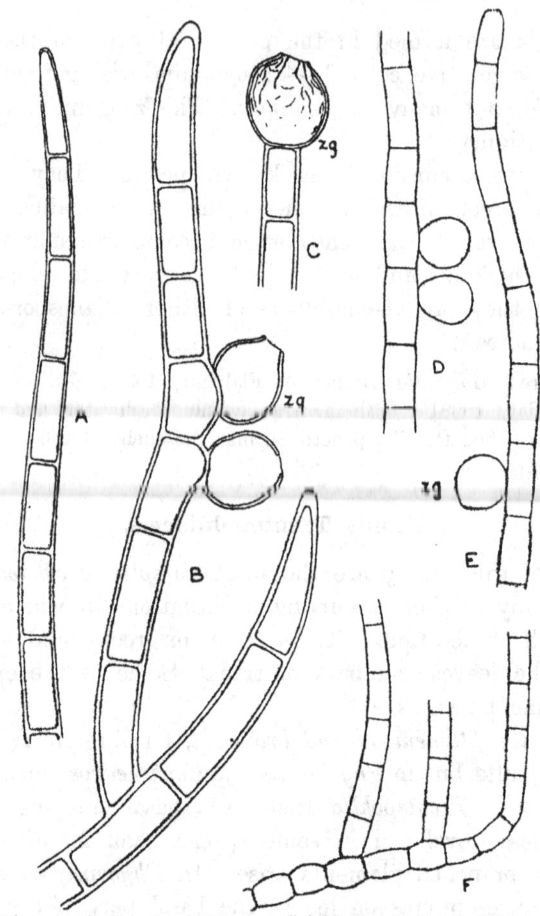

Fig. 199. *A—C*, *Trentepohlia aurea* (L.) Mart. var. *lanosa* Kütz., × 500. *D—F*, *T. calamicola* (Zell.) De Toni, × 500. *zg*, zoogonidangium.

by the distension of only the newly formed layers, the older layers being burst through. In *T. Montis-Tabulæ* var. *ceylanica* all the layers of the cell-wall are burst through at the apex by the extension of the last-formed lamella (fig. 200). In many species of the genus *apical caps* of pectose are secreted at the free end of the apical cell. These caps vary in the extent of their development; they are for the most part absent in species in which the lamellæ of the cell-wall are parallel and the wall at the free end of the apical cell is of approximately the same thickness as the rest of the cell-wall. They are, however, developed in varying degree in those species in which the lamellæ of the cell-walls are divergent. The fewer the lamellæ at the growing extremity of the apical cell, the more complete is the development of the apical cap (West & Hood, '11).

The apical cap is a secretion of the apical cell and is a bonnet-shaped mass of pectose fitting firmly over the delicate end-wall. It is at first homogeneous, consisting of one layer, but in older cells it is composed of two or more layers (fig. 201 *A* and *C*), and sometimes apical cells are surmounted by a successive series of apical caps (fig. 201 *B*). Injury or removal of the cap appears to be followed by the rapid formation of another one. In some species, such as *T. aurea*, the apical cap may become so cumbrous as to be an impediment to apical growth, in which case it is usually displaced by a slight change in the direction of growth of the apical cell (fig. 202 *B*). The original cap is therefore often left in a lateral position some distance behind the growing apex, which develops a new cap. As the growth of the branch proceeds a repetition of this lateral displacement of apical caps results in branches with several lateral excrescences (West & Hood, '11); fig. 202 *C*.

In the division of the apical cell the transverse wall arises as an annular ingrowth from the middle region of the lateral walls. In some species the transverse walls are lamellose like the lateral walls.

Wildeman ('99) has shown that injuries to the thallus of *Trentepohlia* and *Phycopeltis* result in rapid regeneration.

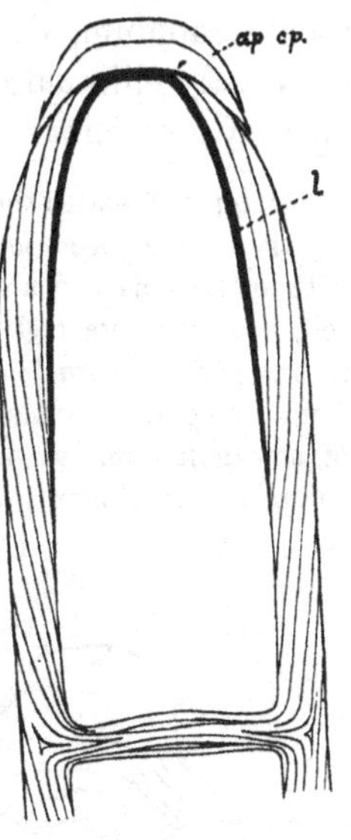

Fig. 200. Semi-diagrammatic figure of the structure of the wall of the apical cell of *T. Montis-Tabulæ* (Reinsch) De Toni var. *ceylanica* W. & G. S. West. The innermost layer of cellulose (*l*) is filled in black. *ap. cp.*, apical cap. (After West & Hood.)

There is normally one nucleus in each cell, but in old cells several nuclei have sometimes been detected. There are several disc-shaped or band-shaped chloroplasts in each cell, without pyrenoids. The green colour is in some species masked by hæmatochrome, some species of all the genera having an orange or red colour. The large quantity of hæmatochrome probably acts as a screen to the chloroplast against too intense light. *Trentepohlia cyanea* Karsten ('91) is bluish in colour and *Phycopeltis nigra* Jennings ('96) is black, but in these and certain other forms the unusual colour is due to the impregnation of the cell-walls with some dark pigment.

Reproduction most commonly occurs by zoogonidia, which are liberated from specially differentiated zoogonidangia. These are solitary or developed in clusters, and are terminal, lateral, or intercalary in position. The terminal zoogonidangium is formed from the apical cell and in consequence the growth of the branch is temporarily arrested. After the escape of the zoogonidia growth may, however, be resumed by the penultimate cell assuming the function of an apical cell and growing upwards through the base of the old zoogonidangium, which it leaves behind as a sort of collar. The zoogonidangia are mostly ellipsoid or ovoid, usually sessile, but sometimes stalked, and they open by a terminal or subterminal pore.

Brand ('10) states that there are three types of zoogonidangia differentiated as follows: (1) the *sessile zoogonidangium*, which may be terminal, lateral or intercalary, and is without any marked annular thickenings on its dividing-wall; it arises from a vegetative cell, is never detached from the filament which bears it, and discharges its zoogonidia *in situ*; (2) the *stalked zoogonidangium*, which is cut off from the end of an outgrowth from a vegetative cell and is therefore either terminal or lateral. It has concentric rings of thickening on its dividing-wall, and these ultimately become a mechanism for its complete detachment, which takes place before the zoogonidia escape; (3) the *funnel*

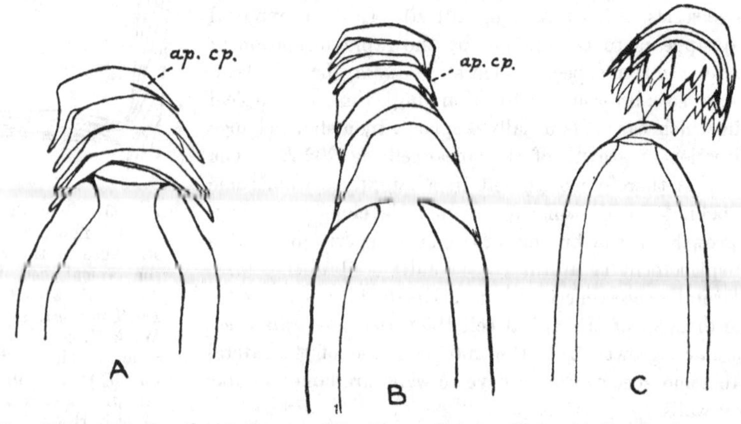

Fig. 201. Three apical cells of *Trentepohlia Montis-Tabulæ* (Reinsch) De Toni var. *ceylanica* W. & G. S. West showing a series of apical caps (*ap. cp.*). In *C* the serrated fringe of a loose cap is well shown. ×800 (after West & Hood).

zoogonidangium, which is always terminal on a branch. The upper part of the terminal cell becomes funnel-shaped by the formation of a subapical constriction and the dilation of the apex, after which it is cut off by a wall on which two annular thickenings are developed, one over the other. These ring-like thickenings eventually cause the detachment of the zoogonidangium before the escape of the zoogonidia.

In *Phycopeltis* the zoogonidangia are borne singly on short stalks of one to six cells, which stand erect from the disc, although the actual supporting-cell may be hooked. In *Cephaleuros* they occur mostly in small clusters at the ends of erect cellular hairs. On becoming wet they rupture and set free the biciliated zoogonidia, but as in *Trentepohlia* they are frequently themselves detached and even distributed by wind before the swarm-spores escape.

Many zoogonidia arise in a zoogonidangium. They are ovoid or pear-shaped, with two cilia, or in some species of *Trentepohlia* with four. It is possible in some cases to discriminate between micro- and macrozoögonidia. When the zoogonidium of *Cephaleuros* has come to rest its protoplast frequently undergoes a rejuvenescence and is then set free by the rupture of the enclosing membrane as a gonidium which can at once germinate (Mann & Hutchinson, '07).

Isogametes are known to occur in a number of different forms. In *Cephaleuros* and *Phycopeltis* they arise from disc-cells, but in *Trentepohlia* the gametangia are found in terminal or intercalary positions on the erect branches, either solitary or in groups; they are very similar to the zoogonidangia, with which it is probable they are often confused. The gametes are biciliated, ovoid and complanate, without a pigment-spot (fig. 92 *B*). They are set free from the gametangia in large numbers and their fusion has been observed in *Trentepohlia* and *Phycopeltis*. The germination of the zygote is not fully known, but is probably direct. Gametes have occasionally been observed to germinate parthenogenetically.

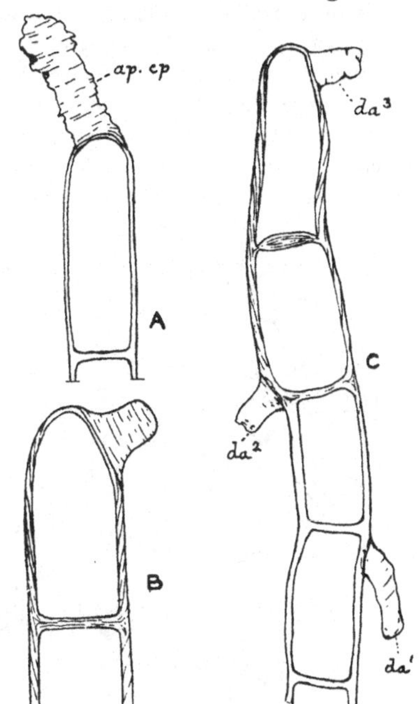

Fig. 202. Apices of three branches of *Trentepohlia aurea* (L.) Mart. *A*, shows a very irregular and much elongated cap (*ap. cp.*). *B*, apical cell with a laterally displaced apical cap. *C*, end of branch showing three successively displaced apical caps (*da*¹, *da*², *da*³). × 800 (after West & Hood).

The genera are: *Trentepohlia* Martins, 1817 [= *Chroolepus* Agardh, 1824; *Nylandera* Hariot, 1889]; *Phycopeltis* Millardet, 1870 [= *Phyllactidium* Kützing, 1849 (in part); *Chromopeltis* Reinsch, 1875; *Hansgirgia* De Toni, 1889]; *Cephaleuros* Kunze, 1828 [= *Mycoidea* Cunningham, 1878; *Phylloplax* Schmidle, 1898; *Weneda* Raciborski, 1900].

The relationships of the three recognized genera were first clearly demonstrated by Karsten ('91). Hariot separated '*Nylandera*' from *Trentepohlia* owing to the fact that certain cells developed hairs, but this character can have no generic value when it is found that some species of both *Cephaleuros* and *Phycopeltis* constantly develop hairs whereas others are quite destitute of them.

The genera *Phycopeltis* and *Cephaleuros* are epiphyllous and are almost exclusively confined to damp tropical countries, whereas those species of *Trentepohlia* which occur on rocks and stones and on the bark of trees are abundant in damp temperate regions and extend far north in Scandinavia. *Trentepohlia aurea* (L.) Mart. is the most widely distributed and conspicuous species in western Europe, often forming large orange-red patches on rocks. *T. jolithus* (L.) Wittr. gives off an odour of violets when moistened. The only true discoidal type native to Europe is *Phycopeltis epiphyton* Mill. (fig. 203 *A*—*C*), which has been found as an epiphyte on the leaves of *Abies pectinata*, *Hedera Helix* and *Rubus* sp. The species of *Phycopeltis* and *Cephaleuros* occur mostly in the damp tropical and subtropical forests of both the New and Old Worlds.

Cephaleuros virescens Kunze (=*Mycoidea parasitica* Cunningham) is a parasite on the leaves of *Camellia*, *Mangifera*, *Rhododendron*, *Thea*, *Croton*, and various Ferns. In North-east India and Assam it causes the 'Red Rust of Tea,' the most serious disease to which the tea-plant (*Thea sinensis*) is liable in that part of the world, and it is as a stem-parasite that it is so destructive (Mann & Hutchinson, '04; '07). The parasite occurs on both leaves and young shoots (fig. 203 *D* and *E*), the latter being mostly infected by zoogonidia from the fructifications of the algal thalli on the leaves. The young shoots are particularly susceptible to attack owing to the rough character of the

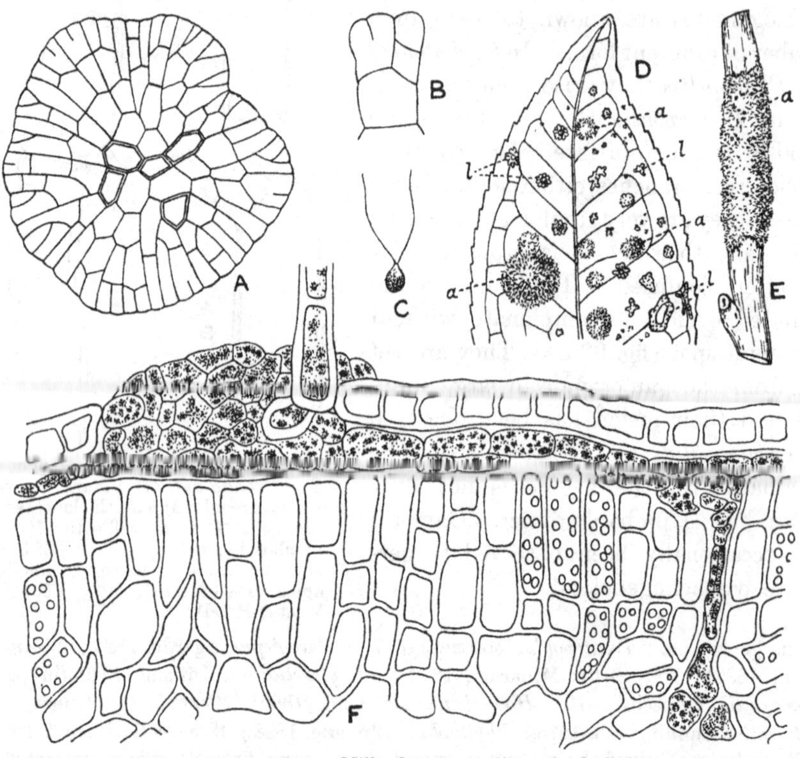

Fig. 203. *A—C*, ***Phycopeltis epiphyton*** Millard. *A*, medium-sized thallus, with the cells in outline only; the five represented with double lines are empty zoogonidangia; *B*, small portion of thallus to show division of peripheral cells; *C*, zoogonidium. *D—F*, ***Cephaleuros virescens*** Kunze. *D*, part of leaf of tea-plant with an epiphytic lichen (*l*) and the parasitic Alga (*a*); *E*, tea-shoot attacked by Alga (*a*); *F*, part of transverse section of tea-leaf showing the penetration of the Alga into the leaf; the algal cells are shaded. *A*, ×300; *B* and *C*, ×900 (after Millard from Wille); *D* and *E*, natural size; *F*, ×about 60 (after Mann & Hutchinson).

bark, in the crevices of which the zoogonidia come to rest and germinate. The parasite is only disastrous in its effect when, owing to want of vitality in the plant attacked, the growth of the Alga is more rapid than the growth of the shoot, in which case the algal filaments penetrate and destroy the tissues of the host. If the shoot is growing faster than the Alga, the latter is removed by exfoliation of the outer tissues and no permanent infection takes place (Mann & Hutchinson, '07).

Some species of the Trentepohliaceæ have become constituents of the thalli of certain Lichens (*vide* p. 141).

The Trentepohliaceæ are a somewhat specialized family of the Ulotrichales related to the Chætophoraceæ possibly through such forms as *Gongrosira* in the Microthamnieæ. The simplest known type is *Trentepohlia umbrina* (Kütz.) Bornet, but this is doubtless a reduced form. The transition from *Trentepohlia* to *Phycopeltis* is seen in those species of *Trentepohlia* belonging to the section '*Heterothallus*' of Hariot, and from *Phycopeltis* to *Cephaleuros* in the section '*Hansgirgia*' of the first-named genus.

Family **Wittrockiellaceæ.**

This family was established by Wille to include the genus *Wittrockiella* described by him in 1909. The thallus consists of erect and slightly branched filaments (fig. 204 *A*) embedded in a tough mucilage derived from the outer

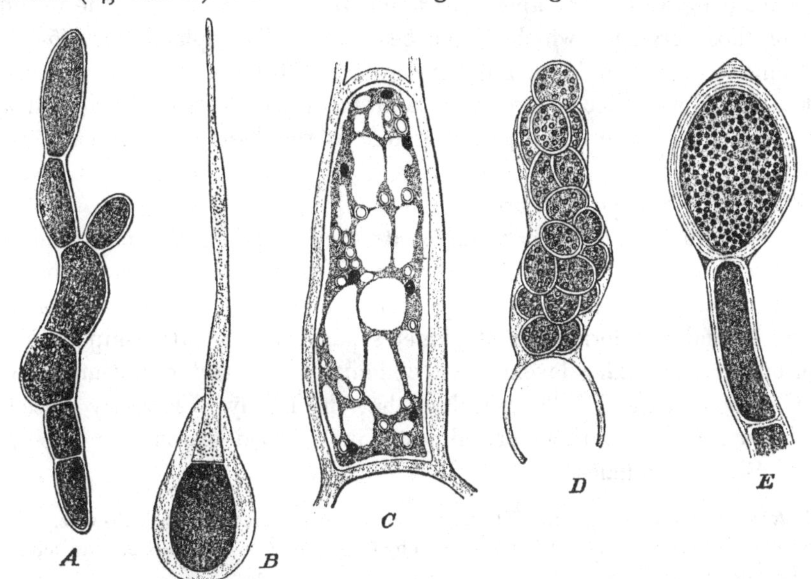

Fig. 204. *Wittrockiella paradoxa* Wille. *A*, vegetative thread with short branches; *B*, germinating akinete which has developed a hair-cell; *C*, single cell fixed and stained to show the reticulate chloroplast with many pyrenoids (with double contours and seven nuclei (darkly shaded)); *D*, formation of aplanospores; *E*, formation of a large akinète. *A* and *B*, ×123; *C*, ×530; *D* and *E*, ×240 (after Wille).

layers of the cell-walls. The branches are few, often unicellular, and they arise from the upper ends of the cells. Numerous thalli are agglutinated to form a cartilaginous stratum on ground which is inundated by salt or brackish water. The cells are mostly inflated, generally globose or ellipsoid, rarely subcylindrical, and often rather irregular. The longest cells are those in the basal part of the stratum and from these cells branches grow downwards to form multicellular rhizoids. From the upper cells hairs are developed. These attain a considerable length, extending beyond the cartilaginous stratum, and are unicellular or rarely bicellular. In the formation of the hairs the outer layers of the thick cell-wall are broken through by the innermost layer, and on the completion of development the hair is cut off from the supporting-cell by a basal wall (fig. 204 *B*).

Each cell possesses a reticulated parietal chloroplast with many pyrenoids (fig. 204 *C*). There are also many oil-drops of variable size, which are green in the inner parts of the thallus, but are of a golden-yellow or orange colour in the outer cells exposed to light.

Reproduction takes place by akinetes and aplanospores. The akinetes are formed only from the terminal cells of the branches, which swell out (up to 60 μ diameter), become rounded, and develop thick walls (fig. 204 *E*). They are finally set free by the dissolution of some of the middle lamellæ of the separating-wall. The aplanosporangia are also formed from the terminal cells of those branches which do not bear hairs. The aplanospores (5—10 μ in diameter) arise in large numbers by free-cell-formation and though at first angular, soon become rounded and furnished with a strong cell-wall (fig. 204 *D*). They are eventually set free by the dissolution of the mother-cell-wall.

Zoogonidia and gametes are unknown.

In the multinucleate character of its 'cells' and in the nature of the chloroplast *Wittrockiella* would appear to be related to the Cladophoraceæ; on the other hand, the branching of the filaments, the accumulation of much oil, and the formation of akinetes suggests a relationship with the Trentepohliaceæ. The development of hairs is a character in common with the Chætophoraceæ. Wille considers that the family is correctly placed in the Ulotrichales, but there are almost equally good reasons for placing it in the Siphonocladiales.

Wittrockiella paradoxa Wille ('09) was originally discovered in brackish-water ditches in southern Norway. It has since been found by G. T. Moore on the coast of Massachusetts (*vide* Collins, '12).

Family **Aphanochætaceæ.**

This is a small family which includes only the one genus *Aphanochæte* A. Braun, 1851 (= *Herposteiron* Nägeli, 1849)[1], which is an epiphyte on larger Algæ and often on aquatic phanerogams. The thallus is creeping in habit and consists of short irregular filaments, with a few slightly attenuated branches. Most of the cells of the thallus possess one or several bristle-like setæ or hairs, cut off from the cell which bears them by a basal septum. The bristles are merely greatly elongated cells, slightly swollen at

[1] Klebahn ('93), Huber ('94) and Nordstedt (in *Botaniska Notiser*, 1906, p. 118) have given cogent reasons for the retention of the generic name *Aphanochæte*.

the base, which never possessed a chloroplast and early lost their scanty protoplasmic contents; they are entirely unseptate, very fragile, and, in all except living plants examined with great care, they are broken off near the base. In the lumen of the bristle there are often plug-like masses of some refractive substance which give the bristle the appearance of being indistinctly articulate. As Fritsch ('02) has shown, staining with Congo red proves that these plugs are not transverse walls.

Each cell contains a massive parietal chloroplast, generally with a conspicuous pyrenoid and numerous small starch-grains.

Asexual reproduction takes place by zoogonidia of which one to four are produced in a mother-cell, the wall of which ruptures and sets them free. They are variable in size, quadriciliated, and with a pigment-spot. On

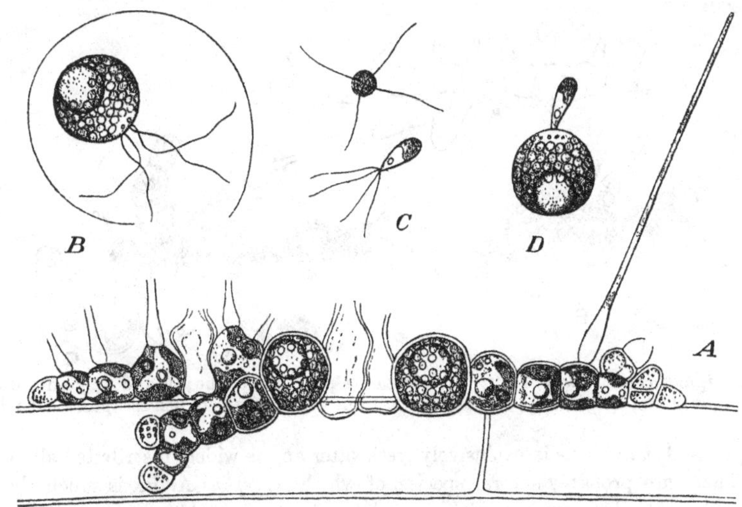

Fig. 205. *Aphanochæte repens* A. Br. *A*, plant with sexual organs, showing epiphytic habit. *B*, escaped quadriciliated oosphere (the circumscribing line represents the extent of a delicate mucous vesicle into which it is at first extruded). *C*, quadriciliated antherozoids. *D*, fertilization of oosphere by antherozoid. All × 600 (after Huber, from Wille).

coming to rest they develop unilaterally into a new plant. Aplanospores are also formed singly from the vegetative cells (G. S. W., '04).

The sexual reproduction of *Aphanochæte* was investigated by Huber ('92), and is of great interest on account of the clear differentiation of both sexual organs and gametes. The oogonia are formed from some of the more central cells of the thallus which are devoid of setæ. These cells grow in size, assume a globular form and become loaded with starch and oil (figs. 205 *A* and 206 *oo*). Only one spherical oosphere is formed within the oogonium; it is quadriciliated and escapes from the oogonium into a delicate hyaline vesicle by the rupture of the upper portion of the wall (fig. 205). The antheridia are developed at the extremities of the branches; they are often almost colourless and usually smaller than the adjacent vegetative cells (figs. 205 *A* and 206 *an*). Each antheridial cell gives origin to one, two (or three ?) antherozoids, which are much smaller than the zoogonidia, pear-shaped, quadriciliated, and with a much reduced chloroplast (fig. 205 *C*). When they first escape they pass into a hyaline vesicle, but the latter undergoes rapid dissolution and sets them free. The antherozoids swim about very rapidly, but the movements of the oospheres are very feeble. The feature of greatest interest in the sexual reproduction of *Aphanochæte* is the fertilization of the oosphere *outside* the oogonium. The antherozoid fuses with the receptive spot of the oosphere by its pointed colourless end (fig. 205 *D*). The oospore surrounds itself with a thick wall and undergoes a period of rest, becoming filled with a red oil. Its germination has not yet been observed.

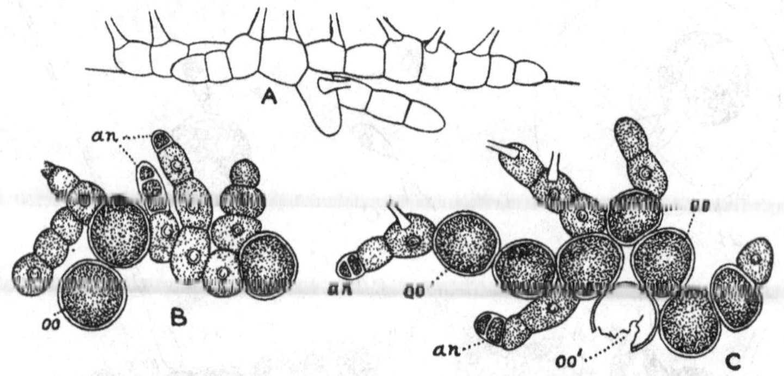

Fig. 206. *Aphanochæte repens* A. Br. *A*, vegetative plant in outline; *B* and *C*, plants with na-theridia (*an*) and oogonia (*oo*); *oo'*, oogonium from which oosphere has escaped. All × 520.

The genus *Aphanochæte* is exclusively freshwater and is widely distributed all over the world. There are probably several species, of which *A. repens* A. Br. is much the most frequent. This Alga is a common epiphyte on larger filamentous Algæ, such as *Œdogonium*, *Cladophora*, *Rhizoclonium*, *Mougeotia*, etc. When attached to the leaves of *Elodea*, to *Lemna*, etc., the thallus is often much branched, the branches of the epiphyte frequently following the contours of the epidermal cells and so forming a reticulum (G. S. W., '99; '04, fig. 19 *B*).

Chodat found that in cultures the setæ were sometimes replaced by branches, a fact which clearly indicates that *Aphanochæte* is nearly related to the Chætophoraceæ.

There does not appear to be any sufficient reason for separating *Gonatoblaste* Huber ('92) from *Aphanochæte*.

Family **Coleochætaceæ.**

In this family there is but one genus—*Coleochæte* Brébisson (1844)—which as regards its sexual reproduction is on a distinctly higher plane than any of the other genera of Green Algæ. The thallus is attached to the stems and leaves of aquatic macrophytes, and is either discoidal or forms cushion-like growths enveloped in mucilage. In the discoidal forms the branches are all procumbent and in some species concrescent (*C. scutata*, *C. orbicularis*), in which case the thallus is a more or less parenchymatous disc one layer of cells in thickness (fig. 207). In *C. soluta* the procumbent filaments are not concrescent and the disc consists of branched filaments radiating from a central point. In other species the ramification is not confined to one plane (*C. irregularis*, *C. pulvinata*), but numerous ascending branches are given off, the whole thallus sometimes forming a hemispherical cushion. Growth is peripheral and due to the divisions of the terminal cells of the branches, which in the discoidal forms are the marginal cells of the disc. In some species the branching is apparently dichotomous, but in others it much resembles that of *Cladophora* or *Stigeoclonium*.

Some of the cells of the thallus are always furnished with sheathed bristles. Each bristle is a long colourless hair issuing from a narrow

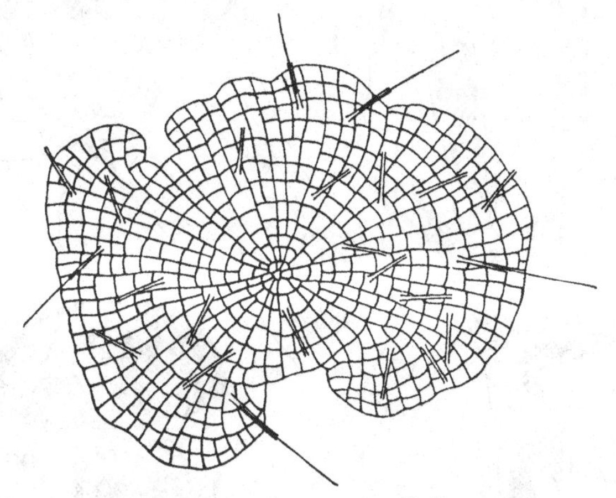

Fig. 207. *Coleochæte scutata* Bréb. ×100.

cylindrical basal sheath (up to 50 μ in length). Pringsheim ('60) originally described these bristles as jointed or septate; and Lambert ('10) has recently asserted that, notwithstanding the fact that this articulation has not been recognized by Jost ('95), Wille ('97), Chodat ('98), West ('04), Oltmanns ('04), or Collins ('09), the bristles of some young *Coleochæte*-plants he examined were distinctly articulate. It is not improbable, however, that this articulation is only apparent, as in *Aphanochæte*, and that the bristles are not truly septate. Lambert found that the bristles attained a length of 4·5 mm.

The most curious species of the genus is *C. Nitellarum* Jost ('95; Lewis, '07), which is endophytic in the outer layers of the wall of species of *Nitella*. In consequence of its habit the cells are greatly flattened with thin cell-walls; they are also of a much more irregular shape than those of any other species of the genus.

Each cell possesses one nucleus, and a single parietal chloroplast with one or, more rarely, two pyrenoids.

Asexual reproduction takes place by means of large ovoid zoogonidia, which are biciliated, without stigma, and with a large parietal chloroplast. They arise singly from the cells of the thallus, more especially from the terminal cells of the branches, and they generally escape from the mother-cell by a circular orifice. On coming to rest the zoogonidia germinate

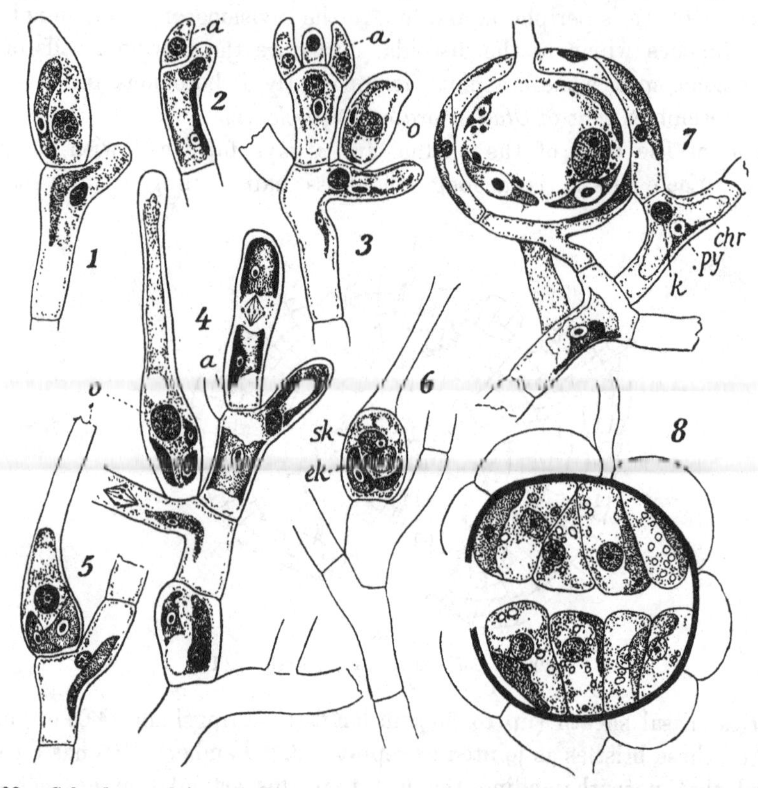

Fig. 208. *Coleochæte pulvinata* A. Br. *1*, young zoogonidangium (?); *2* and *3*, young antheridia (*a*) and oogonium (*o*); *4*, oogonium (*o*) just before the opening of the trichogyne and empty antheridium (*a*); *5*, oogonium (*o*) after opening; *6*, zygote still with distinct male (*sk*) and female (*ek*) nuclei; *7*, zygote which has become surrounded by the 'spermocarp'; *8*, the divisions of the hypnozygote (or oospore after a period of rest). In the vegetative cells in fig. *7*: *k*, nucleus; *chr*, chloroplast; *py*, pyrenoid. (After Oltmanns.)

directly into new plants; those of *C. Nitellarum* penetrate between the lamellæ of the wall of *Nitella*. The first division of the quiescent zoogonidium results in a dorsal and a ventral cell and the disc is developed from the ventral cell.

The sexual reproduction is of a high type. The sexual organs are oogonia and antheridia, and the thalli may be either monœcious or diœcious. The development of the sexual organs was worked out by Pringsheim ('60) in several species, and more recently, and in greater detail, by Oltmanns ('98) in *C. pulvinata.* The antheridia are flask-shaped (fig. 208, *2* and *3 a*) and are usually developed in groups, either terminally on the branches, or, in the discoidal species, sometimes in the vicinity of the oogonium. They are, as a rule, colourless and each one gives origin to a single ovoid antherozoid with two cilia. This antherozoid is in most species colourless, but is green in *C. scutata.* Moreover, in this species the antherozoids apparently arise by the divisions of the contents of ordinary disc-cells. The oogonium is formed by the swelling of the terminal cell of a short branch. In *C. scutata, C. orbicularis,* and others, the adjacent branches continue their growth and the oogonia are gradually enclosed within the disc. In these species the oogonia are depressed with a bottle-neck-like trichogyne, but in *C. pulvinata* the oogonium is bottle-shaped with a long cylindrical trichogyne (fig. 208, *4* and *5 o*). In the discoidal species the oogonia frequently exhibit a zonal arrangement. The young oogonium possesses a relatively large nucleus and one chloroplast. One oosphere is formed within the swollen oogonium and when ready for fertilization it contains a conspicuous chloroplast. Just before fertilization the trichogyne opens at the apex and exudes a small quantity of mucus. According to Lewis ('07) the male nucleus of *C. Nitellarum,* when within the egg-cell, increases in size as it approaches the female nucleus and just before fusion the two nuclei are of approximately the same diameter.

After fertilization the oospore surrounds itself with a wall and grows much in size. When nuclear fusion has taken place the oogonium becomes closely covered with a layer of cortical cells produced by the proliferation of the supporting-cell and other adjacent cells of the thallus. The entire spherical (or spherical-depressed) structure resulting from fertilization is known as a 'spermocarp,' and the cortical cells generally lose their green colour and become brown or brownish red. The spermocarps remain dormant through the winter.

On germination the oospore divides first into octants, and then into 16 or 32 cells. As this division proceeds, the wall of the spermocarp splits into two halves (fig. 208 *8*), and each of the newly formed cells either becomes itself a 'swarm-cell' or gives origin to a 'swarm-cell,' which differs much from the ordinary zoogonidium both in shape and in the attachment of the cilia. When it comes to rest the 'swarm-cell' germinates to form a small asexual plant, often of only a few cells, which is sometimes followed by several asexual generations of equally small plants before a sexual plant is finally developed.

It was Pringsheim ('60) who originally pointed out that in *Coleochæte* there appeared to be an alternation of generations and that the development of the 'swarm-spores' formed on the germination of the oospore into small asexual plants probably represented a rudimentary sporophyte generation. This view was to a great extent upheld till recent years, since the presumed sporophyte was both asexual and post-sexual, but Allen ('05), who studied the cytological details of the germination of the zygote, states that the first division is a reduction-division, in which case the so-called 'sporophyte' is really part of the gametophyte generation (*vide* p. 138).

The branching of the filaments and the general cytological characters indicate that *Coleochæte* originated from the Chætophoraceæ.

There are only about six known species of *Coleochæte* and most of them have a world-wide distribution in fresh water. They are all markedly dorsiventral, occurring as epiphytes on the submerged parts of aquatic phanerogams. *C. scutata* Bréb. has a somewhat irregular disc when fully grown, attaining a diameter of 700—800 μ, and is much the commonest of the discoidal forms. *C. orbicularis* Pringsh. forms almost a circular disc up to 4 mm. in diameter. *C. pulvinata* A. Br. forms more or less hemispherical cushions 2–4 (or even 8) mm. in diameter, and is enveloped in mucilage.

The zoogonidia are generally formed in early summer and the sexual organs in the late summer and autumn, but the time of their formation depends upon both the latitude and the altitude. The oospore germinates in the spring, that is to say, after its winter's rest.

Lambert ('10) has observed some curious unattached zoosporic plants of a species suspected to be *C. scutata*.

LITERATURE CITED

ALLEN, C. E. ('05). Die Keimung der Zygote bei Coleochæte. Ber. Deutsch. Botan. Ges. xxiii, 1905.

ARNOLDI, W. ('13) in Flora, cv, 1913.

ARTARI, A. ('90). Zur Entwickelungsgeschichte des Wassernetzes, Hydrodictyon utriculatum Roth. Bull. de la Soc. Imp. des natur. de Moscou, 1890, no. 2.

ARTARI, A. ('92). Untersuchungen über Entwickelung und Systematik einiger Protococcoideen. Bull. de la Soc. Imp. des natur. de Moscou, 1892.

ASKENASY, E. ('88). Ueber die Entwicklung von Pediastrum. Ber. Deutsch. Botan. Ges. vi, 1888.

BACHMANN, H. ('05). Botanische Untersuchungen des Vierwaldstätter Sees. Ber. Deutsch. Botan. Ges. xxiii, 1905.

BERTHOLD, G. ('80). Zur Kenntnis der Siphoneen und Bangiaceen. Mitteil. der zool. Stat. Neapel, ii, 1880.

BLACKMAN, F. F. ('00). The Primitive Algæ and the Flagellata: An Account of the Modern Work bearing on the Evolution of the Algæ. Ann. Bot. xiv, 1900.

BLACKMAN, F. F. & TANSLEY, A. G. ('02). A Revision of the Classification of the Green Algæ. New Phytologist, i, 1902.

BOHLIN, K. ('97). Die Algen der ersten Regnell'schen Expedition. I. Protococcoideen. Bih. till K. Sv. Vet.-Akad. Handl. Bd. 23, Afd. iii, no. 7, 1897.

BOHLIN, K. ('01). Utkast till de Gröna Algernas och Arkegoniaternas Fylogeni. Upsala, 1901.

BORGE, O. ('06). Süsswasser-Chlorophyceen von Feuerland und Isla Desolacion. Botaniska Studier tillägnade F. R. Kjellman. Upsala, 1906.

BÖRGESEN, F. ('00). A Contribution to the Knowledge of the marine Alga-vegetation on the Coasts of the Danish West-Indian Islands. Botanisk Tidsskrift, xxiii. Kobenhavn, 1900.

BÖRGESEN, F. ('02). The Marine Algæ of the Færöes. The Botany of the Færöes, Part II. Copenhagen, Nov. 1902.

BÖRGESEN, F. ('05). Contributions à la connaissance du genre *Siphonocladus* Schmitz. Bull. Acad. Roy. des Sciences et des Lettres de Danemark, no. 3, 1905.

BÖRGESEN, F. ('07). An Ecological and Systematic Account of the Caulerpas of the Danish West Indies. D. Kgl. Danske Vidensk. Selsk. Skrifter, 7 sér. Section des Sciences, iv, no. 5. Kobenhavn, 1907.

BÖRGESEN, F. ('08). The Dasycladaceæ of the Danish West Indies. Botanisk Tidsskrift, xxviii, 1908.

BÖRGESEN, F. ('11). The Algal Vegetation of the Lagoons in the Danish West Indies. Biol. Arbejder tilegnede Eug. Warming. Nov. 1911.

BÖRGESEN, F. ('13). The Marine Algæ of the Danish West Indies. Part I. Chlorophyceæ. Copenhagen, 1913.

BORNET, E. & FLAHAULT, CH. ('89). Sur quelques Plantes viv. dans le Test calcaire des Mollusques. Bull. Soc. botan. de France, xxxvi, 1889.

BORZI, A. ('83). Studi Algologici I. Messina, 1883.

BRAND, F. ('98). Culturversuche mit zwei Rhizoclonium-Arten. Botan. Centralbl. lxxiv, 1898.

BRAND, F. ('99). Cladophora-Studien. Botan. Centralbl. lxxix, 1899.

BRAND, F. ('01). Ueber einige Verhältnisse des Baues und Wachsthums von Cladophora. Beihefte Botan. Centralbl. Bd. x, Heft 8, 1901.

BRAND, F. ('02 A). Die Cladophora-Aegagropilen des Süsswassers. Hedwigia, xli, 1902.

BRAND, F. ('02 B). Zur näheren Kenntnis der Algengattung *Trentepohlia* Mart. Beihefte z. Botan. Centralbl. xii, 1902.

BRAND, F. ('06). Über die Faserstruktur der Cladophora-Membran. Ber. Deutsch. Botan. Ges. xxiv, 1906.

BRAND, F. ('08). Über Membran, Scheidewände und Gelenke der Algengattung Cladophora. Ber. Deutsch. Botan. Ges. xxvi, 1908.

BRAND, F. ('09). Über die morphologischen Verhältnisse der Cladophora-Basis. Ber. Deutsch. Botan. Ges. xxvii, 1909.

BRAND, F. ('10). Über die Stiel- und Trichtersporangien der Algengattung Trentepohlia. Ber. Deutsch. Botan. Ges. xxviii, 1910.

BRAND, F. ('14). Über die Beziehung der Algengattung Schizogonium Kütz. zu Prasiola Ag. Hedwigia, liv, 1914.

BRAUN, A. ('55). Algarum unicellularum genera nova et minus cognita, etc. Lipsiæ, 1855.

BRUNNTHALER, J. ('13). Die Algengattung *Radiofilum* Schmidle und ihre systematische Stellung. Österr. botan. Zeitschr. no. 1, 1913.

CHATTON, E. ('11) in Bull. Soc. Sci. France et Belgique, sér. 7, xliv, 1911.

CHODAT, R. ('94). Golenkinia genre nouveau de Protococcoidées. Journ. de Botanique, Sept. 1894.

CHODAT, R. ('95 A). Ueber die Entwickelung der Eremosphæra viridis de By. Botan. Zeitung, liii, 1895.

CHODAT, R. ('95 B). Remarques sur le *Monostroma bullosum* Thuret. Bull. Soc. botan. France, xli, 1895.

CHODAT, R. ('97). On the Polymorphism of Green Algæ and the Principles of their Evolution. Ann. Bot. xi, 1897.

CHODAT, R. ('98). Études de biologie lacustre. C. Recherches sur les algues littorales. Bull. de l'Herb. Boiss. vi, no. 6, 1898.

CHODAT, R. ('02). Algues Vertes de la Suisse. Berne, 1902.

CHODAT, R. ('04). Quelques points de nomenclature algologique. I. *Sphærocystis* Chod. ou *Glœococcus* A. Br.? Bull. de l'Herb. Boiss. 2me sér. iv, 1904.

CHODAT, R. ('09). Étude critique et expérimentale sur la Polymorphisme des Algues. Genève, 1909.

CHODAT, R. ('13). Matériaux pour la Flore Cryptogamique Suisse, vol. iv, fasc. 2. Monographies d'Algues en culture pure. Berne, 1913.

CHODAT, R. & HUBER, J. ('95). Recherches expérimentales sur le *Pediastrum Boryanum*. Bull. de la Soc. botan. Suisse, v, 1895.

CHODAT, R. & MALINESCO, O. ('93). Sur le polymorphisme du Scenedesmus acutus. Bull. de l'Herb. Boiss. i, 1893.

CIENKOWSKI, L. ('76). Zur Morphologie der Ulotricheen. Bull. de l'Acad. Imp. d. Sci. de St Petersburg, xxi, 1876.

CLEVE, P. T. ('98). Om Aplanosporer hos Halosphæra. Öfvers. af K. Vet.-Akad. Förh. no. 1, 1898.

COHN, F. ('52). Ueber eine neue Gattung aus der Familie der Volvocineen. Zeitschr. f. wiss. Zoologie, iv, 1852.

COHN, F. ('56). Mém. sur le développement et la mode de reproduction du *Sphæroplea annulina*. Ann. Sci. Nat. bot. 4 sér. v, 1856.

COLLINS, F. S. ('03). The Ulvaceæ of North America. Rhodora, v, 1903.

COLLINS, F. S. ('09; '12). The Green Algæ of North America. Tufts College Studies, Mass. vol. ii, no. 3, 1909; suppl. vol. iii, no. 2, 1912.

CONRAD, W. ('13). Observations sur *Eudorina elegans* Ehrenbg. Recueil de l'Institut botan. Léo Errera, ix, 1913.

CORRENS, C. ('93). Über *Apiocystis Brauniana*. Zimmermann's Beitr. zur Morph. und Physiol. der Pflanzenzelle, III. Tübingen, 1893.

CORRENS, C. ('94). Ueber die Membran von Caulerpa. Ber. Deutsch. Bot. Ges. xii, 1894.

COTTON, A. D. ('06). On some Endophytic Algæ. Journ. Linn. Soc. Bot. xxxvii, July, 1906.

COTTON, A. D. ('11). On the growth of *Ulva latissima* in excessive quantity, with special reference to the *Ulva*-nuisance in Belfast Lough. Seventh Report of the Royal Commission on Sewage Disposal. Appendix IV. 1911.

DANGEARD, P. A. ('98). Mémoire sur les Chlamydomonadinées, ou histoire d'une cellule et théorie de la sexualité. Le Botaniste, vi, 1898.

DANGEARD, P. A. ('00) in Le Botaniste, vii, 1900.

DANGEARD, P. A. ('01). Étude sur la structure de la cellule et ses fonctions. Le *Polytoma uvella*. Le Botaniste, viii, 1901.

DANGEARD, P. A. ('10) in Comptes Rendus, cli, 1910.

DAVIS, B. M. ('04). Oogenesis in *Vaucheria*. Botan. Gazette, xxxviii, 1904.

DAVIS, B. M. ('08). Spore Formation in Derbesia. Ann. Bot. xxii, 1908.

DELF, E. M. ('12). The Attaching Discs of the Ulvaceæ. Ann. Bot. xxvi, April, 1912.

DERBES & SOLIER ('56). Mém. sur quelques points de la physiologie des Algues. Supplément aux Comptes Rendus, 1856.

DESROCHE, P. ('10) in C. R. Soc. Biol. Paris, lxviii, 1910.

DE WILDEMAN, E. ('99). Sur la réparation chez quelques algues. Mém. couronnés etc. par l'Acad. roy. de Belgique, lvii, 1899.

DILL, O. ('95). Die Gattung *Chlamydomonas* und ihre nächsten Verwandten. Jahrb. f. wiss. Botanik, xxviii, 1895.

DODEL, A. ('76). *Ulothrix zonata*, ihre geschlechtliche und ungeschlechtliche Fortpflanzung, usw. Jahrb. f. wiss. Botanik, x, 1876.

ERNST, A. ('02). Siphoneenstudien. I. *Dichotomosiphon tuberosus* (A. Br.) Ernst. Beihefte z. Botan. Centralbl. xiii, 1902.

FREEMAN, E. M. ('99). Observations on Chlorochytrium. Minnesota Botanical Studies. Vol. ii, part 3. 1899.

FRESENIUS, G. ('56—'58). Beiträge zur Kenntniss mikroscopischer Organismen. Abh. Senckenberg Naturf. Ges. Bd. ii, 1856—58.

FREUND, HANS ('07). Neue Versuche über die ungeschlechtliche Fortpflanzung der Algen. Flora, 1907.

FRITSCH, F. E. ('02). Observations on Species of Aphanochæte. Ann. Bot. xvi, 1902.

FRITSCH, F. E. ('03). Observations on the young plants of Stigeoclonium Kütz. Beihefte z. Botan. Centralbl. Bd. xiii, Heft 4, 1903.

FRITSCH, F. E. ('12 A). Freshwater Algæ collected in South Orkneys, etc. Journ. Linn. Soc. Bot. xl, 1912.

FRITSCH, F. E. ('12 B). Freshwater Algæ in the Natural History of the National Antarctic Expedition. Vol. vi, London, 1912.

GARDNER, N. L. ('09). New Chlorophyceæ from California. Univ. California Publ. Botany. Berkeley, 1909.

GAY, F. ('91). Recherches sur le Développement et la Classification de quelques Algues Vertes. Paris, 1891.

GEPP, E. S. ('01). The Genus *Halimeda*. Siboga Expeditie, Monographe lx, Leiden, 1901.

GEPP, E. S. ('04). The Sporangia of Halimeda. Journ. Bot. xlii, 1904.

GEPP, A. & GEPP, E. S. ('04). Rhipidosiphon and Callipsygma. Journ. Bot. xlii, 1904.

GEPP, A. & GEPP, E. S. ('11). The Codiaceæ of the Siboga Expedition, including a Monograph of Flabellarieæ and Udoteæ. Siboga-Expeditie, Monographe lxii, Leiden, Fev. 1911.

GERNECK, R. ('07). Zur Kenntniss nied. Chlorophyceen. Beihefte z. Botan. Centralbl. xxi, 1907.

GIBSON, R. J. H. & AULD, H. P. ('00). Memoirs of the Liverpool Marine Biology Committee. IV. Codium. Liverpool, April, 1900.

GOLENKIN, M. ('99). Algologische Mitteilungen. Ueber die Befruchtung bei *Sphæroplea annulina* und über die Struktur der Zellkerne bei einigen grünen Algen. Bull. de la soc. imp. natur. Moscou, no. 4, 1899.

GOROSCHANKIN ('90—'91). Beiträge zur Kenntniss der Morphologie und Systematik der Chlamydomonaden. I—II. Bull. de la soc. imp. natur. Moscou, 1890-91.

GRIFFITHS, B. M. ('09). On Two New Members of the Volvocaceæ. New Phytologist, viii, April, 1909.

GRINTZESCO, J. ('02). Recherches expérimentales sur la morphologie et la physiologie de *Scenedesmus acutus* Meyen. Bull. de l'Herb. Boiss. 2me sér. ii, 1902.

GRINTZESCO, J. ('03). Contribution à l'étude des Protococcacées. Chlorella vulgaris. Rev. gén. botan. xv, 1903.

HARDY, A. D. ('07). Notes on a Peculiar Habitat of a Chlorophyte, Myxonema tenue. Journ. Roy. Micr. Soc. 1907.

HARDY, A. D. ('10). Association of Alga and Fungus in Salmon Disease. Proc. Roy. Soc. Victoria, xxiii (New Series), 1910.

HARIOT, M. P. ('89; '90). Notes sur le genre Trentepohlia. Journ. de Botanique, 1889—90.

HAZEN, T. E. ('99). The Life History of Sphærella lacustris (Hæmatococcus pluvialis). Mem. Torr. Bot. Club, vi, 1899.

HAZEN, T. E. ('02). The Ulotrichaceæ and Chætophoraceæ of the United States. Mem. Torr. Bot. Club, xi, 1902.

HEERING, W. ('07). Die Süsswasseralgen Schleswig-Holsteins. Siphonales. Jahrb. der Hamburgisch. wiss. Anstalten, xxiv, 1906. Hamburg, 1907.

HEIDINGER, W. ('08). Die Entwicklung der Sexualorgane bei Vaucheria. Ber. Deutsch. Botan. Ges. xxvi, 1908.

HIERONYMUS, G. ('84). Ueber Stephanosphæra pluvialis Cohn, usw. Cohn's Beiträge zur Biol. der Pflanzen, iv, 1884.

HIERONYMUS, G. ('92). Ueber *Dicranochæte reniformis* Hieron., eine neue Protococcacea des Süsswassers. Cohn's Beiträge zur Biol. der Pflanzen, v, 1892.

HOWE, M. A. ('01). Observations on the algal genera Acicularia and Acetabulum. Bull. Torr. Bot. Club, xxviii, June, 1901.

HUBER, J. ('93). Contributions à la connaissance des Chætophorées épiphytes et endophytes, et de leurs affinités. Paris, 1893.

HUBER, J. ('94). Sur l'*Aphanochæte repens* et sa reproduction sexuée. Bull. soc. botan. France, xli, 1894.

JANET, C. ('12). Le Volvox. Limoges, 1912.

JANSE, J. M. ('90). Die Bewegungen des Protoplasma von Caulerpa prolifera. Pringsheim's Jahrbüch. f. wiss. Botan. xxi, 1890.

JENNINGS, A. V. ('96). On two new species of Phycopeltis from New Zealand. Proc. Roy. Irish. Acad. ser. 3, vol. iii, no. 5, 1896.

JOHNSON, L. N. ('93). Observations on the zoospores of Draparnaldia. Botan. Gazette, xviii, 1893.

JOST, L. ('95). Beiträge zur Kenntniss der Coleochaeten. Ber. Deutsch. Botan. Ges. xiii, 1895.

KARSTEN, G. ('91). Untersuchungen über die Familie der Chroolepideen. Ann. du Jard. bot. de Buitenzorg, x, 1891.

KEEBLE, F. & GAMBLE, F. W. ('07). The Origin and Nature of the Green Cells of *Convoluta roscoffensis*. Q. J. M. S. li, part 2, 1907.

KLEBAHN, H. ('93). Zur Kritik einiger Algengattungen. Pringsh. Jahrbüch. f. wiss. Botan. xxv, 1893.

KLEBAHN, H. ('99). Die Befruchtung von *Sphæroplea annulina* Ag. Festschrift für Schwendener. Berlin, 1899.

KLEBS, G. ('81). Beiträge zur Kenntniss neiderer Algenformen. Botan. Zeitung, xxxix, 1881.

KLEBS, G. ('83). Über die Organisation einiger Flagellatengruppen, usw. Unters. a. d. botan. Inst. Tübingen. Bd. i, Leipzig, 1883.

KLEBS, G. ('91). Ueber die Bildung der Fortpflanzungszellen bei Hydrodictyon utriculatum Roth. Botan. Zeitung, 1891.

KLEBS, G. ('92). Flagellatenstudien. Zeitschr. f. wiss. Zool. lv, 1892.

KLEBS, G. ('96). Bedingungen der Fortpflanzung bei einigen Algen und Pilzen. Jena, 1896.

KLEIN, L. ('88). Vergl. Untersuchungen über Morphologie und Biologie der Fortpflanzung bei der Gattung Volvox. Ber. Naturforsch. Ges. zu Freiburg i. B. v, 1888.

KLEIN, L. ('89). Morphologische und Biologische Studien über die Gattung Volvox. Pringsh. Jahrbüch. f. wiss. Botan. xx, 1889.

KOFOID, C. A. ('98). Plankton Studies II. On Pleodorina illinoisensis, a New Species from the Plankton of the Illinois River. Bull. of the State Lab. of Nat. Hist. Urbana, Illinois. v, 1898.

KOFOID, C. A. ('99). Plankton Studies III. On Platydorina, a New Genus of the Family Volvocideæ. *Ibid.* v, 1899.

KUCKUCK, P. ('07). Über den Bau und die Fortpflanzung von Halicystis und Valonia. Botan. Zeitung, lxv, 1907.

KUFFERATH, H. ('13). Contribution à la physiologie d'une Protococcacée nouvelle. Rec. de l'Instit. botan. Léo Errard, ix, 1913.

KÜHN, J. ('78). Über eine neue parasitische Alge, Phyllosiphon Arisari. Sitzungsber. der naturf. Ges. Halle, 1878.

KÜTZING, F. T. ('43). Phycologia generalis oder Anatomie, Physiologie und Systemkunde der Tange. Leipzig, 1843.

KÜTZING, F. T. ('49). Species Algarum. Lipsiæ, 1849.

KÜTZING, F. T. ('45—'66). Tabulæ phycologicæ. Bd. i—xv. Nordhausen, 1845—1866.

LAGERHEIM, G. ('89). Studien über die Gattungen Conferva und Microspora. Flora, lxxii, 1889.

LAGERHEIM, G. ('92 A). Über einige neue Arten der Gatt. *Phyllosiphon* Kühn. La Nuova Notarisia, ser. iii. Padova, 1892.

LAGERHEIM, G. ('92 B). Über die Fortpflanzung von Prasiola. Ber. Deutsch. Botan. Ges. x, 1892.

LAGERHEIM, G. ('92 C). Die Schneeflora des Pichincha. Ber. Deutsch. Botan. Ges. x, 1892.

LAGERHEIM, G. ('94). Ueber die Entwickelung von *Tetraëdron* Kütz. und *Euastropsis* Lagerh., eine neue Gattung der Hydrodictyaceen. Tromso Museums Aarshefter, xvii, 1894.

LAMBERT, F. D. ('10 A). Two New Species of Characium. Tufts College Studies, vol. iii, no. 1. Mass. May. 1910.

LAMBERT, F. D. ('10 B). An unattached zoosporic form of Coleochæte. *Ibid.* 1910.

LAMBERT, F. D. ('12). *Didymosporangium repens*; a New Genus and Species of Chætophoraceæ. *Ibid.* vol. iii, no. 2, 1912.

LEMMERMANN, E. ('98). Beiträge zur Kenntniss der Planktonalgen. I. Hedwigia, xxxvii, 1898.

LEMMERMANN, E. ('04). Das Plankton Schwedischer Gewässer. Arkiv för Botanik utgifv. af K. Svenska Vet.-Akad. Bd. 2, no. 2, 1904.

LETTS, E. A. ('13). On the Occurrence of the Fresh-water Alga (*Prasiola crispa*) on Contact Beds and its Resemblances to the Green Seaweed (*Ulva latissima*). Journ. Roy. Sanitary Instit. xxxiv, no. 10, 1913.

LEWIS, J. F. ('07) in Johns Hopkins Univ. Calendar. Notes Biol. Lab., March, 1907.

LIVINGSTON, B. E. ('05) in Botan. Gazette, xxxix, 1905.

MAIRE, R. ('08) in Bull. soc. botan. France, lv, 1898.

MANN, H. H. & HUTCHINSON, C. M. ('04). Red Rust: a serious Blight of the Tea Plant. Bull. no. 4 of the Indian Tea Association, Calcutta, 1904.

MANN, H. H. & HUTCHINSON, C. M. ('07). Cephaleuros virescens, Kunze: the 'Red Rust' of Tea. Botan. Ser. of Agric. Research Institute, Pusa, vol. i, no. 6, 1907.

MEYER, A. ('95). Über den Bau von Volvox aureus Ehrenb. und V. globator Ehrenb. Botan. Centralbl. lxiii, 1895.

MEYER, A. ('96). Die Plasmaverbindungen und die Membranen von Volvox. Botan. Zeitung, lvi, 1896.

MEYER, K. ('06) in Bull. soc. imp. natur. Moscow, xix, 1906.

MEYER, K. ('13). Über die Microspora amœna (Kütz.) Rab. Ber. Deutsch. Botan. Ges. xxxi, 1913.

MINAKATA, K. ('08) in Nature, Nov. 26th, 1908, p. 99.
MOORE, G. T. ('01). New or little known Unicellular Algæ. II. Eremosphaera viridis and Excentrosphaera. Botan. Gazette, xxxii, Nov. 1901.
MURRAY, G. ('93). On Halicystis and Valonia. Phycological Memoirs, Part II, May, 1893.
MURRAY, G. ('95). An Introduction to the Study of Seaweeds. London, 1895.
MURRAY, G. & BOODLE, L. A. ('89). A Systematic and Structural Account of the Genus *Avrainvillea.* Journ. Bot. xxvii, 1889.
NADSON, G. ('00). Die perforierenden (kalkbohrenden) Algen und ihre Bedeutung in der Natur. Scripta botan. Horti Petropolit. xviii, 1900.
NÄGELI, C. ('49). Gattungen einzelliger Algen. Zurich, 1849.
OLTMANNS, F. ('95). Über die Entwickelung der Sexualorgane bei Vaucheria. Flora, lxxx, 1895.
OLTMANNS, F. ('98). Die Entwickelung der Sexualorgane bei Coleochæte pulvinata. Flora, lxxxv, 1898.
OLTMANNS, F. ('04). Morphologie und Biologie der Algen. Jena, 1904.
OVERTON, E. ('89). Beitrag zur Kenntniss der Gattung Volvox. Botan. Centralbl. xxxix, 1889.
PASCHER, A. ('05) in Flora, xcv, 1905, p. 95.
PASCHER, A. ('06) in Archiv f. Hydrobiol. u. Planktonkunde, i, 1906, p. 433.
PASCHER, A. ('06 B). Über die Zoosporenreproduktion bei *Stigeoclonium.* Oesterr. botan. Zeitschr. 1906.
PASCHER, A. ('07). Studien über die Schwärmer einiger Süsswasseralgen. Bibliotheca Botanica, Heft 67, Stuttgart, 1907.
PASCHER, A. ('12). Zur Kenntnis zweier Volvokalen. Hedwigia, Bd. lii, 1912.
PEEBLES, F. ('09). The life-history of *Sphaerella lacustris.* Centralbl. Bakt. u. Parasitenk. xxiv, 1909.
PETERSEN, J. B. ('11). On Tufts of Bristles in Pediastrum and Scenedesmus. Botanisk Tidsskrift. Bd. xxxi, 1911.
POWERS, J. H. ('05). New Forms of Volvox. Trans. Amer. Micr. Soc. xxvii, 1905.
PRINGSHEIM, N. ('58). Beiträge zur Morphologie und Systematik der Algen. III. Die Coleochaeten. Pringsh. Jahrbüch. f. wiss. Botan. ii, 1858.
RAUWENHOFF, N. W. P. ('88). Recherches sur le *Sphæroplea annulina* Ag. Archiv. néerland. des Sci. exact. et nat. xxii, 1888.
REINKE, J. ('88). Einige neue braune u. grüne Algen der Kieler Bucht. Ber. Deutsch. Botan. Ges. vi, 1888.
REINKE, J. ('89). Atlas deutscher Meeresalgen. 1889.
REINSCH, P. ('67). Die Algenflora des mittleren Theils von Franken. Nürnberg, 1867.
SCHERFFEL, A. ('08 A). Asterococcus n.g. superbus (Cienk.) Scherffel und dessen angebliche Beziehungen zu Eremosphaera. Ber. Deutsch. Botan. Ges. xxvi a, 1908.
SCHERFFEL, A. ('08 B). Einiges zur Kenntnis von Schizochlamys gelatinosa A. Br. Ber. Deutsch. Botan. Ges. xxvi a, 1908.
SCHERFFEL, A. ('10). Rhaphidonema brevirostre nov. spec., zugleich ein Beitrag zur Schneeflora der Hohen-Tatra. Beiblatt zu den Botanikai Közlemén. Heft 2, 1910.
SCHMIDLE, W. ('97). Vier neue, von Prof. Lagerheim in Ecuador gesammelte Baumalgen. Ber. Deutsch. Botan. Ges. xv, 1897.
SCHMIDLE, W. ('03). Bemerkungen zu einigen Süsswasseralgen. Ber. Deutsch. Botan. Ges. xxi, 1903.
SCHMITZ, F. ('78). Über grüne Algen aus dem Golf von Athen. Sitzungsber. der naturf. Ges. zu Halle, Nov. 30, 1878.

SCHMITZ, F. ('79). Über die Zellkerne der Thallophyten. Sitzungsber. d. Niederrhein. Ges. in Bonn, 1879.

SCHRÖDER, B. ('98) in J. B. Schles. Gesellsch. Vaterl. Cult. 1898, Zool., bot. sect.

SCHRÖDER, B. ('02). Untersuchungen über die Gallertbildungen der Algen. Verh. d. Nat.-med. Ver. z. Heidelberg, 1902.

SCOURFIELD, D. J. ('08) in The Essex Naturalist, xv, 1908.

SEWARD, A. C. ('98). Fossil Plants. Vol. i. Camb. Univ. Press, 1898.

SHAW, W. R. ('94). Pleodorina, a New Genus of the Volvocineæ. Botan. Gazette, xix, 1894.

SMITH, G. M. ('13). Tetradesmus, a new four-celled cœnobic Alga. Bull. Torr. Bot. Club, xl, no. 2, Febr. 1913.

SOLMS-LAUBACH, H. GRAF ZU ('91). Fossil Botany. Oxford, 1891. [English Translation by H. E. Garnsey.]

STAHL, E. ('79). Über die Ruhezustände der Vaucheria geminata. Botan. Zeitung, xxxvii, 1879.

STEINMANN, G. ('99). Ueber Boueïna eine fossile Alge aus der Familie der Codiaceen. Ber. der naturforsch. Ges. zu Freiburg i. B. xi, 1899.

STOCKMAYER, S. ('90). Ueber die Algengattung Rhizoclonium. Verhandl. der k. k. zool.-bot. Ges. in Wien, 1890.

SVEDELIUS, N. ('06). Über die Algenvegetation eines ceylonischen Korallenriffes mit besonderer Rücksicht auf ihre Periodizität. Botan. Studier tillägnade F. R. Kjellman. Upsala, 1906.

SVEDELIUS, N. ('07) Ecological and Systematic Studies of the Ceylon Species of Caulerpa. Ceylon Marine Biol. Reports, no. 4, 1906 (1907).

TEODORESCO, E. C. ('05). Organisation et développement du *Dunaliella*, nouveau genre de Volvocacée-Polyblépharidée. Beihefte z. Botan. Centralbl. xviii, 1905.

THURET, G. ('50). Recherches sur les zoospores des algues. Ann. Sci. Nat. Botan. 3e sér. xiv, 1850.

THURET, G. ('78). Études algologiques. 1878.

TILDEN, J. E. ('97). Some New Species of Minnesota Algæ which live in a Calcareous or Siliceous Matrix. Botan. Gazette, xxiii, 1897.

TIMBERLAKE, H. G. ('01). Starch-Formation in Hydrodictyon utriculatum. Ann. Bot. xv, Dec. 1901.

TIMBERLAKE, H. G. ('02). Development and Structure of the Swarmspores of Hydrodictyon. Trans. Wisconsin Acad. Sci. xiii, 1902.

TOBLER, F. ('11) in Flora, ciii, 1911, pp. 78–87.

WEBER VAN BOSSE, A. ('90). Études sur les algues de l'Archipel Malaisien. II. Ann. Jard. Bot. de Buitenzorg. iii, 1890.

WEBER VAN BOSSE, A. ('98). Monographie des Caulerpes. Ann. du Jardin bot. de Buitenzorg, xv, 1898.

WELSFORD, E. J. ('12). The Morphology of Trichodiscus elegans, gen. et sp. nov. Ann. Bot. xxvi, Jan. 1912.

WENT, F. A. F. C. ('89). Les modes de reproduction du Codium tomentosum. Nederlandsch. kruidkundig Archief, 5te Deel, 1889.

WESENBERG-LUND, C. ('03). Sur les *Ægagropila Sauteri* du lac de Sorö. Overs. over det Kgl. Danske Videnskab. Selsk. Forhandl. no. 2, 1903.

WESENBERG-LUND, C. ('04). Studier over de Danske Söers Plankton. Dansk. Ferskr.-Biol. op. 5, Kjöbenhavn, 1904.

WEST, G. S. (G. S. W. '99). The Alga-flora of Cambridgeshire. Journ. Bot. Febr.—July, 1899.

West, G. S. (G. S. W. '04). A Treatise on the British Freshwater Algæ. Cambridge Univ. Press. 1904.

West, G. S. (G. S. W. '06) in the Victorian Naturalist, xxiii, May, 1906.

West, G. S. (G. S. W. '07). Report on the Freshwater Algæ, including Phytoplankton, of the Third Tanganyika Expedition. Journ. Linn. Soc. Bot. xxxviii, Oct. 1907.

West, G. S. (G. S. W. '08). Some Critical Green Algæ. Journ. Linn. Soc. Bot. xxxviii, Jan. 1908.

West, G. S. (G. S. W. '09). The Algæ of the Yan Yean Reservoir: a Biological and Œcological Study. Journ. Linn. Soc. Bot. xxxix, 1909.

West, G. S. (G. S. W. '10). Some New African Species of Volvox. Journ. Quekett Micr. Club, ser. 2, xi, Nov. 1910.

West, G. S. (G. S. W. '12 a). The Freshwater Algæ of the Percy Sladen Memorial Expedition in South-West Africa, 1908—1911. Ann. S. African Museum, ix, part ii, 1912.

West, G. S. (G. S. W. '12 b). Algological Notes. V—IX and X—XIII. Journ. Bot., Nov. 1912.

West, W. & West, G. S. (W. & G. S. W. '96). On some New and Interesting Freshwater Algæ. Journ. Roy. Micr. Soc. 1896.

West, W. & West, G. S. (W. & G. S. W. '03). Notes on Freshwater Algæ. III. Journ. Bot., Febr. and March, 1903.

West, W. & West, G. S. (W. & G. S. W. '06). A Comparative Study of the Plankton of Some Irish Lakes. Trans. Roy. Irish Acad. xxxiii, section B, part ii, 1906.

West, W. & West, G. S. (W. & G. S. W. '11). Reports on the Sci. Investigations of the British Antarctic Expedition 1907—9. Biol. vol. I, part vii (Freshwater Algæ). London, 1911.

West, W. & West, G. S. (W. & G. S. W. '12). On the Periodicity of the Phytoplankton of some British Lakes. Journ. Linn. Soc. Bot. xl, 1912.

West, G. S. & Hood, Olive E. ('11). The Structure of the Cell-wall and the Apical Growth in the genus *Trentepohlia*. New Phytologist, x, 1911.

Whitting, F. G. ('93). On Chlorocystis Sarcophyci—A New Endophytic Alga. Murray's Phycological Memoirs, part ii, May 1893.

Wildeman, E. de ('99). Sur la réparation chez quelques Algues. Mém. couronnés et autres Mém. publ. par l'Acad. roy. de Belgique, 1899.

Wille, N. ('97). Conjugatæ u. Chlorophyceæ in Engler & Prantl, Die natürlichen Pflanzenfamilien. I Teil. Abteilung 2, 1897.

Wille, N. ('01 a; '03; '10). Algologische Notizen. Nyt Magazin f. Naturvidenskab. VI, xxxix, 1901; IX—XIV, xli, 1903; XVI—XXI, xlviii, 1910.

Wille, N. ('01 b). Studien über Chlorophyceen. I—VII. Vid.-selsk. Skrifter. I. Math.-natur. Klasse, 1900 (Christiania, 1901).

Wille, N. ('02). Mittheilungen über einige von C. E. Borchgrevink auf dem antarctischen Festlande gesammelte Pflanzen. Nyt Mag. f. Naturvidenskab. Kristiania, xl, 1902.

Wille, N. ('06). Algologische Untersuchungen an der biol. Stat. in Drontheim. I—VII. Kgl. Norske Videnskab. Selsk. Skrifter. no. 3, 1906.

Wille, N. ('08). Zur Entwicklungsgeschichte der Gattung Oocystis. Ber. Deutsch. Botan. Ges. xxvi a, 1908.

Wille, N. ('09 a). Über *Wittrockiella* nov. gen. Nyt Magazin for Naturvidenskab. Bd. xlvii, 1909.

Wille, N. ('09 b). Chlorophyceæ und Conjugatæ in Engler & Prantl, Die natürlichen Pflanzenfamilien. Nachträge zu I. Teil. Abteilung 2, Leipzig, 1909.

WILLE, N. ('12). Om Udviklingen af Ulothrix flaccida Kütz. Svensk Botanisk Tidskrift. Bd. vi, H. 3, 1912.

WILLE, N. ('13). Algologische Notizen. XXII. Studien in Agardhs Herbarium 1—7. Nyt Magazin f. Naturvidenskab. li, 1913.

WITTROCK, V. B. ('66). Försök till en Monographi öfver algslägtet Monostroma. Upsala, 1866.

WITTROCK, V. B. ('77). On the Development and Systematic Arrangement of the Pithophoraceæ. Nova Acta Reg. Soc. Sc. Upsala, ser. iii, 1877.

WOLLENWEBER, W. ('07). Das Stigma von Hæmatococcus. Ber. Deutsch. Botan. Ges. xxv, 1907.

WOLLENWEBER, W. ('08). Untersuchungen über die Algengattung Hæmatococcus. Ber. Deutsch. Botan. Ges. xxvi, 1908.

YAMANOUCHI, S. ('13). Hydrodictyon africanum, a New Species. Botan. Gazette, lv, Jan. 1913.

ZOPF, W. ('95). Cohn's Hæmatochrom, ein Sammelbegriff. Biol. Centralbl. xv, 1895.

ZUMSTEIN, H. ('99) in Pringsheim's Jahrbüch. f. wiss. Botan. xxiv, 1899.

Division II. AKONTÆ

The name 'Akontæ' was first given to this group of the Chlorophyceæ by Blackman & Tansley ('02) and by them regarded as equivalent to the older name 'Conjugatæ.' The name is here retained for one of the four primary divisions of the Chlorophyceæ, and is applied to *that group of the Green Algæ which is characterized by the complete absence of ciliated reproductive cells.* Thus, neither zoogonidia nor ciliated gametes occur in any member of the group.

In 1904 Oltmanns extended the scope of the Akontæ to include both the Conjugatæ and the Bacillarieæ, but, since there is no evidence of any close phylogenetic relationship between these two groups, this arrangement cannot be upheld. The analogy between the structure of the cell-wall in Diatoms and Desmids, as instituted by Oltmanns, does not stand the test of enquiry (*vide* p. 119).

The only Green Algæ at present known which can rightly be placed in the Akontæ are those belonging to the Conjugatæ and under this heading the general characters of the group can be best discussed.

The Conjugatæ is here placed as an order of the Akontæ, since there may possibly be other Algæ in existence which should be included in this division, and it is also a convenient means of retaining the old name.

Order 1. **CONJUGATÆ.**

The order Conjugatæ is one of the best defined and most natural groups of the Green Algæ, embracing only two families. In the Zygnemaceæ the thallus consists of unbranched filaments of cylindrical cells, but in the Desmidiaceæ the plants are unicellular and generally exhibit a remarkable specialization of form. Some genera of the Desmidiaceæ have become secondarily filamentous, but the filaments are mostly rather fragile and easily become dissociated into their individual cells.

In all members of the Conjugatæ there is a firm wall of cellulose and in addition all these Algæ are remarkable for the great development of the mucilaginous pectose constituents of the cell-wall. There is often either a conversion of the outer cellulose layers into mucilage or a continuous exudation of mucilage, until, in many instances, the gelatinous envelope is of much greater bulk than the individual plant. The unicellular forms not infrequently occur embedded in a mass of transparent jelly formed by the coalescence of their outer mucilaginous coverings. All Conjugates, with the possible exception of *Sirogonium sticticum*, are slimy to the touch.

One of the most conspicuous features of the Conjugatæ is the large size and definite form of the chloroplasts. There are from one to about eight or twelve in each cell and they exhibit great variety in form and disposition. *Each chloroplast contains one or more pyrenoids, which are often very conspicuous and frequently arranged in a symmetrical manner.* Starch is the principal food-reserve. The nucleus is in nearly all cases central in position.

In a number of widely scattered types of the Conjugatæ the cell-sap is coloured purple or violet by a soluble pigment first investigated by Lagerheim ('95) and named by him *phycoporphyrin.* Among Conjugates in which this pigment is present may be mentioned *Mesotænium violascens* De Bary, *M. purpureum* W. & G. S. West, *Mougeotia capucina* (Bory) Ag. and *Pleurodiscus purpureus* (Wolle) Lagerh.

Multiplication takes place by cell-division, which in the Desmidiaceæ is the usual method of propagation. Even in the filamentous Zygnemaceæ the filaments often break up into single cells or short chains of cells, which by further divisions quickly form new filaments. In many species of *Spirogyra* this fragmentation is facilitated by a special mechanism in the transverse walls (*vide* p. 349).

Asexual reproduction sometimes occurs by the formation of aplanospores, both in the Zygnemaceæ and the Desmidiaceæ, but is for the most part rather unusual. Resting 'cysts,' consisting of one or several cells with thick walls, are often formed in *Zygnema.*

Sexual reproduction of a low type occurs by the conjugation of non-ciliated isogametes of relatively large size. The ordinary vegetative cell becomes a gametangium, usually without change of form, and *gives origin only to one gamete.* The latter is derived from the whole contents of the gametangium except in *Mougeotia* and *Pyxispora,* in which some of the protoplast is left unused within the gametangium. In the Zygnemaceæ the gametes unite anisogamously within one of the gametangia or isogamously within the conjugation-tube which joins the two gametangia, but in the Desmidiaceæ (with one solitary exception) the zygote is formed isogamously between the two empty gametangia. The zygospore is in all cases a resting spore.

The sexuality of the Conjugatæ is of a low type and is much less evident in some Conjugates than in others. In the Desmidiaceæ the fusion of the gametes is usually a mere gamogenesis of equal and similar reproductive cells. In *Spirogyra* and many species of *Zygnema* the actual fusion of the gametes takes place within one of the gametangia, which may be to some extent differentiated, either by its swollen character or by the width of its contribution to the conjugation-tube. This gametangium is usually regarded as female in contrast to the emptied one, which is designated the male gametangium. Thus, although the gametes are morphologically indistinguishable, they must be physiologically differentiated, and there is often an obvious morphological differentiation of the gametangia. In the conjugation of most species of *Spirogyra* it is usual for one filament to be completely emptied while the other is filled with zygospores, the physiological differentiation of gametes being uniform in each filament; but in lateral conjugation and the rarer instances of cross-conjugation physiological differentiation of the gametes occurs in the same filament.

In *Sirogonium* and *Temnogametum* the gametangia are clearly differentiated from the vegetative cells, being short cells specially cut off for reproductive purposes.

All the Conjugatæ are inhabitants of fresh water and, so far as has been ascertained, in no single instance has a Conjugate succeeded in adapting itself to a marine life[1]; moreover, the majority of species live only in still water, the few exceptions including *Spirogyra fluviatilis* Hilse, *Sirogonium sticticum* Kütz. and some Desmids. Oltmanns' statement that the Conjugatæ are with few exceptions cosmopolitan is a very erroneous one, since the Conjugatæ, and especially the family Desmidiaceæ, show more decided geographical peculiarities than any other Green Algæ (W. & G. S. W., '07; G. S. W., '07 and '09).

Concerning the classification of the Conjugatæ there have been various suggestions in recent years, but many of the proposed changes are of a rather speculative character and can scarcely be upheld after a careful and detailed consideration of the actual facts.

It was Kützing who first used the group-names Desmidieæ and Zygnemaceæ, although in his *Species Algarum* ('49) they are found widely separated, their close affinity not having been realized. De Bary ('58) divided the Conjugatæ into the Mesocarpeæ, Zygnemeæ and Desmidieæ, but Rabenhorst ('68) recognized only the Desmidieæ and Zygnemeæ.

Palla ('94) was the first to attempt a complete revision of the Zygnemaceæ and he based his groups of the Spirogyraceæ, Mougeotiaceæ and Zygnemaceæ upon the nature of the chloroplasts. Whatever the merits of this classification, and they are certainly great, Palla was in error in placing his Mougeotiaceæ between the very closely allied Spirogyraceæ and Zygnemaceæ.

W. & G. S. West ('97) set up the Pyxisporeæ as a sub-family of the Zygnemaceæ and

[1] *Cosmarium salinum* Hansg. has been described as living in brackish water, and several other small species of *Cosmarium* occasionally occur in limited numbers in water in which brackish Diatoms are living. The larger species of *Spirogyra* may also be cultivated in water of weak salinity.

another family, the Temnogametaceæ, to include the genus *Temnogametum*, but neither of these group-names can be upheld.

Blackman & Tansley ('02) put forward a classification largely based upon the suggestions of Palla. The Conjugatæ were divided into two sections, the Desmidioideæ and the Zygnemoideæ, the latter containing Palla's three families in the order Spirogyraceæ, Zygnemaceæ and Mougeotiaceæ.

Schmitz ('03) reverted to the three families Desmidiaceæ, Mesocarpaceæ and Zygnemaceæ originally suggested by de Bary. His Mesocarpaceæ is, however, a most unnatural group since it includes both *Pyxispora* and '*Zygogonium.*' Moreover, he revives the name '*Mesocarpus*' in a sense quite different from its original application.

West (G. S. W., '99), from general morphological considerations, and Lütkemüller ('02), from a detailed study of the structure of the cell-wall, were agreed upon the classification of Desmids, and Lütkemüller instituted the two sub-families Saccodermæ and Placodermæ. (Consult G. S. W., '04).

Oltmanns ('04) not only included both the Conjugatæ and the Bacillarieæ in the Akontæ, but he also divided the Conjugatæ into the Mesotæniaceæ, Zygnemaceæ and Desmidiaceæ. The suggested association of the Diatoms with the Conjugatæ has already been discussed (*vide* p. 119) and it need only be repeated here that there is no evidence to show that the two groups are in any way nearly related. The three families of the Conjugatæ proposed by Oltmanns were defined as follows:

Mesotæniaceæ. Unicellular; cell-wall in one entire piece; conjugation taking place without the dislocation of the wall into two equal pieces; zygote on germination produces four embryos.

Zygnemaceæ. Filamentous; zygote on germination produces only one embryo, which has a rudimentary rhizoid.

Desmidiaceæ. Unicellular; cell-wall in two pieces; cells more or less constricted in the middle; zygote on germination produces two embryos.

The present author is unable to accept Oltmanns' proposed families; neither could the late Dr Lütkemüller (consult G. S. W., '15). The 'Mesotæniaceæ' is equivalent to the tribe Spirotæniæ of the sub-family Saccodermæ clearly defined by Lütkemüller, and the complete removal of these Algæ from the rest of the Desmidiaceæ does not appear to be in keeping with their true affinities. In this work the Desmidiaceæ, for reasons set out at length later on, are regarded as highly specialized and not primitive.

Wille ('09) introduced into the Zygnemaceæ a new sub-family—the Zygogonieæ—to include the genus *Zygogonium*, but a careful survey of the morphological characters and reproduction of '*Zygogonium ericetorum*' lends no support to the maintenance of such a sub-family or even of the genus '*Zygogonium.*' (Consult p. 346; also West & Starkey, '14; '15.)

In the present volume only two families of the Conjugatæ are recognized.

Fam. *Zygnemaceæ.* Filamentous types with cylindrical cells.

Fam. *Desmidiaceæ.* Unicellular and mostly greatly specialized types. Filamentous condition secondarily acquired in a few genera.

Family **Zygnemaceæ.**

This family of Conjugates has a world-wide distribution and includes some of the most abundant of freshwater Algæ. The thallus is an unbranched filament consisting of a single series of cylindrical cells. Rare instances of branching are known, usually limited to short lateral outgrowths consisting of a few cells. Such rudiments of branches have been observed in *Zygnema*[1] and *Mougeotia* (W. & G. S. W., '98; Pascher, '07).

The cell-wall consists mostly of cellulose in one continuous piece, with its outer surface more or less thickly clothed with mucus, the latter attaining its maximum development in *Zygnema anomalum*, in which it forms a conspicuous sheath four times the diameter of the filament. This outer mucous coat sometimes shows evidence of a fibrillar structure perpendicular to the cell-wall, although the latter has no pores as in so many of the Desmidiaceæ.

In *Spirogyra* the protoplast consists of a thin lining layer of cytoplasm and a small centrally placed mass in which the nucleus is lodged, the two being connected by protoplasmic strands which are frequently branched. The approximately central position of the nucleus is, however, a constant feature throughout all members of the family and the general character of the protoplasmic strands traversing the great sap-vacuole depends upon the form and disposition of the chloroplasts in the various genera. The nucleus may be globular, ellipsoid or lenticular (even in different species of the same genus), and when complanate the flattened sides are always parallel to the plane of division in *Spirogyra* and parallel to the plane of the chloroplast in *Mougeotia*. It has, as a rule, one comparatively large nucleolus, but two, or even three, nucleoli may sometimes be observed.

The most striking feature of all the Algæ belonging to the Zygnemaceæ is furnished by their chloroplasts. These are either axile or parietal, plate-like and solitary, as in *Mougeotia*, star-shaped and binate, as in *Zygnema*[2], or disposed as spiral bands, as in *Spirogyra*.

In many of the Zygnemaceæ numerous refractive globules of small size often occur in the cells. These have been shown to contain much tannin and are termed tannin-vesicles.

Rhizoid-like organs of attachment (haptera) are of frequent occurrence in the young plants of a few species of *Spirogyra* and *Mougeotia*, but have not been noticed in any of the other genera of the Zygnemaceæ. *Spirogyra adnata* and *Sp. fluviatilis* are usually attached to a substratum, and Delf ('13) has recently given a description of the attaching organs of a species which may be identified with *Sp. adnata*. The character of the attaching organs varies much in the different species in which they have been found. Sometimes they are simple rhizoidal outgrowths, but more often they are branched,

[1] Although but rarely observed, branches have been known for a long time to occur in *Zygnema ericetorum*; they may consist of ten or even fifteen cells.

In a small form of this species, described from the West Indies as '*Z. pachydermum* var. *confervoides*' (W. & G. S. W., '94), longitudinal septa of an incomplete character have been observed.

[2] The only exception is *Zygnema ericetorum*. Consult p. 346.

and they may be either lateral or terminal. In *Spirogyra* they may arise by a modification of a conjugation-tube which has been protruded by a cell some distance removed from those cells actually engaged in conjugation (West, '91; Borge, '94). The chloroplasts of those cells which have developed rhizoids

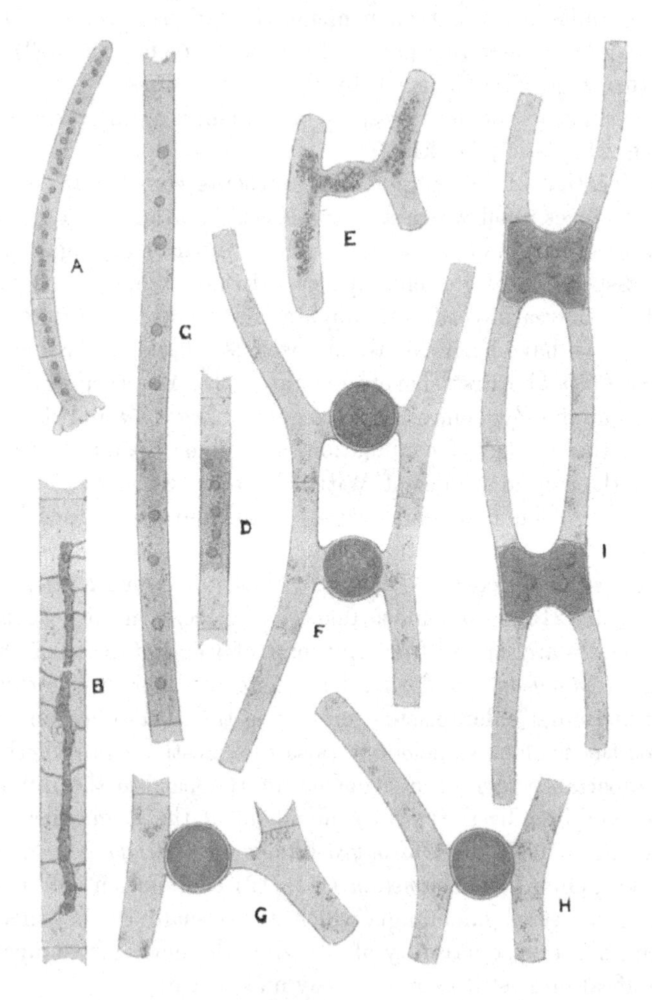

Fig. 209. *A*, young filament of *Mougeotia* sp. showing basal organ of attachment, ×100. *B*, *Mougeotia capucina* (Bory) Ag. showing the edge of the plate-like chloroplast, ×430. *C*, *M. viridis* (Kütz.) Wittr., vegetative cells, ×445. *D—H*, *M. parvula* Hass., ×445. *I*, *M. gracillima* (Hass.) Wittr., ×445.

are often irregular and they frequently degenerate. In *Spirogyra* the nucleus may also disappear, this taking place as a rule before the degeneration of the chloroplasts begins.

In some instances the rhizoids may be formed as a result of the stimulus of contact, as in the profusely branched haptera described by Delf ('13) in an attached *Spirogyra*, but in other cases, such as the replacement of a conjugation-tube by a branched rhizoid, the stimulus is obviously of another kind.

Fragmentation of the filaments is of frequent occurrence in the Zygnemaceæ and is quite a normal phenomenon in various species of *Mougeotia* and in certain of the narrower species of *Spirogyra* (consult p. 349).

The mature zygospore has typically three membranes.

On the germination of the zygospore only a single young plant (sporeling) is formed (fig. 210 *C—E*; fig. 218 *G*).

The classification of the Zygnemaceæ presents certain difficulties. The outstanding features upon which a classification could be based are obviously the method of formation of the gametes, the manner of conjugation, and the nature and disposition of the chloroplasts. In all except a very few of the more recent works treating of the Conjugatæ the methods of gamete-formation and of conjugation have been exclusively used as a basis of classification. It was De Bary ('58) who first sharply separated the Mesocarpeæ (type genus *Mougeotia*) from the Zygnemeæ (typical genera *Zygnema* and *Spirogyra*) on differences in their mode of conjugation, and this separation was further supported by the careful studies of Wittrock ('72; '78) on spore-formation in the Mesocarpeæ. The taxonomic value of the mode of formation of the gametes and of the nature of the conjugation has, however, been seriously discounted by recent discoveries, amongst which the genus *Pyxispora* (W. & G. S. W., '97; fig. 216 *A—C*) must take first place. In this *Zygnema*-like genus the gametes are formed from only part of the protoplast of the gametangium, as in *Mougeotia*, and therefore there are now two known genera with widely dissimilar chloroplasts each having the *Mougeotia*-type of conjugation. This fact is alone sufficient to raise the question as to whether or not too much importance has been attached in the past to the differences in conjugation exhibited by the various members of the Zygnemaceæ. When it is also found that in the *Spirogyra*-like genus *Sirogonium* and in the *Mougeotia*-like genus *Temnogametum* (fig. 212) conjugation only takes place between the gametes of gametangia which are specially cut off, then it must be confessed that the advisability of utilizing the mode of conjugation as a basis of classification is still more seriously questioned.

A consideration of all the known facts concerning the modes of conjugation in the various genera of the Zygnemaceæ causes one to enquire if there are not other characters which can be utilized as a basis of classification, and it is here that the suggestions of Palla ('94) are most helpful. Palla, who was afterwards supported by Blackman & Tansley ('02), suggested that the chloroplast-characters rather than the modes of conjugation should be utilized as a basis of distinction between certain definite groups of filamentous

Conjugates. This basis of distinction also has its weaknesses as well as its merits. It must be realized that without their chloroplasts the genera *Spirogyra* and *Zygnema* are absolutely indistinguishable, and that in the absence of the chloroplast-characters the distinction between the two genera could not be maintained. This fact is significant when considered alongside the conditions prevalent among many of the Desmidiaceæ (the only other family of Conjugates), in which axile and parietal chloroplasts of diverse character often occur in closely allied species of the same genus and sometimes in different individuals of the same species.

Since all the evidence indicates that the Zygnemaceæ should remain intact as a family, it is here suggested that it be subdivided into the three sub-families Mesocarpeæ, Zygnemeæ and Spirogyreæ, this subdivision being based upon chloroplast-characters and not on the method of conjugation. Of these the Mesocarpeæ is the lowest group and the Spirogyreæ the most advanced.

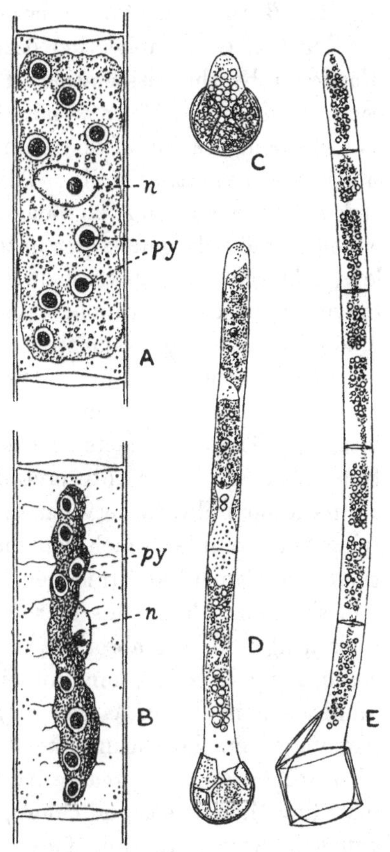

Fig. 210. *A* and *B*, *Mougeotia* sp., ×500; *A*, cell showing the full face of the chloroplast; *B*, cell showing edge of chloroplast. *C* and *D*, germination of the zygospore of *Mougeotia parvula* Hass., ×390. *E*, germination of the zygospore of *M. laetevirens* (A. Br.) Wittr., ×190. *n*, nucleus; *p*, pyrenoid. (*C*—*E*, after De Bary.)

Sub-family MESOCARPEÆ. The Algæ of this sub-family are the narrowest and most delicate of the filamentous Zygnemaceæ. The thickness (3—41 μ) of the filaments and relative length (2—35 times the diameter) of the cells vary between wide limits, and the cell-wall is relatively thin. Each cell contains a single chloroplast, usually in the form of an axile plate (fig. 210 *A* and *B*), which may extend from end to end of the cell or only occupy the median portion (fig. 209 *D*).

The chloroplast contains from 2 to 14 conspicuous pyrenoids, generally arranged in a linear series. It is in some species of *Debarya* (*D. desmidioides*, fig. 213 *G*—*K*; *D. cruciata*) that there are only two pyrenoids, and in some of the thicker species of *Mougeotia* the pyrenoids are more or less scattered (fig. 210 *A*). The chloroplasts of adjacent cells usually lie in the same plane, so that a whole filament of cells may present the full breadth of the chloroplasts or the filament may be in a position such that only the edge of the chloroplasts can be seen (fig. 210 *B*).

The action of light in causing the rotation of the plate-like chloroplasts of *Mougeotia* has been known for a long time. In diffused daylight they place themselves at right-angles to the direction of the incident rays, but the edge of the plate is directed towards strong sunlight. It was shown by Lewis ('98) that the chloroplast occupies on an average 30 minutes to rotate through 90°.

The nucleus is in most cases elliptical in its broadest outline and much compressed, with one flattened side closely adpressed to the chloroplast (fig. 210 *B n*). It contains a large nucleolus.

Vegetative multiplication occurs frequently in nearly all the species of *Mougeotia* by the dissociation of the filaments into their constituent cells, each of which may form a new filament by subsequent rapid cell-division. The dismemberment of the filament is primarily caused by a split in the transverse separating walls. This split, which becomes lenticular in shape and is occupied by mucilaginous material probably derived from the middle lamella, gradually gets larger until it has the form of a thick double-convex lens. At this stage increase in the turgidity of the cells causes the firmer, previously concave, terminal walls to become convex, the area of attachment gradually becoming less until the cells fall apart as the mucilage disappears.

Of the known genera of the Mesocarpeæ only *Mougeotia* is of general world-wide occurrence, and in this genus the conjugation is of peculiar interest. The exact facts of this conjugation have been known for a long time, although their interpretation has been controversial. Conjugation occurs almost always between the cells of the different filaments which are lying side by side. Each cell puts out a protuberance on the side towards the other filament and this meets with a similar protuberance from one of the opposite cells. The ends of the protuberances come into close contact, the separating walls are absorbed, and an open tube is formed placing the two conjugating cells in communication with each other. This channel of communication is known as the *conjugation-tube*. Each conjugating cell is a gametangium, and, during the development of the protuberances and their ultimate fusion, the greater part of the protoplast of each gametangium, including the nucleus and chloroplast, passes into the conjugation-tube. The gametes, which are thus formed from only part of the protoplast of the gametangium, fuse in the conjugation-tube to form a zygospore. Nuclear fusion does not occur until some time afterwards. The subsequent happenings are entirely peculiar to the genus *Mougeotia*. Instead of at once surrounding itself with a new wall this zygospore becomes cut off from the gametangia by partition-walls (two to four in number according to the details of conjugation), so that at first the wall of the zygospore consists of a part or the whole of the walls of the conjugation-tube together with newly formed partition-walls.

Subsequently new continuous layers are deposited internally and the wall becomes differentiated much as in the Zygnemeæ. The relics of the gametangia containing the disintegrating remains of the unused portions of the original protoplasts remain firmly attached to the spore for a long time.

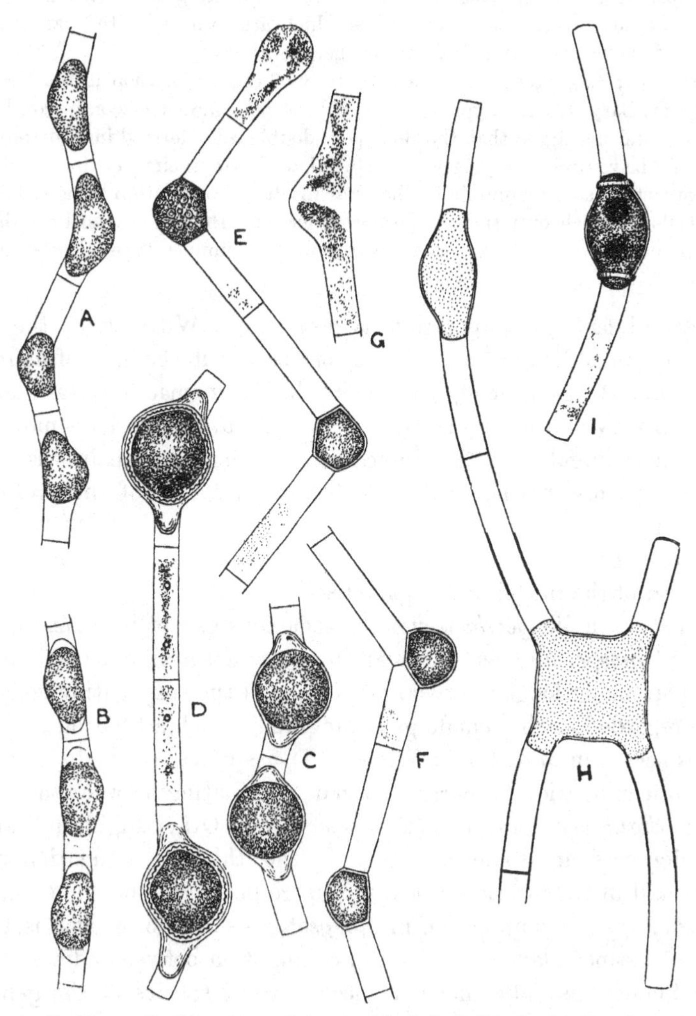

Fig. 211. *A—D*, *Mougeotia* (*Gonatonema*) *tropica* W. & G. S. West, showing four stages up to the completion of the aplanospores. *E—G*, *Mougeotia* (*Gonatonema*) *ventricosa* (Wittr.) Collins. *H* and *I*, *Mougeotia producta* W. & G. S. West. All ×468.

The peculiar conjugation of *Mougeotia* has been interpreted in various ways. De Bary ('58) regarded the H-shaped cell formed by the completion of the conjugation-canal as constituting the zygospore, but that it differed from all other zygospores of the Conjugatæ in not being contracted. He considered that this zygospore existed for a very short time as such, but that after the fusion of the 'chlorophyllaceous bodies (not the whole of the protoplasmic mass)' in the conjugation-tube had been accomplished, the zygospore then became divided by two or four septa into three or five cells, of which the central one was a hypnospore, 'rich in chlorophyllaceous protoplasm (and later in oil), whilst the two or four lateral cells, containing no chlorophyllaceous protoplasm are sterile and soon going to die.' According to De Bary, therefore, the Mesocarpeæ have spores of two kinds, zygospores which are formed by the growing together of the conjugating cells, without contraction, and which do not repose, and hypnospores which are formed by the partition of the zygospore and which rest for a time before germinating.

Wittrock ('72; '78), after a most careful study of the conjugation in the Mesocarpeæ, agreed with De Bary that the hypnospores could not be regarded as zygospores, but at the same time he did not agree that the H-shaped 'double cells' formed immediately on conjugation could be regarded as zygospores. He followed Pringsheim ('77) in considering the H-shaped structure as a 'connubium,' the final result of conjugation being the formation of several cells of which only the central one is fertile. He agreed with Pringsheim that it was a sporocarp, with only one carpospore, and of a simpler type than occurs in any other Algæ.

The actual facts of conjugation as detailed by Wittrock in the various types of the genus *Mougeotia* are quite correct, but in the light of more recent and exact knowledge of conjugation in the Zygnemaceæ as a whole, and especially in view of the conjugation of *Pyxispora*, it is perhaps more correct to regard the conjugation in the Mesocarpeæ as a fusion of a pair of isogametes, the gametes themselves being derived from only a part of the protoplast of the gametangium. The fertile spore—the 'carpospore' of Wittrock—may therefore be regarded as a zygospore, since we know that its nucleus results from a fusion of the nuclei of the gametes.

Conjugation in *Mougeotia* is strictly isogamous except in *M. tenuis* (Cleve) Wittr. (= *Plagiospermum tenue* Cleve) where there is a slight sexual difference, the zygospore being lodged very largely in one of the conjugating cells, which might be regarded as the female gametangium. Such a type of conjugation sometimes occurs in *M. calcarea* Wittr. ('72), a species in which are found all the types of conjugation formerly used in discriminating between the 'genera' *Mougeotia, Staurospermum* and *Plagiospermum.* Other slight indications of sexual differences are sometimes seen in the thicker conjugation-tubes of some cells and in the location of the zygospore nearer to one gametangium.

The usual type of conjugation in *Mougeotia* is scalariform; that is, between the cells of distinct filaments. Lateral conjugation between adjacent cells of the same filament is only known to occur in one species of the genus, viz. *M. genuflexa* (Dillw.) Ag.

Parthenospores are sometimes formed, usually in isolated filaments which have had no opportunity of conjugating, although in most cases among others which have conjugated. The cells send out protuberances as if going to take part in the formation of a conjugation-canal, and most of the protoplast, including the chloroplast, moves into the protuberance, after which a wall appears cutting it off as a parthenospore. Occasionally the parthenospore is formed by the concentration of most of the protoplast in the median part of the cell, after which two somewhat oblique walls appear cutting off this central portion from the remaining distal parts.

Twin parthenospores have been seen by Wittrock ('72) in *Mougeotia genuflexa*, and irregularities are sometimes met with in the conjugation of various species of *Mougeotia*. Cases have been observed in which the terminal cell of a filament has entered into conjugation with another cell through its free end, no conjugation-tube being developed, and rare instances occur in which three cells, each belonging to distinct filaments, have entered into conjugation (W. & G. S. W., '98). Equally rare are the hybrid examples in which conjugation has occurred between species of *Mougeotia* of different thickness.

Twin zygospores have been observed in *Mougeotia capucina* (Bory) Ag., this phenomenon being strictly comparable with that prevailing in certain species of Desmids, such as *Closterium lineatum* and *Penium didymocarpum* (W. & G. S. W., '98).

On the germination of the zygospore the outer coats are burst through by an outgrowth clothed in the inner membrane (fig. 210 *C*). This outgrowth becomes cylindrical, rapidly elongates and ultimately forms the new filament by repeated cell-division. In the *Craterospermum*-section of *Mougeotia* the outgrowth makes its exit from the spore by the removal of a lid (fig. 210 *E*) and it attains a considerable length before any transverse wall appears in it, such walls not making their appearance until there have been formed four chloroplasts and four nuclei.

The genus *Gonatonema* was established by Wittrock ('78) to include two Algæ with the vegetative characters of *Mougeotia* but in which only aplanospores (?) were formed. Some four or five other species of this kind have since been found and aplanospores have also been seen in certain true species of *Mougeotia* (W. & G. S. W., '07; consult fig. 211 *H* and *I*). Spore-formation begins in *Gonatonema* by a distention of the cell in the middle. This distention, however, is not equal all round the cell, but is stronger on one side, the cell bending at the same time like a knee. The formation of spores almost always takes place simultaneously in all the cells of a filament, the cells bending alternately to the right and to the left, so that the filament assumes a zig-zag line (fig. 211 *A*, *E* and *F*). Wittrock stated that in *Gonatonema ventricosum* the spores were formed without any preceding act of conjugation and he also gave cogent reasons for regarding them as neutral and not as parthenospores. As yet the detailed cytology of the spore-formation in *Gonatonema* has not been worked out, but as described originally by Wittrock ('78) in *G. ventricosum*, and as subsequently observed both in that species and in *G. Boodlei* (W. & G. S. W., '97 A), there is in the early stages of spore-formation a more or less complete division of the protoplast into two parts which subsequently fuse together (fig. 211 *G*). The behaviour of the nucleus during this process may prove to be of great interest and would probably decide whether the spores should really be regarded as aplanospores or not. It is possible that some are zygospores and others parthenospores, but in the present state of our knowledge they may for the time being be regarded as aplanospores.

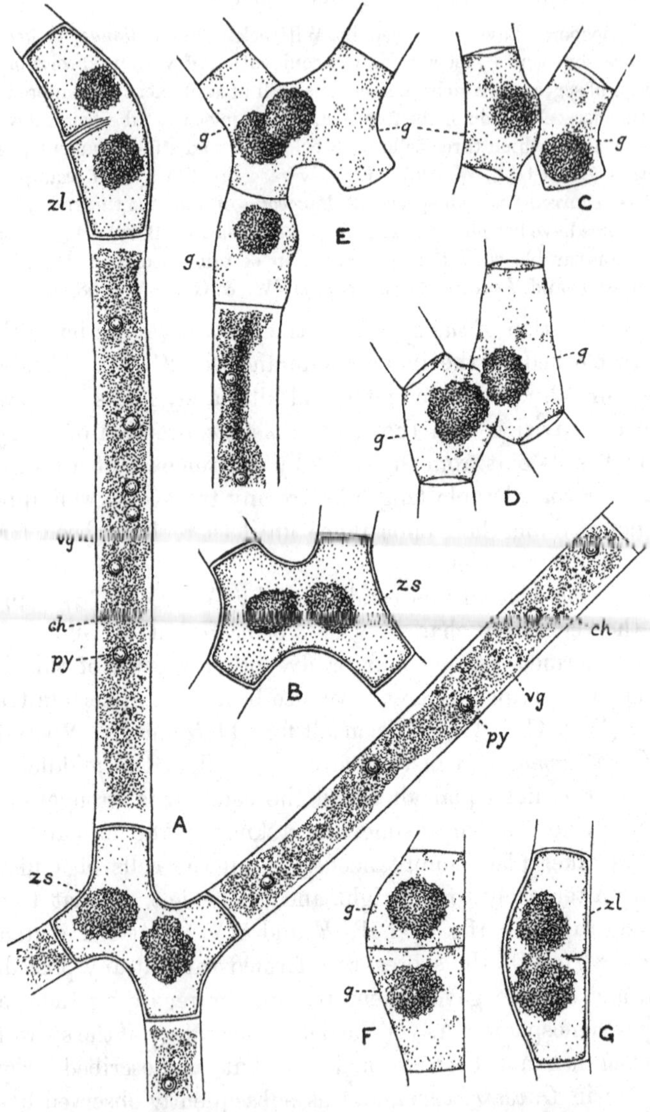

Fig. 212. *Temnogametum heterosporum* W. & G. S. West. *A* shows completed scalariform and lateral conjugation in the same filaments; *B*, zygospore formed by scalariform conjugation; *C* and *D* show early stages in the fusion of the gametangia; *E* is an abnormal case in which two adjacent gametangia were cut off in one filament, one of them having conjugated in a scalariform manner; *F*, early stage in lateral conjugation; *G*, completed zygospore. *ch*, chloroplast; *g*, gametangium; *py*, pyrenoid; *vg*, vegetative cell; *zl*, zygospore produced by lateral conjugation; *zs*, zygospore produced by scalariform conjugation. All ×455.

On the whole it seems highly probable that *Gonatonema* is specialized and not primitive. In all its characters it suggests a set back from *Mougeotia* and in view of the conditions prevailing in *M. producta* (fig. 211 *H* and *I*), it is possibly unwise to separate *Gonatonema* as a genus distinct from *Mougeotia*.

The genus *Temnogametum* (W. & G. S. W., '97), which is exclusively tropical, one species (*T. heterosporum*; fig. 212) being known from West Central Africa and one (*T. Ulmeana*) from Brazil, is unique among the Mesocarpeæ in the fact that the sexual cells are specially cut off. The vegetative filaments are precisely similar to those of *Mougeotia*, but the gametangia are short cells cut off from the more elongated vegetative cells. They are cut off either singly, in which case conjugation is scalariform (fig. 212 *A—E*), or in pairs, when conjugation is lateral (fig. 212 *A*, *F* and *G*).

Of considerable interest is *Debarya* (Wittrock, '72), another genus in which the vegetative cells are precisely like those of *Mougeotia*, but in which each gamete is formed from the entire protoplast of the gametangium. Conjugation is scalariform and the zygospore is lodged in all cases in the conjugation-canal between the gametangia. In *D. calospora* the chloroplast may sometimes be destitute of pyrenoids.

In most species of *Debarya* the walls of the gametangia undergo a pecular thickening during the fusion of the gametes and the formation of the zygospore (G. S. W., '04; '07; '09). In extreme cases the relics of the gametangia become four solid processes which remain permanently attached to the zygospore as in *D. Hardyi* (fig. 213 *A—F*) and *D. desmidioides* (fig. 213 *G—K*). In *D. Hardyi* the thickening begins as soon as the gametes commence to pass into the completed conjugation-tube. 'The terminal transverse walls increase greatly in thickness by the deposition of layer after layer of cellulose, and a slight thickening of the side walls also occurs. As this thickening goes on the cavity of the gametangium is gradually reduced and a hemispherical, or sometimes a bluntly conical mound of cellulose projects into the emptying gametangium. The metamorphosis of the gametangium, which might almost be described as a "solidification," keeps pace with the receding of the gametes, and when the latter have completely coalesced in the wide conjugating-tube, the proximal ends of the four solid processes project as four rounded buttons into the cavity of the zygospore. The mature zygospore possesses four cylindrical truncate horns, each of which has arisen without external change of form from one half of a gametangium, the latter having become *solid* by the deposition of an internal thickening of cellulose' (G. S. W., '09); consult fig. 213, *B—F*. In *D. africana* G. S. West there is no solidification, but the walls of the gametangia increase considerably in thickness, the thickening being evenly laid down except for terminal pits. In *D. glyptosperma* (De Bary) Wittr. the relics of the gametangia become very clear and refractive, exhibiting the delicate striations parallel to the transverse walls which are very noticeable in *D. Hardyi*. They have all the appearance of solidity, although the actual details of the change have not been followed out.

The genera of the Mesocarpeæ are: *Mougeotia* Agardh, 1824; *Debarya* Wittrock, 1872; *Gonatonema* Wittrock, 1878; *Temnogametum* W. & G. S. West, 1897.

Species of *Mougeotia* occur abundantly in almost all parts of the world and are especially frequent in the more elevated regions of temperate countries, in which areas they only rarely conjugate, passing the winter in a vegetative state. The smallest are *M. elegantula*

Wittr. (diam. 4 μ) and *M. minutissima* Lemm. (diam. 2·5—3 μ); the largest is the American species *M. crassa* (Wolle) De Toni (diam. 50 μ).

The 'knee-joint' connections of *Mougeotia genuflexa* (Dillw.) Ag. [= *M. mirabilis* (A. Br.) Wittr.] are deserving of special mention. In north temperate countries this is a common species and although rarely seen in conjugation, contact regularly takes place between neighbouring filaments by means of genuflexions of the cells. These 'knee-joints' of contact have all the outward appearance of the commencement of scalariform conjugation, but Nieuwland ('09 A) has brought forward evidence to show that this is not the case. These contact cells appear to be simply glued together and no direct communication is ever established between them. Nieuwland's observations indicated that filaments thus connected very readily break up into their individual cells, and, since this results in an

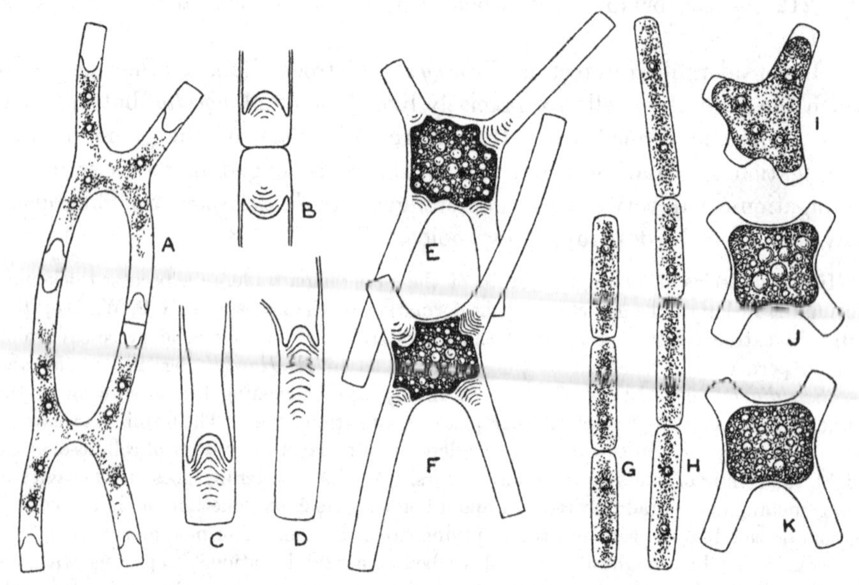

Fig. 213. *A—F, Debarya Hardyi* G. S. West. *A*, conjugating filaments; *B—D*, extremities of gametangia during conjugation showing gradual filling up of internal cavity; *E* and *F*, completed zygospores. *G—K, D. desmidioides* W. & G. S. West. *G* and *H*, small portions of long vegetative filaments; *I*, conjugating cells; *J* and *K*, completed zygospores. *A, E* and *F*, ×470; *B—D*, ×940; *G—K*, ×488.

immediate increase in the quantity of *Mougeotia*, presumably the dissociated cells divide rapidly to form new filaments. Nieuwland regards this 'knee-joint' connection of the filaments as 'a phenomenon preliminary to the process of vegetative multiplication.' The physiological significance of the temporary connection may be an interchange of soluble substances through the apposed walls of the joint, but the whole question requires detailed investigation.

In *Mougeotia gracillima* and *M. parvula* the chloroplast is occasionally constricted and in several instances the present author has observed *two* chloroplasts in each cell, with the nucleus located between them. This condition is a near approach to the cytological structure of the Desmid *Gonatozygon Kinahani*, but the latter has characters which easily distinguish it from any species of *Mougeotia*.

Sub-family Zygnemeæ. The Algæ included in this sub-family consist of unbranched filaments of cells similar to those of the Mesocarpeæ, but usually of slightly larger size and with much shorter cells. The cell-wall is also stronger and thicker, the strongest walls occurring in *Zygnema ericetorum* (fig. 214 *C*; fig. 215). Each cell possesses a centrally placed nucleus, embedded in a small amount of cytoplasm (fig. 215 *D* and *E*; fig. 216 *E*), and two

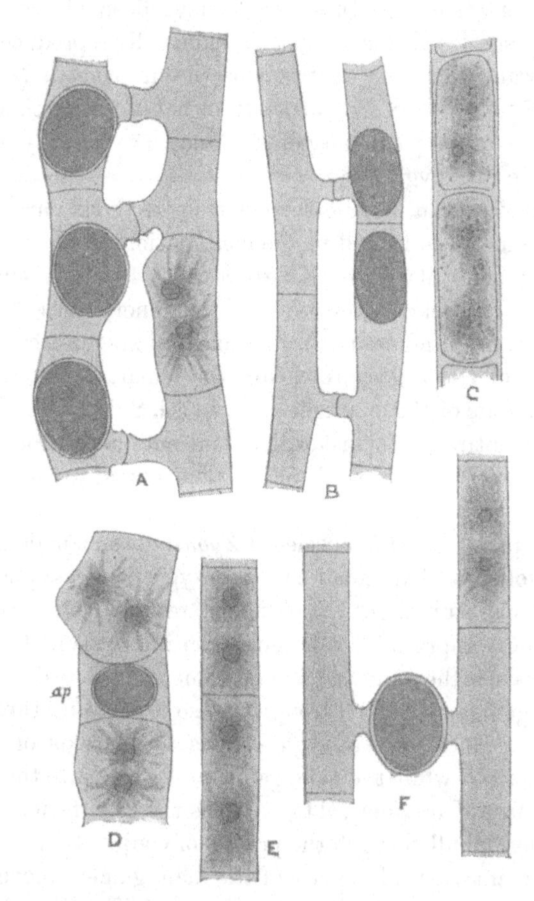

Fig. 214. *A*, *Zygnema stellinum* (Vauch.) Ag., ×430. *B*, *Z. Vaucherii* Ag. var. *stagnale* (Hass.) Kirchn., ×430. *C*, *Z. ericetorum* (Kütz.) Hansg., ×415. *D*, *Z. leiospermum* De Bary, ×430. *E*, *Z. insigne* (Hass.) Kütz., ×330. *F*, *Z. Ralfsii* (Hass.) De Bary, ×430. *ap*, aplanospore.

chloroplasts (except in *Zygnema ericetorum*), the latter being axile in *Zygnema* and *Pyxispora* but parietal in *Pleurodiscus* (fig. 216 *D*). Each chloroplast is furnished with a central pyrenoid.

In *Zygnema* the two chloroplasts are generally star-shaped, having numerous radiating branched processes (fig. 216 *E*). *Z. ericetorum* is a notable exception in which there is but one deeply constricted chloroplast, often twisted at the middle, each half having a central pyrenoid (fig. 215 *D* and *E*). Moreover, in this species the two halves of the chloroplast, although irregularly lobed, are not produced into slender branched processes.

Multiplication by fragmentation of the vegetative filaments is very unusual in *Zygnema*.

Asexual reproduction by aplanospores is more frequent in *Zygnema* than in any other Conjugates except those which have been placed in the genus *Gonatonema*. One species, *Z. spontaneum*, is habitually reproduced by aplanospores, although scalariform conjugation does occur (consult Nordstedt, '78; W. & G. S. W., '97; '07; G. S. W., '09). It is not unlikely that the aplanospores are in some cases parthenospores. They are usually globular and rather smaller than the zygospores.

Conjugation in *Zygnema* is almost exclusively of the scalariform type and the zygospore is sometimes lodged in the conjugation-canal, species in which this occurs having been placed by Kützing ('43; '49) in a separate genus, *Zygogonium*. This character is, however, of no generic importance, since in *Z. spontaneum*, a tropical and subtropical species, all stages occur between the complete inclusion of the zygospore within the female gametangium and its lodgment in the middle of the conjugation-tube (G. S. W., '09). Each gamete is formed from an entire protoplast which contracts away from the cell-wall and becomes a somewhat ellipsoidal mass. The gametes then glide slowly into the conjugation-tube or one of them glides through that tube into the opposite gametangium. In a few species of *Zygnema* the female gametangium is considerably swollen and in some the ripe zygospores assume a dark blue colour. The zygospore undergoes a period of rest and on germination the young plant shows no appreciable differentiation into base and apex.

During conjugation the male and female gametes behave differently. The axis of the male gamete rotates through 90° so that a line through the two chloroplasts and the nucleus is at right-angles to the axis of the filament. This position is retained while the male gamete advances into the conjugation-canal (Dangeard, '09; Kurssanow, '11), whereas the orientation of the female gamete remains normal all through the process of conjugation. On the fusion of the gametes the anterior chloroplast of the male gamete moves to one side so that the male and female nuclei come into contact. The time elapsing before the fusion of the nuclei is apparently variable, being much shorter in some species than in others. According to Kurssanow ('11) the male chloroplasts disintegrate immediately after the formation of the zygote-walls. The nucleus (fusion nucleus) of the zygote thereupon divides twice, three of the nuclei degenerating and one assuming the characters of the true zygote-nucleus. Kurssanow states that there is no conjugation of two of the secondary nuclei (*i.e.* two of the four produced by division of the fusion-nucleus of the zygote) such as Chmielewski describes in *Spirogyra*. It must be confessed, however, that the published statements regarding the cytology of conjugation in *Zygnema* are not at all in agreement with one another, and that much further investigation is still required.

It is necessary to consider the case of *Zygnema ericetorum* (Kütz.) Hansg. separately. This Alga has a world-wide distribution and in temperate countries it lives equally well in

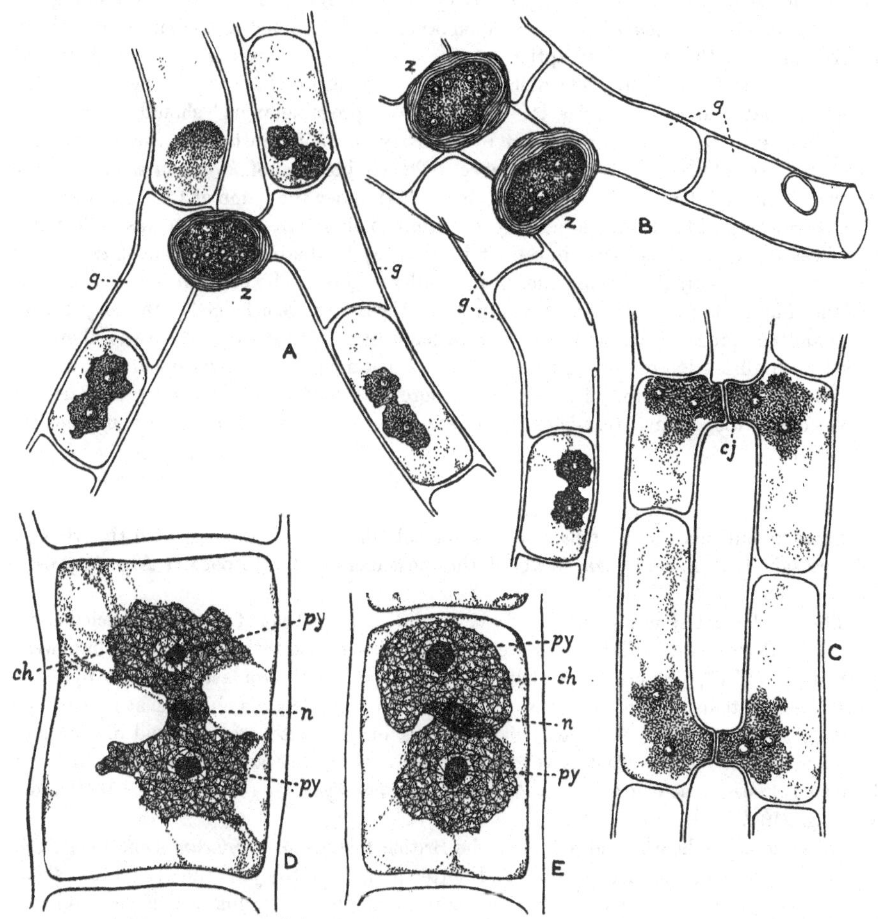

Fig. 215. *Zygnema ericetorum* (Kütz.) Hansg. *A* and *B*, fully conjugated specimens; *C*, an early stage in conjugation; *D* and *E*, stained specimens to show the chloroplast and nucleus. *ch*, chloroplast; *cj*, conjugation-canal; *g*, gametangium; *n*, nucleus; *py*, pyrenoid; *z*, *zygospore*. *A*—*C*, ×425; *D* and *E*, ×1500.

Note. *A*—*C* were originally described under the name of *Zygnema pachydermum* W. & G. S. West.

water or on damp heaths or peaty moors, not infrequently assuming a purple or violet colour owing to the appearance of phycoporphyrin in the cell-sap. In 1858 De Bary attempted to put the genus *Zygogonium* upon a new footing, including in it *Z. ericetorum* and *Z. didymum*[1], the generic character being the formation of 'progametangia' which ultimately fuse to form the zygospore. He illustrated this by two figures of conjugating filaments of *Z. didymum*. Various authors, notably Wille ('09) have accepted De Bary's conception of the genus *Zygogonium*.

In the first place it should be mentioned that the sole character upon which Kützing ('43) separated *Zygogonium* from *Zygnema* Agardh ('24) was the lodgment of the zygospore in the conjugation-canal, and since recent evidence proves this to be a character of no taxonomic value (*vide* G. S. W., '09) De Bary was in error in trying to establish a genus under the name of '*Zygogonium*' based upon other characters such as were never mentioned by Kützing. With the one exception of *Z. ericetorum* no species included in Kützing's *Zygogonium* could have any place in De Bary's genus of that name. Secondly, it is essential that the evidence for the existence of 'progametangia' should be carefully examined, since the two figures given by De Bary in 1858 illustrating these structures constitute the sole record of their existence. The conjugation of *Zygnema ericetorum* was described and well figured by W. & G. S. West ('94) under the name of '*Z. pachydermum*.' West & Starkey ('14; '15) have recently re-examined this material, which was collected in the West Indies, and find that it cannot in any way be distinguished from *Z. ericetorum*, the cytological characters being identical in all respects. The conjugation was in every instance like that of any other *Zygnema* in which the zygospore is lodged in the conjugation-canal and no 'progametangia' were observed (consult fig. 215 *A—C*). There is, therefore, as yet no confirmation of the formation of progametangia in *Z. ericetorum*, and it is quite possible that the two figures, first published more than half a century ago by De Bary and since copied into many text-books, are abnormalities and do not represent healthy conjugation.

West & Starkey ('14; '15) made numerous cultures of *Z. ericetorum*, extending over a period of two years, and many of these cultures were subaërial, but all attempts to induce conjugation failed, although aplanospores were obtained in abundance. On the whole, it may be said that *Z. ericetorum* is about the most inert and unresponsive Alga at present known.

The cytological structure of *Z. ericetorum* differs from that of all other species of the genus. Wille ('97) stated that there were two chloroplasts, each with one pyrenoid, but twelve years later (Wille, '09) he states that there is one axile chloroplast with one pyrenoid. Both these statements are erroneous. There is but a single axile chloroplast in each cell, deeply constricted and sometimes twisted in the middle, with one pyrenoid in each lobe. The lobes of the chloroplast have an irregular outline, but are not produced into the branched processes so characteristic of most species of *Zygnema* (compare fig. 215 *D* and *E* with fig. 216 *E*).

On some of the heaths and moors of the British Islands *Z. ericetorum* sometimes fulfils an important function. In the drier and hotter periods of the year thickly-matted sheets of this Alga, often many square feet in extent, are found covering extensive patches of almost bare sand or peat, round such plants as *Drosera*, *Carices*, etc. These mats of *Zygnema* have great absorptive capacity, greedily taking up water, and in this way they regulate the moisture of the surface soil, the thriving of some of the smaller phanerogams depending to a great extent on the presence of the *Zygnema*[2].

[1] The present author has examined Rabenhorst's original specimens of *Zygogonium didymum* (issued in Rabenhorst's *Alg. Exsic.* no. 182) and finds them identical in every respect with *Zygnema ericetorum*.

[2] This phenomenon is much more evident in some parts of the tropics and attention was first called to it by Welwitsch in the *Journal of Travel and Natural History*, vol. i, 1868 (consult p. 31).

The genus *Pleurodiscus* is peculiar in the possession of parietal discoidal chloroplasts, two of which are present in each cell. There is only one known species—*Pl. purpureus* (Wolle) Lagerh.—in which the cell-sap is coloured purple with phycoporphyrin. Conjugation takes place as in *Zygnema* and the zygospore is formed in the female gametangium. *Pyxispora*, a genus known only from West Africa (*vide* W. & G. S. W., '97), bears the same relationship to *Zygnema* that *Mougeotia* does to *Debarya*, the gametes being formed from only part of the protoplast of the gametangium (fig. 216 *A* and *B*). In this genus the outer wall of the ripe zygospore splits along an equatorial crack (fig. 216 *C*).

The genera are: *Zygnema* Agardh, 1824 [inclus. *Zygogonium* Kützing, 1843; *Zygogonium* De Bary, 1858; Wille, '97; '09]; *Pleurodiscus* Lagerheim, 1895; *Pyxispora* W. & G. S. West, 1897.

The doubtful genus *Mesogerron* Brand ('99) does not appear to belong to the Conjugatæ but rather to some group of the Ulotrichales.

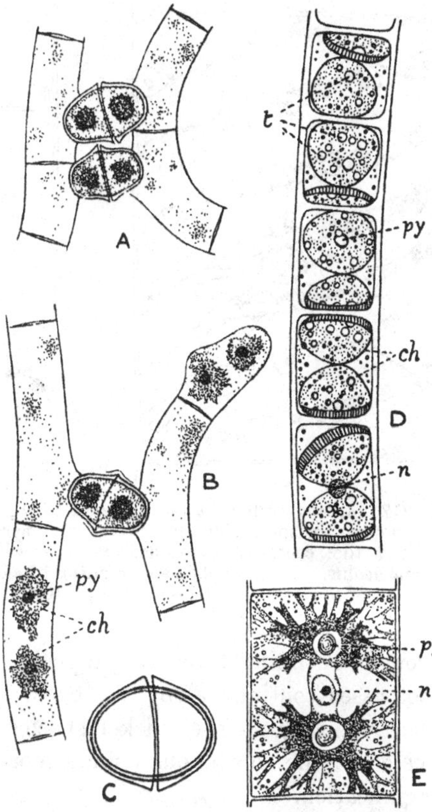

Fig. 216. *A—C*, *Pyxispora mirabilis* W. & G. S. West; *A* and *B*, conjugated specimens, ×520; *C*, zygospore, ×1000. *D*, vegetative filament of *Pleurodiscus purpureus* (Wolle) Lagerh., ×600 (after Lagerheim); *E*, single vegetative cell of *Zygnema* sp., stained and mounted in Venetian turpentine, ×500. *ch*, chloroplast; *py*, pyrenoid; *n*, nucleus; *t*, tannin globules.

Sub-family SPIROGYREÆ. This sub-family includes the largest of the Zygnemaceæ, some species of *Spirogyra* attaining a diameter of over 160 μ. The distinguishing character of the group lies in the band-like parietal chloroplasts, which are for the most part spirally twisted. *Spirogyra* is the best known and most conspicuous of the Conjugatæ, its cells exhibiting great variability both in their diameter and relative length. The cell-wall is very firm and the mucous coat, which is well developed, especially in the larger species, renders the plants very slimy to the touch.

The chloroplasts of *Spirogyra* are disposed in the lining layer of cytoplasm in the form of spiral bands, and they vary in number from one to twelve or fourteen in each cell. In some species they are coiled into very close spirals, but in others they are practically straight and longitudinal; in some, as in

S. neglecta (Hass.) Kütz., their margins are quite smooth and there is a regular axile series of pyrenoids; in others, as in *S. nitida* (Dillw.) Link or *S. porticalis* (Vauch.) Cleve, the chloroplasts are very broad, with serrated margins and scattered pyrenoids. Between these two extremes there is every gradation, and the character of the chloroplasts always remains constant for any particular species, even though the number of them may vary in different cells of the same filament. It has been ascertained by Kolkwitz that the chloroplasts grow in length in the direction of the coils by both apical and intercalary growth; and therefore as this is obliquely to the surface of the cell-wall, there is a gliding motion of the spiral bands through the lining layer of cytoplasm. The chloroplasts of any one cell in most cases continue along the exact spiral line of those in adjacent cells, a fact which gives the thicker species with several chloroplasts the appearance of having continuous spirals, more especially when viewed under a low magnification.

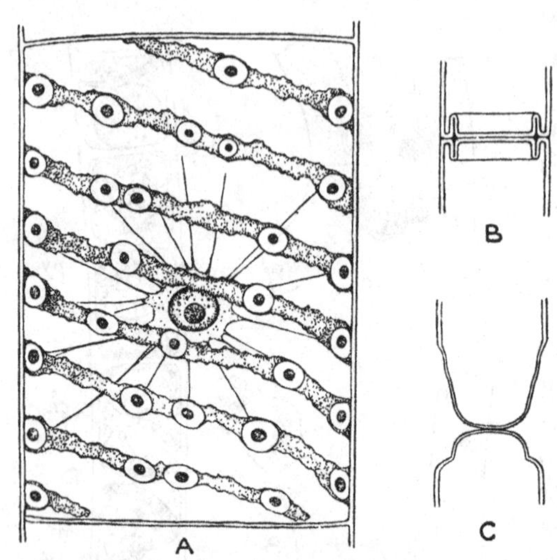

Fig. 217. *A*, single cell of *Spirogyra crassa* Kütz., × 350, showing the spiral chloroplasts with their pyrenoids and the centrally suspended nucleus with its nucleolus. *B*, the replicate extremities of adjacent cells of *Sp. tenuissima* (Hass.) Kütz., × 1000. *C*, the same with the infoldings everted.

The nucleus, which in some species is transversely ellipsoid or complanate and in others globose, is in all cases suspended in the median part of the cell, embedded in a small amount of cytoplasm from which radiate numerous protoplasmic strands to the lining layer (fig. 217 *A*). Many of these strands reach the lining layer exactly opposite a pyrenoid, and, since the starch-grains are laid down around the pyrenoids, this fact has been brought forward as a reason for supposing the nucleus to have a direct influence on starch-formation. In some species, such as *Spirogyra majuscula* Kütz. and *Sp. pellucida* (Hass.) Kütz., the nucleus is clearly visible in the living cells. Gerassimoff ('97; '00) has found that the cells are sometimes binucleated or they may contain a 'compound nucleus.'

In cases of this kind the abnormal nuclear structure is owing to adverse conditions arising during nuclear division. If sudden cooling occurs at the very commencement of nuclear division it may result in the production of two cells, one with an extraordinarily large nucleus and one without any nucleus; if the cooling takes place at a later stage then both nuclei may be in one cell the other being again entirely without[1].

There is usually one large nucleolus (fig. 217 *A*), but two, or even three may sometimes occur. The nucleolus contains much chromatin.

Any cell of the filament may undergo division and from the time Strasburger ('82) first gave an account of the nuclear division, many investigators have studied the mitosis in *Spirogyra*. Considerable differences of opinion exist with regard to the actual details and especially as to the number of chromosomes which arise from the nucleolus, Van Wisselingh ('00) stating that only two of them arise in that way, whereas Berghs ('06) and Moll ('08) state that all the chromosomes arise from the nucleolus[2]. The spindle is broad and cylindrical and the balance of evidence is in favour of its purely cytoplasmic origin[3].

Amitotic division of the nucleus has been induced by Gerassimoff ('00), and also by Nathansohn ('00) by subjecting the filaments to anæsthetics, such as chloroform and ether, of a strength of 0·5—1 per cent. in the culture medium. Gerassimoff also obtained the same result by cooling the filaments below freezing-point for a time. These results should be regarded, however, as pathological.

On the division of binucleated cells the daughter-cells are again binucleated through several generations.

In the division of the cell the new transverse wall arises as a delicate annular ingrowth from the lateral walls and by gradual extension finally becomes a complete partition. As the new wall gradually thickens the original partition becomes the middle lamella.

Vegetative propagation sometimes occurs by the fragmentation of the filaments either into individual cells or short chains of cells, which then form new filaments by cell-division. It is in the slender species of *Spirogyra*, and more especially those in which the cells possess 'replicate extremities,' that this method of multiplication is most frequently observed. The 'replicate extremity' consists of an annular infolding of the end-wall (fig. 217 *B*), which facilitates fragmentation by becoming everted (fig. 217 *C*).

Reproduction occurs by the conjugation of isogametes which always fuse in one of the gametangia. In the formation of a gamete the protoplast contracts away from the wall, receding from the end-walls first. It soon becomes more or less ellipsoidal, after which one gamete of each conjugating pair begins very gradually to glide from its mother-cell (the male gametangium) through the conjugation-canal into the opposite cell (the female gametangium).

[1] The cells without nuclei may live for some weeks and are able to store starch.

[2] For a brief but comprehensive summary of the various views on nuclear division in *Spirogyra* consult Lutman ('11).

[3] It is probable that the discrepancies in the various accounts of nuclear division in *Spirogyra* are to be attributed to defective methods of fixation and the consequent misinterpretation of facts.

One of the conjugating cells is thus completely emptied while the other lodges the zygospore (fig. 218 *C*). The gametes may coalesce immediately on contact or for a brief period they may lie side by side in the gametangium before fusing.

The fully-formed zygospore is generally ellipsoidal, but it may be spherical, and sometimes it is distinctly compressed, as in *Spirogyra maxima*, *Sp. majuscula* and others. It provides itself with a thick wall of three coats (*vide* De Bary, '58, and others). The outer coat consists mostly of cellulose and is the first-formed wall of the zygospore. The middle coat, which is generally stronger and thicker, gives no cellulose reaction, and is apparently partially chitinized. It is usually of a brown colour, and in many cases is furnished with small, irregular and anastomosing ridges, or with scrobiculations. The inner coat is the most delicate and consists of pure cellulose.

The fusion of the nuclei is delayed for some time after the formation of the zygospore. Overton ('88) and Klebahn ('88) each stated that the fusion did not take place for weeks or even months after the complete formation of the zygospore, but Chmielewski ('90) declared that fusion occurred quite soon after the union of the gametes, and that it was followed by two successive mitotic divisions resulting in four nuclei. Two of these decay, the other two uniting to form the 'secondary nucleus' of the zygospore[1]. Trondle ('07) states that the nuclei of the gametes lie side by side for about seventeen to twenty-one days, and then fuse. He confirms Chmielewski's observations on the double mitotic division with the formation of four nuclei and the subsequent fusion of two of them. The whole process is, however, at present rather obscure. Karsten ('09) has also observed the double mitosis of the zygote-nucleus and states that the first division is heterotypic.

Concerning the behaviour of the chloroplasts of *Spirogyra* during conjugation there is much difference of opinion. Overton ('88) stated that the chloroplasts of the two gametes disorganized and became amalgamated, whereas Chmielewski ('91) states that only the chloroplast of the male gamete disorganizes, that of the female gamete remaining intact. There appears to be good evidence that in some species both chloroplasts disintegrate during conjugation, and it may be that the behaviour of the chloroplasts during conjugation is not precisely uniform. In *Sirogonium* the chloroplasts of both gametes disintegrate at the very commencement of conjugation.

The mature zygospores are usually brown or sometimes brownish-red in colour, and most of the starchy reserves are usually converted into a fatty oil.

On the germination of the zygospore much starch appears and the oil vanishes. The two outer coats are ruptured, generally at one end of the spore, and the inner coat is protruded as a short outgrowth. Nuclear- and cell-division soon occur and the outgrowth becomes divided by a transverse wall (fig. 218 *G*), one cell sometimes developing organs of attachment and

[1] Compare with Kurssanow's statement concerning *Zygnema* (*vide* p. 344)

containing little or no chlorophyll, whereas the other cell increases in size develops one or more chloroplasts, and by repeated divisions forms a filament. Except in one or two species of *Spirogyra* the distinction between base and apex is soon lost and the filaments float freely in the water; thus the slight rhizoidal development which sometimes occurs is of little practical value since the filaments almost immediately become free-floating. In some cases organs of attachment (haptera) are developed subsequently from older cells, and in *Spirogyra adnata* several of the adjacent basal cells may participate in the formation of attaching organs (Delf, '13).

It has been disputed that *Spirogyra* exhibits any sexuality, but there are many facts which suggest a differentiation of sex. The cell containing the zygospore becomes in some species greatly inflated, a distention which is never observed in the emptied cell. That part of the conjugation-tube developed from the cell in which the zygospore is lodged is very often thicker and shorter than the part developed from the opposite cell, and in some cases (*e.g. Sp. punctata* Cleve, *Sp. tenuissima* Hassall, *Sp. orientale* W. & G. S. West) it may not be developed at all or is obliterated by the distention of the cell. The location of the zygospore in that gametangium which shows these special characters indicates that there is a differentiation of sex similar to that which occurs in *Zygnema*.

In most cases the cells of one filament become all of the same sex and two filaments only are concerned in scalariform conjugation, but three, four, five, six, or even more may occasionally take part in one conjugating example (consult fig. 218 *A*). Statistics show that in these cases the female filaments predominate (a condition frequently termed 'polygamy'), the relatively shorter length of many of the cells of a male filament often permitting of the formation of sufficient male gametes to supply two female filaments.

Lateral conjugation, or conjugation between adjacent cells of the same filament, is much more frequent in *Spirogyra* than in *Zygnema*, and is the general rule in certain of the more slender species, such as *Sp. tenuissima*, *Sp. inflata* (fig. 218 *D*), *Sp. quadrata* and others. It also occurs in large species like *Sp. majuscula*. In these cases a differentiation of sex must take place amongst the cells of the same filament, some becoming male and others female. In rare instances scalariform conjugation takes place between filaments with this promiscuous differentiation of sex, perfectly normal zygospores being formed in the female cells of each filament. This type of conjugation has been termed 'cross-conjugation' (*vide* W. & G. S. W., '98, t. v, f. 81).

Irregularities are frequently met with in conjugating filaments. Zygospores are occasionally observed which have been produced by the fusion of three gametes, two male and one female (W. W., '91; Borge, '91; W. & G. S. W., '98), but such attempts are usually abortive (fig. 218 *B*; consult also W. & G. S. W., '98; Schmula, '99; etc.). Copeland ('02) states that in these cases the nucleus of the abortive male cell is situated against the wall opposite to and remote from the conjugation-canal. Gerassimoff has observed the conjugation of two female cells with one male cell and the formation of the zygospore by the fusion of the male gamete with one female gamete, a parthenospore being formed by the other female gamete. Owing to sudden changes of the physiological conditions it not infrequently happens that conjugation is brought to an abrupt termination before the fusion of the gametes has taken place. In such cases parthenospores are often formed, either one in each gametangium or two in the female gametangium (*vide* Rosenvinge, '83; W. W., '91; Hansgirg, '88; W. & G. S. W., '98; etc.). The discovery by Gerassimoff of binucleated cells in the Zygnemaceæ may perhaps afford an explanation of some of these irregularities of conjugation, since he states that in the conjugation of binucleated cells parthenospores were sometimes formed.

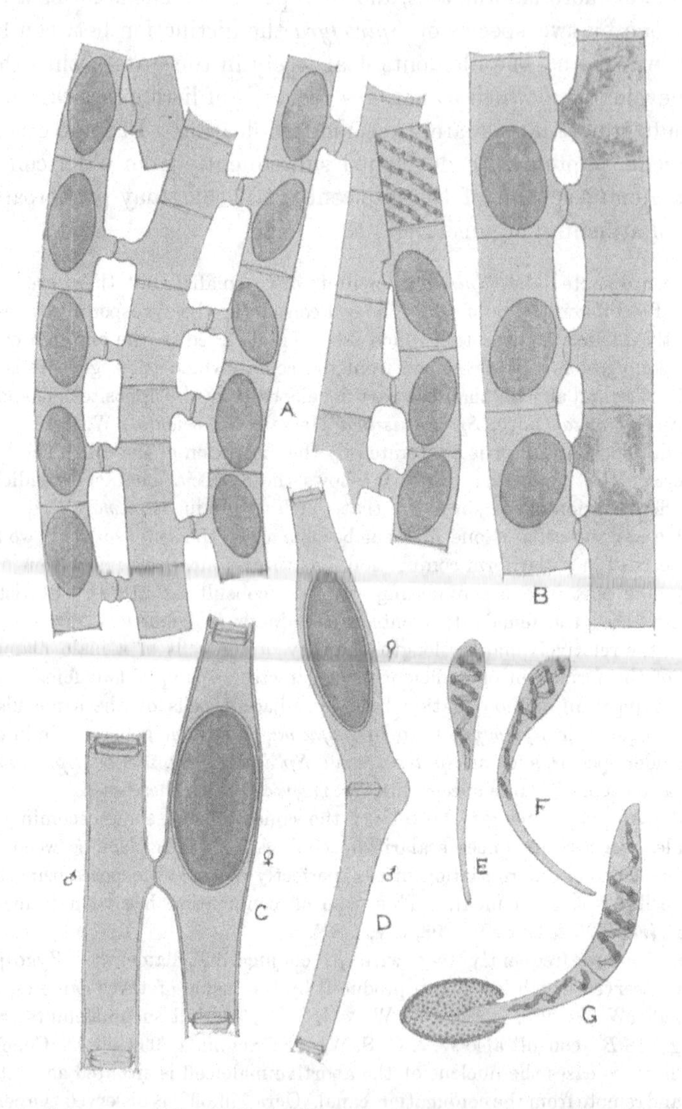

Fig. 218. *A*, *Spirogyra nitida* (Dillw.) Link, showing scalariform conjugation between six filaments, ×75. *B*, *Sp. setiformis* (Roth) Kütz., showing frustrated attempts at conjugation between two male cells and one female cell, ×90. *C*, *Sp. Spreeiana* Rabenh., ×390. *D*, *Sp. inflata* (Vauch.) Rabenh., lateral conjugation, ×390. *E* and *F*, zygospores of *Sp. velata* Nordst. germinating immediately after their formation and before the differentiation of the wall of the zygospore, ×165. *G*, germination of zygospore of *Sp. velata* after normal period of rest, ×230.

Klebs found that still water and bright illumination were favourable to conjugation, whereas running water and an excess of available food-materials were inimical to it. Thus, if a *Spirogyra* be transferred from a dilute nutrient solution to rain-water or tap-water and placed in bright light conjugation is induced. Conjugation is also said to be induced by placing the filaments in a 2—4 per cent. solution of cane sugar and exposing to sunlight. The student must not, however, hope to be uniformly successful in attempts to induce conjugation, since many such attempts fail hopelessly, probably because of the lack of the requisite combination of operative factors. Klebs also found that the formation of parthenospores could be induced by placing conjugating filaments in a 60 per cent. sugar solution. In these cases of apogamy the parthenospores are usually of the same form as the zygospores and they germinate in the same way. Parthenospores are not infrequently produced under natural conditions and have been recorded by numerous observers.

The case of *Spirogyra mirabilis* (Hass.) Petit is peculiar and may be the result of degeneration, the formation of spores being apparently analogous to that which occurs in *Gonatonema*. In spore-formation the cells become swollen in the median part and the protoplast divides into two parts which subsequently fuse together (Petit, '80). The cytology of this process would be of great interest, but it still requires investigation.

In the genus *Sirogonium*, the sterile threads of which differ little from *Spirogyra* except in the somewhat irregular course of the chloroplasts, the conjugation is comparable with that of *Temnogametum*. Certain cells of the filaments become bent into a knee-like form, after which they attach themselves to similar bent cells of other filaments, the attachment being rendered firm by a ring of tough mucilage. The protoplasts of the attached cells now divide and partition-walls appear, cutting off either one or two smaller cells from the remaining cell, which is the gametangium. Dissolution of the wall occurs at the point of attachment of the gametangia and the zygospore is formed in the female gametangium.

The genera of the Spirogyreæ are: *Spirogyra* Link, 1820, and *Sirogonium* Kützing, 1843 [= *Choaspis* S. F. Gray, 1821[1]].

Species of *Spirogyra*, which as a genus is one of the commonest laboratory types studied among the Algæ, occur mostly in still waters and very largely in ponds and ditches, where they often form floating masses buoyed up by numerous bubbles of oxygen. These flocculent masses gradually change colour from green to yellow, mostly owing to the death of a large proportion of the filaments as a result of exposure to light of too great an intensity, although sometimes owing to almost universal conjugation. Some species habitually occur in very slow streams and rivers. *Sirogonium sticticum* Kütz. generally occurs attached to stones over which the water is quickly running. It is not infrequent in limestone districts and is noteworthy for the comparatively small amount of external mucus.

[1] It is to be recommended at the next International Botanical Congress that the generic name *Sirogonium* be retained.

Family **Desmidiaceæ.**

The Algæ included in this family are remarkable for their great diversity of form and their wonderful symmetry. Indeed, the group includes some of the most beautiful of microscopic objects. Desmids are unicellular plants and the greater number of them lead a solitary existence. Certain of them are, however, associated in colonies and others are more or less closely united into longer or shorter filaments. They are essentially free-floating Algæ and frequently occur in great abundance in small ponds, in the quiet margins of rocky lakes, in *Sphagnum*-bogs, and in other favourable localities.

Most Desmids exhibit a more or less distinct constriction into two perfectly symmetrical halves; each half is termed a *semicell* and the narrower part connecting the two semicells is known as the *isthmus*. The angle resulting on either side from the constriction or narrowing of the cell is known as the *sinus*. The depth of the constriction varies very much even in different species of the same genus (compare fig. 219 *A* and fig. 219 *D*) and most of the very deeply constricted forms have a linear sinus. None of the true Saccoderm Desmids are constricted, but of the vast majority of Desmids, which belong to the Placodermæ, few are without a median constriction of greater or less depth[1].

One of the most striking features of the family is the extraordinary complexity of the cell-outlines. The margin of the cell is often deeply lobed or incised, and the exterior of the cell-wall is frequently covered with granules, spines, wart-like thickenings, or other protuberances, most of which are arranged in some definite pattern. The greatest complexity of cell-outline is met with in species of *Micrasterias* (fig. 223), *Euastrum* (fig. 220) and *Staurastrum* (figs. 221 and 222). Some Desmids are more or less cylindrical (consult figs. 227, 230 and 238), many others are compressed in the plane of the front view (consult figs. 219, 220 and 223), and those of the genus *Staurastrum*, when seen from the end view, are triangular, quadrangular, or with radiating processes (consult figs. 221 and 222). In *Closterium* the cells are subcylindrical, attenuated from the middle towards each end, and curved or sublunate (fig. 229 *A* and *B*; fig. 231 *A—E*). In some species of this genus the cell-wall is longitudinally striated, the striations representing internal thickenings of the wall. External to the firmer portion of the cell-wall, which consists chiefly of cellulose, are layers of mucilaginous pectose compounds. The latter often form a thick mucilaginous coat completely surrounding the individual (fig. 228 *B—E*), or, in the case of some colonial forms, entirely enveloping the colony. It is by means of this mucilaginous envelope that many Desmids adhere to other larger aquatic plants and that

[1] With the exception of one or two species, such as *Closterium subcompactum* W. & G. S. West ('02), the species of *Closterium* are destitute of a median constriction. Several species of *Penium* are likewise unconstricted.

others increase their floating capacity in the plankton; and occasionally, when the conditions have been favourable for rapid multiplication, enormous numbers of individuals occur embedded in masses of jelly. Sometimes the secreted mucus is very tough. In some species of the genus *Spondylosium* (fig. 224, *I*, *M* and *O*) the cells are united into filaments by mucous threads passing between their apposed ends, and the filaments break much more readily across the isthmus of a cell than at the points of apical attachment.

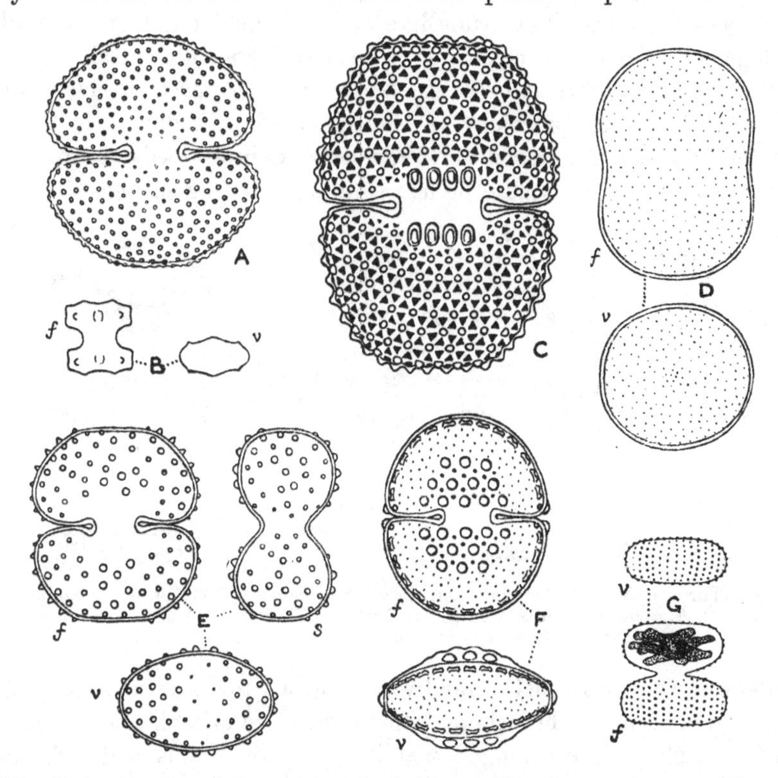

Fig. 219. Various species of *Cosmarium*. *A*, *C. Botrytis* Menegh. var. *depressum* W. & G. S. West, ×430; *B*, *C. Regnesi* Reinsch var. *montanum* Schmidle, ×1200; *C*, *C. decoratum* var. *dentiferum* W. & G. S. West, ×520; *D*, *C. pseudoconnatum* Nordst., ×520; *E*, *C. præmorsum* Bréb., ×520; *F*, *C. Prainii* W. & G. S. West, ×520; *G*, *C. Pappekuilense* G. S. West, ×500. *f*, front view of cell; *s*, side view; *v*, vertical or end view.

The colonial Desmids mostly occur as filaments (figs. 224 and 225), the degree of firmness of attachment of the cells being very variable. It is in *Gonatozygon* (fig. 238 *C* and *D*) that the attachment is most fragile, so much so that individual cells of this genus are much more frequently met with than filaments. The mere adhesion by tough mucus secreted between apposed apices of cells is in a few cases supplemented by the development of overlapping apical outgrowths of the cell-wall, as in *Onychonema* (fig. 225). In *Sphærozosma* there are also supplementary button-like outgrowths, which in the case of *S. vertebratum* and *S. Aubertianum* are the sole means of attachment of the cells. Colonial species sometimes occur in genera in which the species are normally solitary, such as *Pleurotænium coronulatum*, *Pl. perlongum*, and *Micrasterias foliacea* (fig. 239 *B*). The latter is a tropical and subtropical species with remarkable tooth-like apical connections, which result in very rigid filaments of cells. The apical attachment of this extraordinary species was first accurately described and figured by Johnson ('94). In *Cosmocladium* the *Cosmarium*-like cells are held together by stalks of mucilage, and the entire colony is quite small, branched in character, and free-floating (fig. 226 *A* and *B*). In *Oocardium*, the most curious of all the genera of Desmids, the colony is hemispherical in shape, 1 to 2 mm. in diameter, and occurs attached to calcareous rocks in the beds of streams. It consists of a number of more or less parallel, radiating strands of mucus of

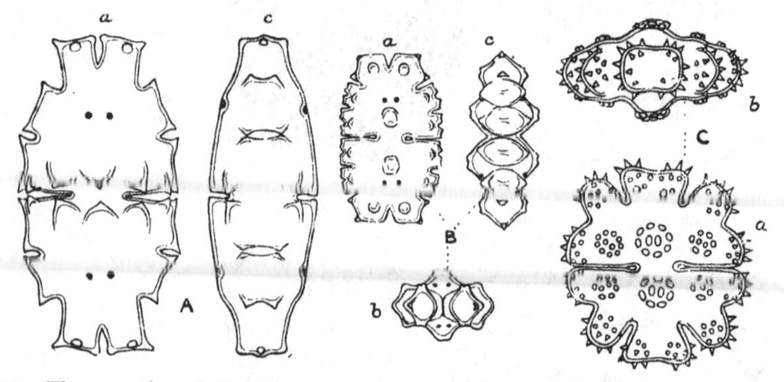

Fig. 220. Three species of *Euastrum*. *A*, *E. asperum* Borge, ×468; *B*. *E. serratum* Joshua, ×468; *C*, *E. inermius* (Nordst.) Turn. var. *burmense* W. & G. S. West, ×468. *a*, front view of cell; *b*, vertical or end view; *c*, side view.

considerable thickness, each strand widening out towards the surface of the colony and occasionally branching. In the free end of each mucous strand is lodged a single cell, so arranged that its longitudinal axis is at right angles to the axis of the mucous strand (fig. 226 *D*—*F*).

The minute structure of the cell-wall was first studied by Klebs ('85) who examined the nature of the mucilaginous outer coat and also demonstrated in certain species the presence of pores passing right through the wall. Hauptfleisch ('88) also contributed some information on this subject, more especially with regard to the extent and structure of the mucilaginous envelope. Some years later Lütkemüller ('94) described very minutely the pores in the cell-wall of *Closterium*, and eight years later (Lütkemüller, '02) published an important account of the general structure of the cell-wall in Desmids. A further paper by the same author emphasized the importance of the structure of the cell-wall in formulating any sound scheme of classification of Desmids (*vide* Lütkemüller, '05).

In a few Desmids—those belonging to the Spirotænieæ among Saccoderm Desmids—the cell-wall consists of a single layer of cellulose, all in one piece, and of a homogeneous structure. In the remaining Saccoderm Desmids, of the Gonatozygæ, the cell-wall is also quite continuous in one piece, but there is a differentiated outer layer. The great majority of Desmids belong to the sub-family Placodermæ, in which the cell-wall is composed of two well differentiated layers of varying thickness. Micro-chemical tests show that the inner layer is practically structureless and consists of pure cellulose. The outer layer is stronger and thicker, consisting of a ground substance of cellulose

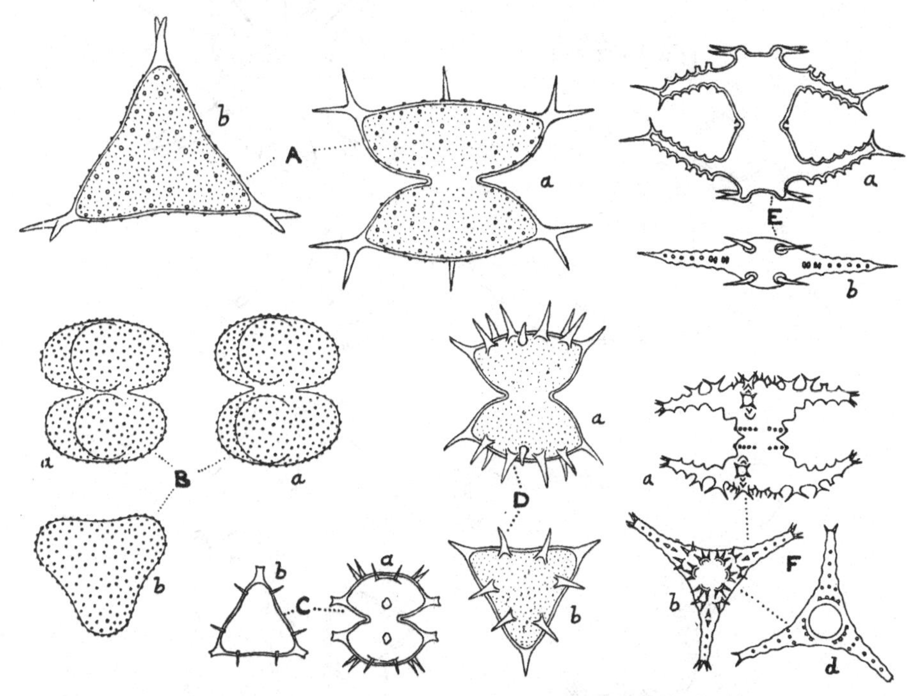

Fig. 221. Various species of *Staurastrum*. *A*, *St. pelagicum* W. & G. S. West, ×520; *B*, *St. turgescens* De Not., ×520; *C*, *St. monticulosum* Bréb. var. *pulchrum* W. & G. S. West, ×520; *D*, *St. pungens* Bréb., ×520; *E*, *St. saltans* Joshua, ×520; *F*, *St. cyclacanthum* W. & G. S. West, ×520. *a*, front view of cell; *b*, vertical or end view; *d*, basal view of semicell.

impregnated with various other substances, amongst which compounds of iron have been demonstrated. It is in some species of *Closterium* and *Penium* that the iron compounds are most prominent, and this is presumably directly concerned with the yellow-brown colour of the cell-wall so often seen in these genera.

The pores pass through both layers of the cell-wall and in the outer layer each pore is surrounded by a cylindrical tube-like structure which does not consist of cellulose. To these parts of the cell-wall Lütkemüller gave the name of 'pore-organs' and he considered them to be directly concerned in mucus-formation, a view which is probably correct since there is no doubt that mucus is passed outwards through the fine pores. The contents of the pore-canal often terminate on the inner surface of the cell-wall in a button-shaped or lens-shaped swelling, and from the outer end of the pore-canal there often extends a delicate flower-like or club-shaped mass of tough mucus through which the canal passes (fig. 226 *I*). In many cases this projecting

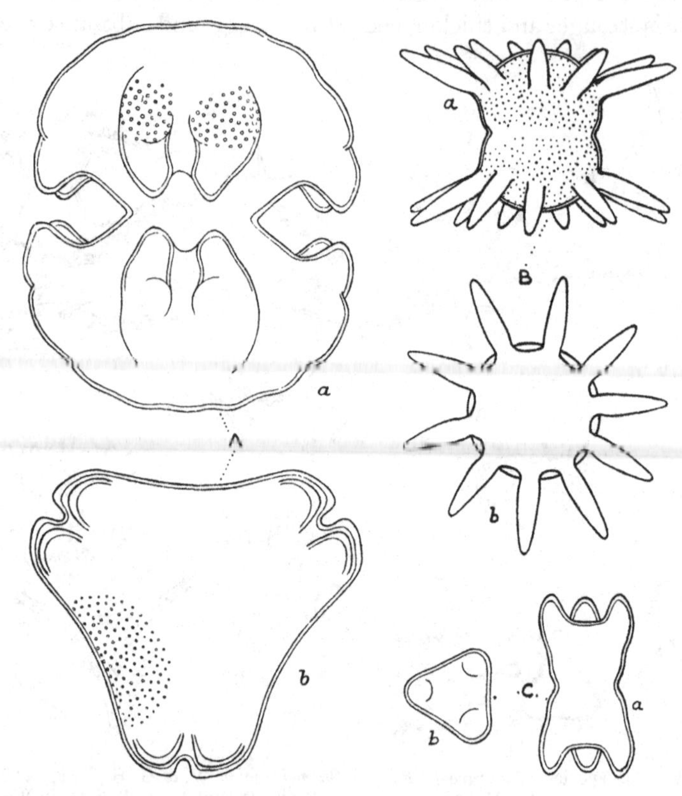

Fig. 222. Various species of *Staurastrum*. *A*, *St. Burkillii* W. & G. S. West, ×468; *B*, *St. victoriense* G. S. West, ×450; *C*, *St. unguiferum* Turn. var. *inerme* (Turn.) W. & G. S. West, ×468. *a*, front view of cell; *b*, end or vertical view.

mucous process is bifurcated (consult fig. 228 *D* and *E*[1]) and it is frequently sufficiently resistant to reagents to stand out clearly after fixation. It may also be yellow or even yellow-brown in colour, presumably owing to the deposition of iron salts, and may therefore be more or less permanent in character. More often this mucous process is entirely wanting or is replaced by a small perforated button (fig. 226 *H*). In most of the larger Desmids

[1] Consult also W. & G. S. West ('04—'11), t. 113, f. 15; t. 117, f. 5.

the pores are evenly distributed over the whole surface of the wall, except at the isthmus, where they are always absent, and in many cases there are numerous smaller pores between the large ones. It is highly probable that pores are present in the cell-wall of all Placoderm Desmids. In the striated species of *Closterium* the pores are in longitudinal rows between the striations. In granulated and verrucose species of *Cosmarium* the pores are in many instances arranged in groups around the individual granules, and in some genera there are often larger pores at the extremities of the cell, which are usually situated in local thickenings of the cell-wall. In *Cosmocladium* and in many of the filamentous genera of Desmids, such as *Hyalotheca*, *Gymnozyga*, *Onychonema*, etc., the pores are not evenly distributed, but have a definite disposition in certain parts of the wall (consult figs. 225 *B* and 226 *G*).

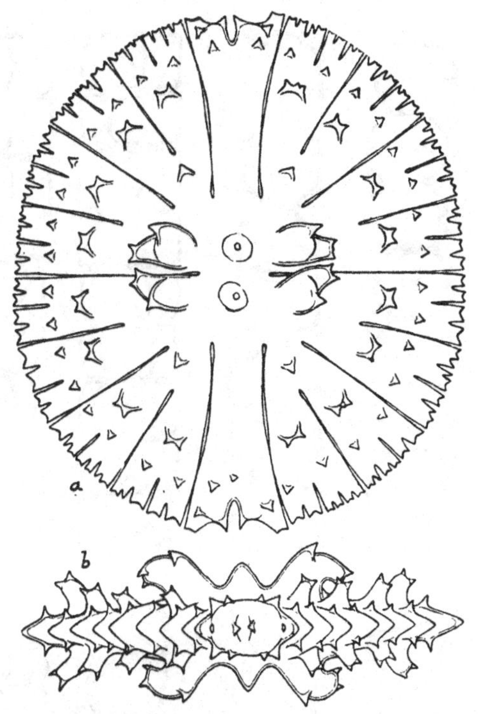

Fig. 223. *Micrasterias Thomasiana* Arch. var. *pulcherrima* G. S. West. *a*, front view of cell; *b*, end or vertical view. ×315.

External to the firm layers of the cell-wall is the mucus which is secreted through the pores. In the great majority of Desmids this mucus is small in amount and is easily diffluent, but in some it is copious and forms a mucous mass entirely enveloping the individual (fig. 228 *B*—*E*), or, in some filamentous species, it forms a cylindrical mucous sheath of variable width surrounding and enclosing the filament. In most filamentous forms of Desmids the apical attachment of the cells is by a tough mucus which acts as a cementing substance. The outer mucilaginous coat often exhibits a prismatic or fibrillar structure (fig. 228 *C*) and is frequently the home of rod-shaped bacteria or of minute epiphytic Flagellates. Schröder ('02) investigated the nature and amount of the mucus secreted by certain Desmids, obtaining much information by placing living Desmids in indian ink or sepia, the exact limitations of the secreted mucus being thus revealed.

It has been known for a long time that some Desmids exhibit irregular and jerky movements. These movements were first investigated by Stahl

('79) and afterwards by Klebs ('85), Schröder ('02) and others. In various species of *Closterium* one end of the cell becomes temporarily fixed and the whole cell then performs swinging movements with the fixed end as a pivot.

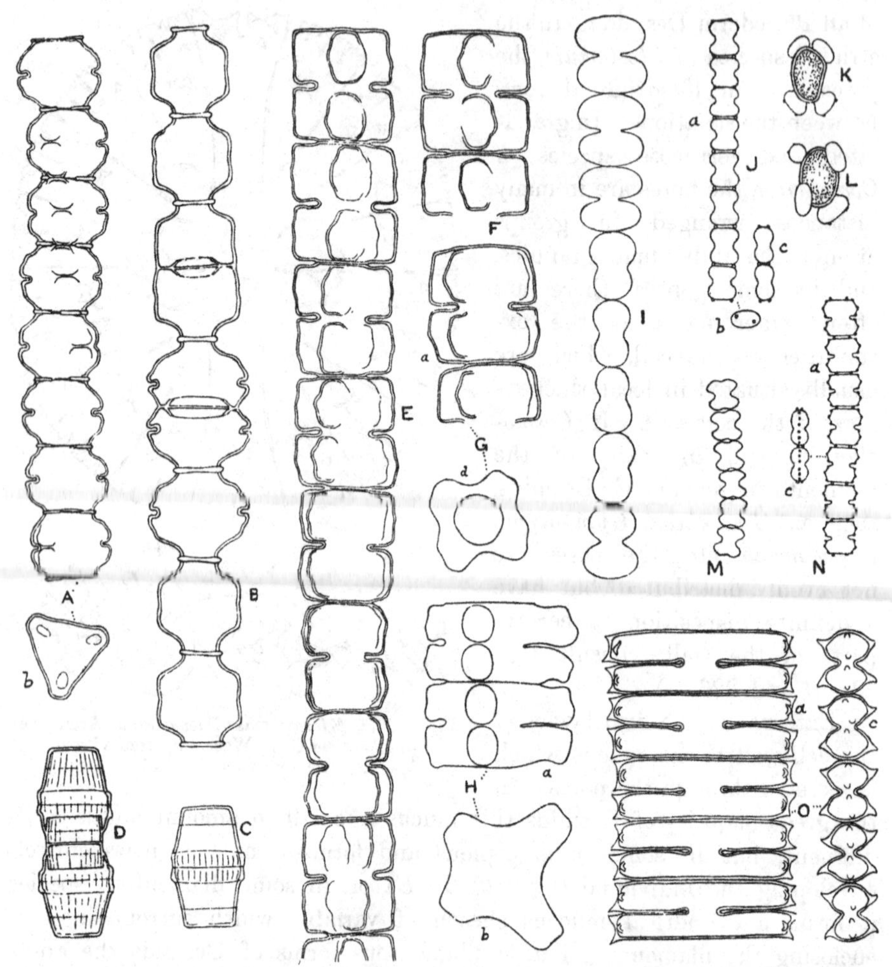

Fig. 224. Various filamentous types of Desmids. *A*, *Desmidium occidentale* W. & G. S. West. *B*, *D. coarctatum* var. *cambricum* W. West. *C* and *D*, *Gymnozyga moniliformis* Ehrenb.; *C*, single cell; *D*, dividing cell. *E—G*, *Phymatodocis Nordstedtiana* Wolle. *H*, *Phy. irregularis* Schmidle. *I*, *Spondylosium ellipticum* W. & G. S. West. *J—L*, *Sphærozosma excavatum* Ralfs; *K* and *L*, zygospores. *M*, *Spondylosium secedens* De Bary. *N*, *Sphærozosma granulatum* Roy & Biss. var. *trigranulatum* W. & G. S. West. *O*, *Spondylosium rectangulare* (Wolle) W. & G. S. West. *a*, front view; *b*, vertical or end view; *c*, side view; *d*, basal view of semicell. *A* and *B*, ×333; *C—O*, ×433.

Sooner or later the other pole becomes similarly attached and further oscillatory movements are carried out. Sometimes, as Stahl showed in *Cl. moniliferum*, the cell swings completely over through 180° and then becomes attached by the other pole, the movement being repeated. This reversal of position occupies from 6 to 35 minutes according to the temperature, the change of position occurring more rapidly as the temperature is increased. At 33° C. the reversal occupied from 6 to 8 minutes. In a number of other genera, such as *Cosmarium, Euastrum, Micrasterias*, etc., the movements are mostly quite irregular and spasmodic, although sometimes of a gliding character. In greatly compressed Desmids the flattened sides may be kept towards the incident light. The movements have been chiefly ascribed

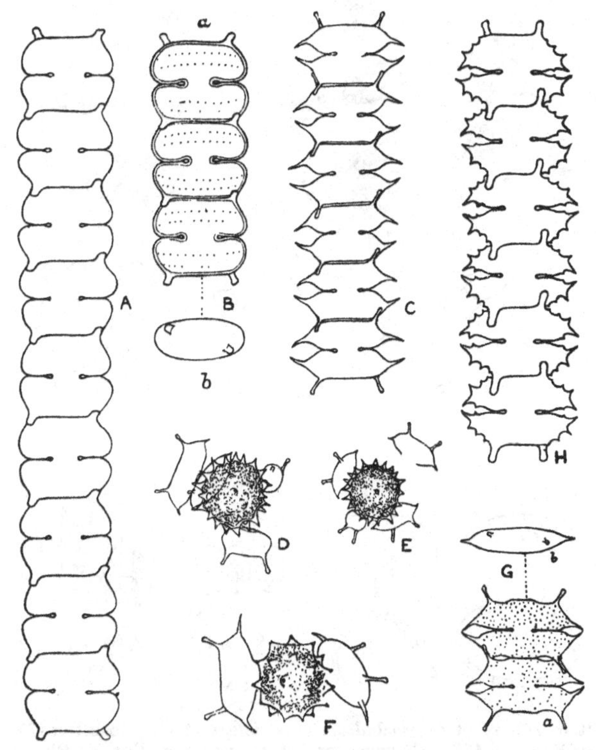

Fig. 225. *A* and *B*, *Onychonema compactum* W. & G. S. West. *C—F*, *O. læve* Nordst.; *D* and *E*, zygospores; *F*, aplanospore. *G*, *O. læve* Nordst. var. *latum* W. & G. S. West. *H*, *O. uncinatum* G. C. Wallich. *a*, front view; *b*, vertical or end view. *E*, ×360; all the remainder ×468.

to heliotropic and geotropic responses, but there is little doubt that in most cases the spasmodic nature of the movement is owing to the secretion of an irregular stalk of mucilage through the larger pores at the extremity of the cell (fig. 228 *A*), and that, moreover, a substratum is necessary for the movement to be carried out. If a sediment of organic detritus containing a number of living Desmids be exposed to the light, in a few hours the Desmids will have moved towards the incident light, having collected in clusters surrounded by abundant mucilage.

In some species of the genus *Pleurotænium*, such as *Pl. trochiscum* (fig. 227 *A*) and *Pl. doliiforme* (fig. 227 *B*), the cell-wall possesses large internal pits. These are thin places in the cell-wall and except near the poles of the cell are mostly of a rectangular shape.

Every Desmid, like every Diatom, consists of two halves, one of which is

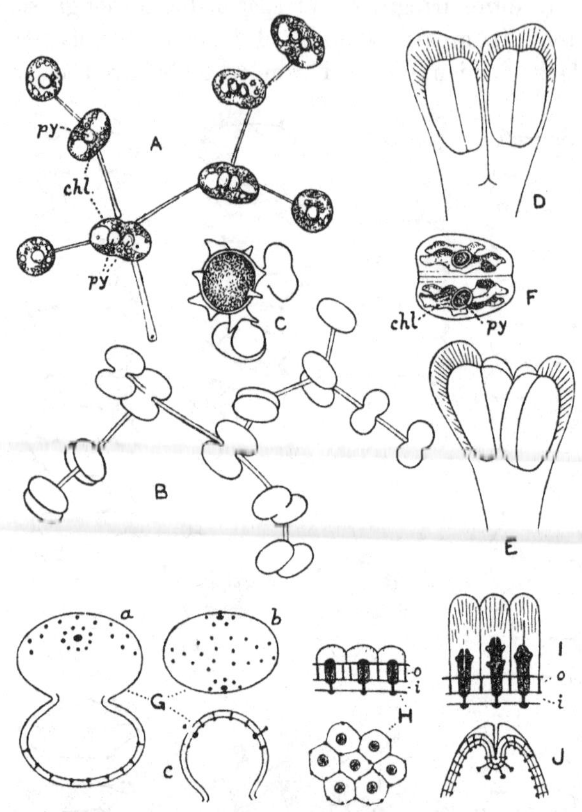

Fig. 226. *A*, part of colony of *Cosmocladium constrictum* (Arch.) Josh., ×475. *B*, colony of *C. pulchellum* Bréb., ×475. *C*, zygospore of *C. perissum* Roy & Biss., ×475. *D—F*, *Oocardium stratum* Näg. ×730; *F* shows the nature of the chloroplasts. *G*, *Cosmocladium saxonicum* De Bary showing disposition of pores in cell-wall, ×1200. *H*, pore-organs and mucous secretion in *Cosmarium turgidum* Bréb.; the upper figure is in optical section; ×1600. *I*, section of cell-wall of *Xanthidium armatum* (Bréb.) Rabenh. showing the pore-organs and secreted mucus, ×1600. *J*, apex of semicell of *Tetmemorus granulatus* (Bréb.) Ralfs showing the pore-organs, ×800. (*D—J*, after Lütkemüller; *G—J*, specimens treated with fuchsin and potassium acetate.) *chl*, chloroplast; *o*, outer layer; *i*, inner layer of cell-wall; *py*, pyrenoid; *a*, front view; *b*, end or vertical view; *c*, side view of semicell.

at least one generation older than the other, and may be many generations older. These halves are technically known as the 'semicells.' In the Placoderm Desmids the wall is composed of two pieces joined at the isthmus, the suture consisting of the overlapping bevelled edges of the two halves; but in the Saccoderm Desmids the wall is all in one piece and no suture is discernible though still one half of it is older than the other half. In a few species of *Closterium* and *Penium* the wall is composed of more than two pieces owing to the development of one or more girdle-bands subsequent to cell-division (consult p. 369).

The protoplast of the Desmid varies much, but there is always a nucleus of moderate size, usually embedded in a small band of cytoplasm traversing the median part of the cell and, in the constricted types, lying in the isthmus. The nucleus consists of a fine reticulum, with few chromatin granules, but with a large compound nucleolus[1]. There is the usual lining layer of cytoplasm, but the general disposition of the rest of the cytoplasm depends entirely on the nature and arrangement of the chloroplasts. In those Desmids with large axile chloroplasts the sap-vacuoles are much reduced and they are often confined to one at each extremity of the cell; in others with parietal chloroplasts one large vacuole usually occupies the central part of each semicell. Owing to transparency of the cell-wall the circulation of the protoplasm is often seen extremely well, especially in the larger species of *Closterium, Netrium,* etc. In the genera *Gonatozygon* and *Closterium,* and in certain species of *Pleurotænium* and *Penium,* there is a conspicuous terminal vacuole at each extremity of the cell containing one or more moving granules. These granules, which may be quite irregular in form or of some definite shape, exhibit rapid vibratory movements and in the genus *Closterium* have been shown to be minute crystals of gypsum.

The chloroplasts of Desmids present a very interesting study[2]. They are always of comparatively large size and usually of some definite form; in most cases they are situated in an axile position in the cell or semicell, although less frequently they take the form of parietal cushions or bands. In the cells of *Spirotænia* (fig. 230 *A* and *B*), *Mesotænium* (fig. 230 *E* and *F*; fig. 231 *F—I*) and *Roya* there is only one chloroplast and, except for a few species of *Spirotænia,* the nucleus consequently occupies an asymmetrical position. The same is true of some forms of *Cosmocladium* (fig. 226 *A*), *Gonatozygon* and *Cosmarium*[3]. The majority of Desmids possess two axile chloroplasts, symmetrically arranged one in each semicell (figs. 229 *A* and *B, D—F*; fig. 230 *H, J* and *K*), although in many species of *Cosmarium* there are two more or less distinct axile chloroplasts in each semicell. In *Pleurotænium,* and in a few species of *Cosmarium, Staurastrum* and *Xanthidium* there are

[1] The nuclear structure is well seen in various species of *Gonatozygon, Closterium, Roya* and *Micrasterias.* For nuclear structure in *Closterium* consult Lutman ('10).

[2] But for the untimely death (in 1912) of Dr J. Lütkemüller of Vienna we should have had by now an exhaustive study of the chloroplasts of Desmids. Much of his work and most of his drawings were communicated to the present author, who deplores the loss of a valued correspondent and of the unwritten work founded upon many years of patient study.

[3] As for example *Cosmarium subtile* (W. & G. S. West) Lütkem. and *C. subtilissimum* G. S. West.

several parietal cushion-like chloroplasts in each semicell, but the parietal disposition is not constant. In many species the chloroplasts are very variable in character and disposition, numerous intermediate states existing between

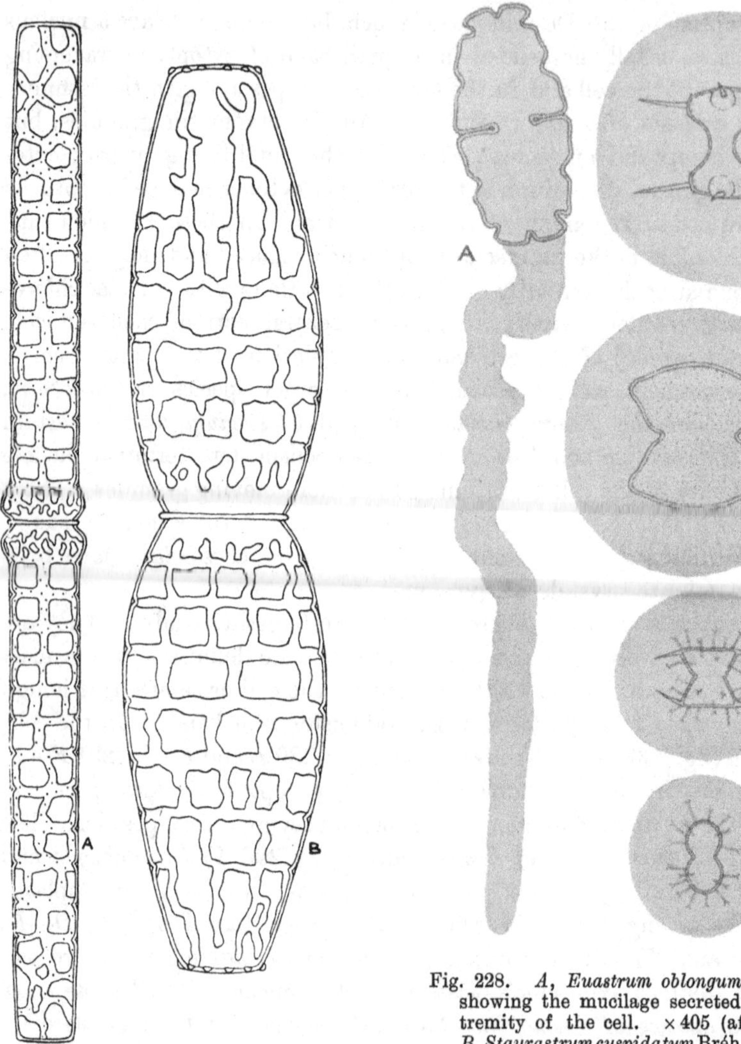

Fig. 227. *A*, *Pleurotænium trochiscum* W. & G. S. West. *B*, *Pl. doliiforme* W. & G. S. West. Both ×433.

Fig. 228. *A*, *Euastrum oblongum* (Grev.) Ralfs showing the mucilage secreted from one extremity of the cell. ×405 (after Schröder). *B*, *Staurastrum cuspidatum* Bréb. var. *maximum* W. & G. S. West, ×360. *C*, *St. subpygmæum* W. West, ×468. *D* and *E*, *Arthrodesmus Incus* (Bréb.) Hass. var. *Ralfsii* W. & G. S. West forma *latiuscula*, ×450.

the truly axile and truly parietal conditions. This is especially noticeable in certain species of *Xanthidium*, in *Staurastrum grande* Bulnh. and *St. brasiliense* Nordst. var. *Lundellii* (*vide* W. & G. S. W., '04—'11). There is little doubt that the axile chloroplast was the primitive one in Desmids as a whole and that the parietal condition has been acquired by a few scattered species in various genera[1]. In Desmids in which the cell is deeply lobed or incised the form of the chloroplasts frequently follows the external configuration of the cell.

Pyrenoids are present in the chloroplasts of all Desmids. In most species there are one or two pyrenoids in each semicell, but in the large flattened forms

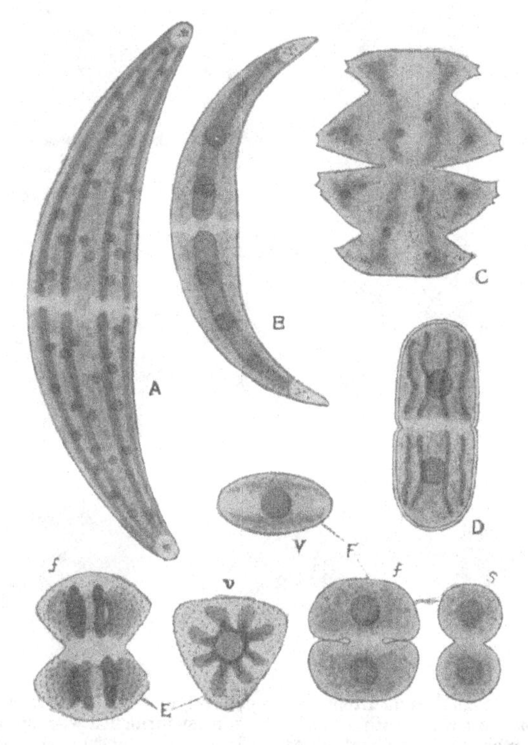

Fig. 229. Desmids showing chloroplasts. *A*, *Closterium Ehrenbergii* Menegh., × 184. *B*, *Cl. Leibleinii* Kütz., × 334. *C*, *Micrasterias oscitans* Ralfs var. *mucronata* (Dixon) Wille, × 184. *D*, *Cosmarium cucurbitinum* (Biss.) Lütkem., × 435. *E*, *Staurastrum punctulatum* Bréb. var. *Kjellmani* Wille, × 435. *F*, *Cosmarium subtumidum* Nordst. var. *Klebsii* (Gutw.) W. & G. S. West, × 435. *f*, front view of cell; *s*, side view; *v*, vertical or end view.

of *Euastrum* (fig. 220 *A*) and *Micrasterias* (fig. 223), and in the elongated cells of *Closterium* (fig. 229 *A* and *B*; fig. 231 *A—D*), *Pleurotænium*, *Tetmemorus*,

[1] All attempts to split up the genera *Cosmarium*, *Staurastrum* and *Xanthidium* on the basis of axile and parietal chloroplasts have merely resulted in confusion. Such divisions are purely artificial, resulting in the association of species which have no near relationship with each other. In those few species with a tendency to the parietal disposition of chloroplasts it is by no means uncommon to find axile chloroplasts in one semicell and parietal chloroplasts in the other semicell of the same individual.

etc., there may be many pyrenoids in each chloroplast[1]. In certain cases pyrenoids are subject to variation in number and disposition, but in the great majority of Desmids they are remarkably constant in their relative position. In many instances they have a conspicuous starch sheath (consult fig. 231 *A—D*) and Lutman ('10) has shown that all the starch found in the cell of *Closterium* is originally formed around the pyrenoids.

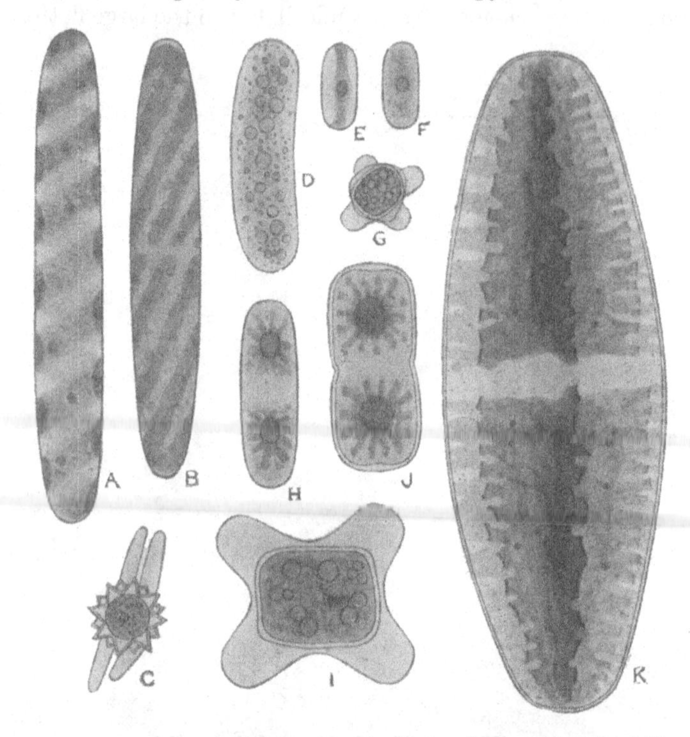

Fig. 230. *A*, *Spirotænia condensata* Bréb. with a parietal, spirally twisted, band-like chloroplast, × 334. *B*, *Sp. obscura* Ralfs, with an axile chloroplast furnished with spirally twisted ridges, × 435. *C*, zygospore of *Sp. truncata* Arch., × 250 (after Archer). *D*, *Mesotænium De Greyi* W. B. Turner, × 435. *E* and *F*, *M. macrococcum* (Kütz.) Roy & Biss., × 334; note the plate-like chloroplast seen from the flat side in *F* and from the edge in *E*. *G*, zygospore of *M. chlamydosporum* De Bary, × 334. *H* and *I*, *Cylindrocystis Brébissonii* Menegh.; *H*, vegetative cell; *I*, zygospore; × 435. *J*, *Cosmarium diplosporum* (Lund.) Lütkem., × 435. *K*, *Netrium Digitus* (Ehrenb.) Itzigsh. & Rothe, showing axile chloroplasts with radiating serrated plates, × 435.

If Desmids are kept living in small glass vessels for some time, and therefore under abnormal conditions, curious cytological changes occur, resulting in the formation of large vacuoles which previously did not exist. These vacuoles generally contain numbers of minute moving granules which are somewhat different in appearance from those normally present in the apical vacuoles of *Closterium*. As many as six large vacuoles have been

[1] In a few Desmids, among which may be mentioned *Spirotænia acuta* Hilse, *Cosmarium subtile* (W. & G. S. West) Lütkem. and *Cosmocladium constrictum* (Arch. Josh., there may be *only one pyrenoid in each cell.*

noticed in a single semicell of *Pleurotænium nodulosum* var. *coronatum* (G. S. W., '99), each one being partially filled with an incessantly moving mass of minute granules, which move freely in the vacuole, but always collect towards its base. These granules are of a pale yellow colour and appear brown in a thin stratum; but when present in immense numbers they give the Desmid almost a black appearance. The development in quantity of these moving granules is a pathological condition and is associated with the gradual disintegration of the chloroplasts. At the same time the cell-sap in the large vacuoles very often becomes coloured violet with phycoporphyrin, a pigment which occurs normally in the cell-sap of very few Desmids. This pathological state can be seen frequently in *Penium*, *Cosmarium*, *Euastrum*, *Micrasterias*, *Staurastrum* and other genera.

The normal method of increase in the Desmidiaceæ is a vegetative multiplication by simple cell-division, which occupies about a day in the smaller and simpler types, but several days in the larger species with more complex outlines. The essential points of the cell-division of *Closterium* were well described by Fischer ('83) and more recently a very detailed account of the process as it occurs in *Closterium Ehrenbergii* and *Cl. moniliferum* has been given by Lutman ('11). The details of cell-division are not quite the same in all Desmids and it is a curious fact that so far they have only been completely worked out in the unconstricted forms, whereas fully 90 per cent. of known Desmids are more or less deeply constricted.

In the more usual type of Desmid with a median constriction the first step in cell-division is a division of the nucleus and, as this approaches completion, there is an elongation of the isthmus causing a slight separation of the semicells. The elongated isthmus generally becomes slightly swollen and across its median part a new cell-wall is laid down[1]. The parts of the isthmus on either side of this new wall are destined to grow into the new semicells. They rapidly become more turgid and soon grow into adult semicells. The lobulation or ornamentation of the species gradually makes its appearance on the new semicells as they assume the normal size. As a rule the two new semicells, and therefore the daughter-cells resulting from division, remain attached until they are full-grown. Chloroplasts appear in the new semicells at a comparatively early age, but at present there is no exact information concerning their origin. The pyrenoids in many cases certainly arise *de novo*.

In the unconstricted genus *Closterium*[2], the only genus of Desmids in which cell-division has been minutely studied, the process of division, although similar in essential points, differs in details from that which occurs in constricted Desmids. Lutman ('11) has shown that in *Closterium Ehrenbergii* and *Cl. moniliferum* the first visible appearance of division in any individual

[1] The details of the early stages of cell-division yet require careful investigation.

[2] There are several slightly constricted species of this genus; notably *Closterium Braunii* Reinsch ('67) and *Cl. subcompactum* W. & G. S. West ('02).

cell is the pinching in of the two chloroplasts at a point about a third the distance from the middle to the extremity, and that the division of the chloroplasts is entirely distinct from that of the cell. This constriction occurs altogether inside the plasmatic membrane of the cell and is probably the

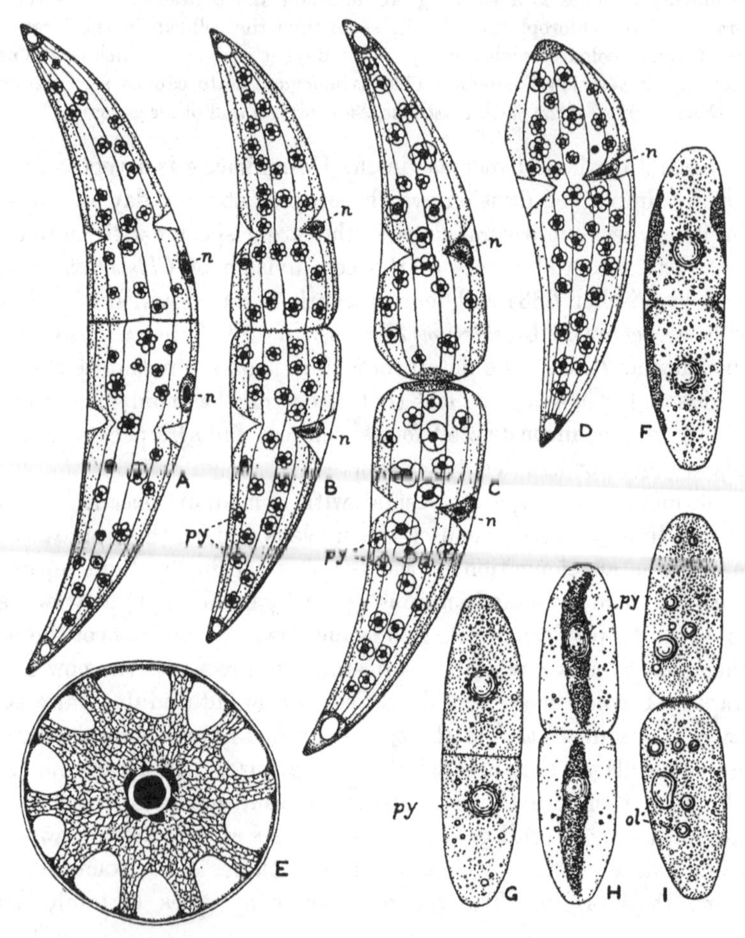

Fig. 231. *A—D*, stages in the division of *Closterium Ehrenbergii* Menegh., × 272 (after Lutman). *E*, tranverse section of *Cl. moniliferum* (Bory) Ehrenb., showing ten peripheral ridges on the chloroplast and one of the axial row of pyrenoids, × about 800 (after Lutman). *F—I*, stages in the division of *Mesotænium caldariorum* (Lagerh.) Hansg., × 900. *n*, nucleus; *ol*, oil globules; *py*, pyrenoid.

constriction of a membrane which forms the outer layer of the chloroplast. No nuclear changes are visible at this stage. During the time the two chloroplasts are dividing, the nucleus divides and after reconstruction of the daughter-nuclei, the latter can be seen moving out to the surface of the chloroplast and making their way along the convex margin of the cell to their new positions (fig. 231 *A*). Each nucleus arrives in the notch of the constricting chloroplast before that division is completed (fig. 231 *B—D*). The new cross-wall is formed at right angles to the old walls (fig. 231 *A*) very much as in *Spirogyra*. It grows in from the periphery of the cell during the metaphase of nuclear division and cuts across the spindle fairly at its centre, being a third of the way across by the time the fibres have disappeared. The daughter-nuclei have been reconstructed and are moving to their new positions before the new wall is completed. The young semicells are cone-shaped (fig. 231 *D*), but they soon acquire their proper characters. Lutman states that in these two species of *Closterium* cell-division and nuclear-division represent at least a two-night process, the chloroplasts dividing the first night and the nucleus probably on the second.

In *Mesotænium caldariorum*, also an unconstricted Desmid, cell-division is precisely as in the Placoderm Desmids (G. S. W., '15). After the division of the nucleus a new cell-wall is laid down in an exactly transverse plane (fig. 231 *F*) in a manner very similar to that which occurs in the Zygnemaceæ. There is only one chloroplast in the cells of *Mesotænium* and the new transverse wall cuts right across it. On the completion of this wall there is no trace of a constriction of the cell (fig. 231 *G*). The new semicells are now developed as in other Desmids. The middle lamella of the new transverse wall gradually disappears from the periphery inwards, and during its disappearance, which probably results from its conversion into mucilage, the part of the new wall belonging to each semicell begins to bulge outwards, assuming a greater and greater convexity (fig. 231 *H* and *I*). This finally results in the separation of the daughter-cells.

As a rule all growth ceases in a Desmid after the new semicell is fully formed and has acquired its distinctive specific characters, the cell then consisting of two halves of different generations. [Consult the scheme of normal cell-division in *Closterium* given in fig. 232 *I*, showing four generations, the semicells of each generation marked s, s_1, s_2, s_3.] In certain species of *Penium* and *Closterium*, however, there is a further growth subsequent to completed cell-division. This consists in the development of an elongated median girdle-band, which is a cylindrical piece of cell-wall intercalated between the old and the new semicells. The general scheme of this type of division was first portrayed in detail by Lütkemüller ('02) and is diagrammatically represented in fig. 232 *II*. The scheme depicted is an ideal one, representing the growth of a new girdle-band after each division, but as a matter of fact this rarely happens, and many divisions may take place without the addition of a new girdle-band. In the genus *Closterium* each division results in a new junction of the old and new cell-walls, the optical effect of which is the addition of a new transverse suture (consult fig. 232 *III* and *IV*). In no known species of *Closterium* are more than two girdle-bands ever developed, but in *Penium spirostriolatum* and *P. spirostriolatiforme* the cell-wall may contain several girdle-bands of different generations, the greatest number occurring in the last-named species, which is so far known to occur only in Ceylon (W. & G. S. W.,

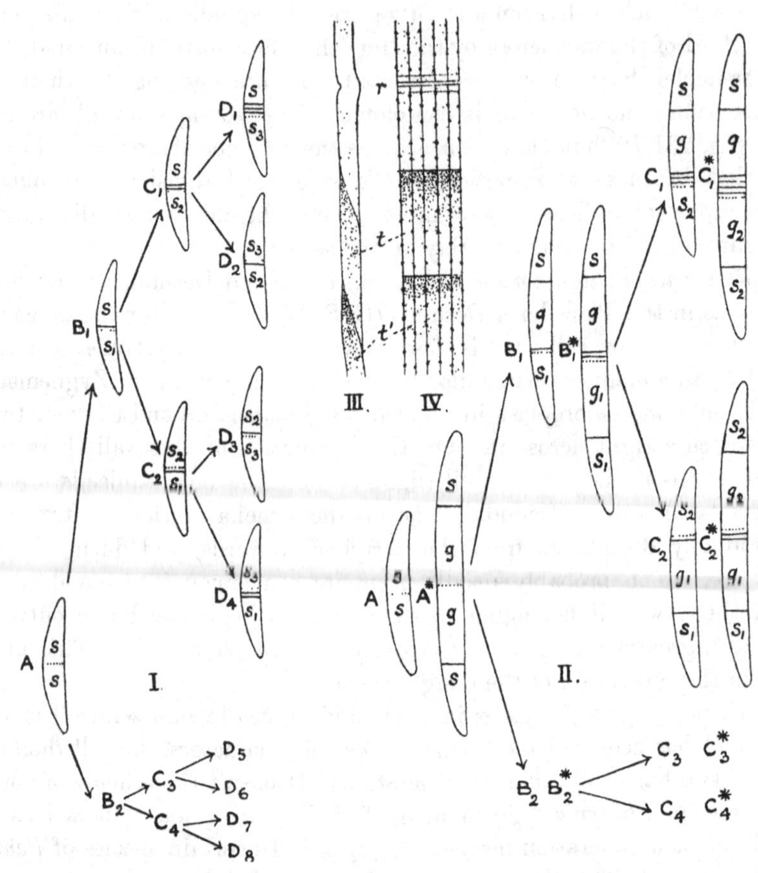

Fig. 232. *I*, diagrammatic scheme depicting four generations (*A*, *B*, *C* and *D*) in the usual type of *Closterium*; *s*, s_1, s_2 and s_3 refer to the semicells of successive generations. *II*, similar scheme depicting three generations of the type of *Closterium* which develops a girdle-band; *A*, *B* and *C*, cells of successive generations immediately after cell-division; *A**, *B** and *C**, corresponding generations after development of girdle-bands; *s*, s_1 and s_2 refer to semicells, and *g*, g_1 and g_2 to girdle bands of successive generations. *III* and *IV*, *Closterium turgidum* Ehrenb. var. *giganteum* Nordst., ×1200; *IV*, surface view of the median part of the cell; *III*, longitudinal section of same; *r*, the 'ring-furrow' in the middle of the cell; *t* and *t'*, short transverse segments of cell-wall, resulting from cell-divisions, showing the bevelled sutures. (All modified from Lütkemüller.)

'02). The girdle-bands (or intercalary bands) are not comparable with those of Diatoms (consult p. 119).

Asexual reproduction takes place very occasionally by the formation of aplanospores (fig. 233 *E*—*G* and *H*—*J*).

These have been observed in *Spondylosium nitens* (Wallich) Arch. (Wallich, '60; Turner, '93), *Closterium Cornu* Ehrenb. (Lagerheim, '86), *Hyalotheca neglecta* Racib. (W. & G. S. W., '98), *H. dissiliens* (Sm.) Bréb. (W. & G. S. W., '07) and *Cosmarium bioculatum* Bréb. (Nieuwland, '09).

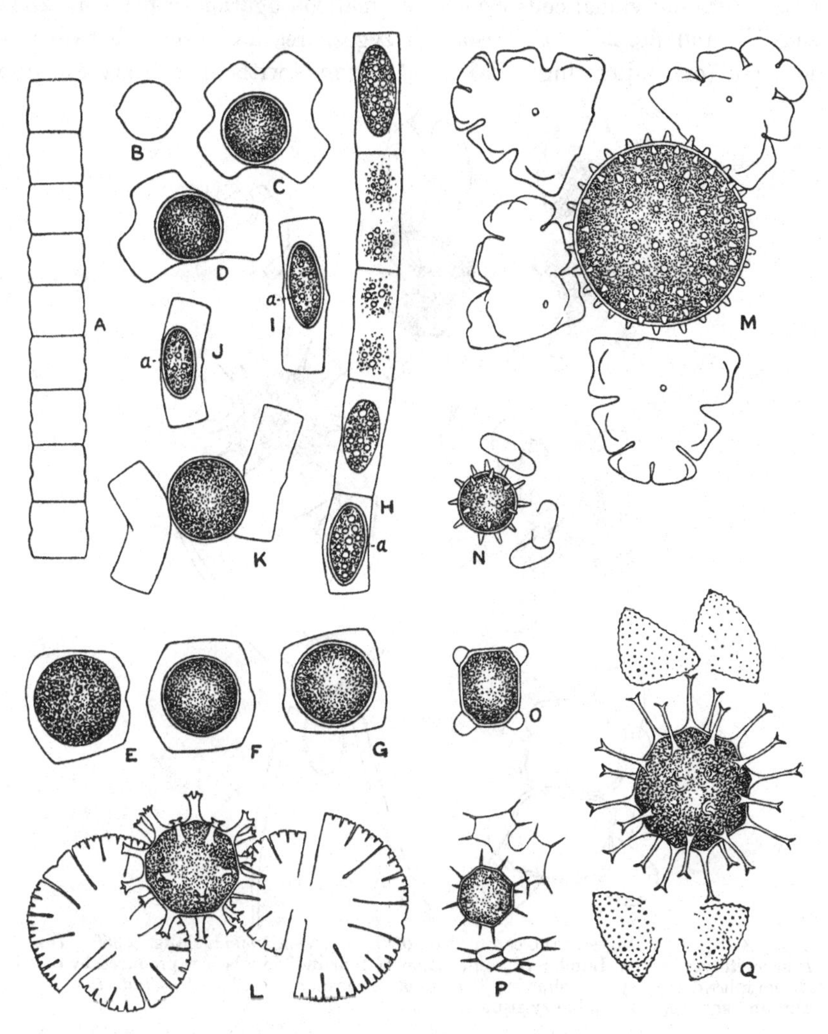

Fig. 233. Zygospores and aplanospores of Desmids. *A—G*, *Hyalotheca dissiliens* (Sm.) Bréb.; *A*, vegetative filament; *B*, end view of cell; *C* and *D*, zygospores, ×338; *E—G*, aplanospores, ×468. *H—K*, *H. neglecta* Racib.; *H—J*, aplanospores (*a*); *K*, zygospore, ×430. *L*, *Micrasterias denticulata* Bréb., ×100. *M*, *Euastrum oblongum* (Grev.) Ralfs, ×180. *N*, *Cosmarium bioculatum* Bréb., ×425. *O*, *Penium suboctangulare* W. West, ×328. *P*, *Arthrodesmus octocornis* Ehrenb., 425. *Q*, *Staurastrum granulosum* (Ehrenb.) Ralfs, ×468.

Sexual reproduction of a low, and probably degenerate, type occurs by the conjugation of two individual cells and the fusion of two large equal gametes to form a zygospore. The conjugation is similar to that in the Zygnemaceæ and with the solitary exception of *Desmidium cylindricum*, there is no definite conjugation-tube. Except in *Desmidium Baileyi*, and to a much less-marked degree in *Desmidium Swartzii* and *D. aptogonum*, all the filamentous Desmids dissociate into individual cells prior to normal conjugation (consult fig. 233 *C*, *D* and *K*; and fig. 235 *J*). Also, the zygospores are formed between the gametangia (or conjugating cells) in all Desmids with the solitary exception

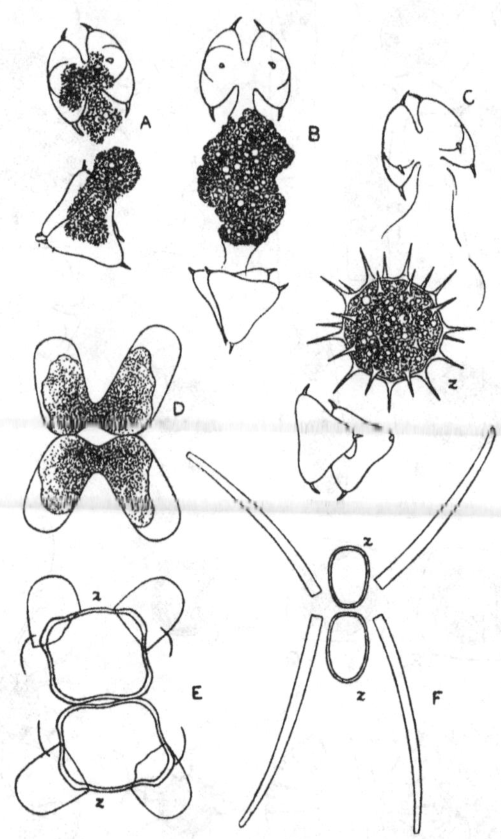

Fig. 234. *A—C*, three stages in the conjugation of *Staurastrum Dickiei* Ralfs, × 356. *D* and *E*, *Penium didymocarpum* Lund.; *D*, conjugation of four individuals just produced by division; *E*, completed conjugation showing the double zygospore, × 464. *F*, *Closterium lineatum* Ehrenb., showing the double zygospore, × 100. *z*, zygospore.

of *Desmidium cylindricum* Grev., in which the zygospore is formed within the female gametangium and a distinct conjugation-tube is developed (fig. 235 *J*). The two conjugating cells, which are sexually indistinguishable, approximate and usually become enveloped in mucus. In the Cosmarieæ the longitudinal axis of one conjugating cell is very often at right-angles to that of the other (fig. 234 *A* and *B*; fig. 235 *J*) and at the commencement of conjugation the semicells of each individual come apart at the isthmus. Each gamete issues from the dislocated isthmus as a green protoplasmic vesicle containing the nucleus and the more or less disintegrated chloroplasts. The fusion of the gametes results in a zygospore which develops a cell-wall with three distinct layers. The inner layer consists of cellulose and is thin and colourless; the middle layer is brown and firm and possibly cutinized; the outer layer consists mostly of cellulose and may be quite smooth (fig. 233 *K* and *O*; fig. 235 *J*) or covered with variously arranged warts or spines (fig. 233 *L*—*N*, *P*—*Q*; fig. 234 *C*).

Sometimes more than two cells have participated in the formation of a zygospore, the latter having been formed by the union of three (W. W., '91 A; W. & G. S. W., '97 A) or even four (Turner, '93, t. x, f. 16 *e*) gametes. Cytological details are entirely lacking.

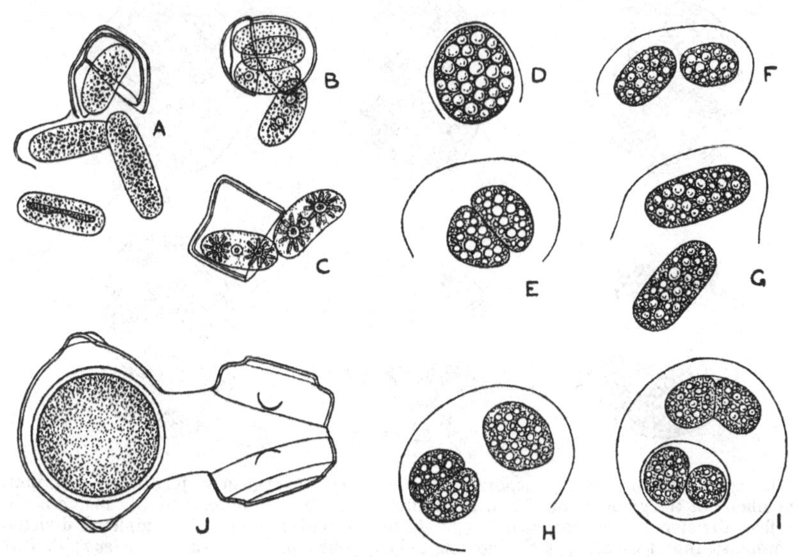

Fig. 235. *A*, germination of zygospore of *Mesotænium chlamydosporum* De Bary, showing four embryos, × 390 (after De Bary). *B* and *C*, germination of zygospore of *Cylindrocystis Brébissonii* Menegh.; *B*, with four embryos; *C*, with two embryos, × 390 (after De Bary). *D*—*I*, germination of zygospore of *Mesotænium caldariorum* (Lagerh.) Hansg., × 500; *D*—*G*, the usual germination with the formation of two embryos; *H* and *I*, the exceptional germination with the formation of four embryos. *J*, zygospore of *Desmidium cylindricum* Grev., × 500.

In many cases conjugation takes place immediately after vegetative division and before the young semicells have arrived at maturity, the conjugation actually occurring between the two daughter-cells which have resulted from the division of the original mother-cell. The present author has observed this to take place in *Micrasterias denticulata*, *Penium didymocarpum* (*vide* figs. 233 *L* and 234 *D* and *E* and several species of *Closterium*), and the same phenomenon was observed many years ago by Archer. Conjugation between adjacent cells (lateral conjugation) in the filamentous species of Desmids has been recorded in the genera *Sphærozosma* and *Spondylosium*, but such cases are exceptional and very rare.

In a few cases *double zygospores* are normally produced. The best known species are *Closterium lineatum* Ehrenb. (fig. 234 *F*), *Cl. Ralfsii* var. *hybridum* Rabenh., *Cosmarium diplosporum* (Lund.) Lütkem., and *Penium didymocarpum* Lund. (fig. 234 *D* and *E*). In the last-named species conjugation occurs between four cells resulting from two consecutive divisions of one individual (G. S. W., '04). Each of the two zygospores is formed by the union of a distinct pair of gametes. In the two species of *Closterium* in which double zygospores occur, *each semicell produces a gamete* and each of the two zygospores results from the fusion of a pair of gametes, one from a semicell of each individual. The cytology of this extraordinary conjugation is not yet worked out and would obviously prove of great interest. In *Spirotænia condensata* Bréb. two distinct and separate zygospores are produced from a single pair of conjugating cells (Archer, '67).

A double zygospore of *Closterium rostratum* Ehrenb. has been recorded by Lagerheim ('86), but although this species conjugates more frequently than any other *Closterium*, this is the only recorded instance of a double zygospore, which must therefore be regarded as abnormal.

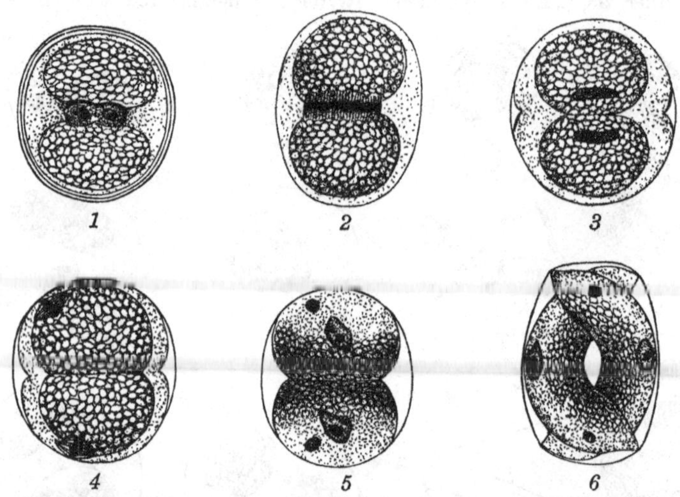

Fig. 236. Germination of the zygospore of *Closterium* sp. *1*, zygospore just before germination, the nuclei of the gametes not having yet fused; *2*, the first mitosis of the fusion-nucleus; *3*, first division of nucleus completed; *4*, the second mitosis; *5*, completed division of protoplast into two cells, each showing a large nucleus and a small nucleus; *6*, further stage in germination, the cells beginning to assume a definite shape. All × 308. (After Klebahn, from Oltmanns.)

Klebahn ('90) has observed the formation and germination of parthenospores in a species of *Cosmarium*, and Borge ('13) has also noted the production of parthenospores in abortive attempts at conjugation in *Cylindrocystis Brébissonii* Menegh.

The zygospore rests for some time before germination. The first observations of importance on the germination of the zygospore were those of De Bary ('58), and further details with regard to certain genera have been furnished by Millardet ('70) and especially by Klebahn ('90). The nuclei of the gametes lie side by side for some time before fusion (figs. 236 *1* and 237 *A*). During the conjugation of most Desmids the chloroplasts are quite disintegrated, but in the ripe zygospore there is evidence in some cases that two chloroplasts are reorganized (fig. 237 *A* and *B ch*). Much more detailed investigation is, however, required on this point. Soon after the fusion of the two nuclei (in the Placoderm Desmids) the fusion-nucleus divides mitotically *twice*, and after the completion of the second mitosis the protoplast also divides forming two bi-nucleated daughter-cells (figs. 236 *5* and 237 *D*). In each of these cells the nuclei become differentiated into a 'large nucleus' and a 'small nucleus,' and during subsequent development the small nucleus gradually disappears (fig. 237 *E*—*G*), the large nucleus assuming a median position and becoming the nucleus of the fully developed cell (fig. 237 *H*). It seems probable that in the Placoderm Desmids two embryos are formed from each zygospore (consult fig. 236 *6* and 237 *E*), but in the Saccoderm Desmids there is considerable variability in the number of embryos which are produced on the germination of the zygospore. The observations

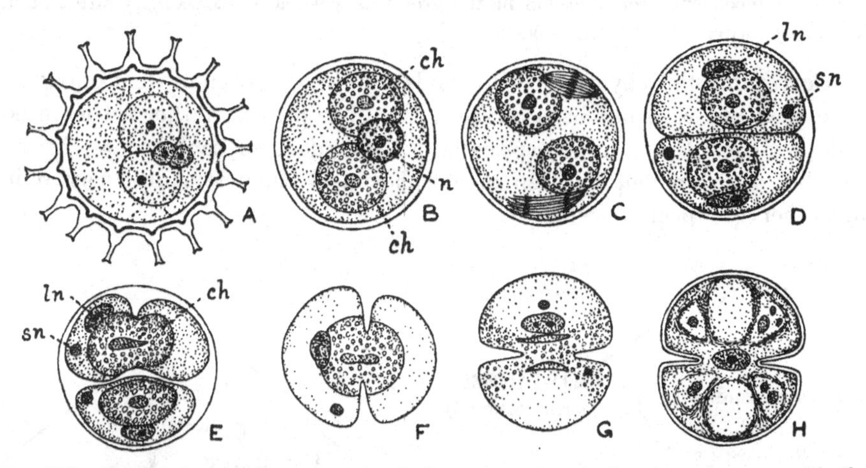

Fig. 237. Germination of the zygospore of *Cosmarium* sp. *A*, the ripe zygospore with the nuclei of the gametes as yet unfused. *B*, commencement of germination. *C*, second mitosis of the fusion-nucleus. *D*, complete division into two cells, each with a large and small nucleus. *E*, upper cell showing commencement of median constriction. *F*, one of the two cells formed by germination; large and small nuclei both in one semicell. *G*, further stage with small nucleus gradually disappearing. *H*, final stage, the cell having assumed its proper shape and the large nucleus having become median in position. *ch*, chloroplast; *n*, fusion-nucleus; *ln*, large nucleus; *sn*, small nucleus. All ×350 (after Klebahn).

of De Bary tend to show that in the Gonatozygæ only one embryo is produced precisely as in the Zygnemaceæ. In the Spirotænieæ there may be either two or four. Four may be the general rule in this sub-family, but in various species of *Mesotænium* and in *Cylindrocystis Brébissonii* the number of embryos may be only two, and in *Mesotænium caldariorum* two is the usual number (G. S. W., '15; also consult fig. 235 *C*—*G*).

The observations on the cell-division and germination of the zygospore of *Mesotænium caldariorum* are of particular interest in view of the attempt by Oltmanns to establish within the Conjugatæ a third family, the Mesotæniaceæ (consult p. 331). The removal of the Desmidian Conjugates embraced in Oltmanns' Mesotæniaceæ (= Lütkemüller's Spirotænieæ of the Saccodermæ) from the Desmidiaceæ appears to be contrary to their affinities.

As pointed out elsewhere (G. S. W., '15), there is no essential *family* difference between a *Mesotænium* and a *Closterium*. The fact that there is no obvious line of junction between the new and the old semicells in the Spirotænieæ is of little importance, since the development of the new semicells is the same as in *Closterium* (compare fig. 231 *A* and *D* with fig. 231 *F* and *I*) and other Desmids. Moreover, Oltmanns' further contention that in those Desmids which he would place in the 'Mesotæniaceæ' the germination of the zygospore results in the production of four embryos is not entirely supported by facts.

Only one true case of hybridization has been observed amongst Desmids and in that case the development of the zygospore was not followed out (Archer, '75). In this case a zygospore had been produced by the conjugation of two cells, one of which was *Euastrum Didelta* (Turp.) Ralfs and the other *Euastrum humerosum* Ralfs. All other recorded cases of hybrids are conjectural and most of them are obviously forms produced by ordinary vegetative division. It is unlikely that many true hybrids can exist among the Desmidiaceæ, since sexual reproduction in the group as a whole is exceedingly rare and in very many forms is not known to occur.

Concerning the phylogeny of Desmids and their position in the Conjugatæ more than one view has been put forward. The present author is convinced that they constitute a highly specialized family, degenerate in sexual characters, but with an amplification of morphological characters unsurpassed in any other group of unicells.

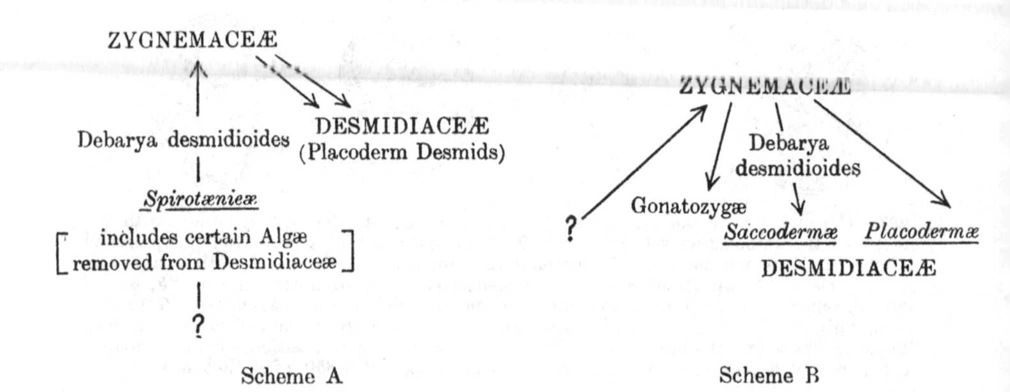

Three possibilities may be discussed in considering the relationships between Desmids and other Conjugates. The first is that the unicellular Desmids are primitive and that the Zygnemaceæ have been derived from them. This hypothesis may be dismissed, since there is strong evidence that many Desmids have originated from filamentous ancestors and no definite evidence that any Desmid has arisen from a unicellular ancestor.

The second possibility is that the Desmidiaceæ are not a natural family and that the Spirotænieæ, and even the Gonatozygæ, should be removed from them. It might then be suggested that the Spirotænieæ (= the Mesotæniaceæ of Oltmanns) arose from unicellular ancestors and in course of time gave rise to such forms as *Debarya desmidioides*, through which and similar forms the Zygnemaceæ were evolved. The great bulk of the Desmidiaceæ (all the Placoderm Desmids) would then be derived from the Zygnemaceæ by retrogression and specialization (consult scheme A).

The third hypothesis is that the Desmidiaceæ are a homogeneous group and that all Desmids primarily arose from filamentous ancestors. There is little doubt that the great majority of Desmids have arisen in this way, and since the various members of the Spirotænieæ, which is the chief group of Saccoderm Desmids, are more closely allied to the Placoderm Desmids than to the Zygnemaceæ there appears to be little reason for supposing the family Desmidiaceæ (in its widest sense) to be a mixed assemblage. This third hypothesis is the one supported by a considerable amount of evidence and it seems on the whole to be the most probable (consult scheme B).

There are many cogent reasons for regarding the Desmidiaceæ as a specialized family of Conjugates which has originated from filamentous ancestors. The loss of the filamentous condition has been accompanied by the development of complex morphological characters, and this has gone on hand in hand with the loss of sexual differentiation of the conjugating cells (G. S. W., '99; '04). The structure of the cell-wall of the Placoderm Desmids, which is so much more complex than anything exhibited in the Zygnemaceæ, and the elaboration of the chloroplasts, clearly indicate that *the Desmidiaceæ are not primitive.*

It has been previously mentioned that *Desmidium cylindricum* is the only known Desmid in which the zygospore is lodged in the female gametangium (consult p. 372 and fig. 235 *J*) and the very occasional reversion to this type of conjugation in *Hyalotheca dissiliens* (Joshua, '82; Boldt, '88) is of considerable significance. It is in *Desmidium cylindricum* and in the presumed 'abnormal' cases of conjugation in *Hyalotheca dissiliens* that one is probably witnessing the type of conjugation which was prevalent in the ancestors of the Desmidiaceæ. Another Desmid of great interest is *Debarya desmidioides* (W. & G. S. W., '03), since it stands in a somewhat intermediate position between *Mesotænium* and such genera as *Mougeotia* and *Debarya.*

The average Desmid must be regarded as a unit of a dismembered filament. The complete individuality of the cell is the only *real* distinction between Desmids and other Conjugates. Lutman ('11) from his observations on the division of *Closterium Ehrenbergii* states that 'the position of the young transverse wall would seem to indicate that the pointed ends are secondarily formed, and that *Closterium* was originally a filamentous alga, which has developed the habit of breaking up into single cells.'

Desmids exhibit a considerable tendency towards the secondary assumption of the filamentous condition. This has resulted in the production, from unicells, of filamentous genera, such as *Streptonema, Onychonema, Sphærozosma, Desmidium,* etc., in which special apical connecting processes are for the most part conspicuously developed (*vide* fig. 224 *A*, *J* and *N*; and fig. 225 *A—H*). The same tendency sometimes occurs in certain species of genera which are normally unicellular. Thus, short filaments of cells have been figured in such species as *Cosmarium obliquum* Nordst. ('73), *C. moniliforme* (Turp.) Ralfs, *C. Regnellii* Wille (W. & G. S. W., '96), *Euastrum binale*

(Turp.) Ehrenb. (W. & G. S. W., '98) and *Staurastrum inconspicuum* Nordst. (Börgesen, '01). Certain of the tropical species of *Pleurotænium* form longer or shorter filaments, and *Micrasterias foliacea* Bail. is a true filamentous form of a typically unicellular genus with complex cell-outlines (fig. 239 *B*).

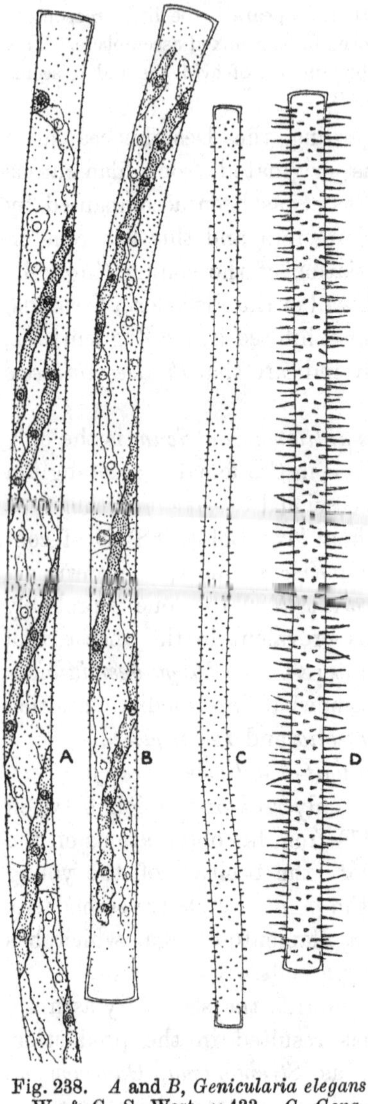

A B C D

Fig. 238. *A* and *B*, *Genicularia elegans* W. & G. S. West, ×433. *C*, *Gonatozygon monotænium* De Bary var. *pilosellum* Nordst., ×416. *D*, *G. aculeatum* Hastings, ×416.

The general evolution of the Placoderm Desmids is fairly clear (*vide* G. S. W., '04), but the genus *Roya* has to be removed from the sub-family Placodermæ. Lütkemüller ('10) has recently shown that the species of this genus are Saccoderm Desmids and that *Roya* is a close ally of *Mesotænium*.

It is highly probable that the complexity of outline of the Desmid, which is so frequently accompanied by a defensive armour of spines and spinous processes, has been acquired as a means of defence against the attacks of small aquatic animals. After the loss of the filamentous condition it became necessary for the solitary and unprotected individuals to acquire some other means of defence, and presumably the present morphological complexity is the result. It is a notable fact that those species which occur on wet rocks and in other situations in which Amœbæ, Oligochætes, Tardigrades and Crustacea are either absent or very scanty, especially at high elevations, usually possess a comparatively simple outline; whereas those species occurring in deep bog-pools, in the plankton, and at the quiet margins of deep lakes, in which localities such enemies abound, are generally possessed of a more complicated, and in many cases of a formidable, exterior. These characters acquired by the unicell are not only protective against the depredations of aquatic animals, but are also useful as anchors in the time of floods.

There are several thousand known species of Desmids, almost all of which can be readily identified by their external morphological features. The majority of Desmids have three principal axes of symmetry at right angles to one another, and for this reason it is usually necessary to examine them in three positions. The most important aspect is the *front view*, in which the plant is observed in that plane containing the two longest axes. The other important positions are the *vertical* (or *end*) view and the *lateral* (or *side*) view.

Desmids are subject to some amount of variation, but only within certain limits, and one of the most extraordinary facts concerning these unicellular plants is the constancy of the ornamentation of the cell-wall. Variations occur mostly where the conditions have allowed of very rapid multiplication by cell-division, such prolific increase resulting in peculiar, and certainly abnormal, physiological conditions. Similar conditions may supervene in pure cultures with consequent modification of specific characters.

In pure cultures of Diatoms, which are much more easily obtained than in the case of Desmids, similar loss of specific characters is far more noticeable, but it has been shown experimentally that if these degenerate individuals are transferred to fresh culture-solutions in which the physiological conditions are more normal, the proper specific characters quickly reappear during subsequent divisions.

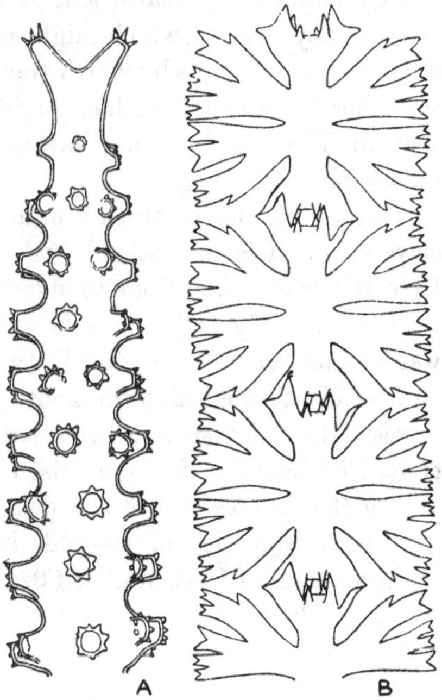

A B

Fig. 239. *A*, a variety of *Triploceras verticillatum* Bail., from Australia, ×416. *B*, portion of a filament of a Burmese form of *Micrasterias foliacea* Bail. in which the toothed polar lobes of the cells are not so firmly interlocked as usual, ×433.

Our knowledge of the geographical distribution of Desmids, although very incomplete, has advanced considerably in recent years. We know now that the family as a whole exhibits geographical peculiarities of a more striking character than those shown by any other group of Green Algæ. Although numerous species are ubiquitous there are many others confined to definite land-areas of the earth's surface. Thus, one is able to discriminate between definite Indo-Malayan types, African types, American types, Arctic types, and so on.

The only continental area from which Desmids are practically absent is the Antarctic continent[1], in which the conditions of existence are more severe than in the Arctic areas and would soon prove fatal. The chief reason for the absence of Desmids, however, is probably the great distance of the Antarctic continent from the nearest land. This may be fully realized when it is

[1] Only one Desmid is known from the Antarctic continent, viz.: a small species of *Cosmarium* closely allied to *C. Cucurbita* Bréb. This was recorded and described by Fritsch ('12) under the name of '*Penium* sp.'

borne in mind that the direct transference by natural means from one country to another distant country of any living Desmid is in most cases an utter impossibility, since desiccation, or in many cases even partial drying is quickly followed by death, and submergence in sea-water is equally fatal. Moreover, zygospores, which might possibly withstand the entailed vicissitudes if circumstances arose by which they could be transferred from one country to the other (such as by the long flight of a wading bird), are so rarely produced that distribution by their means across an expanse of ocean is almost impossible.

Borge ('92) has recorded a number of sub-fossil Desmids from the glacial clays of the Isle of Götland, and a few subfossil forms of existing species have also been noted from an ancient peat deposit near Filey in E. Yorkshire.

Desmids only thrive in soft water, and they are very numerous in peaty water which has a trace of acidity and in the almost pure water of rocky lakes on the old geological formations. There are a few terrestrial or partially terrestrial species, such as *Mesotænium caldariorum* and sometimes *Cylindrocystis Brébissonii*. There are also others which live in well aërated positions among Bryophytes on wet and dripping rocks.

The classification of Desmids is on a very sound basis, largely owing to the researches of Lütkemüller ('02; '05), and the work of the present author (G. S. W., '99; etc.) mostly verifies Lütkemüller's conclusions. There are two well-defined sub-families in which the genera are arranged as follows:

Sub-family I. SACCODERMÆ.

Cell-wall unsegmented and without pores, soluble in an ammoniacal solution of cupric oxide. Point of division of cells somewhat indefinite and unknown previous to the actual division. The young half of the cell is sometimes developed obliquely and its walls are absolutely continuous with the walls of the older half.

Tribe 1. **Gonatozygeæ.** Cells elongate, cylindrical and unconstricted, solitary or forming loose filaments. Cell-wall with a differentiated outer layer of which the small roughnesses or spines form a part. *Gonatozygon* De Bary, 1856; *Genicularia* De Bary, 1858.

Tribe 2. **Spirotænieæ.** Cells solitary, relatively short and unconstricted[1]. Cell-wall a simple sac, destitute of a differentiated outer layer. The cell becomes adult by periodical growth. *Cylindrocystis* Menegh., 1838; *Spirotænia* Bréb., 1848; *Netrium* Näg., 1849; em. Lütkem., 1902; *Mesotænium* Näg., 1849 [inclus. *Ancylonema* Berggren, 1870]; *Roya* W. & G. S. West, 1896.

Sub-family II. PLACODERMÆ.

Cell-wall segmented, with a differentiated outer layer, only slightly soluble in an ammoniacal solution of cupric oxide. Cell-division follows a fixed type, with the interpolation of younger semicells between the old ones. The younger portions of the cell-wall are joined to the older portions by an oblique surface.

[1] All the supposed constricted species of *Cylindrocystis* have now been shown to belong to the genus *Cosmarium*.

Tribe 3. **Penieæ.** Cells short or of moderate length, straight, cylindrical, sometimes with a slight median constriction. Cell-wall without pores. Point of division of cell sometimes variable. The cell may arrive at maturity by the development of girdle-bands. *Penium* Bréb., 1844; em. Lütkem., 1905.

Tribe 4. **Closterieæ.** Cells elongate, sometimes very long, generally curved; symmetrical in one longitudinal plane only. Cell-wall usually with pores. The cell may arrive at maturity by the development of girdle-bands. At each extremity of the cell is a terminal vacuole with gypsum crystals. *Closterium* Nitzsch, 1817.

Tribe 5. **Cosmarieæ.** Cells exhibit great variety of form. Cell-wall consists of two thin, firm layers, always furnished with pores. Girdle-bands are never developed, the cell becoming adult very soon after division by growth of the young semicell to maturity.

Section *a.* The point of division, where the new and old parts of the cell-wall are obliquely fitted together, remains plane during division.

† Cells solitary. *Docidium* Bréb., 1844; em. Lund., 1871; *Pleurotænium* Näg., 1849; *Triploceras* Bailey, 1851; *Ichthyocercus* W. & G. S. West, 1897; *Tetmemorus* Ralfs, 1844; *Euastridium* W. & G. S. West, 1907; *Euastrum* Ehrenb., 1832; *Micrasterias* Agardh, 1827; *Cosmarium* Corda, 1834; *Xanthidium* Ehrenb., 1834; *Arthrodesmus* Ehrenb., 1838; *Staurastrum* Meyen, 1829.

†† Colonial or filamentous. *Cosmocladium* Bréb., 1856; *Oocardium* Näg., 1849; *Sphærozosma* Corda, 1835; *Onychonema* Wallich, 1860; *Spondylosium* Bréb., 1844; *Phymatodocis* Nordst., 1877: *Hyalotheca* Ehrenb., 1841.

Section *b.* The point of division of the cell, where the new and the old parts of the cell-wall are obliquely fitted together, develops a girdle-like thickening or ingrowth, which projects both ways into each of the old semicells during division.

Streptonema G. C. Wallich, 1860; *Desmidium* Agardh, 1824; *Gymnozyga* Ehrenb., 1840.

The largest genera are *Cosmarium* and *Staurastrum.* A very valuable bibliographical work on the family has been published by Nordstedt ('96; '08).

LITERATURE CITED

Archer, W. ('67). On the conjugation of *Spirotænia condensata* (Bréb.) and *Spirotænia truncata* (Arch.). Quart. Journ. Micr. Sci. n.s. vii, July, 1867.

Archer, W. ('75) in Quart. Journ. Micr. Sci. 1875, pp. 414, 415.

Berghs, J. ('06). Le noyau et la cinèse chez le *Spirogyra.* La Cellule, xxiii, 1906.

Blackman, F. F. & Tansley, A. G. ('02). A Revision of the Classification of the Green Algæ. New Phytologist, i, 1902.

Boldt, R. ('88). Desmidieer från Grönland. Bih. till K. Sv. Vet.-Akad. Handl. Bd. xiii, Afd. 3, no. 5, 1888.

Borge, O. ('91). Sibiriens Chlorophyllophycé-Flora. Bih. till K. Sv. Vet.-Akad. Handl. Bd. xvii, no. 2, 1891.

Borge, O. ('92). Subfossila sötvattensalger från Gotland. Botaniska Notiser, 1892.

Borge, O. ('94). Über die Rhizoidenbildung bei einigen fadenförmigen Chlorophyceen. Upsala, 1894.

Borge, O. ('13). Beiträge zur Algenflora von Schweden. 2. Botaniska Notiser, 1913.

Börgesen, F. ('01) in the Botany of the Færöes, Copenhagen, 1901, p. 235, t. 8, f. 4.

CHMIELEWSKI, V. F. ('90). Materialien zur Morphologie und Physiologie des Sexualprozesses bei niederen Pflanzen. Arbeit. Ges. der Naturf. d. Charkower Univ. xxv, 1890.

CHMIELEWSKI, V. F. ('91). Eine Notiz über das Verhalten der Chlorophyllbänder in den Zygoten der Spirogyra-Arten. Botan. Zeitung, xlviii, 1891.

COPELAND, E. B. ('02). The Conjugation of *Spirogyra crassa* Kg. Bull. Torr. Bot. Club, xxix, 1902.

DE BARY, A. ('58). Untersuchungen über die Familie der Conjugaten. Leipzig, 1858.

DELF, E. M. ('13). Note on an Attached Species of Spirogyra. Ann. Bot. xxvii, April, 1913.

FISCHER, A. ('83). Ueber die Zelltheilung der Closterien. Botan. Zeitung, xli, 1883.

FRITSCH, F. E. ('12). 'Freshwater Algæ' in the Reports of the National Antarctic Expedition. Natural History, vol. vi, 1912.

GERASSIMOFF, J. J. ('97). Ueber die Kopulation der zweikernigen Zellen bei *Spirogyra*. Bull. de la soc. impér. de naturalistes de Moscou, 1897.

GERASSIMOFF, J. J. ('00). Ueber die Lage und die Function des Zellkerns. *ibid.* 1900.

HANSGIRG, A. ('88) in Hedwigia, Heft 9 u. 10, 1888, t. x, f. 6.

HAUPTFLEISCH, P. ('88). Zellmembran und Hüllgallerte der Desmidiaceen. Inaug.-Dissertation Univ. Greifswald, 1888.

JOHNSON, L. N. ('94). On some species of Micrasterias. Botan. Gazette, xix, 1894.

JOSHUA, W. ('82). Notes on British Desmidieæ. Journ. Bot. xx, 1882.

KARSTEN, G. ('09). Die Entwicklung der Zygoten von *Spirogyra jugalis* Ktzg. Flora, xcix, 1909.

KLEBAHN, H. ('88). Über die Zygosporen einiger Conjugaten. Ber. Deutsch. Botan. Ges. vi, 1888.

KLEBAHN, H. ('90). Studien über Zygoten. I. Die Keimung von Closterium und Cosmarium. Jahrbüch. f. wiss. Botan. xxii, 1890.

KLEBS, G. ('85). Ueber Bewegung und Schleimbildung der Desmidiaceen. Biol. Centralbl. v, 1885.

KÜTZING, F. T. ('43). Phycologia generalis oder Anatomie, Physiologie und Systemkunde der Tange. Leipzig, 1843.

KÜTZING, F. T. ('49). Species Algarum. Lipsiae, 1849.

LAGERHEIM, G. ('86). Algologiska bidrag. I. Botaniska Notiser, 1886.

LAGERHEIM, G. ('95). Ueber das Phycoporphyrin, einen Conjugatenfarbstoff. Vidensk.-Selsk. Skrifter. I. Mathem.-naturv. kl. Kristiania, 1895.

LEWIS, F. J. ('98). The Action of Light on Mesocarpus. Ann. Bot. xii, 1898.

LÜTKEMÜLLER, J. ('94). Die Poren der Desmidiaceen Gattung Closterium Nitzsch. Oesterr. botan. Zeitschr. xliv, 1894.

LÜTKEMÜLLER, J. ('02). Die Zellmembran der Desmidiaceen. Cohn's Beiträge zur Biologie der Pflanzen, viii, 1902.

LÜTKEMÜLLER, J. ('05). Zur Kenntnis der Gattung *Penium* Bréb. Verhandl. der k. k. zool.-botan. Ges. Wien, lv, 1905.

LÜTKEMÜLLER, J. ('10). Zur Kenntnis der Desmidiaceen Böhmens. Verhandl. der k. k. zool.-botan. Ges. Wien, lx, 1910.

LUTMAN, B. F. ('10). The Cell Structure of Closterium Ehrenbergii and Closterium moniliferum. Botan. Gazette, xlix, 1910.

LUTMAN, B. F. ('11). Cell and Nuclear Division in Closterium. Botan. Gazette, li, 1911.

MILLARDET ('70). De la germination des zygospores dans les genres Closterium et Staurastrum etc. Mém. soc. des sci. nat. de Strasbourg, vi, 1870.

MOLL, J. W. ('08). Die Fortschritte der mikroskopischen Technik seit 1870. Progressus Rei Botanicæ, ii, 1908.

NATHANSOHN, A. ('00). Physiologische Untersuchungen über amitotische Kerntheilung. Jahrbüch. f. wiss. Botan. xxxv, 1900.

NIEUWLAND, J. A. ('09). Resting Spores of Cosmarium bioculatum, Bréb. The Midland Naturalist, Univ. Press, Notre Dame, Indiana. April, 1909.

NIEUWLAND, J. A. ('09 A). The 'Knee-Joints' of Species of Mougeotia. The Midland Naturalist, Univ. Press, Notre Dame, Indiana. October, 1909.

NORDSTEDT, O. ('73). Bidrag till kännedomen om Sydligare Norges Desmidiéer. Acta Univers. Lund, ix, 1873.

NORDSTEDT, O. ('78). De Algis aquæ dulcis et de Characeis ex insulis Sandvicensibus a Sv. Berggren 1875 reportatis. Lund, 1878.

NORDSTEDT, O. ('96). Index Desmidiacearum citationibus locupletissimus atque Bibliographia. Lund, 1896.

NORDSTEDT, O. ('08). Index Desmidiacearum. Supplementum. Lund, 1908.

OLTMANNS, F. ('04). Morphologie und Biologie der Algen. Jena, 1904.

OVERTON, E. ('88). Über den Konjugationsvorgang bei Spirogyra. Ber. Deutsch. Botan. Ges. vi, 1888.

PALLA, E. ('94). Ueber eine neue, pyrenoidlose Art und Gattung der Conjugaten. Ber. Deutsch. Botan. Ges. xii, 1894.

PASCHER, A. A. ('07). Über auffallende Rhizoid- und Zweigbildungen bei einer *Mougeotia*-Art. Flora, Bd. xcvii, 1907.

PETIT, P. ('80). Spirogyra des environs de Paris. Paris, 1880.

PRINGSHEIM, N. ('77). Ueber die Sprossung der Mossfrüchte und den Generationswechsel der Mallophyten. Jahrbüch. f. wiss. Botan. xi, 1877.

REINSCH, P. ('67). Die Algenflora des mittleren Theilcs von Franken. Nürnberg, 1867.

ROSENVINGE, L. K. ('83). Om *Spirogyra grœnlandica* n. sp. og dens Parthenosporedannelse. Öfvers. af K. Vet.-Akad. Förhandl. no. 8, 1883.

SCHMIDT, M. ('03). Grundlagen einer Algenflora der Lüneburger Heide. Inaugural-Dissertation, Univ. Göttingen, 1903.

SCHRÖDER, B. ('02). Untersuchungen über Gallertbildungen der Algen. Verhandl. des Naturhist.-Med. Vereins zu Heidelberg, vii, 1902.

SCHMULA ('99). Ueber abweichende Copulation bei Spirogyra nitida (Dillwyn) Link. Hedwigia, Bd. xxxviii, 1899.

STAHL, E. ('79). Über den Einfluss des Lichtes auf die Bewegung der Desmidiaceen. Verhandl. d. phys.-medizin. Ges. zu Würzburg, xiv, 1879.

STRASBURGER, E. ('82). Ueber den Theilungvorgang der Zellkerne und das Verhältniss der Kerntheilung zur Zelltheilung. Bonn, 1882.

TRÖNDLE, A. ('07). Ueber die Kopulation und Keimung von Spirogyra. Botan. Zeitung, xxv, 1907.

TURNER, W. B. ('93). The Freshwater Algæ (principally Desmidieæ) of East India. K. Svenska Vet.-Akad. Handl. xxv, no. 5, Stockholm, (1892) 1893.

WALLICH, G. C. ('60). Desmidiaceæ of Lower Bengal. Ann. Mag. Nat. Hist. ser. 3, v, 1860.

WEST, G. S. (G. S. W., '99). On Variation in the Desmidieæ and its Bearings on their Classification. Journ. Linn. Soc. Bot. xxxiv, 1899.

WEST, G. S. (G. S. W., '04). A Treatise on the British Freshwater Algæ. Cambridge University Press, 1904.

WEST, G. S. (G. S. W., '07). Report on the Freshwater Algæ, including Phytoplankton, of the Third Tanganyika Expedition. Journ. Linn. Soc. Bot. xxxviii, 1907.

WEST, G. S. (G. S. W., '09). The Algæ of the Yan Yean Reservoir, Victoria ; a Biological and Œcological Study. Journ. Linn. Soc. Bot. xxxix, 1909.

West, G. S. (G. S. W., '15). Algological Notes. XIV—XVII. Journ. Bot. March, 1915.

West, G. S. & Starkey, Clara B. ('14). *Zygnema ericetorum* and its Position in the Zygnemaceæ. Report Brit. Assoc. Birmingham Meeting, 1913 (1914).

West, G. S. & Starkey, Clara B. ('15). A Contribution to the Cytology and Life-History of *Zygnema ericetorum* (Kütz.) Hansg., with Some Remarks on the 'Genus' *Zygogonium*. New Phytologist, xiv, 1915.

West, W. (W. W., '91). Sulla conjugazione delle Zignemee. La Notarisia, vi, no. 23, Feb. 1891.

West, W. (W. W. '91 a). A Contribution to the Freshwater Algæ of West Ireland. Journ. Linn. Soc. Bot. xxix, 1891.

West, W. & West, G. S. (W. & G. S. W., '94). On some Freshwater Algæ from the West Indies. Journ. Linn. Soc. Bot. xxx, 1894.

West, W. & West, G. S. (W. & G. S. W., '96). On Some North American Desmidieæ. Trans. Linn. Soc. Bot. ser. 2, v, December, 1896.

West, W. & West, G. S. (W. & G. S. W., '97). Welwitsch's African Freshwater Algæ. Journ. Bot. 1897.

West, W. & West, G. S. (W. & G. S. W., '97 a). A Contribution to the Freshwater Algæ of the South of England. Journ. Roy. Micr. Soc. 1897.

West, W. & West, G. S. (W. & G. S. W., '98). Observations on the Conjugatæ. Ann. Bot. xlv, 1898.

West, W. & West, G. S. (W. & G. S. W., '02). A Contribution to the Freshwater Algæ of Ceylon. Trans. Linn. Soc. Bot. ser. 2, vol. vi, part 3, 1902.

West, W. & West, G. S. (W. & G. S. W., '03). Notes on Freshwater Algæ. III. Journ. Bot. Febr. and March, 1903.

West, W. & West, G. S. (W. & G. S. W., '04—'11). Monograph of the British Desmidiaceæ. Ray Society, vol. i, 1904; vol. ii, 1905; vol. iii, 1908; vol. iv, 1911.

West, W. & West, G. S. (W. & G. S. W., '07). Fresh-water Algæ from Burma, including a few from Bengal and Madras. Ann. Roy. Botan. Gard. Calcutta, vol. vi, part ii, 1907.

Wille, N. ('97; '09). Conjugatæ in Engler & Prantl, Die natürlichen Pflanzenfamilien. I. Teil, Ab. 2, Leipzig, 1897; Nachträge zu I. Teil, Ab. 2, Leipzig, 1909.

Van Wisselingh, C. ('00). Ueber Kerntheilung bei *Spirogyra*. Flora, lxxxvii, 1900.

Wittrock, V. B. ('72). Om Gotlands och Ölands Sötvattens-Alger. Bih. till K. Sv. Vet.-Akad. Handl. i, no. 1, 1872.

Wittrock, V. B. ('78). On the Spore-formation of the Mesocarpeæ and especially of the New Genus *Gonatonema*. Bih. till K. Sv. Vet.-Akad. Handl. v, no. 5, 1878.

Wolle, F. ('87). Freshwater Algæ of the United States. Bethlehem, Pa. 1887.

Division III. STEPHANOKONTÆ

The group-name 'Stephanokontæ' was suggested by Blackman & Tansley ('02) for those Algæ in which the motile reproductive cells possess a crown of cilia round the clear anterior end. As a divisional name it stands on an equal footing with the Isokontæ, Akontæ and Heterokontæ. The Algæ included in it are filamentous, with a great uniformity of morphological characters, and they constitute one of the most distinctive orders of the Chlorophyceæ, not only because of the peculiar nature of their ciliated reproductive cells, but also on account of the unique type of division of their vegetative cells.

Order 1. **ŒDOGONIALES.**

In this order the thallus consists of fixed, simple or branched filaments, with a wide range of thickness in the various species. There are only three genera.—*Œdogonium*, *Bulbochæte* and *Œdocladium*, the first-named having simple filaments and the two others branched filaments. The following account of the structure and reproduction refers to the genera *Œdogonium* and *Bulbochæte*, the third genus *Œdocladium* being dealt with separately.

In the young stages of these plants there are well-marked organs of attachment, although many species of *Œdogonium* float freely when adult. The distinction between base and apex is, however, never lost and can be discerned in every individual cell in the filament. In almost all cases the apical part of the cell is somewhat wider than the basal part and very often slightly tumid. (Consult fig. 240 *A—C*; fig. 246 *B*; etc.) There is also a characteristic rigidity about a filament of *Œdogonium* which precludes any such graceful bending as may be noticed in the flexible filaments of *Spirogyra* or *Ulothrix*, and in many of the smaller species there is an inequality in the growth of the individual cells which gives rise to irregular twists and turns in the filament[1]. The basal cell invariably differs from the other cells; it is

[1] This feature reminds one of the similar unequal growth of the segments of a *Rhizoclonium*-filament (*vide* p. 267).

generally somewhat tumid with a basal holdfast (fig. 244; fig. 245 *A*, *C—H*), but in a few of the smaller species of *Œdogonium* it is depressed-ellipsoid (fig. 246 *B—D*). The apical cell is often bluntly rounded, but frequently possesses a conical cap (fig. 244 *I*). In some species the cap is apiculate, in which case this character is shown in the very first cell developed from the zoogonidium (fig. 245 *A*); and in *Œdogonium ciliatum* (Hass.) Pringsh. the terminal cell is greatly elongated to form a delicate hair. In the genus *Bulbochæte* nearly all the cells, except at the points of insertion of the branches, bear at their dilated upper extremities a laterally placed bristle. This bristle is hollow, with a swollen bulbous base, and may be more than 1 mm. in length (fig. 242; fig. 251). The terminal cell of each branch is always furnished with a bristle.

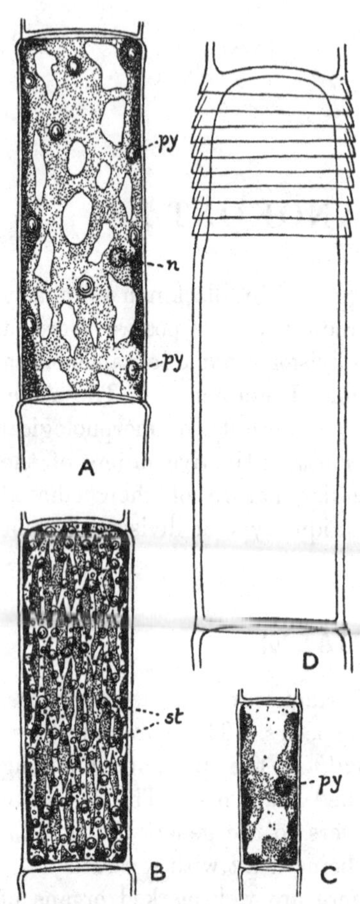

Fig. 240. *A*, *B* and *C*, cells of different species of *Œdogonium* to show the variable character of the parietal chloroplast. *D*, cell of *Œ. giganteum* Kütz. showing a succession of apical caps. *n*, nucleus; *py*, pyrenoid; *st*, starch. All ×430.

The cell-wall consists of a thick cellulose layer outside which is a thin cuticle (fig. 241 *B*). In several species of *Bulbochæte* and in one or two species of *Œdogonium* the exterior of the wall is finely granulate.

The protoplast is largely disposed as a parietal layer in which the chloroplast is embedded, and the cells are uninucleate. The nucleus may be more or less centrally located or it may occupy a parietal position, and in most cases it possesses a prominent nucleolus. The chloroplast is somewhat variable in character, but is always parietal and generally in the form of a reticulum (fig. 240 *A*). In the smaller species of *Œdogonium* the network is much reduced, but in the larger species it may be rather complex and the pieces of the reticulum often have a tendency to become separate and distinct (fig. 240 *B*). The chloroplast is furnished with one or more pyrenoids, the number depending largely upon the size of the cells. In the smaller species there is generally only one pyrenoid (fig. 240 *C*), but in the larger species there may be many pyrenoids distributed through the reticulum. Sometimes there are no pyrenoids, but large numbers of small starch grains (fig. 240 *B*), and in the autumn the cells of many species of *Œdogonium* are frequently packed with moderately large starch-grains.

Many of the cells in a filament of *Œdogonium* may exhibit transverse striations at their upper extremities. This feature, which is most obvious in the larger species, is due to a succession of 'apical caps' resulting from repeated cell-division (fig. 240 *D*).

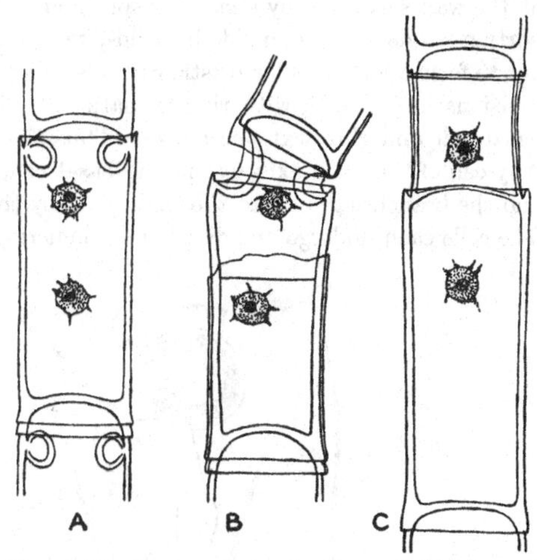

Fig. 241. Three stages in the cell-division of *Œdogonium Borisianum* (Le Cl.) Wittr. × about 620 (after Hirn).

Growth is not apical, but for the most part intercalary, and any cell except the basal one may divide. The details of cell-division have been worked out by various authors and notably in recent years by Hirn ('00) and Kraskovits ('05). It is preceded by the division of the nucleus, during the mitosis of which there is formed an intranuclear spindle. At this period there arises a ring-like thickening on the inner side of the wall at the apex of the cell (fig. 241 *A*). This ring is a circular cushion, the central portion of which consists either of mucilaginous material (Hirn) or of cell-wall substance containing a greater amount of water than the ordinary cell-membrane (Wille), while the peripheral layer (towards the interior of the cell) consists of cellulose. The latter becomes intimately concrescent with the old wall above and below the ring. The first rupture is probably that of the cuticle, which is torn irregularly (fig. 241 *B*), after which the old cell-wall undergoes a circular split and the peripheral layer of the ring becomes the new intercalary piece of cell-wall (fig. 241 *B* and *C*). The new transverse wall arises as a cell-plate between the daughter nuclei and gradually extends outwards to the old wall. This is similar to the condition in higher plants and quite different from the in-creeping transverse walls in *Spirogyra* or *Cladophora*. The upper part of the old wall is pushed upwards as an 'apical cap' by the gradual extension of the new intercalary wall and the number of 'apical caps' indicates the number of divisions any cell has undergone. It is not improbable that the cuticle of the new piece of cell-wall is derived from the material which formed the central part of the ring-thickening.

The plants of *Bulbochæte* are branched (fig. 242), often very profusely, and are invariably attached.

On the germination of the zoogonidium in *Bulbochæte* there is formed a basal cell similar to the rhizoidal basal cell in many species of *Œdogonium.* This remains the only active cell in the primary axis and is therefore the only one capable of division. The first division of the basal cell is of a simple type, without any ring-formation. The upper part of the wall separates by a circular split and, as the new cell grows upwards, the detached portion is pushed to one side in a lid-like manner (fig. 246 *J*). This first new cell grows out to form a long tubular bristle with a swollen bulb-like base (fig. 246 *J br*). Subsequent divisions are normal, with ring-formations, and each new cell is intercalated between the basal cell and the next one above. Thus the upper cell of a branch is the oldest one. Any cell of the main axis, except the basal cell, is a potential basal cell of a side branch, and the branches grow like the main axis by the divisions of their respective basal cells. The cells each undergo one division without ring-formation, the new cell breaking through the mother-cell-wall laterally at its upper end and forming a tubular bulbous bristle (fig. 246 *K*). Should the cell then become the basal cell of a branch subsequent divisions occur with ring-formation and the axis of the branch is deflected in the direction of the axis of the bristle. The cells of lateral branches of the first order become the basal cells of branches of the second order and so on.

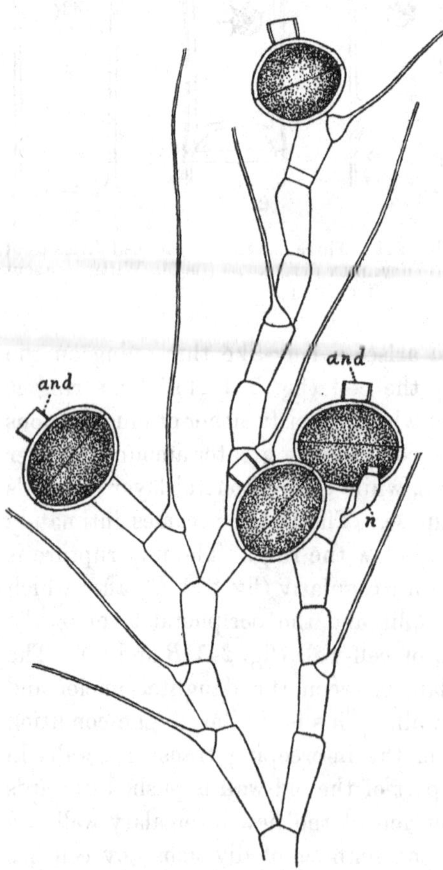

Fig. 242. *Bulbochæte minuta* W. & G. S. West. ×453. *and*, androgonidangium; *n*, nannandrium or dwarf male.

Asexual reproduction takes place by means of zoogonidia of comparatively large size, which are formed singly from the vegetative cells. Very good and reliable accounts of the liberation of the zoogonidia were given by Braun ('49) and by Pringsheim ('58), and since that time some further details have been added. In the vegetative cell about to become a zoogonidangium the protoplast undergoes a rejuvenescence and gradually contracts away from the wall. The cell-wall then splits transversely at or near the upper end of the cell and the rejuvenated protoplast, enclosed in a delicate hyaline vesicle, emerges from the opening (fig. 243 *A* and *B*). At first the emerging zoogonidium is rather irregular in shape, but it quickly becomes rounded (fig. 243 *B*) and in many cases pyriform (fig. 89). Around its anterior pole, which is colourless and sometimes extended to form a protuberance, there is a circle of numerous short cilia. This unique zoogonidium sometimes possesses a pigment-spot. Its exit from the mother-cell only occupies a few minutes, since the rapid disappearance of the delicate vesicle enables it to swim quickly away. On coming to rest it attaches itself by its anterior hyaline end, loses its cilia and develops a cell-wall. From this point onwards germination conforms to two very distinct types, although in both types the cell formed directly from the zoogonidium remains as a differentiated basal cell.

Freund has stated that zoogonidia can be produced in *Œdogonium pluviale* by transferring the filaments from a cane-sugar solution to dilute Knop's solution. They are often formed in numbers in both the spring and the autumn when *Œdogonium*-filaments are brought from outside temperatures to the temperature of the laboratory.

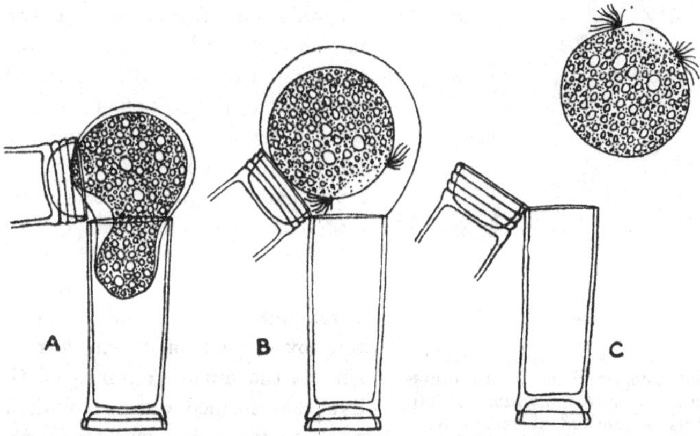

Fig. 243. Formation and escape of the zoogonidium in *Œdogonium concatenatum* (Hass.) Wittr. × 276 (after Hirn).

In the first and most usual type of germination the fixed end of the cell puts out an organ of attachment, which may be simple (fig. 244) or branched (fig. 245 *A*, and *E*—*H*). The nature of this hapteron depends to a great extent upon the particular species and so also does the degree of tumidity of the cell. In most species the basal cell is more or less tumid, but in a few it remains almost cylindrical. The hapteron is in some cases an attaching disc and in others a simple or branched rhizoid. The protoplast extends into the rhizoid and its branches, but as a general rule the chlorophyll does not. The disc-development is generally in those young plants which are attached to a definite substratum (fig. 244 *D*, *E* and *H*), whereas when the young plants remain free-floating and unattached the rhizoidal development is usually most pronounced, and often only a single unbranched rhizoid is developed (fig. 244 *F*, *G* and *I*). This unicellular plant has a distinct apical cap, often convex or conical, and sometimes furnished with an apiculus.

Wille ('87) stated that his observations led him to the conclusion that only those young plants which had become attached were able to grow and divide, the unattached unicellular plants invariably forming zoogonidia again. This is very often the case (fig. 245 *A*), but both Fritsch ('02) and West (G. S. W. '04) have observed short free-floating young filaments (*vide* fig. 244 *G* and *I*). In *Œ. fonticola* A. Br. a basal disc is developed and in those plants which become attached the margins of this disc never become more than crenulate (fig. 245 *C* and *D*), whereas in unattached unicellular plants the disc becomes much lobed and branched (fig. 245 *E*—*H*; G. S. W., '12).

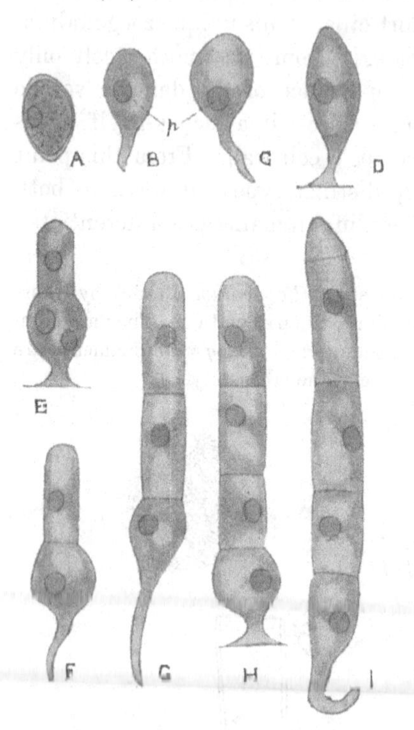

Fig. 244. Various stages in the development of the zoogonidium in an undetermined species of *Œdogonium*. ×460. This is the usual method of germination with the formation of a basal rhizoid.

It is not unusual for the young unicellular plants to liberate their contents as a zoogonidium after a day or two, and this may be repeated through several generations (*vide* fig. 245 *A*). During this process the apical cap either falls off or remains as a hinged lid. After a time the young plants become two-celled. Poulsen ('77) was the earliest investigator to give a precise account of this first division and Fritsch ('02 B) largely confirms his observations. The cell-wall in the apical region becomes thickened by the development of an inner secondary membrane the lower part of which forms a cellulose ring at the upper extremity of the cell, very like that formed during cell-division in the adult filament, but whereas in the latter the outline of the ring makes two acute angles with the inner limit of the cell-wall, in the unicellular plant there is only the lower acute angle, the upper part of the ring going over into the inner layer of the wall above in a gradual curve. A circular split is formed in the outer part of the ring, and by the growing out of the contents of the old cell the ring is gradually stretched until it forms the lateral wall of the new cell. The upper end of the old part of the cell-wall, which now forms the first apical cap, is in some species thrown off, but in others it is permanently retained. After the new lateral wall has been considerably stretched and elongated the first transverse wall appears as a cell-plate. Subsequent divisions generally take place normally as in the cells of the adult filament.

In the second type of germination the basal cell swells out and becomes either depressed-globose or somewhat hemispherical owing to a slight flattening of the under-side by which it is attached (fig. 246). Attention was first directed to this mode of development by Lemmermann ('98) and subsequently both Scherffel ('01) and Fritsch ('04) have given a more extended account of it. This basal cell has abundant chlorophyll and starch-grains. It also possesses a small apical cap in the middle of the upper surface. On the first division the apical cap is detached and through the aperture is protruded a more or less cylindrical outgrowth clothed in a cellulose wall. This is the second cell of the now bicellular plant which either pushes the cap on one side like a hinged lid (fig. 246 *G*) or carries it upwards on its apex (fig. 246 *F* and *H*). This type of growth is the result of the new substance of the cell-wall not arising in the form of a ring but as a hemispherical layer in the upper part of the rounded basal cell.

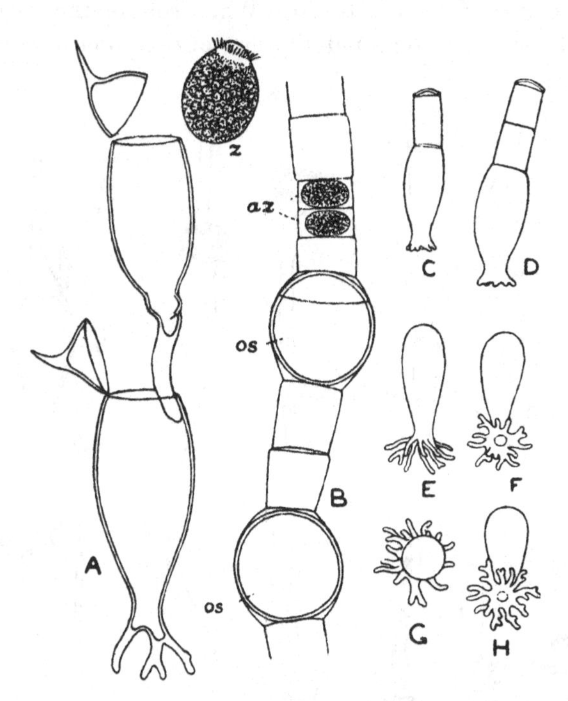

Fig. 245. *A*, three generations of *Œdogonium rivulare* (L. Cl.) A. Br., two unicellular plants and an escaping zoogonidium (*z*), × 310. *B—H*, *Œ. fonticola* A. Br., × 330. *B*, female plant with two oogonia containing oospores (*os*) and two androzoogonidangia (*az*) each containing a single immature androzoogonidium; *C* and *D*, young male plants which were attached; *E—H*, developing androzoogonidia which were unattached, showing the much-branched basal disc.

In the first type of germination the actual attachment of the basal cell to the substratum appears to be by a brown cementing substance which is either ferric oxide or some ferric salt (Fritsch, '02), but in the second type of germination there is evidence that the depressed basal cell is attached by a tough mucus.

Sexual reproduction in the Œdogoniales is of a high type and there is a greater specialization of the male and female organs than in any other Green Algæ. The sexual organs are oogonia and antheridia. The oogonium arises as a result of the division of a vegetative cell, and in *Œdogonium* the lower cell forms a supporting cell while the upper cell becomes the oogonium. This upper cell is very often one which, owing to repeated previous divisions, possesses a number of apical caps (*vide* figs. 247 *F*; 249 *B* and *C*; 250 *A*), and after division it contains most of the protoplast of the original mother-cell. The supporting cells of some species of *Œdogonium* are decidedly swollen [consult fig. 249 *C* of *Œ. lautumniarum* Wittr. and fig. 250 *A* of *Œ. Borisianum* (Le Cl.) Wittr.]. The oogonia are ovoid, globose or depressed-globose in outward form and in *Œdogonium* they occur singly at intervals along the filament or in series of from 2 to 10. When seriate the oogonia arise in basipetal succession by the repeated divisions of the lower or supporting cell.

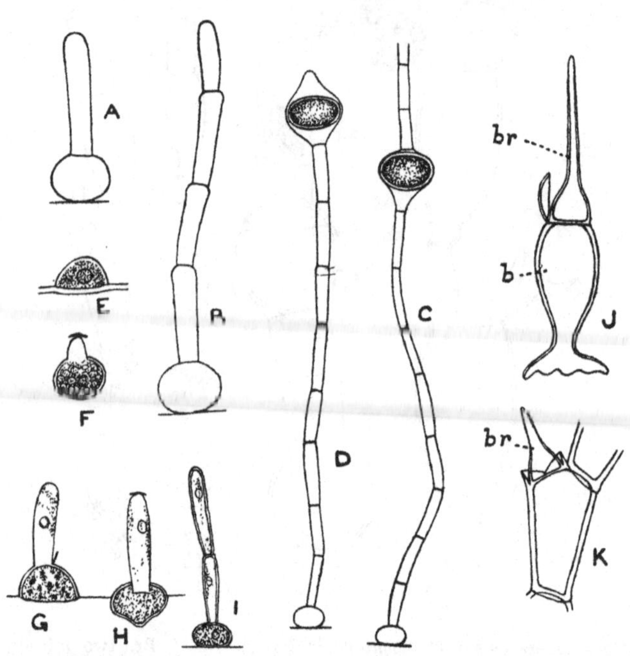

Fig. 246. *A* and *B*, young plants of *Œdogonium Howardii* G. S. West, × 520. *C* and *D*, adult plants of *Œ. inconspicuum* Hirn, × 500. *E—H*, developmental stages (from the zoogonidium) of *Œ. rufescens* Wittr. var. *Lundellii* (Wittr.) Hirn, × 372. *I*, young plant of *Œ. Vireeburgense* Hirn, × 372. (*E—I*, after Scherffel.) Note the form of the basal cell in all these figures. *J*, young plant of *Bulbochæte intermedia* De Bary showing the basal cell (*b*) and the growing bristle (*br*) which is the first cell formed from the basal cell. *K*, part of a branch of *B. intermedia* showing the first division of an intercalary cell resulting in the formation of a laterally placed bristle (*br*). *J* and *K*, × 300 (after Hirn).

In the genus *Bulbochæte* the formation of the oogonium is not so simple as in *Œdogonium* since it arises as the result of a double division. The first division results in a transverse wall which cuts off a supporting cell from a primary oogonium-cell (fig. 247 *A pr*). The latter bulges outwards at its upper end, the new wall bursting through the old mother-cell-wall and forming a swelling under the bulbous base of the bristle. This swelling is the young developing oogonium and about the time it attains its full size the second division occurs. The new transverse wall cuts off the swollen oogonium from the lower part of the primary oogonium-cell, which becomes a second or upper supporting cell (fig. 247 *B s*). Each oogonium has thus two supporting cells, of which the lower corresponds to the supporting cell in *Œdogonium*. No further division can take place in either of these cells.

The protoplast of each oogonium becomes rounded off to form a single oosphere or egg-cell, the chloroplast disintegrating and the egg-cell assuming an intensely green colour except at the receptive spot.

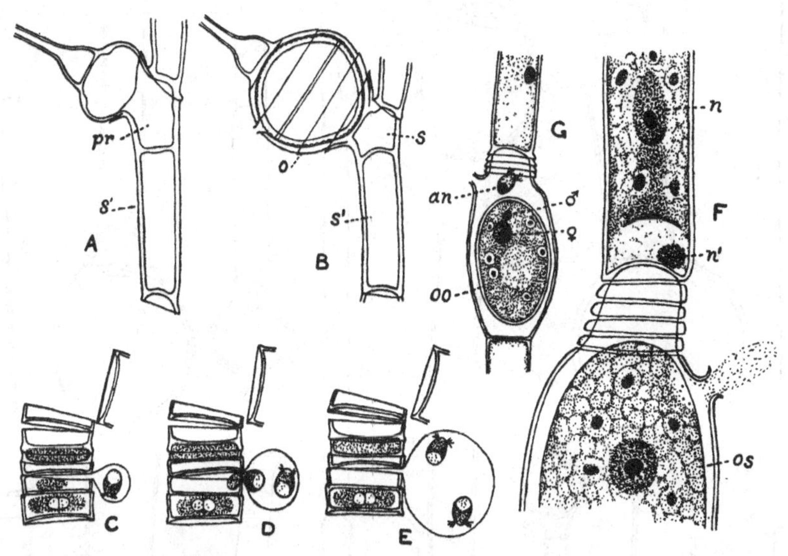

Fig. 247. *A* and *B*, development of the oogonium in *Bulbochæte setigera* (Roth) Ag., ×262; *pr*, primary oogonium-cell; *s* and *s'*, upper and lower supporting cells; *o*, oogonium. *C—E*, formation and escape of the antherozoids in *Œdogonium Landsboroughi* (Hass.) Wittr., ×262. *F* and *G*, *Œ. Boscii* (L. Cl.) Wittr.; *F*, upper part of oogonium with oosphere (*os*) showing receptive spot and exudation of mucilage from opening of oogonium; also division of upper cell prior to formation of another oogonium (*n*, nucleus of future oogonium; *n'*, nucleus of future supporting cell), ×720; *G*, oogonium with oospore (*oo*), but with the male (♂) and female (♀) nuclei not yet fused; *an*, a superfluous antherozoid which has entered the oogonium, ×300. (*A—E*, after Hirn; *F* and *G*, after Klebahn.)

There are three distinct sections of the genus *Œdogonium* characterized by the place of development of the antheridia. In the first type the antheridia arise in the same filament as the oogonia and the plants are *monœcious* (fig. 248). The other two types are *diœcious*. In one the antheridia are developed in male filaments similar in character and but little inferior in size to the female filaments. Species in which this occurs are *diœcious macrandrous* (fig. 249). In the remaining type the male plants are very small and are attached to the female filaments in the vicinity of the oogonia. Such species are said to be *diœcious nannandrous* (fig. 250).

In the monœcious species of *Œdogonium* and *Bulbochæte*, and in the diœcious macrandrous species of *Œdogonium*, the antheridium arises by a division of a vegetative cell, in which the first dividing wall is laid down near the apex of the cell. It is the upper small daughter-cell which becomes an antheridial cell, and in some of the monœcious species it may remain the only antheridial cell, thus constituting a unicellular antheridium. In most

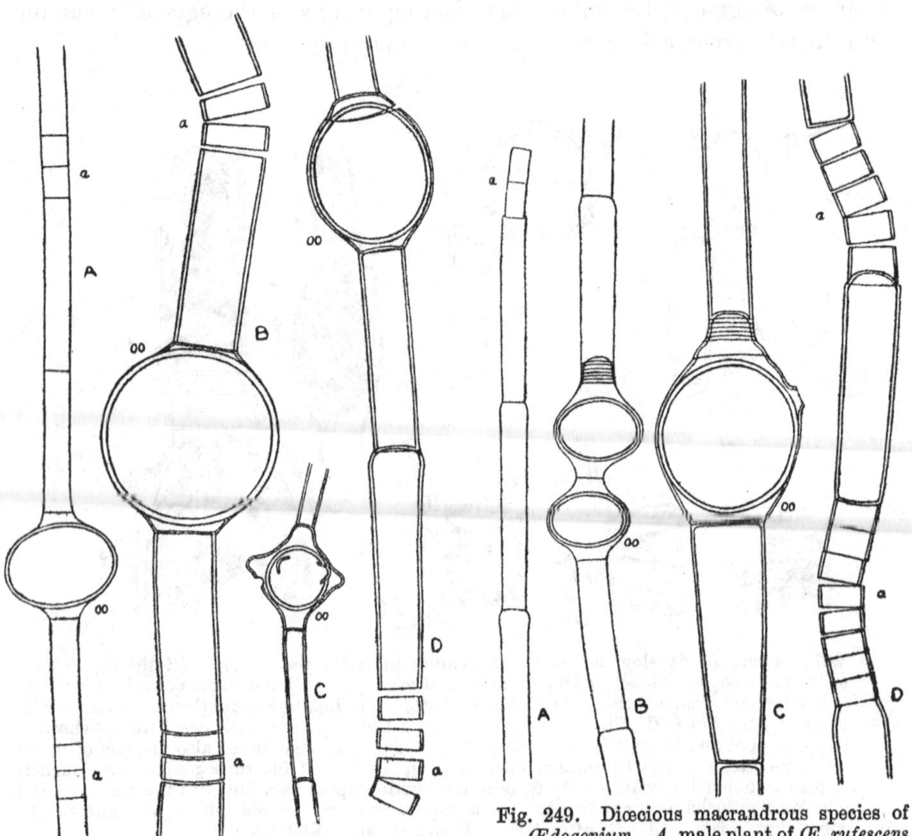

Fig. 248. Monœcious species of *Œdogonium*. *A*, a form of *Œ. obsoletum* Wittr. *B*, *Œ. zig-zag* Cleve var. *robustum* W. & G. S. West. *C*, *Œ. Itzigsohnii* De Bary var. *minor* W. West. *D*, *Œ. Ahlstrandii* Wittr. *a*, antheridium; *oo*, oogonium. All × 460.

Fig. 249. Diœcious macrandrous species of *Œdogonium*. *A*, male plant of *Œ. rufescens* Wittr.; *B*, female plant of same. *C*, female plant of *Œ. lautumniarum* Wittr.; *D*, male plant of same. *a*, antheridium; *oo*, oogonium. All × 460.

cases, however, the antheridium consists of more than one cell, the antheridial cells arising either by the continued division of the lower daughter-cell or by division of the first-formed antheridial cell.

The number of cells composing the antheridium is fairly constant for any one species, and in the monœcious species the position of the antheridium relatively to the oogonium is also a character of specific importance.

The antherozoids are not always formed in quite the same way. In some of the smaller species of *Œdogonium* the protoplast of the antheridial cell forms one antherozoid. This is exceptional, as in general two antherozoids arise in each antheridial cell (fig. 247 *C—E*). In some species of *Œdogonium*, especially the diœcious macrandrous species, they are formed by a vertical division of the protoplast, the two antherozoids being side by side in the

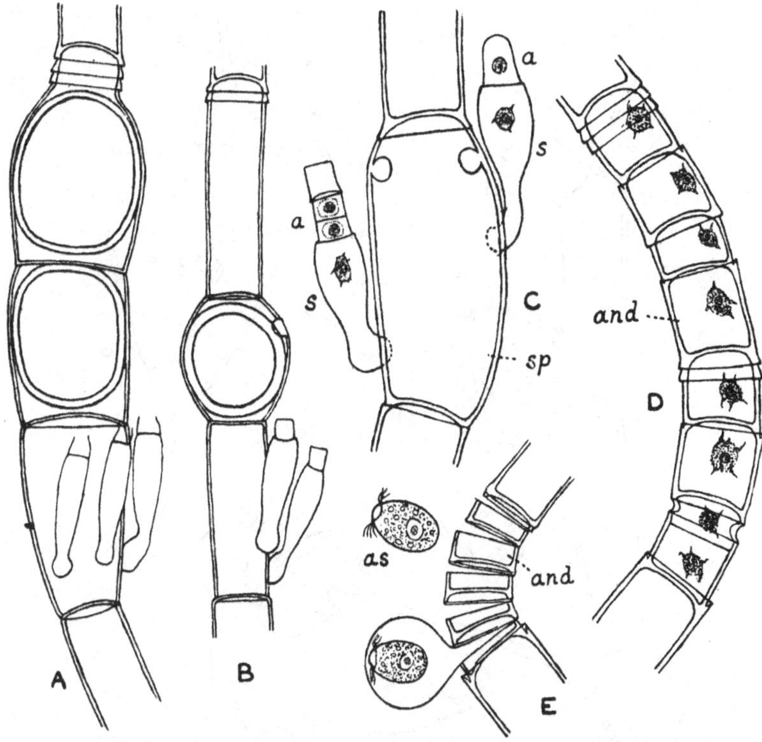

Fig. 250. Diœcious nannandrous species of *Œdogonium*. *A*, *Œ. Borisianum* (Le Cl.) Wittr., ×400. *B*, *Œ. crassiusculum* Wittr. var. *idioandrosporum* Nordst. & Wittr., ×220. *C* and *D*, *Œ. concatenatum* (Hass.) Wittr., ×300; *C*, supporting cell of oogonium (which is just about to be formed by cell-division) carrying two nannandria; *D*, chain of androgonidangia. *a*, antheridium; *and*, androgonidangium; *s*, supporting cell of antheridium; *sp*, supporting cell of oogonium. *E*, androgonidangia (*and*) of *Œ. Braunii* Kütz. showing escape of androgonidia (*as*). (*C—E*, after Hirn.)

antheridial cell, but in many others and in *Bulbochæte* the division is transverse so that the antherozoids are one over the other.

In the diœcious nannandrous species the male plants are very small and are epiphytic on the female plants. They are developed from special motile cells known as *androgonidia* (or *androspores*) which are produced in *androgonidangia* (fig. 250 *D* and *E and*). The latter, which are often called androsporangia, are short cells, solitary or in chains, and they are sometimes a little narrower than the filament in which they are produced. The first androgonidangium always arises by the division of a vegetative cell and the others are formed in a manner very similar to the development of a multicellular antheridium. In fact, there is often a very close resemblance between the latter and a chain of androgonidangia. Sometimes the androgonidangia occur in the same filaments as the oogonia (gynandrosporous forms) or they may only be formed in separate filaments (idioandrosporous forms), and in

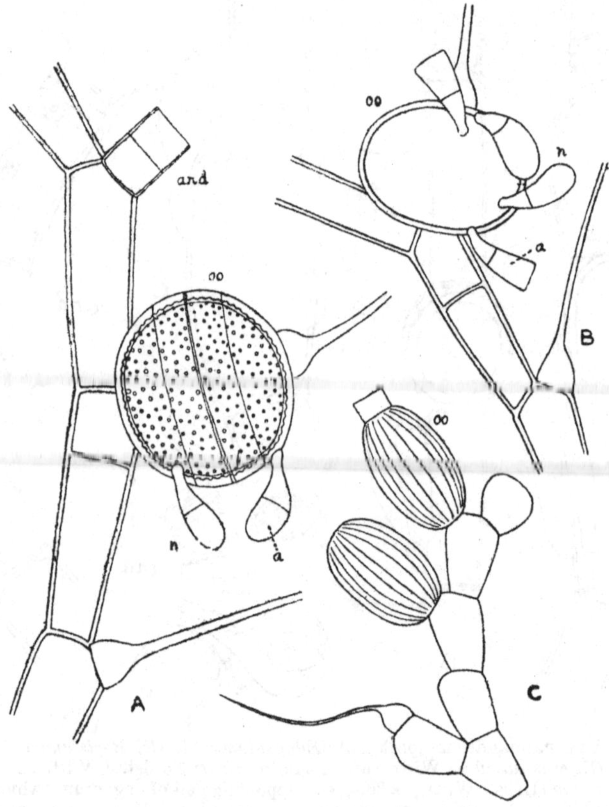

Fig. 251. *A, Bulbochæte subintermedia* Elfv. *B, B. Nordstedtii* Wittr. *C, B. nana* Wittr. *a*, antheridium; *and*, androgonidangium; *n*, nannandrium; *oo*, oogonium. All ×495.

the former case they may, like the antheridia of the monœcious species, occupy more or less constant positions relative to the oogonia.

The androgonidia come to rest either on the wall of the oogonium or on the supporting cells and on germination each produces a 'dwarf male' or *nannandrium* (fig. 250 *A—C*). These tiny male plants are sometimes unicellular, the antherozoids being produced within the single cell. More often they are bicellular or pluricellular, in which cases the antheridia may arise in one of two ways. In many species a transverse wall appears in the originally unicellular dwarf male and the upper cell becomes the antheridium (*antheridium interior*; fig. 251 *A* and *B*); in other species the antheridium may consist of one or more cells which arise by the division of the first-formed cell in a manner characteristic of cell-division in *Œdogonium* (*antheridium exterior*; fig. 250 *C a*).

When the oosphere is ready for fertilization there is formed in it a clear pellucid space, the *receptive spot*, opposite the opening in the wall of the oogonium. The latter opens in two distinct ways, either by a pore or by a circular split all round the wall. In both cases the opening may be superior, median or inferior. The antherozoid enters by the opening and sexual fusion results in the oospore (fig. 247 *G*). The wall of the ripe oospore consists of from one to three layers and not infrequently it exhibits various sculptures. In *Œdogonium* it is sometimes clothed with short spines. The oospore is a resting spore and a short time after the formation of its wall the chlorophyll disappears, the protoplast becoming largely filled with a fatty oil in which there is commonly dissolved a brown or red pigment. It is liberated by the decay of the wall of the oogonium. On germination the thick walls are burst open and the protoplast, now green again, divides into four cells each of which becomes a swarm-cell of the typical *Œdogonium*-type (fig. 252). After swarming for a time the motile cells come to rest and begin to grow into new filaments. In some cases these filaments are asexual and they give rise to several other asexual generations before forming a sexual plant. It has been suggested that the four swarm-cells formed on the germination of the fertilized egg represent a rudimentary sporophyte generation, but this is doubtful and there is no cytological evidence in support of it.

Œdogonium fonticola A. Br., which has male plants of small size, is apparently a species intermediate between the truly diœcious macrandrous forms and those with large nannandria. It thus sheds much light on the possible evolution of the nannandrous types (G. S. W., '12). It seems not improbable that the macrandrous species arose from the monœcious species by the physiological differentiation of the zoogonidia, some of which gave rise only to female filaments and others only to male filaments. Thus there would have arisen *androzoogonidia* and *gynozoogonidia*. In the nannandrous species there is a further differentiation between these two types of zoogonidia in point of view of size, the androzoogonidia being as a rule not more than half the size of the gynozoogonidia, and intermediate between the gynozoogonidium and the antherozoid. All these motile bodies exhibit

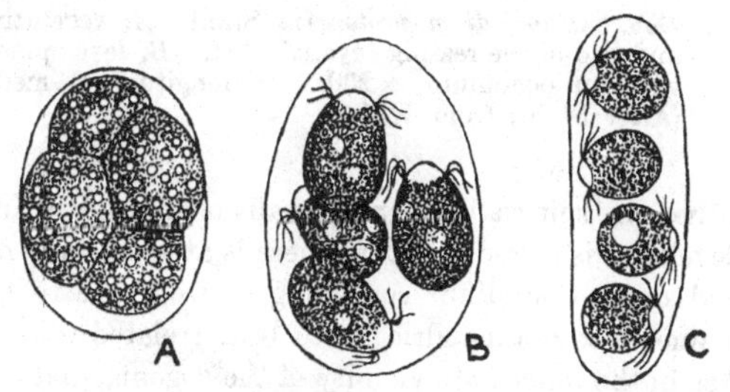

Fig. 252. *A* and *B*, germination of the oospore of *Œdogonium pluviale* Nordst. *C*, the four swarm-spores produced on the germination of the oospore of *Bulbochæte elachistandra* Wittr., × 350. (*A* and *B*, after Jurányi; *C*, after Pringsheim.)

precisely similar morphological characters. How have the androzoogonidia of the nannandrous species arisen? It is not at all probable that they have arisen from the antherozoids of the monœcious species by an increase in size of the antherozoid mother-cells, because two antherozoids almost invariably arise in each cell of the antheridium by the division of the protoplast, whereas the entire protoplast of the androzoogonidangium forms a single androzoogonidium. Seeing that this is also the case in the formation of both the androzoogonidia and gynozoogonidia of the macrandrous species, and in the gynozoogonidia of the nannandrous species, it is reasonable to suppose that the small androzoogonidium of the latter has arisen merely by a reduction in size, which has gone on hand in hand with a greater sexual differentiation. The specialization has become such that the androzoogonidia are attracted to the vicinity of the oogonia and only germinate

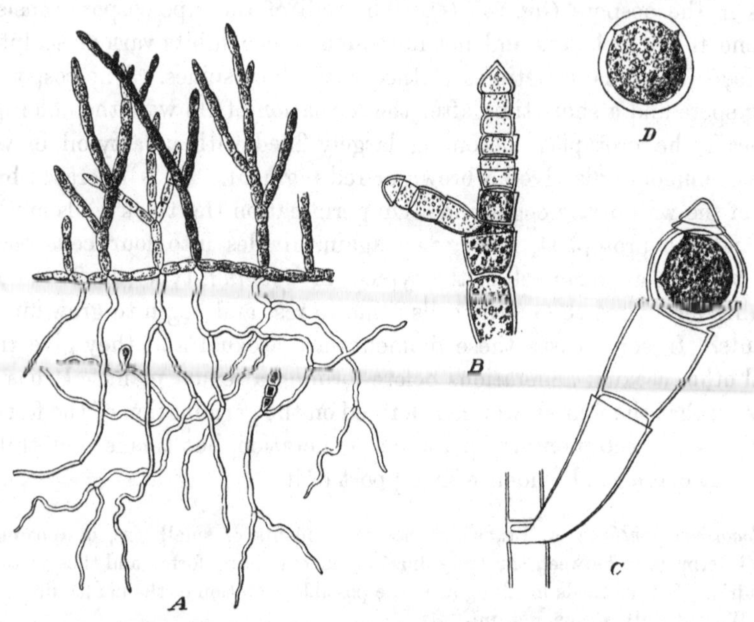

Fig. 253. *Œdocladium protonema* Stahl. *A*, vegetative plant showing the colourless rhizoids and two of the resting 'cysts,' × 41. *B*, development of a male branch, × 330. *C*, a branch with an oogonium, × 230. *D*, longitudinal median section through the zygote, × 230. (After Stahl, from Wille.)

either on their walls or on the walls of the supporting cells. The male plant which is then developed is so reduced that there is at most only one vegetative cell, or sometimes none, and one antheridium consisting of from one to five cells. The reduction of the male filaments to 'nannandria' is to be correlated with the fact that the antherozoids are set free in the immediate vicinity of the oogonia, and, therefore, fewer of them are required in order to ensure fertilization. Moreover, vegetative cells are unnecessary except as a support for the single antheridium, for which one cell easily suffices (G. S. W., '12).

Hirn ('00) is also in agreement with the view just expressed as to the origin of the dwarf males of the nannandrous species of *Œdogonium* by the reduction and greater specialization of the androzoogonidia of the macrandrous forms, but Pascher ('06) states that the nannandrous forms have not arisen from the macrandrous forms.

The genus *Œdocladium*, of which there is only one known species, *Œdocladium protonema* Stahl ('91), is a subaërial Alga forming a stratum on damp ground. There is a creeping portion of the thallus, firmly fastened to the substratum by means of numerous colourless rhizoids. Standing up from the creeping part of the thallus are erect branches, which are themselves branched (fig. 253 *A*). The cell division is like that of *Œdogonium*, but growth in length of the branches is mostly the result of the activity of the apical cell. The latter has a conical cap, but is not piliferous. The zoogonidia have the typical circlet of cilia. The plants are monœcious and the oogonia are globose. Propagation may occur by resting 'cysts' of one or several cells formed on the underground rhizoids.

Family **Œdogoniaceæ.**

This is the only family of the Œdogoniales.

The genera are: *Bulbochæte* Agardh, 1817; *Œdogonium* Link, 1820; *Œdocladium* Stahl, 1891. The family was first monographed by Wittrock ('74) and more recently on a more elaborate scale by Hirn ('00). *Bulbochæte* only occurs in the comparatively still waters of pools and lakes or in quiet bog-pools. It is less abundant than *Œdogonium*, species of which are found in both still and running water. There are about 240 known species of *Œdogonium*, almost all of which are recognized by the characters of the sexual organs. Rather less than half the species are diœcious nannandrous and most of the remainder are monœcious. In the smaller genus *Bulbochæte* most of the species are diœcious nannandrous and the rest monœcious. In the latter the antheridia may be either erect or patent.

In the genus *Œdogonium* there is considerable range in size, the smallest species, *Œ. angustissimum* W. & G. S. West, having filaments only 2μ in thickness, whereas those of *Œ. fabulosum* Hirn attain a diameter of 85μ.

LITERATURE CITED

Blackman, F. F. & Tansley, A. G. ('02). A Revision of the Classification of the Green Algæ. New Phytologist, 1902.

Braun, A. ('49). Betrachtungen über die Erscheinung der Verjüngung in der Natur. 1849–50.

Fritsch, F. E. ('02 a). Algological Notes. II. The Germination of the Zoospores in Oedogonium. Ann. Bot. xvi, June, 1902.

Fritsch, F. E. ('02 b). The Structure and Development of the young plants in Œdogonium. Ann. Bot. xvi, Sept. 1902.

Fritsch, F. E. ('04). Some points in the structure of a young Œdogonium. Ann. Bot. xviii, Oct. 1904.

Hirn, K. E. ('00). Monographie und Iconographie der Œdogoniaceen. Acta Soc. Sci. Fennicæ, xxvii, no. 1, 1900.

Hirn, K. E. ('06). Studien ueber Œdogoniaceen. I. Acta Soc. Sci. Fennicæ, xxxiv, no. 3, 1906.

Kraskovits, G. ('05). Ein Beitrag zur Kenntnis der Zellteilungsvorgänge bei Oedogonium. Sitzungsber. der k. Akad. der Wissensch. Wien, Mathem.-naturw. Klasse, Bd. cxiv, Abt. I, 1905.

Lemmermann, E. ('98). Algologische Beiträge. IV. Süsswasseralgen der Insel Wangerooge. Abh. Nat. Ver. Bremen, xiv, 1898.

Pascher, A. ('06). Über die Zwergmännchen der Œdogoniaceen. Hedwigia, xlvi, 1906.

Poulsen, V. A. ('77). Om sværmsporens spiring hos en art af slægten *Oedogonium*. Botanisk Tidsskrift, ser. 3, bd. ii, Kjobenhavn, 1877.

Pringsheim, N. ('58). Beiträge zur Morphologie u. Systematik der Algen. I. Morphologie der Oedogonien. Pringsh. Jahrbüch. f. wiss. Botan. i, 1858.

Scherffel, A. ('01). Einige Beobachtungen über Oedogonien mit halbkugeliger Fusszelle. Ber. Deutsch. Botan. Ges. xix, 1901.

Stahl, E. ('91). *Oedocladium protonema.* Pringsh. Jahrbüch. f. wiss. Botan. xxxiii, 1891.

Tuttle, A. H. ('10). Mitosis in *Œdogonium.* Journ. Exp. Zool. ix, 1910.

West, G. S. (G. S. W., '12). Algological Notes. X—XIII. Journ. Bot. Nov. 1912.

Wille, N. ('87). Algologische Mittheilungen. V. Über das Keimen der Schwärmsporen bei *Oedogonium.* Pringsh. Jahrbüch. f. wiss. Botan. xviii, 1887.

Wittrock, V. B. ('74). Prodromus Monographiæ Oedogoniearum. Upsaliæ, 1874.

Division IV. HETEROKONTÆ

It is to Luther ('99) that we owe the establishment of the group Heterokontæ to embrace a varied assortment of Green Algæ which were previously scattered throughout the different families of that section of the Chlorophyceæ now well-established as the Isokontæ.

Our knowledge of these Algæ has increased very much during the past few years and it is now possible to discuss their inter-relationships with some degree of certainty. *They are of a prevailing yellow-green colour owing to the presence of a greater amount of xanthophyll than occurs in other groups of the Chlorophyceæ. The chromatophores are parietal, often discoidal, and are destitute of pyrenoids.* There are usually several or many in each cell. *Starch is absent and the stored product of photosynthetic activity is a fatty oil.* In other respects the cytological details are similar to those described for the Isokontæ.

The group contains unicellular, colonial, filamentous and cœnocytic types, and it is possible to institute a comparison between these and corresponding types in the Isokontæ.

Vegetative multiplication occurs in a few genera by the dissociation of large colonies into smaller aggregates which soon increase in size by further cell-division.

Asexual reproduction takes place by *ovoid or pear-shaped zoogonidia, which are furnished with two cilia of unequal length.* The latter are often attached rather to one side of the anterior extremity of the zoogonidium. Moreover, the zoogonidia possess, as a rule, more than one chromatophore. Aplanospores are frequently formed in some genera, either singly or several in each cell.

Gamogenesis occurs in a number of genera and the isogametes are biciliated. It is not improbable that exact observations and accurate methods will demonstrate that the ciliation of the gametes is similar to that of the zoogonidia, but so far such observations as have been made are few in number and not sufficiently precise[1].

[1] It should be remembered that the second short cilium was not clearly demonstrated until 1898, previous to which date the zoogonidia and gametes of these Algæ were almost invariably described as possessing only one cilium. The short cilium is usually carried in a backward direction, pressed more or less closely against the body of the motile cell.

It is a noteworthy fact that there are no highly developed reproductive organs, or even differentiated reproductive cells, in any family of the group.

Bohlin, in two excellent papers ('97 ; '01 A), strongly emphasized the importance of maintaining the group of the Heterokontæ, but in completely separating the group from all other Green Algæ he was just as much in error as was Wille in separating the Conjugatæ from the rest of the Chlorophyceæ. Blackman & Tansley ('02) and also West (G. S. W., '04) supported Bohlin's views, the former even to the extent of including the Vaucheriaceæ in the Heterokontæ (*vide* p. 249).

Concerning the classification of the Heterokontæ there is no doubt that the old order 'Confervales,' which embraced all but the Flagellate forms, was much too wide in its scope and cannot in future be retained. A number of suggestions have recently been put forward by Pascher ('13 B ; '14) and it is necessary that they should be carefully considered. In his first paper in 1913 he subdivided the Heterokontæ into five groups which he regarded as equivalent to corresponding groups in the Chlorophyceæ (= Isokontæ of this work). They may be briefly summarized as follows:

Heterokontæ	*Chlorophyceæ* (of Pascher)
Heterochloriadales	Volvocales
Heterocapsales	Tetrasporales
Heterocapsaceæ	
? Botryococcaceæ	
Mischococcaceæ	
Heterococcales	Protococcales
Chlorobotrydaceæ	
Sciadiaceæ	
Heterotrichales	Ulotrichales
Tribonemaceæ	
Heterosiphonales	Siphonales

In his later paper, in 1914, Pascher widely separated the Heterokontæ from the 'Chlorophyceæ,' placing the latter along with the Conjugatæ in a new primary group, the Chlorophyta, and the former, along with the Chrysophyceæ and Bacillariales, in another new group, the Chrysophyta. The correlated scheme of arrangement of the 'Chrysophyta,' 'Pyrrophyta' (which includes the Peridinieæ and the Cryptomonadales) and 'Chlorophyta,' as set forth by Pascher, is very methodical, but the scheme must be considered in the light of facts. There is little, if any, reason for separating the Heterokontæ so decisively from the remainder of the Green Algæ that two of the intervening groups are the 'Phæophyta' (which is presumably a new name for the Phæophyceæ) and the 'Pyrrophyta.' Moreover, a large part of the scheme depends upon the supposed validity of the 'Volvocales,' 'Tetrasporales' and 'Protococcales' as independent orders. In this work a division of this kind cannot be admitted, since the complete separation of such groups would be to a great extent artificial. The so-called 'Volvocales' and 'Tetrasporales' are merely sections of the Protococcales. The 'Volvocales' have never emerged from the Protococcales; nor have the 'Tetrasporales,' although the tetrasporine tendency has resulted in the evolution of Algæ of greatly increased complexity and definiteness of form (*vide* p. 157).

For the classification of the Heterokontæ to be consistent with that of the Isokontæ adopted in this work it is necessary to include Pascher's Heterochloriadales and Heterocapsales in the Heterococcales, which thus becomes in the Heterokontæ the precise equivalent of the Protococcales in the Isokontæ. The Heterotrichales and the Heterosiphonales (including *Botrydium* only) are legitimate and well-founded groups. It is highly probable, as both Luther and Bohlin contend, that the Heterokontæ were evolved along an independent line from the rest of the Green Algæ, having arisen from Flagellate forms such as *Chloramœba* (fig. 254) and passed through low intermediate types like *Leuvenia* Gardner ('10) and *Chlorosaccus* Luther ('99), in which the dominant phase has become non-motile.

Fig. 254. *Chloramœba heteromorpha* Bohlin. × 600 (after Bohlin).

A good general account of the group and a synopsis of many of the species has been given by Heering ('06), but the more recent suggestions concerning the classification of the group do not, of course, occur in his work.

Order 1. HETEROCOCCALES.

The Algæ included in this order of the Heterokontæ are *unicellular or colonial.* The cells vary much in outward form and in the way in which they are associated to form colonies.

In the more primitive types (*Chlorosaccus, Stipitococcus*) the cell-wall is very thin, but in the more advanced types (*Chlorobotrys, Botrydiopsis, Ophiocytium*) the wall is firm and strong. In some genera, such as *Chlorobotrys* and *Pelagocystis*, there is a great development of mucilage. *Mischococcus* has a branched colony which is almost unique amongst the Green Algæ, and the shapeless colonies of *Botryococcus* are even yet somewhat of a puzzle. In *Ophiocytium* the cells are elongate and curved or coiled, and in the *Sciadium*-section of the genus the daughter-cells, which are developed from zoogonidia or aplanospores, remain attached in an umbellate manner to the open apex of the old mother-cell, the colony being technically of the nature of a cœnobium. The cells contain from one to many chromatophores, the greatest number occurring in *Botrydiopsis.*

In the Heterococcales it would appear that there is nothing corresponding to the family Volvocaceæ of the Protococcales. The only forms predominantly motile, such as *Chloramœba* and *Vacuolaria*, are distinctly Flagellate in character, being naked and amœboid, and cannot rightly be placed in the Heterococcales. Pascher ('13 B) tabulates his 'Heterochloriadales' as more or less equivalent to his 'Volvocales,' but this is not in agreement with known facts and the two genera included in his new group, namely, *Chloramœba* and *Stipitococcus*, have not much in common with each other.

The families Chlorosaccaceæ and Chlorotheciaceæ are respectively almost exactly equivalent to the Palmellaceæ and the Planosporaceæ among the Isokontæ. In *Ophiocytium* is seen the probable starting-point of the principal filamentous type, and the striking similarity between the structure of the cell-wall in this genus and in *Tribonema* (consult fig. 262) is particularly significant.

Multiplication occurs in various families by cell-division and by the breaking up of the larger colonies. Reproduction occurs in many cases by zoogonidia and sometimes, as in *Ophiocytium*, aplanospores are chiefly formed.

Family **Chlorosaccaceæ.**

This is the most primitive family of the unicellular and colonial Heterokontæ, occupying much the same position as the Palmellaceæ among the Tetrasporine Protococcales. The non-motile condition is the dominant one, and in *Chlorosaccus* and *Racovitziella* the cells are embedded in a copious mucus. Cell-division takes place abundantly in this state. In *Stipitococcus* (fig. 255) and *Peroniella* the cells are attached by delicate stalks to the filaments of larger Algæ, and in many cases the length of the stalk is determined by the thickness of the mucous investment of the host. It seems probable that in both genera the stalk is the modified longer cilium. In *Mischococcus* the cells are globular and are united to form small branched colonies by thick tubular stalks of mucilage in the distal ends of which the cells are embedded (fig. 256). Each cell possesses from one to four chromatophores. Reproduction is by zoogonidia, which in *Chlorosaccus* and *Mischococcus* have each a pair of chromatophores. The unequal cilia are attached at the anterior extremity. Aplanospores are known to be formed in *Peroniella*.

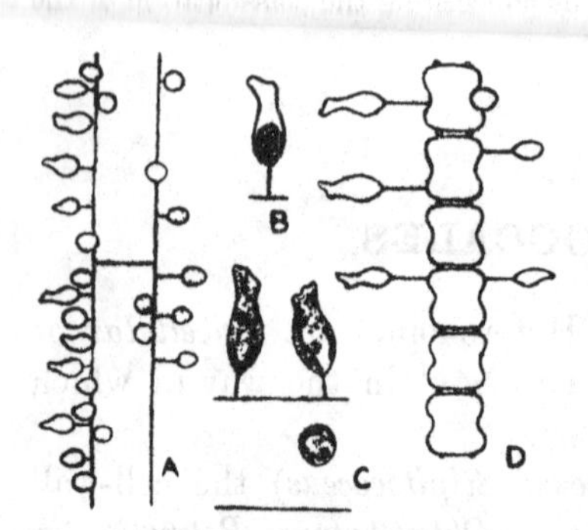

Fig. 255. *Stipitococcus urceolatus* W. & G. S. West. *A—C*, epiphytic on a filament of *Mougeotia*; *D*, epiphytic on *Sphærozosma excavatum*. *A* and *D*, ×500; *B* and *C*, ×780.

Isogametes occur in *Mischococcus*, in which genus the zygospore on germination divides in two directions in one plane, forming an epiphytic cushion, all the cells of which are situated on short, broad, mucilaginous stalks. This condition might almost be regarded as a palmella-state.

The genera are: *Peroniella* Gobi, 1887; *Stipitococcus* W. & G. S. West, 1898; *Mischococcus* Nägeli, 1849; *Chlorosaccus* Luther, 1898; *Racovitziella* De Wildeman, 1900 [= *Tetrasporopsis* Lemmermann & Schmidle (according to Wille); *Dictyosphæropsis* Schmidle, 1903].

Peroniella and *Stipitococcus* might be compared with *Physocytium* among the Palmellaceæ, and similarly a comparison might be instituted between *Mischococcus* and *Prasinocladus*.

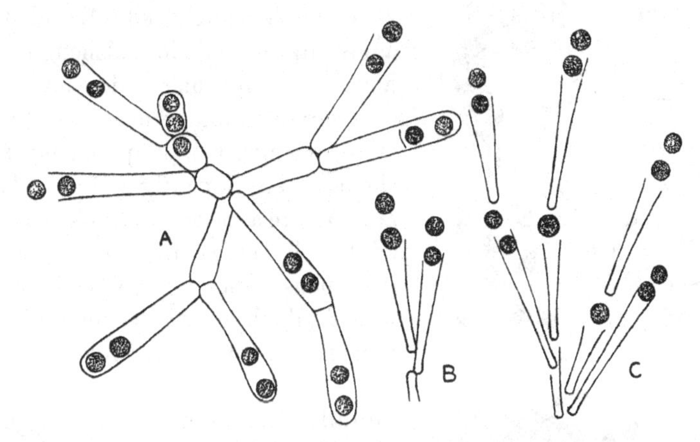

Fig. 256. *Mischococcus confervicola* Näg. *A*, entire colony; *B* and *C*, parts of other colonies. All × 500.

Family **Botryococcaceæ**.

It is highly probable that the Botryococcaceæ form a natural family and the genera included in it seem to belong to the Heterokontæ rather than to the Isokontæ. All the forms are colonial, the smallest colonies being those of *Stichoglœa* and the largest those of *Botryococcus*. The colonies are, as a rule, free-floating and they may occur in great numbers in the freshwater plankton. In *Stichoglœa* they usually consist of four or eight cells, enveloped in abundant but rather indistinct mucilage, each group of four cells being disposed in a somewhat irregular cruciform manner, and the poles of the cells being connected by firmer and thicker mucilaginous strands. There are two parietal chromatophores in each cell. A less-known genus is *Askenasyella* in which the colonies, although larger than in *Stichoglœa*, are still comparatively small. The cells are rounded or pear-shaped and arranged in a more or less radiating manner in the enveloping mucus. In each cell there is a single greatly hollowed chromatophore.

Botryococcus is the most important genus of the family and in some respects it is the most curious genus in the whole of the Chlorophyceæ. Its characters long remained obscure and to some extent they are yet very puzzling. It is a difficult Alga to examine with accuracy, with the result that it has been described under several generic names. It occurs as aggregates of botryoidal groups of cells, each group being so closely encased in a peculiar envelope that in the common species *Botryococcus Braunii* nothing can be seen of the structure of the colony by direct observation (fig. 257 *A* and *B*). Chodat ('96) has given the most complete account of *B. Braunii* yet published. Each botryoidal aggregate is rounded although slightly hollowed on its inner side, and it consists of a variable number of cells (20 to 30, or more) arranged as a peripheral layer within the outer envelope (fig. 257 *C*). The cells are obovate and somewhat elongated, each adult cell being embedded in a sort of gelatinous cupule. The outer envelope, which is doubtless a secretion of the underlying cells, is of a most irregular character, being folded, wrinkled and often produced into all manner of irregular lobes, processes and spines. This membrane stains strongly with fuchsin, but with chlor-zinc-iodine there is no colouration. Strand-like continuations of the membrane join together the various botryoidal groups of the colony.

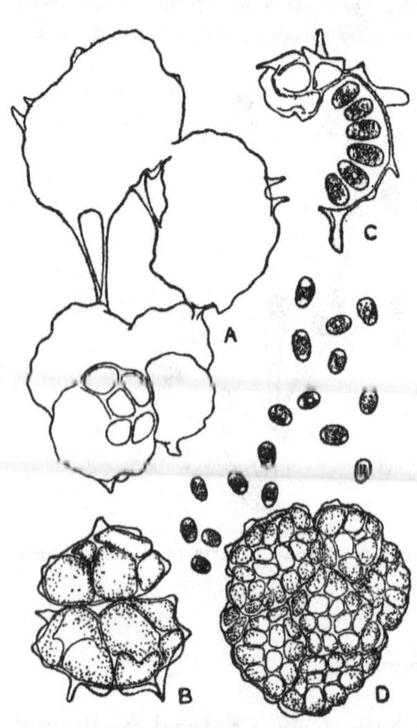

Fig. 257. *Botryococcus Braunii* Kütz. *A*, outline of medium-sized colony; *B*, smaller colony; *C*, part of single botryoidal group in section to show the cells within the outer envelope; *D*, a colony from which many cells have been extruded by pressure. All ×450.

Each cell contains one cup-shaped chloroplast, but this does not as a rule extend to the outer extremity of the cell. Small granules of starch have been detected in the cells and a variable quantity of oil is produced in each cell. This oil may sometimes pass out of the cells, in which case it adheres to the botryoidal groups and their connecting strands and adds to the general obscurity of the structure of the colony. In the late summer and autumn the whole colony may assume a brick-red colour owing to the formation of a pigment which is dissolved in the oil. The cells multiply exclusively by longitudinal division.

No motile state is known in any of the Botryococcaceæ.

The genera are: *Botryococcus* Kützing, 1849 [=*Ineffigiata* W. & G. S. West, 1897; em. 1903; *Botryomonas* Schmidle, 1899; *Botryodictyon* Lemmermann, 1903]; *Stichoglœa* Chodat, 1897 [inclus. *Oodesmus* Schmidle, 1902]; *Askenasyella* Schmidle, 1902 [=*Actinobotrys* W. & G. S. West, 1905].

Botryococcus Braunii Kütz. is a very abundant Alga, occurring in bogs, ditches, tanks, water-butts, ponds and lakes, but it is in the plankton of lakes that it attains its greatest development. It is equally abundant in both temperate and tropical regions, and whereas in bogs and pools the colonies rarely exceed 80—100 μ in diameter, in the plankton they may attain a diameter of 1 mm. (or even more). The large colonies are very oily and in calm weather they float in great numbers at the surface of the water. In the late summer, when the oil becomes brick-red, the waters of an entire lake may become tinged with a red colour. This red colouration of the waters of lakes owing to a vast abundance of colonies of *Botryococcus Braunii* has been observed in England, Switzerland and Central Africa.

Family **Chlorotheciaceæ**.

The Algæ of this family are all unicellular epiphytes, mostly gregarious in habit. The cells are subglobose, ovoid, ellipsoid or sometimes rather elongated and acuminate at the apex. They are attached by a stalk of variable length at the base of which is a more or less conspicuous disc. The stalk and disc form the basal part of the cell-wall, so that the attachment of the cells is of an entirely different character from that which occurs in *Peroniella* and *Stipitococcus*. Each cell contains from two to many parietal chromatophores. The two known genera of the family were described by Borzi and it is to him, especially in his later work (Borzi, '95), that we owe our knowledge of their life-histories. Cell-division does not occur. On the reproduction of *Chlorothecium* there is an increase in the size of the cell followed by a division of the protoplast which results in the formation of a number (24 or more) of aplanospores. These are liberated by the dissolution of the mother-cell-wall and at once become zoogonidangia, setting free either two or four zoogonidia. The latter come to rest and grow into new plants, or they may be facultative gametes and conjugate in pairs. In *Characiopsis* reproduction usually occurs by zoogonidia, of which four or eight are formed in each cell. At other times several or many aplanospores may be formed and, as in *Chlorothecium*, on liberation they immediately become gametangia which set free two or four gametes. In both genera the zygospores do not rest, but at once produce zoogonidia.

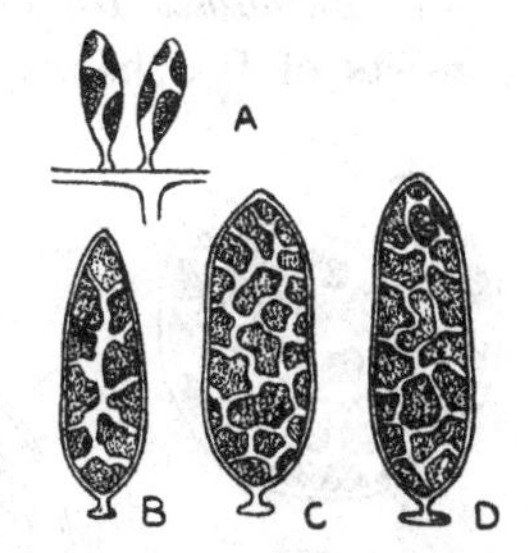

Fig. 258. *A*, *Characiopsis minuta* (A. Br.) Borzi. *B—D*, *Ch. turgida* W. & G. S. West. All × 500.

The genera are: *Chlorothecium* Borzi, 1885, and *Characiopsis* Borzi, 1895. Species of *Characiopsis* are not rare, but very little is known concerning *Chlorothecium* other than the observations of its original describer. Printz ('14) has recorded it from Norway. The only species is *Chlorothecium Pirottæ* Borzi and should it be shown that the cells of this Alga are able to give rise directly to zoogonidia then the genera *Characiopsis* and *Chlorothecium* would have to be united. As it is, the distinction between them is very slender. Borzi figures the zoogonidia with one cilium, as he also does in *Mischococcus*, but it is probable that the shorter cilium was overlooked.

Characiopsis greatly resembles the genus *Characium* in the Protococcales and the two genera have not only been greatly confused in the past, but they are by no means properly understood at the present time. (Compare Lemmermann, '14, and Printz, '14.)

Family **Chlorobotrydaceæ.**

This family includes a number of free-floating unicellular Algæ with firm cell-walls. In *Botrydiopsis, Pseudotetraëdron* and *Centritractus* the cells are solitary, in *Chlorobotrys* they occur in fairly regular groups of 2, 4 or 8 enveloped in mucilage, and in *Polychloris* they are so aggregated that they are sometimes angular by compression. The firmness of the cell-wall is undoubtedly a character of the family and, in *Chlorobotrys*, Bohlin ('01 B) states that the wall contains silica although it is not brittle. The amount of silica must, however, be very small; less, even, than in some Desmids[1].

In *Pseudotetraëdron* (*vide* Pascher, '13 A) and in *Centritractus* the cell-wall consists of two halves, one of which slightly overlaps the other. It would also appear from Bohlin's account of specimens from the Azores that the wall of the cylindrical cysts of *Chlorobotrys regularis* consists of two halves, but in the vegetative cells of *Chlorobotrys* the wall is continuous.

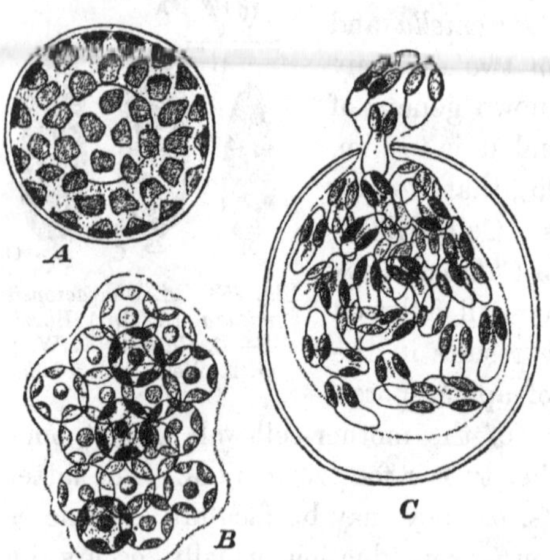

Fig. 259. *Botrydiopsis arrhiza* Borzi. *A*, vegetative cell; *B*, formation of aplanospores; *C*, formation of zoogonidia. All ×600 (after Borzi, from Wille).

The cells are globose except in *Pseudotetraëdron* and *Centritractus*; in the first-named genus they are rectangular when seen from the front and compressed when seen from the side or end. In this genus the cells are also furnished with four long bristles, one at each angle of the front view. In *Centritractus* the cells are cylindrical with somewhat swollen conical extremities, each of which is furnished with a long bristle. There are two to many parietal

[1] *Chlorobotrys* is very abundant in the bogs of the British Islands (from which situations it was first described as *Chlorococcum regulare* W. West) and the present author has found that boiling in fuming nitric acid dissolves the walls more easily than in some species of *Pleurotænium* and *Euastrum*.

chromatophores in each cell, fewest in *Centritractus* and most in *Botrydiopsis* and *Polychloris*. The chromatophores are sometimes discoidal, but at other times cushion-like. The cells are uninucleate.

Multiplication takes place by cell-division in three directions in *Chlorobotrys* and *Polychloris*, and in the former it is the usual method of propagation, motile cells being unknown. In this genus the families of 4, 8 or 16 cells are often very symmetrical, but beyond this number they become irregular. During division the contiguous walls of the daughter-cells are at first much flattened, but afterwards they gradually become more convex. In *Pseudotetraëdron* and *Chlorobotrys* 'cysts' occur, which on germination produce one or two individuals. In *Botrydiopsis* a number of spherical aplanospores are sometimes formed in each mother-cell (fig. 259 *B*). It is only in *Botrydiopsis* and *Polychloris* that zoogonidia occur, a large number being formed in each cell (fig. 259 *C*). These zoogonidia possess two or many chromatophores and in *Botrydiopsis* it has been shown that they possess two unequal cilia. In *Botrydiopsis* the aplanospores may at once become zoogonidangia, or, after a period of rest, they may become gametangia. The gametes have only one chromatophore and a pair of almost equal cilia.

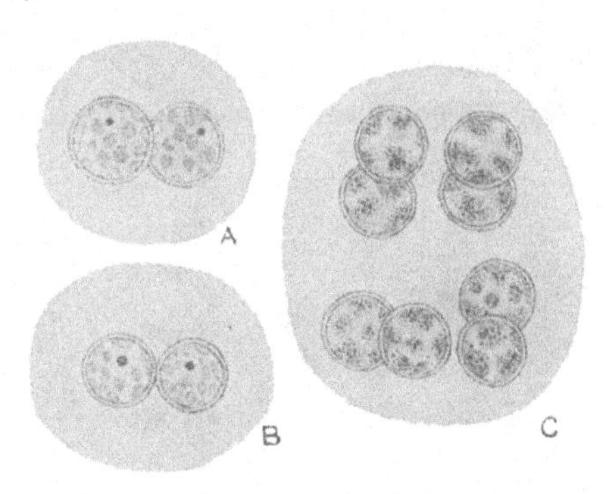

Fig. 260. *Chlorobotrys regularis* (West) Bohlin. Three colonies, ×450.

The genera are: *Botrydiopsis* Borzi, 1889; *Polychloris* Borzi, 1892; *Chlorobotrys* Bohlin, 1901; *Centritractus* Lemmermann, 1900; *Pseudotetraëdron* Pascher, 1913.

Chlorobotrys mostly occurs in *Sphagnum*-bogs. The other genera are usually found in small ponds and ditches, although *Centritractus* has been observed in the plankton.

It is possible that *Pelagocystis* Lohmann, 1903 [=*Clementsia* Murray, 1905] should be placed in this family. It is an Alga of the marine plankton occurring as small colonies, which are enveloped in a rounded mass of mucus. The cells are mostly arranged in pairs, or in fours, and each group shows a more or less distinct lamellation of the mucus immediately surrounding it. The cell-wall is firm, and there is apparently only one parietal chromatophore and an abundance of oil as a food-reserve. Division of the cells takes place in three directions and the larger colonies dissociate into smaller ones. The general disposition of the cells in the colony is very similar to that in *Chlorobotrys* although the colonies attain a larger size.

Family **Ophiocytiaceæ**.

The Algæ of this family are either free-floating or attached and they differ from all other members of the Heterococcales in the greatly elongated cells. The principal genus is *Ophiocytium*, in which the cells are generally many times longer than their diameter and almost always curved, in very many cases being spirally contorted (fig. 261 *A*). There is, as a rule, a distinct base and apex to each cell, the base being produced into a stalk or a short spine and the apex more or less distinctly swollen (fig. 261 *A*). Sometimes both ends of the cell are similar, either blunt or spined (fig. 261 *H* and *I*). There is at first a single nucleus, but in the larger and more elongate cells several nuclei have been detected. The chromatophores are rather large, parietal and (in optical section) H-shaped, being arranged in a series from end to end of the cell. Pyrenoids do not occur, but in some species oil-globules are a conspicuous feature of the cells. No vegetative division occurs and reproduction so far as is definitely known is entirely asexual. It takes place normally by the division of the protoplast to form ellipsoidal aplanospores (up to 16 in number), which escape from the apical end of the mother-cell by the detachment of a lid. More rarely zoogonidia are produced, 8 in each cell, and these also escape from the apex of the cell. The existence of gametes has been suspected, but is doubtful. In the attached forms, which were at one time placed in the genus *Sciadium*, the zoogonidia usually come to rest on the rim of the empty tube-like mother-cell and there grow into a colony of adult cells. Another generation may be formed in a like manner from each of these cells and thus a curious umbellate colony may be built up.

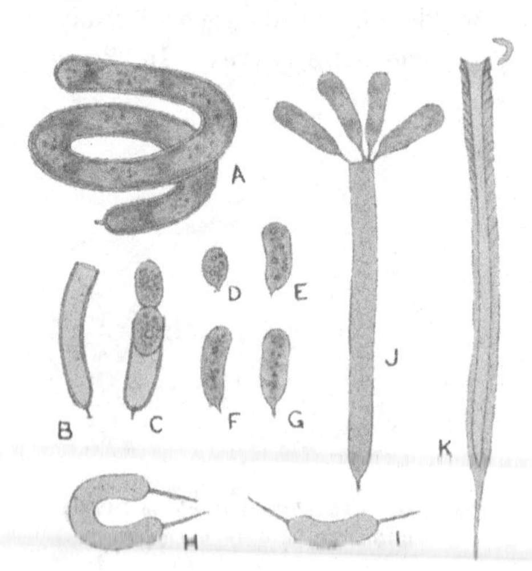

Fig. 261. *A*, *Ophiocytium majus* Näg. *B—G*, aplanospore-formation and germination of aplanospores in *O. cochleare* (Eichw.) A. Br. *H* and *I*, *O. bicuspidatum* (Borge) Lemm. forma *longispina* Lemm. *J*, *O. Arbuscula* (A. Br.) Rabenh. *K*, *O. graciliceps* (A. Br.) Rabenh., after treatment with potassium hydrate. *A—J*, × 450; *K*, × 570 (after Bohlin).

The structure of the cell-wall was carefully investigated by Bohlin ('97), who found that it was essentially similar to that exhibited by *Tribonema* (*vide* fig. 262).

In *Bumilleriopsis brevis* (Gerneck) Printz the cells are comparatively short, bent, and irregularly cylindrical, with parietal discoidal chromatophores. Moreover, the cell never has more than one nucleus.

The genera are: *Ophiocytium* Nägeli, 1849 [inclus. *Sciadium* A. Br., 1855] and *Bumilleriopsis* Printz, 1914. Species of *Ophiocytium* are common in both temperate and tropical countries, very often occurring in small ponds and pools in which there is a deficiency of aëration.

It seems probable that the genus *Actidesmium* Reinsch (1891) should be included in the Ophiocytiaceæ. This Alga forms free-floating colonies in which the primary mother-cells radiate from a central point. These cells are elongate-fusiform and each sets free 16 zoogonidia from its distal end. The latter come to rest on the open end of the mother-cell and there grow into new cells, the whole colony being reminiscent of *Ophiocytium Arbuscula*. The cells possess a parietal chromatophore and the food-reserve is oil.

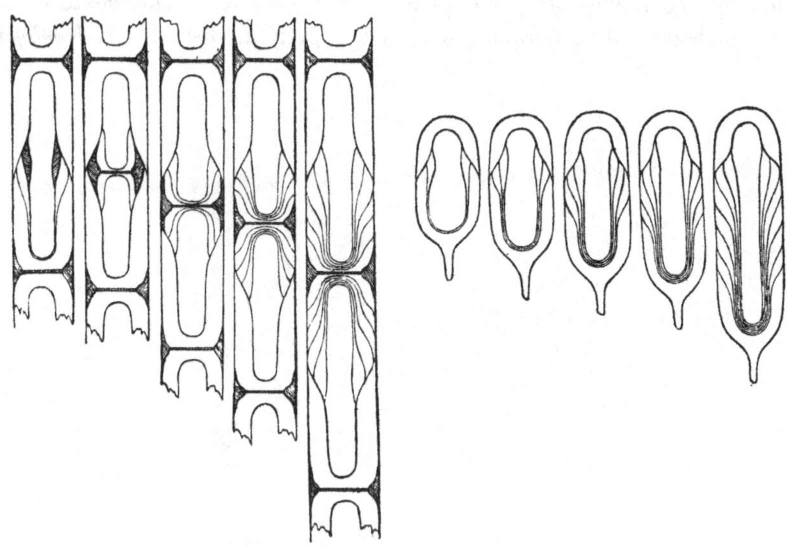

Fig. 262. Diagrams to illustrate the structure of the cell-wall in *Tribonema* and *Ophiocytium*. The series on the left represents cell-division and the growth of the cell-wall in *Tribonema*. On the right the growth of the wall in *Ophiocytium* is depicted. (After Bohlin.)

Order 2. **HETEROTRICHALES.**

All the Algæ of this order are *filamentous*, with or without a little enveloping mucus. Almost all forms of asexual reproduction occur within the order and gamogenesis has also been observed.

There is only one established family—the Tribonemaceæ—which is more or less equivalent to the Ulotrichaceæ among the Isokontæ.

Family **Tribonemaceæ.**

First established in 1904 (G. S. W., '04) this family is here limited to filamentous unbranched types of the Heterokontæ. *Tribonema* is one of the most abundant of the genera of Green Algæ, *T. bombycina* having a world-wide distribution. In *T. affinis* and other species the filaments are exactly cylindrical, but in *T. bombycina* the cells are normally a little barrel-shaped and not infrequently a trifle irregular in their growth (fig. 263 *A*). The apical cell may have a conical extremity or, as in *T. affinis*, it may be apiculate. The cell-wall is very firm and has a definite structure, readily breaking up into H-pieces in the genus *Tribonema*, but to a much less marked degree in *Bumilleria*. Each H-piece consists of a transverse wall with a cylindrical piece on either side, and the whole is composed of a number of apposed layers of pectic compounds (fig. 263 *G*). Each cell is thus bounded by the halves of two H-pieces. The cells are uninucleate or very rarely binucleate. The chromatophores are parietal discs in *T. bombycina*

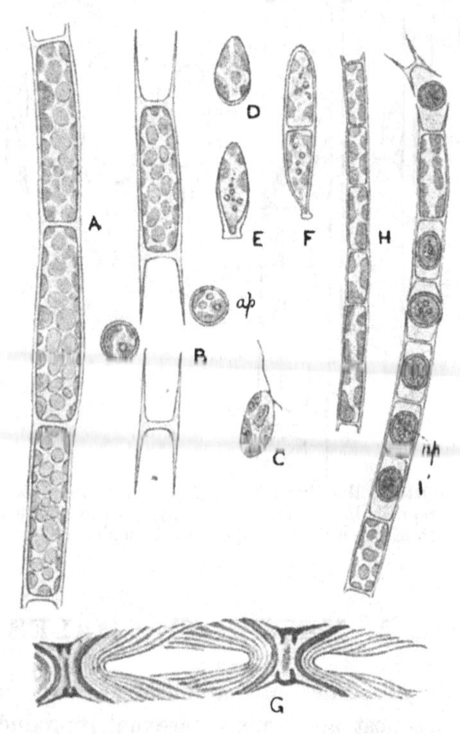

Fig. 263. *A—G*, ***Tribonema bombycina*** **(Ag.) Derb. & Sol.; *A*, part of vegetative filament; *B*, showing aplanospores (*ap*); *C*, zoogonidium; *D—F*, germination of zoogonidium and formation of young plants; *G*, showing structure of cell-wall after treatment with potassium hydrate. *H* and *I*, *T. bombycina* forma *minor* (Wille) G. S. West. *G*, × 570 (after Bohlin); *A—F*, *H* and *I*, × 450.**

and most other species (consult fig. 263 *A* and *B*). *It is this discoidal character of the parietal chromatophores which at once distinguishes the common species of* Tribonema *from all other Green Algæ.* In *T. affinis* the chromatophores, although parietal, are few in number and very irregular in shape. The stored reserve is oil, which is scattered in small globules through the cell.

Asexual reproduction takes place by the formation of globular or ellipsoid aplanospores, which escape by the breaking up of the filament (fig. 263 *B* and *I*); also by zoogonidia with a pair of unequal cilia and two or more parietal chromatophores (fig. 88 *C*; fig. 263 *C*). Either one or two zoogonidia may arise from a single cell; the inequality of the cilia was first demonstrated by Bohlin ('97). In the escape of the aplanospores and zoogonidia the H-pieces of the cell-wall fall apart, thus causing a dislocation of the filaments.

Gametes are only rarely produced. They are isogamous although conjugation is to some extent anisogamous, since it is stated that one gamete comes to rest before the other swarms up to it and fuses with it.

In *Bumilleria* the filaments are slender and somewhat delicate, and the cell-wall is not so distinctly built up of H-pieces; neither do the latter exhibit the special structure which is so characteristic of the wall of *Tribonema.* On the other hand, the filament may possess a by no means insignificant

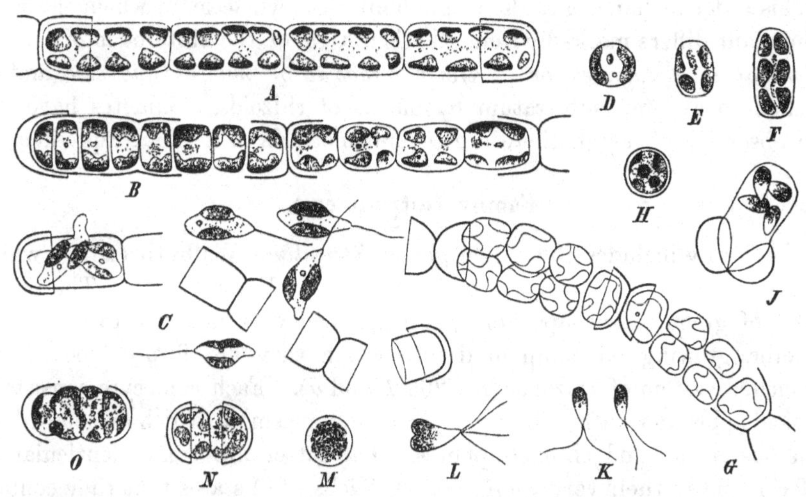

Fig. 264. *Bumilleria sicula* Borzi. *A*, vegetative filament; *B*, filament showing cell-division; *C*, escape of zoogonidia; *D—F*, germination of zoogonidia; *G*, formation of gamete-mother-cells; *H*, an isolated gamete-mother-cell; *J*, formation of gametes; *K*, gametes; *L*, fusion of gametes; *M*, zygospore; *N* and *O*, germination of zygospore. × about 500. (From Wille, after drawings by Borzi.)

N.B. The zoogonidia are depicted with only one cilium, but this is probably an error in the original observations.

mucous sheath, continuous in character and consisting of the pectic constituents of the wall. Much investigation is still required concerning the zoogonidia and gametes of this genus.

The genera are: *Tribonema* Derbes & Solier, 1856 [= *Conferva* as defined by Lagerheim, 1888]; *Bumilleria* Borzi, 1895. For full reasons for the abandonment of the generic name '*Conferva*' consult Hazen ('02) or West (G. S. W., '04). *Tribonema bombycina* (with its forma *minor*) is a common Alga in all countries and it occurs in very varied habitats.

Bumilleria is much less frequent than *Tribonema*, but is occasionally found in considerable quantity in small ponds.

The genus *Monocilia* Gerneck, '07 [= *Heterococcus* Chodat, '08] should have a place in the Heterotrichales. It is a small Alga with very short branched filaments, the cells of which have thin walls and a few parietal discoidal chromatophores. Owing to the branched character of its filaments and the homogeneous nature of its thin cell-walls it is not possible to include it in the Tribonemaceæ. It should rather be placed in another family—the **Monociliaceæ**. The food-reserve is oil and the zoogonidia have two unequal cilia. A 'palmella-state' has been observed in this Alga. There are two described species, *M. viridis* Gerneck and *M. flavescens* Gerneck.

Order 3. HETEROSIPHONALES.

This order includes only the single family Botrydiaceæ, in which the form of the plant differs markedly from all other Algæ of the Heterokontæ. *Each individual is a rounded or pyriform cœnocyte of macroscopic size* and is attached to a damp substratum by means of rhizoids. Gametes have not been observed, although there are several methods of asexual reproduction.

Family **Botrydiaceæ**.

The family includes the single genus *Botrydium* Wallroth (1815), which was first accurately investigated by Rostafinski & Woronin ('77). The plants consist of green pear-shaped or spherical cœnocytes, about 1 to 2 mm. in diameter, growing on damp mud into which they are firmly 'rooted' by a branched system of rhizoids (fig. 265 *1* and *5*). Each cœnocyte consists of a vesicular bladder with a lining layer of cytoplasm in which are embedded numerous nuclei and chromatophores. The latter are small, lenticular or fusiform, and in their very young stages, Klebs ('96) states that they contain 'pyrenoid-like' bodies. The chromatophores are situated in the outer part of the lining cytoplasm and in full-grown plants they are usually arranged in several layers, the nuclei being internal to them. Starch is not formed. The rhizoids possess protoplasmic contents and many nuclei, but normally they contain no chromatophores. Wager has observed mitotic division of the nuclei in the rhizoids and states that the chromatic substance appears to reside wholly in the nucleolus.

Reproduction is purely asexual and may occur in a variety of ways, depending largely upon the conditions of environment, any change of conditions usually resulting in a corresponding variation of the reproductive process. Rostafinski & Woronin ('77) worked out the different methods of asexual reproduction and the final result in every case is the production either of zoogonidia or of 'cysts' (sometimes termed 'aplanospores'), the latter often becoming hypnocysts. The whole vesicular plant becomes a huge zoogonidangium when submerged in water and the zoogonidia escape in great numbers from an apical opening (fig. 265 *2*). The zoogonidia are small and ovoid in shape, with two chromatophores and two unequal cilia. If the plants are wet, but not submerged, the zoogonidia do not swarm out, but round themselves off as non-motile gonidia, each of which can grow into a new plant. If, on the other hand, the plants become dry, such as when

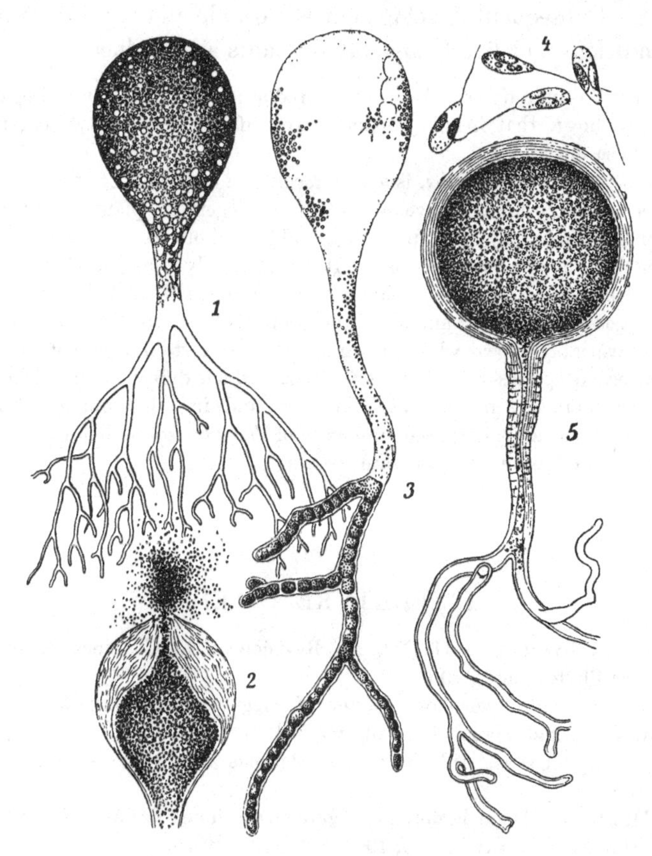

Fig. 265. *1—4*, *Botrydium granulatum* (L.) Grev. *1*, vegetative plant; *2*, the escape of zoogonidia; *3*, the formation of 'cysts' in the rhizoids; *4*, four zoogonidia. *5*, *Botrydium Wallrothii* Kütz. *1—3* and *5*, × about 50; *4*, × about 800. (After Rostafinski & Woronin, from Oltmanns.)

N.B. The zoogonidia are here depicted with only one cilium, but it has been definitely shown that they possess two of unequal length.

exposed to strong sunshine, the cytoplasm with the nuclei and chromatophores passes down into the rhizoids and there becomes divided up into a number of separate portions which surround themselves with new walls and become 'cysts' (fig. 265 *3*). The cysts may develop in more than one way. If submerged in water they form zoogonidia at once; if placed on damp earth in the light each cyst puts out a rhizoid and develops directly into a new plant; if the cysts remain in the mud in the dark they become hypnocysts and may rest for a long time, although when ultimately moistened they at once become zoogonidangia. The zoogonidia always begin development in the same way. On coming to rest they become rounded off and increase in size. The nucleus begins division and there is soon a great increase in the number of both nuclei and chromatophores. At this stage a delicate rhizoid is developed. Subsequent development is variable and depends largely upon external conditions, but finally vegetative plants are produced.

Rostafinski & Woronin described a sexual reproduction by the fusion of isogametes, but Klebs ('96) has shown that in all probability the life-histories of two organisms were confused by those authors.

Botrydium granulatum (L.) Grev. is a very local Alga and is not often observed because the conditions necessary for the development of the vegetative plants only rarely obtain. It is found almost exclusively upon drying-up mud turned out from ditches, canals, etc., or on the drying bottom of a muddy pond. It seems highly probable that some form of resting-spore of *Botrydium*, probably a hypnocyst, is universally distributed in the mud of fresh waters in temperate regions and also in many parts of the tropics, since the vegetative plants almost invariably appear when such mud attains a certain degree of dryness.

There is a second species—*B. Wallrothii* Kütz.—which differs in its thicker lamellate cell-wall, in the characters of the rhizoidal part, and in other minor details. Klebs regarded this species as a hypnosporangium-state of *B. granulatum*, but Iwanoff has given good reasons for its retention as a separate species.

LITERATURE CITED

BLACKMAN, F. F. & TANSLEY, A. G. ('02). A Revision of the Classification of the Green Algæ. New Phytologist, i, 1902.

BOHLIN, K. ('97). Studier öfver några slägten af Alggruppen Confervales Borzi. Bihang till K. Sv. Vet.-Akad. Handl. Bd. xxiii, no. 3, 1897.

BOHLIN, K. ('01 A). Utkast till de Gröna Algernas och Arkegoniaternas Fylogeni. Upsala, 1901.

BOHLIN, K. ('01 B). Étude sur la flore algologique d'eau douce des Açores. Bihang till K. Sv. Vet.-Akad. Handl. xxvii, no. 4, 1901.

BORZI, A. ('95). Studi Algologici. II. Palermo, 1895.

CHODAT, R. ('96). Sur la structure et la biologie de deux Algues pélagiques. Journ. de Botanique, 1896.

CHODAT, R. ('08) in Bull. de l'Herb. Boiss. 1908, p. 81.

DERBES, A. & SOLIER, A. J. J. ('56). Mémoire sur quelques points de la physiologie des Algues, 1856.

GARDNER, N. L. ('10). Leuvenia, a new genus of Flagellates. Univ. of California Publications, Botany, iv, no. 4, May 1910.

HAZEN, T. E. ('02). The Ulotrichaceæ and Chætophoraceæ of the United States. Memoirs Torr. Bot. Club, xi, no. 2, 1902.

HEERING, W. ('06). Die Süsswasseralgen Schleswig-Holsteins usw. 1 Teil: Einleitung.—Heterokontæ. Jahrb. Hamburgischen Wiss. Anstalten, xxiii, 1905 (1906).

Klebs, G. ('96). Bedingungen der Fortpflanzung bei einigen Algen und Pilzen. Jena, 1896.

Lagerheim, G. ('89). Studien über die Gattungen *Conferva* und *Microspora*. Flora, lxxii, 1889.

Lemmermann, E. ('03). Beiträge zur Kenntnis der Planktonalgen. XV. Das Phytoplankton einiger Plöner Seen. Forschungsber. aus der Biol. Stat. zu Plön, x, 1903.

Lemmermann, E. ('14). Algologische Beiträge. XII. Die Gattung Characiopsis Borzi. Abh. Nat. Ver. Brem. xxiii, 1914.

Luther, A. ('99). Ueber *Chlorosaccus* eine neue Gattung der Süsswasseralgen, nebst einigen Bemerkungen zur Systematik verwandter Algen. Bihang till K. Sv. Vet.-Akad. Handl. Bd. 24, Afd. iii, no. 13, 1899.

Murray, G. ('05). On a New Genus of Algæ, Clementsia Markhamiana. Geographical Journal, Febr. 1905.

Pascher, A. ('13 a). Die Heterokontengattung Pseudotetraëdron. Hedwigia, liii, 1913.

Pascher, A. ('13 b). Zur Gliederung der Heterokonten. Hedwigia, liii, 1913.

Pascher, A. ('14). Über Flagellaten und Algen. Ber. Deutsch. Botan. Ges. xxxii, 1914.

Printz, H. ('14). Kristianiatraktens Protococcoideer. Videnskapsselskapets Skrifter. I. Mat.-naturv. Klasse, 1913, no. 6. (Christiania, 1914.)

Rostafinski, J. & Woronin, M. ('77). Über *Botrydium granulatum*. Bot. Zeitung, xxxv, 1877.

Schmidle, W. ('99). Über Planktonalgen und Flagellaten aus dem Nyassasee. Engler's Botan. Jahrbüch. xxvii, 1899.

West, G. S. (G. S. W., '04). A Treatise on the British Freshwater Algæ. Camb. Univ. Press, 1904.

West, W. & West, G. S. ('97). A Contribution to the Freshwater Algæ of the South of England. Journ. Roy. Micr. Soc. 1897.

West, W. & West, G. S. ('03). Notes on Freshwater Algæ. III. Journ. Bot. Febr. & Mar. 1903.

THE OCCURRENCE AND DISTRIBUTION OF FRESHWATER ALGÆ

FRESHWATER Algæ are universal in their occurrence, no moist situation being without some type of Alga. They are found on damp earth, rocks, walls, tree-trunks, etc.; they are met with in all kinds of running water, from the torrent and the waterfall to the slowest river; but it is in the still waters of pools and lakes that they exhibit the greatest diversity and attain their maximum abundance. They occupy, therefore, very varied habitats, and *it is because habitat plays such an important part in both the occurrence and distribution of freshwater Algæ that it is here made the basis of the treatment of the subject*[1].

The ecology of freshwater Algæ is still so much in its infancy that a satisfactory ecological summary is impossible with our present knowledge. The following account of the *occurrence and distribution* of these plants is based largely upon the author's wide experience of them during the past twenty years or more, although the views of other authors have been carefully considered and reference is constantly made to them[2]. In view of our limited knowledge of the factors controlling the distribution and periodicity of freshwater Algæ generalizations are well nigh impossible and certainly very unwise.

Special ecological terms have been avoided as far as possible, since in dealing with freshwater Algæ they are so easily mis-applied. Most of the algal vegetation of fresh waters can be regarded as forming associations of a more or less definite character, the peculiarities of which are the direct result of habitat and the nature and amount of the dissolved salts in the water. The latter is a factor of great importance and is to a large extent

[1] Some of the main features of this chapter were explained by the author in a lecture delivered at University College, London, on Feb. 18th, 1915.

[2] Many of the conclusions put forward by various authors in recent years have been very erroneous and do not stand the test of enquiry. This is mainly owing to deductions having been made from very limited investigations and to want of precise knowledge of the Algæ dealt with. To make any reliable contribution to our knowledge of algal ecology it is essential that the author should have an extensive grasp of the taxonomics of the various groups, since to discuss the distribution and periodicity of *genera*, as is so often done, is in most cases quite futile and merely leads to confusion.

dependent upon the geological formation. In both small and large bodies of water the algal associations often change with the seasons, so that a succession of associations may occur in the same habitat.

In recognition of the primary importance of habitat the subject is dealt with under the following headings:

I. Subaërial Associations.
II. Associations of Irrorated (or dripping) Rocks.
III. Aquatic Associations.

I. Subaërial Associations.

Many freshwater Algæ live under subaërial conditions, having adapted themselves to a life in a damp atmosphere. Most of them are able to survive a considerable period of desiccation, and it is amongst these subaërial types that one meets with what is, perhaps, the nearest approach to an actual plant-formation among freshwater Algæ.

1. *Protococcus*-formation. This is the most general and at the same time most distinctive algal 'formation' in north temperate regions. It consists of a bright green incrustation of *Protococcus viridis* Ag. (= *Pleurococcus vulgaris* auct.) covering the windward side of tree-trunks, branches, walls, palings, etc. It is most conspicuous in those areas in which the annual rainfall exceeds 30 inches, but may be almost equally abundant in low-lying damp areas with a less rainfall, as in the fen-districts of the east of England. The Alga is perennial and in some cases has a slight admixture of *Stichococcus bacillaris* Näg. In parts of Canada the latter may almost entirely replace the *Protococcus*.

2. *Zygnema ericetorum*-formation. Extensive felt-like mats of this Alga occur on the surface of the ground on heaths and moors, more particularly where the soil is peaty. On almost pure peat-soil the algal cells develop phycoporphyrin so that the mats assume a purple colour. In the British Islands *Zygnema ericetorum* occurs at all elevations, but is seen in greatest abundance on peat-moors, such as those of the Pennine Chain. On damp heaths the Alga is not so conspicuous. It is perennial and can be found at all times of the year, although its maximum activity is in the late spring. As a rule it remains throughout the year in a purely vegetative condition, but in the winter months aplanospores may be formed (West & Starkey, '15).

3. *Prasiola*-formation. This is very limited in extent except on certain sea-coasts. *Prasiola crispa*, which is the commonest species, is a perennial Alga able to withstand much desiccation, especially in the '*Hormidium*-state.' It requires an abundance of nitrogen and is a common Alga in towns and villages. It sometimes occurs in quantity in the haunts of sea-birds,

27—2

and in the Antarctic continent it forms a veritable carpet on the sloping ground below the penguin rookeries (W. & G. S. W., '11).

4. *Leaf- and bark-epiphytes.* Prominent amongst these is the genus *Trentepohlia*, the abundant presence of which depends entirely upon rainfall. There are many species of the genus, about two-thirds of the number being

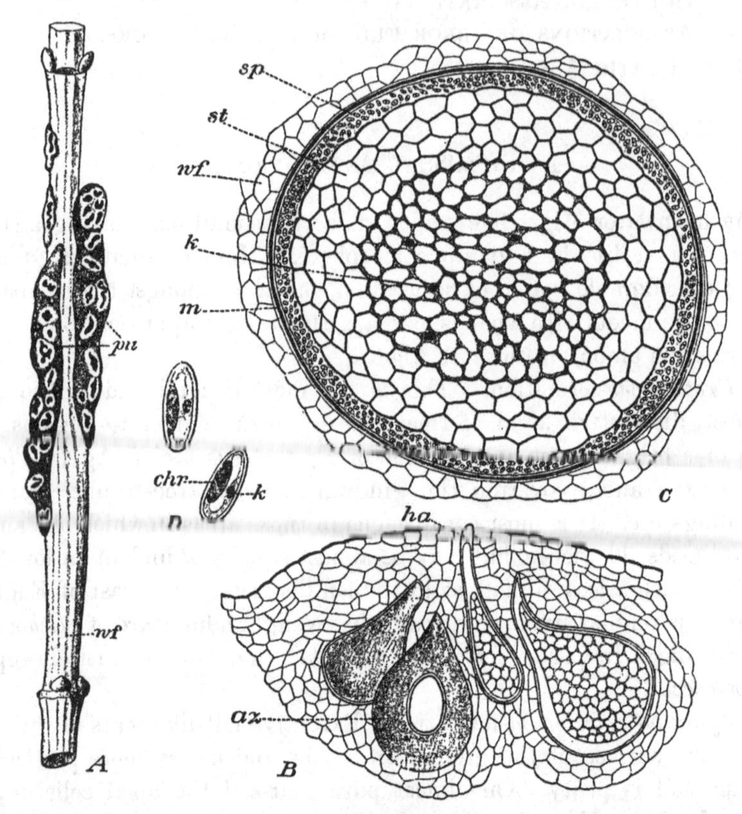

Fig. 266. *Phytophysa Treubii* Weber van Bosse. *A*, stem of *Pilea* (*wf*) with pustules (*pu*) caused by algal parasite; *B*, section through part of one of the galls showing four of the flask-shaped cœnocytes (*az*) of the Alga with their necks (*ha*) projecting outwards; *C*, algal cœnocyte during spore-formation; *k*, nuclei; *m*, wall of cœnocyte; *sp*, aplanospores; *st*, sterile cells; *wf*, some of the surrounding cells of the host-plant; *D*, two aplanospores; *chr*, chromatophore; *k*, nucleus. *A*, nat. size; *B*, × 36; *C*, × 49; *D*, × about 700. (After A. Weber van Bosse, from Wille.)

tropical epiphytes, and it is in the damp tropical woodland areas that species of *Trentepohlia* attain their greatest luxuriance. In temperate countries they are frequent in those districts with an annual rainfall of upwards of 40 inches. In western Europe several species occur on the trunks and branches of trees in woodlands at no considerable altitude. It is a curious fact, however, that *T. aurea*, which is much the commonest species in the

British Islands, occurs almost entirely on the vertical faces of rocks facing the 'drive' of the wind. This species ascends to 1200 feet in the Pennine Chain. In the tropics many species of *Trentepohlia* are leaf-epiphytes.

Belonging to the same family, the Trentepohliaceæ, are the genera *Phycopeltis* and *Cephaleuros*, most of the species of which are epiphyllous. One species of *Phycopeltis* occurs in Europe, but all the known species of *Cephaleuros* are tropical and one of them, *Cephaleuros virescens*, is a destructive parasite (consult p. 310). In some species of all the genera of the Trentepohliaceæ the cells contain a quantity of the red pigment hæmatochrome and are thus able to live in situations where they are exposed to light of strong intensity.

Phyllosiphon (fig. 156) and *Phytophysa* (fig. 266), both of which belong to the Phyllosiphonaceæ, have become partial parasites, the former in the leaves of various genera of the Araceæ and the latter in the stems of *Pilea*, a tropical genus of the Urticaceæ.

In damp tropical areas leaf- and bark-epiphytes are largely composed of the Myxophyceæ. On the bark of trees they often occur amongst a more or less prolific growth of Bryophytes. There are various associations in which the following genera are abundantly represented: *Hapalosiphon*, *Stigonema*, *Scytonema*, *Schizothrix*, *Phormidium*, *Chroococcus* and *Glœocapsa*. Other genera are also represented to a less degree. Fritsch ('07 B and '07 C) has remarked upon the abundance of such associations in Ceylon, but it seems that in certain of the mountainous islands of the West Indies these Blue-green associations attain their maximum development (consult W. & G. S. W., '94; '99) and amongst the tangled mass of threads many other Algæ also occur. In the island of Dominica various genera of desmids, diatoms and of the Protococcales habitually live on trees!

One minute diatom—*Navicula contenta*—is chiefly found as an epiphyte on the leaves of trees.

5. *Miscellaneous 'formations' and associations on rocks and on damp ground.* Among the Blue-green Algæ there are several more or less definite subaërial formations.

In the north-west of Scotland and in the Hebrides there is a *Glœocapsa magma*-formation on the damp ground. It is sometimes sufficient in amount to give a distinct reddish-brown colouration to wide areas.

In West Africa *Porphyrosiphon Notarisii* may carpet the ground extensively, covering wide areas with a reddish-brown felt. In the same part of the world *Scytonema Myochrous* var. *chorographicum* is the cause of the 'pedras negras' of Angola (*vide* Welwitsch, '68; W. & G. S. W., '97).

In temperate climates many kinds of subaërial Algæ occur on the ground and on damp rocks and stones. Much the most frequent Blue-green Alga is *Phormidium autumnale*, which is found almost everywhere on damp ground and stones as a bluish-black stratum. *Symploca muscorum* is frequent on mossy ground and *Nostoc commune* often occurs in quantity on the surface of cultivated land after a period of damp weather.

Various species of *Stichococcus* frequently form a thin green stratum on walls, damp stones and wooden palings, and *Ulothrix æqualis* is frequent on damp shady banks. *Coccomyxa subellipsoidea* occurs as a mucous green stratum on damp rocks and stones, particularly on the softer sandstones. *Mesotænium macrococcum* occurs on damp rocks and *M. caldariorum* sometimes forms a stratum on the ground under the shade of trees. *Vaucheria terrestris* and *V. hamata* are common Algæ on damp soil, forming bright green felts similar in outward appearance to moss-protonema.

Porphyridium cruentum, which is a primitive member of the Bangiales, is a familiar object at the base of damp walls and on the damp flag-stones of cold greenhouses, old churches, etc.

II. Associations of Irrorated Rocks.

There are many associations of Algæ on rocks which are kept constantly wet by trickling water. *Such Algæ are really submerged, but with the maximum of aëration.* It is in damp mountainous regions that algal associations of this kind occur in abundance and they are especially well marked in the mountainous areas of the British Islands where the rainfall varies from 40 to upwards of 100 inches. *The Myxophyceæ (or Blue-green Algæ) are for the most part dominant.* Pure algal strata are not uncommon and often cover several square yards of rock-surface, but mixed associations with Bryophytes are more frequent. These associations are very widespread in the damper parts of both temperate and tropical countries, and it is not too much to state that a great deal of the colouration of the landscape in damp temperate countries results from associations of Blue-green Algæ. The same is true of the damper parts of the tropics, but on the whole to a less degree, not because of the fewer Blue-green Algæ, but because their colouration there enters into a competition with coloured foliage such as does not occur in temperate climates.

It is on the dripping rocks in the deep glens and gulleys of mountainous areas that the Blue-green Algæ are most abundant. Many species of *Nostoc* are found among the wet mosses, certain of the larger species of *Stigonema* occur in tufted masses, and *Scytonema Myochrous* and *S. mirabile* often form thick mats with bristling upstanding branches. Many species of *Phormidium* occur in quantity, sometimes in pure sheets covering many square yards, as may be the case with *Ph. purpurascens* (*vide* W. & G. S. W., '01). Various species of *Glœocapsa* and *Chroococcus* are often found in pure gelatinous masses, although they are frequently abundant among filamentous types.

Many genera other than those mentioned are to be found, sometimes sparingly and occasionally in abundance, giving an extremely varied character to these Blue-green associations.

Desmids are not uncommon on dripping rocks and some are particularly characteristic of such habitats. They usually occur amongst the wet Bryophytes, but they not infrequently form pure gelatinous masses. The following are some of the typical wet-rock desmids: *Mesotænium chlamydosporum, M. De Greyii, M. macrococcum, Cosmarium anceps, C. cymatopleurum* var. *tyrolicum, C. didymochondrum, C. dovrense, C. Etchachanense, C. Holmiense, C. microsphinctum, C. nasutum, C. pseudarctoum, C. speciosum, C. sphalerostichum, C. subexcavatum* var. *ordinatum, C. tumens, Staurastrum Meriani* and *St. pileolatum.*

A few diatoms are also typically wet-rock types, and among them *Melosira arenaria* and *Navicula borealis* deserve special mention. The former often occurs as coarse mats on dripping sandstone rocks and the latter, although abundant in various habitats in high latitudes or at altitudes of over 1000 feet, is in some areas a conspicuous feature of the algal associations which occur among the mosses of wet rocks. *Melosira Roeseana* occurs in similar habitats.

In limestone areas certain of the Blue-green Algæ build up calcareous deposits. *Dichothrix gypsophila* is a notable example (consult p. 34 and fig. 21).

Mention should not be omitted of *Hildenbrandtia rivularis*, which on irrorated rocks forms red incrustations of a very striking character.

III. Aquatic Associations.

The truly aquatic associations of freshwater Algæ may be dealt with under four headings:

A. Swiftly running water.
B. Bogs and swamps.
C. Ponds and ditches.
D. Pools and lakes.

Unlike the formations of terrestrial plants, many aquatic associations vary greatly from season to season. There is usually a marked periodicity, a number of dominant forms succeeding one another in the course of twelve months. Thus, such associations have their different phases and pass through an annual cycle.

In other cases the same association may occur all the year round, with little variation for a number of years. This is well exemplified in *Sphagnum*-bogs, possibly owing to the relatively uniform conditions which obtain in such habitats. Most desmids, for instance, are perennial and, although fewer in numbers in the winter months, a percentage of individuals invariably survives that period in the vegetative state. The same is true of freshwater diatoms, but to a more marked degree, since these organisms are mostly cold-water types.

Where there is a succession of dominant types the cycle is controlled by many diverse factors. There are seasonal factors, such as temperature, intensity of light, amount of dissolved oxygen, etc.; and various other determining factors, among which may be mentioned the geological formation (and its effect upon the chemical composition of the water), altitude, and, in the case of lakes, the nature of the banks, whether marshy or rocky.

A. **Algal Associations of Swiftly Running Water.**

In this category are all those Algæ which inhabit swift rivers, cataracts and waterfalls. Among the most characteristic are certain of the freshwater Rhodophyceæ, the genera *Lemanea*, *Sacheria* and *Chantransia* being especially noteworthy. The two first-named genera occur only in the most rapid torrents and in waterfalls, always where the force of the water is greatest. The most abundant species in temperate countries is *Sacheria mamillosa*, an Alga which often occurs in artificial torrents such as mill-sluices. Species of *Batrachospermum* may occur in running water, but are found abundantly in pools and lakes, and even in bogs.

Of the Chlorophyceæ, several species of *Cladophora* are abundant, more especially *Cl. glomerata* and *Cl. fracta*. *Vaucheria geminata* is frequently abundant where running water overflows rocks and boulders, and also where the waters of a spring irrigate mossy ground. That most extraordinary of all desmids—*Oocardium stratum*—occurs in swift mountain streams in limestone areas, forming small opaque white, encrusted pilules attached by much mucus to submerged rocks and stones. *Microspora amœna* is also a frequent Alga in swiftly running streams, more especially in the early spring or late autumn.

Many attached diatoms occur in running water especially some of the stalked species of *Cymbella* and *Gomphonema*. The most notable is *Gomphonema geminatum* which often forms greyish felt-like masses attached to the rocks of mountain streams and cataracts.

* * * * * * * * *

Here should be mentioned the Algæ of Hot Springs, since they occur in running water, but the water is at a comparatively high temperature and in many cases sulphurous. They are mostly Blue-green Algæ and have been found to occur in water at over 80° C. (*vide* p. 34). In the warm streams flowing from these springs several species of *Rhizoclonium* occur and also *Zygnema ericetorum*.

B. **Algal Associations of Bogs and Swamps.**

All the algal associations of bogs and swamps are mixed associations with a less marked periodicity than in ponds, pools or lakes. They differ much in character according to geological formations, local conditions (such as prevalence of iron salts), altitude, etc.

The commonest type of bog in temperate areas is the *Sphagnum*-bog, although species of *Hypnum* play an important part in upland districts. An average mixed association consists of various species of the Zygnemaceæ in small quantity, certain desmids, the larger species being mostly of the ubiquitous type, many diatoms, a few representatives of the Protococcales, certain of the Blue-green Algæ and sometimes one or two members of the Peridinieæ. Of the Zygnemaceæ, a few species of *Zygnema* and *Mougeotia* are not infrequently met with. The most conspicuous desmids are *Closterium Lunula, Cl. turgidum, Cl. didymotocum, Cl. striolatum, Cl. juncidum, Cl. gracile, Penium margaritaceum, P. spirostriolatum, Euastrum crassum, E. ampullaceum, E. Didelta, E. pectinatum, Micrasterias denticulata, M. rotata, M. papillifera, M. truncata, Xanthidium armatum, Cosmarium tetraophthalmum, C. margaritiferum, C. Cucurbita, Staurastrum hirsutum, St. muricatum, St. Reinschii, St. brachiatum, Tetmemorus granulatus, T. lævis, Hyalotheca dissiliens, Gymnozyga moniliformis* and many others. The diatoms are especially represented by the larger species of *Navicula* (*N. nobilis, N. major, N. viridis, N. Iridis* and others) and various species of *Eunotia* (*E. Arcus, E. majus, E. tetraodon*, etc.). *Stauroneis Phœnicenteron* is very frequent and *Vanheurckia rhomboides* var. *saxonica* is often present in great abundance, sometimes to the comparative exclusion of other types. Of the Protococcales, *Oocystis solitaria* and *Eremosphæra viridis* are the most noteworthy, the latter sometimes occurring in abundance among some of the larger desmids. The only representative of the Heterokontæ of any importance is *Chlorobotrys regularis*, which may occur in considerable quantity. A number of Blue-green Algæ habitually occur in *Sphagnum*-bogs, of which may be mentioned *Stigonema ocellatum, Hapalosiphon hibernicus, Cylindrospermum stagnale*, several species of *Anabæna, Synechococcus major* and *Chroococcus turgidus. Glenodinium uliginosum*, one of the Peridinieæ, is also a common organism amongst submerged *Sphagnum*, and there is evidence to show that it is profoundly influenced by temperature (consult fig. 267; and G. S. W., '09 A).

Cedergren ('13) has suggested that there are two decided subassociations in a *Sphagnum*-bog, a *Sphagnetum desmidiosum* characterized by the abundance of desmids and a *Sphagnetum naviculosum* characterized by the dominance of the larger species of naviculoid diatoms. These suggested subassociations scarcely hold good, however, since the larger desmids and naviculoid diatoms so frequently occur together in abundance, evidently thriving under the same conditions.

In some of the more spongy *Sphagnum*-bogs *Utricularia minor* is abundant[1], and this at once adds to the richness of the alga-flora. It harbours amongst its leaves more desmids and a greater variety of them than almost any other aquatic macrophyte. Also, if the bogs are on the older geological formations they have a much richer and more varied desmid-flora.

Altitude has a decided effect upon the Algæ of a *Sphagnum*-bog. In north temperate latitudes the desmid-flora changes considerably with the

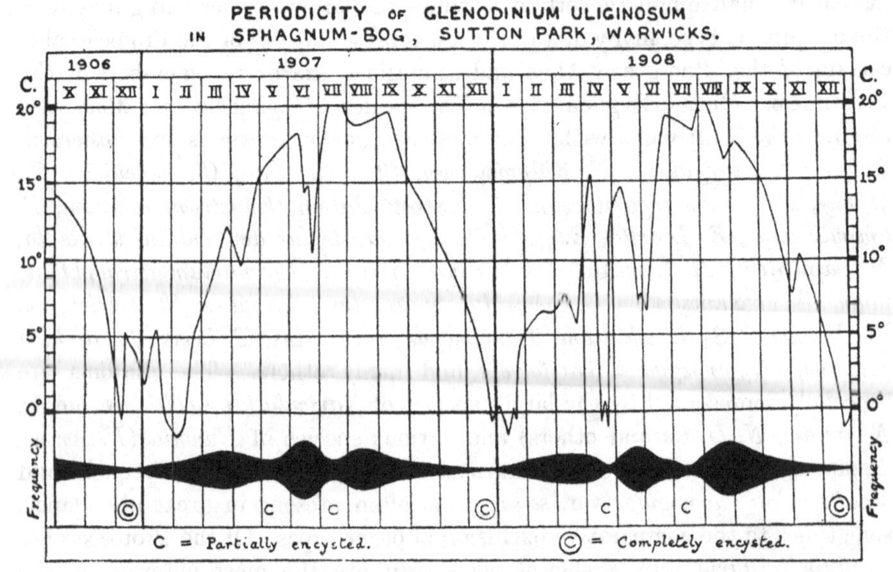

Fig. 267. Periodicity of *Glenodinium uliginosum*. The upper curve represents the water-temperatures from October, 1906 to December, 1908. The lower black line represents by its varying width the relative abundance of *Glenodinium uliginosum* at different seasons of the year.

altitude, additional species, such as *Micrasterias oscitans*, *M. Jenneri*, *Cosmarium Ralfsii*, *C. cœlatum*, and many others, becoming increasingly frequent with increase in altitude up to several thousand feet. In the northern hemisphere increase in latitude has a marked effect on the desmid-flora, many arctic types making their appearance, amongst which may be mentioned *Xanthidium grœnlandicum*, *Euastrum tetralobum*, *Staurastrum acarides*, *St. subsphæricum*, *St. rhabdophorum*, *Cosmarium spetsbergense*, *C. pericymatium*

[1] The author has seen the deep bogs on Cocket Moss, West Yorkshire, on Thursley Common, Surrey, and in parts of Achill Island, Mayo, a sheet of yellow with the flowers of *Utricularia minor*. Such bogs always contain a great abundance of desmids, but the number and character of the species depend largely upon the nature of the geological formation.

and many others. Many of the arctic types are alpines in lower latitudes, but others (such as *Xanthidium grœnlandicum*) are strictly arctic. Diatoms are also influenced in the same way. *Navicula alpina* is strictly a subalpine or montane species in England, but occurs in great abundance in the *Sphagnum*-bogs of the Orkneys and Shetlands almost at sea-level.

In the tropics most of the boggy areas are at considerable altitudes and they contain many of the ubiquitous types of desmids and diatoms so frequent at lower altitudes in temperate regions. In the lowland swampy places of the tropics desmids are often abundant, but the species are to a great extent very different from those of temperate areas and also much more numerous. Quite a number are more or less ubiquitous throughout tropical areas, but very many are certainly restricted to definite geographical areas.

In his observations on the alga-flora of Ceylon, Fritsch ('07 A and B) states that the filamentous types of desmids occur mostly in the lowland pools and he is inclined to attribute this fact to the lower percentage of dissolved oxygen. It seems probable, however, that other factors are concerned in the production of this filamentous tendency. For instance, in the British Islands the largest number of filamentous desmids occurs in the subalpine lakes of the Welsh and Scottish mountains, and, moreover, under conditions of great aëration.

C. **Algal Associations of Ponds and Ditches.**

Ponds and ditches, being only very small sheets of water, exist under very varied conditions and they differ much among themselves in their alga-flora. All exhibit a decided periodicity and irregular factors have much more effect on the alga-flora than is the case in large sheets of water.

The Zygnemaceæ are represented more especially by species of *Spirogyra* and *Mougeotia*. Various species of *Zygnema* occur in an irregular way, but on the whole in small bodies of water this genus only occurs abundantly in rocky pools or where there is a considerable amount of peat. The various species of *Spirogyra* differ in their periodicity. Some are purely vernal, whereas others may be both vernal and autumnal. The autumnal phase has been stated to be the result of the germination of a certain percentage of the zygospores formed in the spring (Fritsch & Rich, '07), but in many instances it results from the persistence through the summer of a number of vegetative filaments of short length and few cells. In temperate countries there is no doubt that conjugation takes place mostly in the vernal phase and depends upon a combination of recurring factors. *Mougeotia* is more particularly a vernal type in ponds and ditches, and reproduction occurs normally in the vernal phase. In temperate areas *Mougeotia viridis* is almost invariably the first species to enter into the fructiferous condition in the spring (G. S. W., '09 B), beginning its spore-formation in the midlands and southern counties of England in March, with a water-temperature of 5°—7° C., and at higher altitudes in the mountainous areas in April or May. Other Conjugates are represented by certain desmids, of which the following are fairly generally distributed in the ponds and ditches of temperate countries: *Closterium Ehrenbergii, Cl. moniliferum, Cl. rostratum, Cl. acerosum, Cosmarium Botrytis, C. granatum, C. Meneghinii*, and *Staurastrum punctulatum*; others, less frequent, are *Closterium peracerosum, Cl. acutum, Cosmarium biretum* and *Staurastrum crenulatum*. In larger ponds, and in others which never dry up, many additional species are found, such as *Cosmarium reniforme, C. humile, C. Boeckii, Closterium Venus* and various others. On the older rocks in the mountainous areas of western Europe roadside ditches and pools very rarely dry up and the species of desmids are often quite different. *Cosmarium speciosum, C. decedens, C. pseudarctoum, Euastrum dubium, Tetmemorus lævis, Arthrodesmus Incus* and many others make their appearance.

In the tropics species of *Spirogyra* are quite as abundant as in temperate regions, but there is a striking absence of the narrower species with the replicate or infolded extremities of the cells (Fritsch, '07 A and B; G. S. W., '12). It is also a noteworthy fact that *Sp. decimina* is much the commonest species. The desmids are for the most part essentially different from those of temperate areas.

The genus *Microspora* is well represented in temperate countries, most of the species being decidedly spring types, although some of them occur in considerable quantity in the late autumn. Vegetative filaments of these species never entirely disappear in the summer from the waters of those ponds and ditches in which they occur. The investigations of Fritsch & Rich ('13) confirm the view that the genus as a whole is a cold-water type and a very vigorous one. In tropical countries *Microspora* is principally an upland type occurring in mountainous areas up to 8000 or 9000 ft. *M. amœna*, which is the largest species, thrives best in running water although it may occur in ditches.

One or two species of *Cladophora* occur in ponds, generally where there is good aëration of the water, but as a rule species of this genus require more aëration than can be obtained in very small bodies of water. *Rhizoclonium hieroglyphicum* may occur in very stagnant ponds, this species being able to adapt itself to almost every kind of environment. *Pithophora* is almost exclusively tropical and subtropical, replacing to some extent the genus *Cladophora* in the smaller sheets of water. Fritsch ('07 A) considers *Pithophora*, with its comparatively thin cell-walls, as better suited than the other

Cladophoraceæ for life in tropical waters, which owing to their higher temperatures contain less dissolved oxygen and carbon-dioxide. It is also possible that the peculiar resting-spores so characteristic of this genus enables it to live in situations liable to rapid desiccation. The Cladophoraceæ of ponds and ditches are generally infested with epiphytes. In temperate climates these are largely diatoms of the genera *Cocconeis* (consult fig. 84), *Epithemia*, *Rhoicosphenia* and *Achnanthes*. In the tropics the filaments of *Pithophora* often carry epiphytes, the most important being a number of the smaller species of *Œdogonium* (generally with depressed-ellipsoid basal cells) and several species of *Endoderma* (consult fig. 194 *D* and *E*).

Many species of *Œdogonium* occur in ponds and ditches, and in the larger ponds a few species of *Bulbochæte* may be found. They are either spring or summer types, depending upon the species, although not a few are found far into the autumn. Sexual reproduction depends very largely upon meteorological conditions and in most cases does not occur every year, the plants being reproduced asexually until the exact conditions for sexual reproduction supervene. The sexual organs may be produced any time between April and September. The filaments are sometimes covered with species of the epiphytic genus *Characium*, and *Aphanochæte repens* is often a common epiphyte.

Of the Ulvaceæ only *Enteromorpha intestinalis* is at all frequent, although at least two species of *Monostroma* occur in small ponds.

Of the Protococcales the Volvocaceæ deserve the first mention. The genera *Carteria*, *Chlamydomonas*, *Pandorina*, *Gonium*, *Eudorina* and *Volvox* all occur abundantly; *Eudorina* is not uncommon, but *Pleodorina* is decidedly rare. The various species of these genera are to a great extent erratic in their occurrence, but on the whole they are cold-water types. Species of *Chlamydomonas* are the most numerous and some of them may be found at any time between September and May in the active motile condition. The Autosporaceæ are well represented by *Scenedesmus*, *Ankistrodesmus*, *Selenastrum*, *Cœlastrum* and other genera, and several species of *Pediastrum* may be abundant. The great majority of the Tetrasporine and Chlorococcine Protococcales are late spring and summer forms, often occurring far into the autumn[1]. Genera of the Micractinieæ, which were at one time thought to be largely plankton organisms, often occur in quantity in small ponds.

In the Heterokontæ *Tribonema bombycinum* (with its forma *minor*) is often abundant and not infrequently mixed with *Microspora floccosa* and

[1] Delf ('15) writing on the 'Algal Vegetation of Ponds on Hampstead Heath,' London, states that the majority of the Protococcales are early spring forms attaining their maximum development in the month of March, but most of the genera mentioned by this author belong to the Volvocine series, a group in which the majority of the species have a preference for comparatively cold water, although they are often very erratic in their appearance.

M. stagnorum; *Tribonema affine* is not uncommon in peaty ponds and ditches. Several species of *Ophiocytium* are very common in stagnant pools, especially *O. majus* and *O. parvulum*. Species of *Tribonema* are not common in the tropics and occur mostly in elevated areas, but the genus *Ophiocytium* is quite as abundant as in any temperate region. Most genera of the Heterokontæ like shaded pools or those which are grass-grown at the margin. *Centritractus belonophorus* and *Botrydiopsis arrhiza* occur as rare constituents of the alga-flora of small ponds, although the former has been found in the plankton of lakes.

The Peridinieæ of small ponds and ditches are not particularly numerous. The commonest form in western Europe is most probably *Peridinium cinctum* although several species of *Gymnodinium* are frequent.

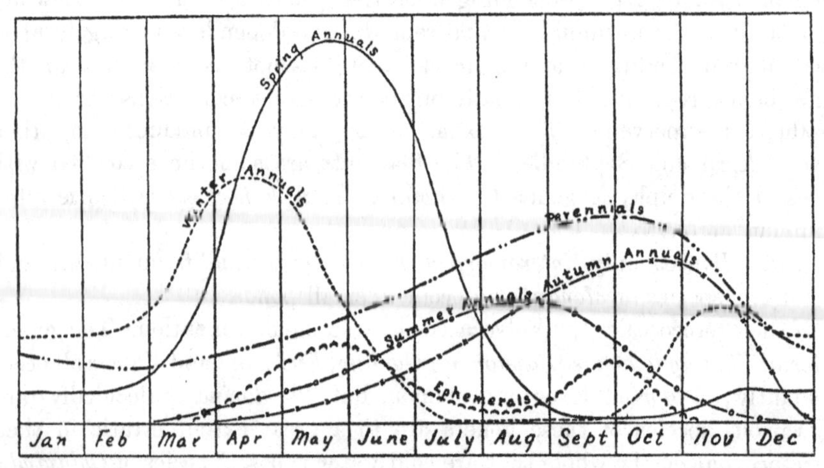

Fig. 268. Estimated relative importance of the several types of algal periodicity and the composition of the alga-flora at any time of the year in the waters of eastern Illinois. The 'irregulars' are not depicted. (After Transeau.)

Fritsch ('06) suggested a division of small pieces of water into two series:

a. *Those containing Cladophoraceæ*, in which the water is to a great extent fairly well aërated.

b. *Those from which Cladophoraceæ are absent*, in which the aëration is much less.

There is no doubt that the presence of perennial species of *Cladophora* has not only a marked effect on the alga-flora of a pond but on the annual cycle of that pond. There is, of course, always a struggle between *Cladophora* and its epiphytes, and this is not much affected by the presence of other Algæ. These epiphytes are, however, not confined to *Cladophora* and often occur in abundance on *Vaucheria*-filaments and on the leaves of aquatic macrophytes.

In all small ponds and ditches with reasonable exposure to sunlight *the general periodicity is much the same: diatoms being the dominant winter types and Zygnemaceæ the dominant spring types (with abundance of diatoms).* It is the summer phase which is so very variable and there is no doubt that meteorological conditions exert a profound influence. Therefore in temperate countries the controlling factors are, and must be, irregular. *The determination of the* LIMITING FACTORS *will probably afford a solution of the whole problem.* It is usually one factor which exercises a decisive influence and this factor is the limiting factor.

In the United States, Transeau ('13) has made a large number of careful observations on the periodicity of the Algæ of small bodies of water in Illinois and has concluded that 'on the basis of their periods of greatest abundance, the duration of the vegetative cycles, and the times of reproduction, freshwater Algæ in general may be divided into seven classes.' These are: winter annuals, spring annuals, summer annuals, autumn annuals, perennials, ephemerals and irregulars. The accompanying chart illustrates some of Transeau's observations.

D. Algal Associations of Pools and Lakes.

So far as can be ascertained there is no zoning in the algal associations of pools and lakes. There are the Algæ around the shores of the lake constituting the *benthos* and those which occur free-floating in the waters of the lake constituting the *plankton.* In the smaller pools many of the benthic species are found in the 'plankton,' which is thus composed of a mixture of forms many of which are not by any means true plankton organisms. This so-called 'plankton' of small pools is often designated as *heleoplankton.* The general alga-flora varies very considerably, depending largely upon the geological formations and the altitude.

1. **Benthos.** The alga-flora of the benthic region of pools and lakes varies much according to the nature of the banks, whether rocky, sandy or marshy. The algal associations occur mostly among various aquatic macrophytes to which many of the species are attached.

In the more reedy lakes *Œdogonium* and *Bulbochæte* are not uncommon, species of the first-named genus sometimes dominating all the filamentous types. Several species of *Coleochæte* are common epiphytes, *C. scutata* being the most abundant. Several species of *Ulothrix* may occur, but not very commonly, and *Chætophora* is distinctly a vernal type. *C. pisiformis* is epiphytic and is generally distributed, whereas the curious colonies of *C. incrassata* (fig. 188 *A*) are unattached and more locally distributed. Other more or less vernal types are represented by various species of *Microspora* (*M. amœna, M. floccosa* and others), and *Tribonema bombycinum* is frequent.

Sometimes members of the Cladophoraceæ occur, more especially attached to rocks and stones.

The Zygnemaceæ are fairly common, *Mougeotia* being more abundant than either *Spirogyra* or *Zygnema*. In the upland and alpine lakes and tarns of the British Islands, in which species of *Mougeotia* and *Zygnema* abound, the maximum vegetative abundance of these genera usually occurs in the late summer and early autumn as the temperature is gradually declining (G. S. W. '09 B). In the littoral alga-flora of the alpine lakes of the Pike's Peak region, Colorado, Shantz ('07) also records the maximum abundance of species of *Mougeotia* and *Zygnema* in September, when the temperature is falling. In these habitats spores are only rarely produced, the winter season being passed in the form of 'cysts.' In the tropics species of both the above-mentioned genera are less abundant than in temperate areas. On the other hand, certain exclusively tropical genera are known to exist, such as *Temnogametum* (fig. 212) and *Pyxispora* (fig. 216 *A—C*).

The desmid-flora in pools and lakes is chiefly dependent upon the geological formation. It is very poor in lakes situated on the newer formations and correspondingly rich in those on the older formations. This is strictly true of western Europe and there is much evidence that the same is true of other parts of the world (W. & G. S. W. '00 B; G. S. W. '09 B). If there is little contamination of the water, that is, if the water is very pure, with only a small amount of dissolved salts, desmids are as a rule fairly numerous, but the abundance of species depends upon the geological formation and the species themselves are dependent to some extent upon the altitude. It would serve no useful purpose to mention specifically any of the desmids which occur in the benthos of lakes. There are about 2000 of them and those of the tropics are for the most part very different from those which occur in subtropical and temperate areas.

Many of the Protococcales are found amongst the aquatic macrophytes of the benthic region of pools and lakes. Various species of *Oocystis, Scenedesmus, Ankistrodesmus, Sorastrum, Pediastrum,* etc., are frequent and often abundant.

The diatoms are largely dependent upon the amount of the dissolved salts in the water and they are in general most abundant in the spring. The zig-zag chains of *Tabellaria flocculosa, T. fenestrata* and *Diatoma elongatum* occur sometimes in quantity, and *Eunotia pectinalis, E. lunaris,* various species of *Cocconema, Navicula, Gomphonema,* etc., are often abundant.

Peridinium inconspicuum has a world-wide distribution.

Of the Blue-green Algæ several species of *Anabæna* occur among aquatic macrophytes at the margins of pools and lakes, and *Merismopedia glauca* is frequent. In alpine lakes species of *Calothrix* and *Stigonema* are common on the rocks of the shore. Some of the Blue-green Algæ, such as species of

Schizothrix and *Glœocapsa*, build up calcareous pebbles in the littoral region of lakes (*vide* p. 35) and other Algæ perforate both shells and stones.

2. **Plankton.** The methods of collection of plankton-organisms and the subsequent examination and estimation of catches, which are much the same as in the case of marine plankton, have been well described by many authors and the student is referred to the works of Apstein ('96), Kofoid ('97) and Bachmann ('11).

Three types of plankton are recognized:—the LIMNOPLANKTON of lakes, the POTAMOPLANKTON of rivers[1], and the CRYOPLANKTON of perpetual snow and ice.

Owing to the limited size of freshwater basins the shore- and bottom-species play a more important part in the plankton than is the case in the sea. In general, fresh waters contain a greater mass of plankton in the same volume of water than does the sea, and, moreover, the plankton is of a more composite character.

LIMNOPLANKTON.

The phytoplankton of lakes may be of great bulk, but more often it is not. On the whole, it is unusual for it to colour the water to any appreciable extent, except in shallow lakes on the more recent geological formations[2]. The greatest amount of phytoplankton occurs in most cases during the autumnal decline in temperature and in the great majority of lakes there are certain more or less well-marked phases in the phytoplankton, each phase dominated by one or more of the constituents.

There is a considerable uniformity in the phytoplankton of lakes which are situated on relatively recent geological formations, especially those of lowland areas. On the other hand, the phytoplankton of the more upland lakes of the older geological formations is usually quite different and, moreover, affords striking contrasts even between lakes in the same area. In many lake-areas it is not an easy matter to compare the phytoplankton of one lake with that of another, since the annual phases of one probably do not correspond with those of the other.

Wesenberg-Lund ('08) has objected to Apstein's suggested grouping of lakes according to the quality of the plankton, such as lakes with Myxophyceæ and lakes with *Dinobryon*. This objection was based, however, only on a knowledge of the lakes of Denmark, which contain plankton of the most monotonous character and are amongst the least interesting in Europe. Apstein was fundamentally correct in his suggested classification of lakes

[1] Potamoplankton only occurs in comparatively slow rivers, and is therefore discussed here and not under the heading of 'Algal Associations of Swiftly-running Water.'

[2] Such colouration of the water is very noticeable in the shallow Danish lakes, in some of the Baltic lakes, and in Lough Neagh and a few other Irish lakes.

according to *quality, i.e.*, according to the nature of the constituents of the plankton, but he had not realized the basic principle underlying this difference in quality. Desmid-plankton was unknown at the time and the great importance of the geological formations of lake-basins had not been recognized.

Wesenberg-Lund also states that pond-plankton (heleoplankton) is more variable from pond to pond than lake-plankton. This may be true in Denmark and in the Baltic lakes, but it is not often true of other regions. Take three of the English lakes situated within a few miles of one another—Ennerdale Water, Wastwater and Windermere; 91 species have been recorded in the phytoplankton of Ennerdale Water, 50 species in that of Wastwater and 65 in Windermere; yet only 15 species are common to all three lakes, and the dominant species are for the most part different (W. & G. S. W., '12). The Central African lakes afford another example of widely differing plankton; 85 species have been recorded in the phytoplankton of Tanganyika, of which 61 do not occur in the other great lakes (G. S. W., '07). Wesenberg-Lund differentiated between the plankton of deep lakes and that of shallow lakes, stating that deep lakes were characterized first and foremost by their enormous diatom maxima and shallow lakes by 'water-bloom.' These statements require careful revision, since they were based upon a limited and very inadequate experience. The deep lakes of North Wales and the deepest lake (Wastwater) in the English Lake District have no great diatom maxima in either spring or autumn. The same is also true of many of the deep Scottish lochs. On the other hand, the great African lake Tanganyika, which on the whole is a deep lake, has a very great diatom-flora. It seems highly probable that the great diatom-maxima of certain lakes have little connection with the depth. 'Water-bloom,' which is caused by a great development of certain species of Blue-green Algæ, is, however, a feature of shallow lakes in lowland areas. It is a conspicuous feature of Lough Neagh (Dakin & Latarche, '13) just as it is of the Danish Lakes. Wesenberg-Lund's statement that Chlorophyceæ are practically absent from deep lakes is very erroneous, since Green Algæ often form upwards of 70 per cent. of the phytoplankton of deep lakes.

The principal groups of Algæ represented in the freshwater phytoplankton are the Myxophyceæ, Peridinieæ, Bacillarieæ and Chlorophyceæ, and a few comments upon the occurrence of each of these groups may prove useful to the student.

MYXOPHYCEÆ.—In those lakes in which Blue-green Algæ are conspicuous, the various species are usually abundant in the warmer part of the year, more particularly in the early part of the autumnal decline in temperature. The lakes of the British Islands and Scandinavia are less dominated by Blue-green Algæ than most other European lakes, only *Cœlosphærium Kützingianum*, *Gomphosphæria Nägeliana* and *Oscillatoria Agardhii* occurring in such abundance as sometimes to dominate the plankton, and then only in contaminated lakes (*i.e.* those with a relatively high percentage of dissolved salts). Myxophyceæ occur very abundantly in many of the shallower lakes of the European continent situated on the more recent geological formations (Schröder, '00; Lemmermann, '03; etc.) and the same is true in N. America (Marsh, '03). *Gomphosphæria lacustris* may sometimes occur in great quantity

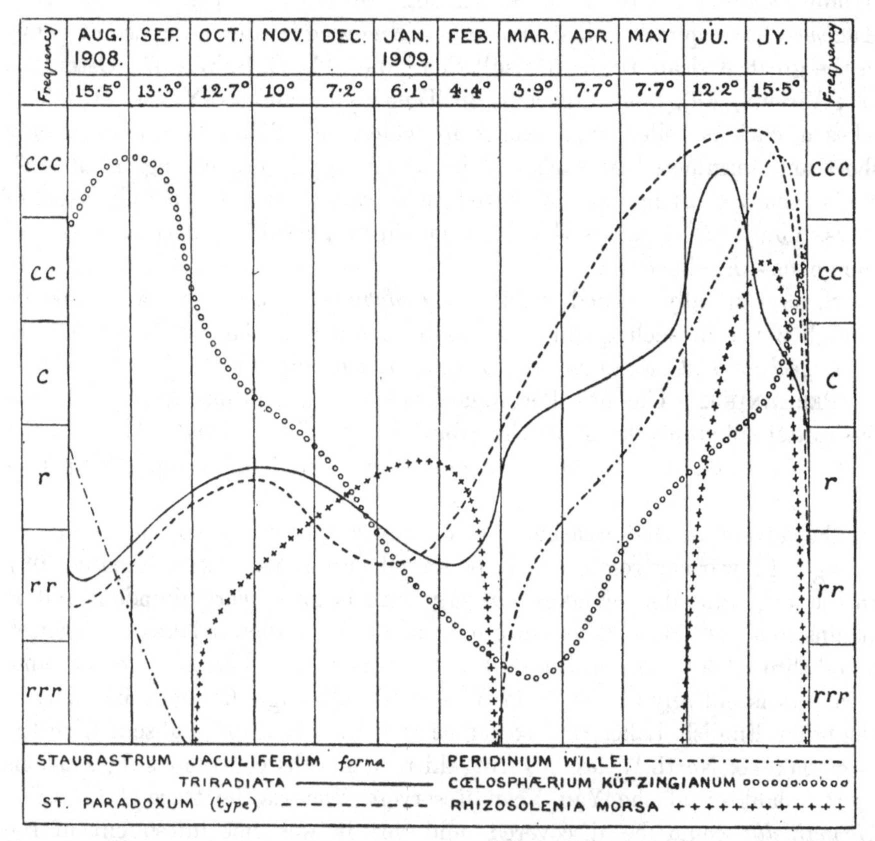

Fig. 269. Chart of the periodicity of five of the dominant constituents of the phytoplankton of Loch Katrine, Scotland. The temperatures are in degrees Centigrade.

and in Lough Corrib in the west of Ireland it completely dominates the summer-plankton (W. & G. S. W., '06). *Chroococcus limneticus* is also of frequent occurrence and in the larger pools may often be quite abundant.

Various species of *Anabæna* (notably *A. Flos-aquæ*, *A. circinalis* and *A. Hassallii*) and *Aphanizomenon Flos-aquæ* often occur in quantity, and they may attain great maxima, along with *Oscillatoria Agardhii* and *Glœotrichia echinulata*, in pools and smaller lakes on the occasions signalized by great masses of 'water-bloom' (*vide* p. 32). This phenomenon, which does not occur in the deep, uncontaminated lakes, yet requires thorough investigation, and it would appear that its complete scientific explanation can only be obtained by simultaneous biological and chemical observations extending over a considerable period of time (W. & G. S. W., '12)[1]. *Anabæna Lemmermanni* sometimes occurs in abundance, its spores forming deep blue-green floating clusters which may give a decided colour to surface water of an entire lake (Lemmermann, '03; W. & G. S. W., '06). Most of the plankton-species of *Anabæna* have spirally coiled filaments. *Anabæna* is also an important genus in the great African Lakes (consult Schmidle, '02; G. S. W., '07; Ostenfeld, '08; Virieux, '13), more especially in Tanganyika and in Nyasa. In these lakes a curious coiled type occurs in which the filaments are extremely short and terminated at each end by a heterocyst (consult fig. 19 *A—E*). Woloszynska ('12) has also observed these same forms in the plankton of lakes in Java and suggests that they should be placed in a special section of the genus—*Anabænopsis*.

Several spirally twisted species of *Lyngbya* also occur in the plankton, though never in such quantity as to be dominant. The most notable are *L. Lagerheimii*, *L. contorta* and *L. circumcreta* (fig. 19 *F—H*).

PERIDINIEÆ.—The one Peridinian which is ubiquitous throughout the freshwater phytoplankton of the world is *Ceratium hirundinella*. In the colder temperate countries it is a summer form, completely disappearing from the plankton in the winter months, entering into an encysted state (fig. 54 *E*) on the advent of cold weather. It usually reappears about the middle of spring. In warmer regions it is perennial (Entz, '04; Lemmermann, '08). In the larger and deeper lakes this organism is rarely very abundant, but it attains to a position of dominance in some of the shallower lakes. Although found throughout the world it does not occur in every lake even in the same area. It is entirely absent from Wastwater, although frequent in many of the other English Lakes, and so far as is known it is quite absent from the larger lakes of North Wales. During thirteen months' continuous observations on the plankton of the Yan Yean Reservoir, Victoria, no trace of *Ceratium hirundinella* could be discovered, and yet it was not infrequent in the Toorourong Reservoir from which the main water-supply of the Yan Yean is derived by aqueduct (G. S. W., '09 B). The organism is exceedingly variable in size and in the length and number of its horns, the many forms having been well described and figured (consult Lemmermann, '04; '10; W. & G. S. W., '05; '06; Bachmann, '07). Two of these forms stand out conspicuously from the others. One is var. *brachyceras* (v. Daday) Ostenfeld ('09), a stunted form characteristic of Victoria Nyanza, and the other a form with a curiously

[1] It has been suggested by Snow ('03) as a result of both observations in nature and cultural experiments that the appearance of 'water-bloom' is possibly due to the presence of an unusual amount of dead organic matter in the water.

deflected first antapical horn which so far is only known from a few lakes in the Outer Hebrides and the west of Ireland (W. & G. S. W., '06; '09 B). In some lakes seasonal form-variations have been observed, such as those described by Wesenberg-Lund ('08) from the Danish Lakes, but these variations have not been observed in the British lakes, although several forms frequently occur simultaneously in the plankton of one lake.

Ceratium cornutum is another summer type, but of less frequent occurrence. It may occur in very small numbers in lakes in which *C. hirundinella* is abundant, but more often it is found in those lakes, generally with very pure water, from which *C. hirundinella* is absent. It is common in the Welsh lakes and in the lake-area of Carnarvonshire it occurs in the well-aërated *Sphagnum*-bogs.

Many species of *Peridinium* are characteristic of lake-plankton. In European areas *P. cinctum* and *P. Willei* (fig. 44) are the most conspicuous, the latter being a summer form with a maximum at some period during the warmer months (W. & G. S. W., '09 A). As in the case of *Ceratium hirundinella*, *Peridinium Willei* is a perennial constituent of the lake-plankton of more southern latitudes. Species of *Peridinium*, like many diatoms, do not attain a universal maximum at one definite period of the year, but the various species reach their greatest vegetative development at different times of the year (W. & G. S. W., '12). In some of the pools of the English Midlands there is a summer species (a var. of *P. cinctum*), a spring species (*P. anglicum*; fig. 51), and a very early spring—almost a winter—species (*P. aciculiferum*; fig. 50), each of which has been shown to form resting-cysts at the close of the vegetative period, even though the vegetative periods are all at different seasons. The observations on the Peridinieæ of Bracebridge Pool in Sutton Park, Warwickshire (consult G. S. W., '09 A) show that temperature is really an important factor in the occurrence of each species.

BACILLARIEÆ.—Only in the more contaminated lakes (*i.e.* those with a relatively high percentage of dissolved salts) do diatoms attain great maxima. Vast maxima occur periodically in the central European Lakes, in the Danish lakes and in a few of the British lakes; also in Tanganyika. The Pennate Diatoms are much more numerous and for the most part much more conspicuous than the Centric Diatoms. In lakes with very pure water, such as the Carnarvonshire lakes, the plankton contains very few diatoms. The evidence at present available shows that, although many of the plankton-species of diatoms occur in greatest quantity in the spring, some of them attain their maxima in the summer and autumn, and several of them have a double maximum, one in the spring and the other in the autumn. An instance of the latter is afforded by *Asterionella gracillima* in Windermere (W. & G. S. W., '09 A), in which the spring maximum is of greater bulk but not so prolonged as the autumn maximum (consult fig. 270).

The double maximum of diatoms in general does not appear to be so marked in the freshwater plankton as in the marine plankton, and the double maximum of the *same species*, as instanced by *Asterionella gracillima, Cyclotella compta, Rhizosolenia morsa* (W. & G. S. W., '12) and *Fragilaria crotonensis* (Wesenberg-Lund, '08), is of great interest.

A number of the freshwater diatoms are perennial constituents of the plankton of certain lakes and can be collected in a living condition all the year round.

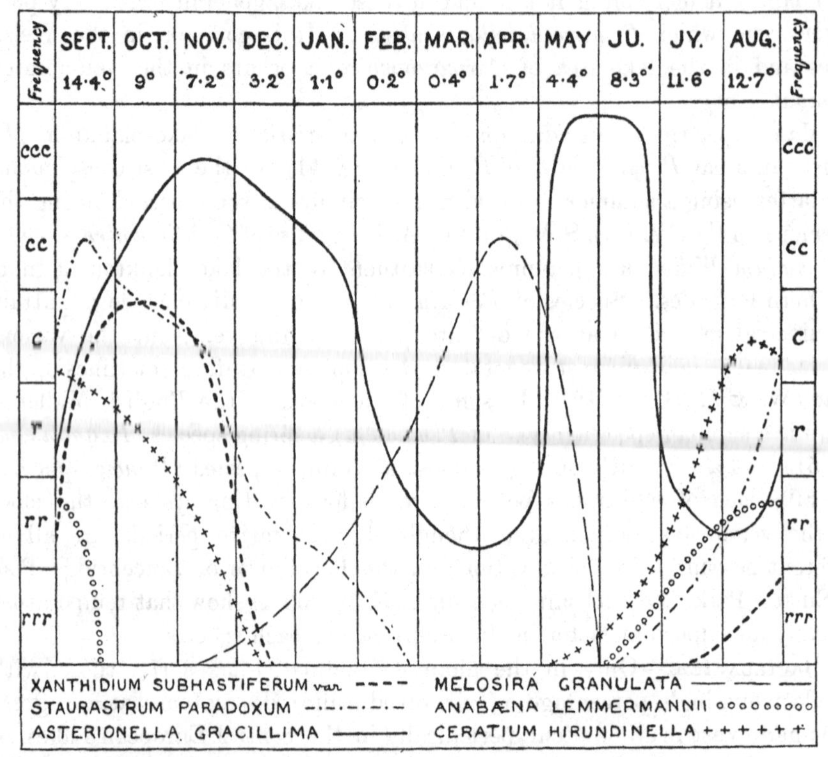

Fig. 270. Chart showing the periodicity of six of the most abundant constituents of the plankton of Windermere, in the English Lake District, from September 1907 to August 1908. The temperatures are in degrees Centigrade.

In the European lakes the most notable forms are *Asterionella*, with a range of form and size which embraces both *A. formosa* (fig. 73 *A*) and *A. gracillima*, and the two species of *Tabellaria*. On the whole, *T. fenestrata* is much more abundant than *T. flocculosa*, except in the English Lake District. The chain-forms of *T. fenestrata* are the most frequently observed, but the star-dispositions of the frustules are common except in the lakes of North Wales. *T. fenestrata* var. *asterionelloides* (fig. 73 *B*) is one of the dominating features of the late spring, the summer and the early autumn plankton of many of the western European lakes (W. & G. S. W., '02; '05; '06; '09 A and B; Lemmermann, '04; Wesenberg-Lund, '04; '08; Bachmann, '11; etc.). Star-dispositions of *T. flocculosa* are very uncommon, but they have been observed by Holmboe ('99) in Norway and by Wesenberg-Lund ('04) in Denmark.

Fragilaria crotonensis is a very characteristic plankton-diatom of the pools and lakes of Europe and North America, but does not often occur in large quantity. In Loch Ruar, in Sutherland, there is a unique variety—var. *contorta*—with curiously twisted short filaments (W. & G. S. W., '05).

The phytoplankton of the large African lakes is peculiar in the almost entire absence of many genera of diatoms which are a dominant feature of the lakes of north temperate areas. *Asterionella, Tabellaria* and *Rhizosolenia* are wanting or of small importance. Four species of the last-named genus have recently been recorded by Woloszynska ('14) from Victoria Nyanza, but star-dispositions of diatoms are practically confined to several species of *Synedra* of the section *Belonastrum*.

Attheya is a characteristic plankton-genus closely allied to *Rhizosolenia* and although frequent in the lakes of continental Europe has not been found in the British lakes.

In all the British lakes and in the large African lakes the genus *Surirella* plays an important part in the phytoplankton. In the British lakes *Surirella robusta* var. *splendida* is the most frequent form, sometimes occurring in great abundance, but *S. biseriata* and *S. linearis* are both general (W. & G. S. W., '05; '09 B). In the great lakes of Africa a number of beautiful species of *Surirella* occur in abundance, many of them, such as *S. Nyassæ, S. Malombæ* and *S. Füllebornii*, being apparently restricted to that part of the world (O. Müller, '05; G. S. W., '07; Ostenfeld, '09; etc.). These African species show many transition stages from one to the other. Certain species of *Cymatopleura* have also become true constituents of the plankton; they are common in the Irish lakes (W. & G. S. W., '06) and in Victoria Nyanza (G. S. W., '07), and Wesenberg-Lund ('04) records *Cymatopleura elliptica* as being a typical plankton-organism in one of the Danish lakes.

The Centric diatoms of the plankton are only conspicuous in certain lakes. Apart from *Rhizosolenia*, various species of *Melosira* are often well represented. In the European area *M. granulata* may occur in such quantity as to give a *Melosira*-phase to the plankton and the same is true of the Yan Yean Reservoir in Australia. The records of this species show such erratic vegetative periods[1] in relation to water-temperatures that it is not improbable that various species have been recorded under the name of '*Melosira granulata*' or that there are different biologic forms of this diatom. In Windermere,

[1] *Rhizosolenia morsa* is apparently as erratic as *Melosira granulata* in its occurrence. It frequently has a double maximum, the autumn maximum being the larger (W. & G. S. W., '12).

in the English Lake District, *M. granulata* has a large maximum in April with a water-temperature of 1·7° C.; in Loch Lomond it attains its maximum from May to June (temp. 5°—13·3° C.); in Lough Neagh the *Melosira* maximum corresponds closely with that of Windermere, occurring from February to March. In the Yan Yean Reservoir the maximum is attained in the middle of the warm period with a water-temperature of 21° C.; moreover, auxospore-formation took place at approximately the period of highest temperature. This is in striking contrast to the behaviour of *Melosira islandica* (a closely allied species to *M. granulata*) recorded by Ostenfeld & Wesenberg-Lund ('06) in the plankton of Thingvallavatn, Iceland, in which auxospore-formation occurred from December to January with a water-temperature of 1°—2° C. In the European area some lakes are '*Melosira*-lakes' whereas others are not, and this difference occurs in lakes in close proximity. Similarly, of the large African lakes, Nyasa and Victoria Nyanza are both '*Melosira*-lakes,' whereas Tanganyika is not. In the two first-named lakes there are many species (*M. nyassensis, M. ikapoensis, M. ambigua, M. italica, M. argus, M. granulata, M. Agassizii, M. Schroederi*, etc.), almost all of which are represented by several varieties. In fact, the *Melosira*-flora is both phenomenal and dominant in the early spring (consult O. Müller, '04; Ostenfeld, '09; Woloszyńska, '14). In most cases the *Melosira*-filaments are approximately straight, but in several instances Lemmermann, Volk, Ostenfeld, and others, have recorded the occurrence of spirally coiled filaments.

The various species of *Cyclotella* are important as plankton-forms, especially in many European lakes (*vide* Bachmann, '11), and *Stephanodiscus Astræa* is sometimes very abundant both in European and African lakes. *St. Niagaræ* is dominant at one period of the year in certain Canadian lakes. *Coscinodiscus lacustris* is a very noteworthy feature of the plankton of Lough Neagh in Ireland (W. & G. S. W., '02; '06; Dakin & Latarche, '13).

CHLOROPHYCEÆ.—The only groups of Green Algæ of importance in the plankton are the Protococcales and the Desmidiaceæ, and almost all the different forms attain their maximum vegetative abundance during the autumnal fall in temperature.

Of the Protococcales several genera are more or less important. Several species of *Pediastrum*, and especially the greatly perforated varieties of *P. simplex* and *P. duplex*, may occur in abundance, although the usual pond-forms are not common except in the shallower and more lowland lakes. *Sphærocystis Schroeteri* is frequent in some lakes, but quite absent from others. Species of *Scenedesmus* and *Crucigenia*, although not occurring in any great quantity, are often characteristic of the plankton of certain lakes. The same is true of several species of *Oocystis*. On the other hand, *Dictyosphærium pulchellum* may occasionally occur in great abundance. A few species of *Cœlastrum* are common and *Cœlastrum reticulatum* is mostly confined to the plankton. Species of *Tetradesmus* and *Ankistrodesmus* occur in some lakes, but never in quantity. *Ankistrodesmus Pfitzeri* is a characteristic plankton-form in European lakes. *Closteriopsis longissima* is frequent in many lakes, and *Kirchneriella lunaris* and *K. obesa* are often notable constituents of the phytoplankton. Nearly all these members of the

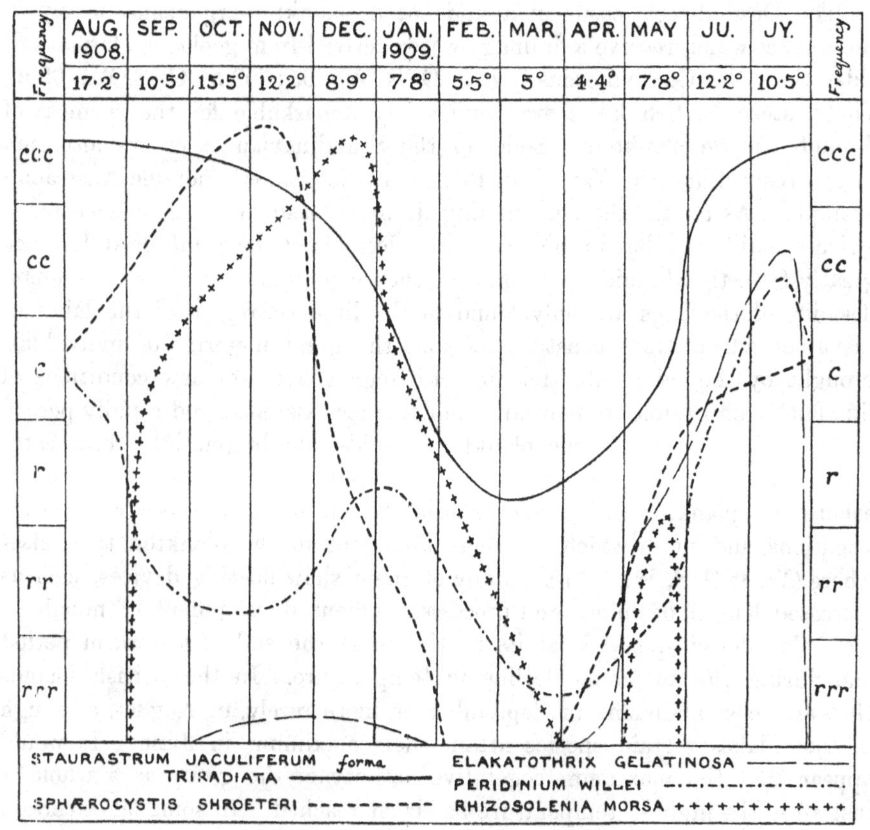

Fig. 271. Chart of the periodicity of five of the most conspicuous constituents of the phytoplankton of Wastwater, in the English Lake District. The temperatures are in degrees Centigrade.

Protococcales occur in greatest amount in the slightly contaminated lakes and pools. From Victoria Nyanza Woloszynska ('14) has described a number of other plankton-genera, such as *Schmidleia* and *Schroederiella*, but they are only scarce plankton-constituents. *Elakatothrix gelatinosa* is a rare Alga which apparently occurs in very pure water, being known from the lakes of Norway and from Wastwater (W. & G. S. W., '09 A). Two other species occur in ponds and lakes in N. America, Europe and Asia.

Botryococcus Braunii (which the present author regards as a member of the Heterokontæ; consult p. 406) is a ubiquitous plankton-form, although by no means confined to the plankton. It is much more abundant in some lakes than in others and at times forms a yellow-green scum on the surface of the water. This scum may subsequently become of a yellowish-red or brick-red colour, owing to the development of a pigment, and the entire lake may be coloured by it.

The Desmidiaceæ occur in abundance as plankton-organisms mostly in those lakes which receive a drainage-water derived from geological formations older than the Carboniferous (W. & G. S. W., '05; '09 B; G. S. W., '09 B). The western British lakes are particularly remarkable for the richness of desmids in the plankton. Some of the Scandinavian lakes are also rich in this respect and the Yan Yean Reservoir, Victoria, is a notable Australian example. As a rule, the surrounding drainage-basins of such lakes contain a rich desmid-flora, but in most cases the desmids of the plankton differ very greatly from the desmid-community of the surrounding area. The common desmids of the bogs are only found in the limnetic region of the lakes as casual or adventitious constituents, and the great majority of individuals brought by the rains into this limnetic region, with its new conditions of life, find it impossible to maintain their further existence, and rapidly perish. On the other hand, the true plankton-desmids may be put into three categories: those which are exclusively confined to the plankton, those which are exclusively plankton-varieties of species which frequently occur in other situations, and those which are more abundant in the plankton than elsewhere (W. & G. S. W., '09 B). Many of them show floating devices, such as increased length of spines and processes, copious development of mucilage, etc. The desmid-phase is at its maximum at the end of the warm period and during the autumnal decline in temperature. In the British Islands they are most abundant in September or, more rarely, in August, although in some lakes certain species attain their maximum in June. It would appear that the maximum vegetative activity of the group as a whole is just after the highest temperature has been reached. In some lakes certain species are perennial, but these also have their maximum at the same season as the others (W. & G. S. W., '12).

There are a few other Green Algæ which are sometimes of importance in the phytoplankton. A form of *Tribonema bombycinum* occurs in Lough Neagh in fairly large quantities and also in the Danish lakes. All these lakes are shallow and very different in character from the deep rocky lakes of mountainous areas, in which *Tribonema* does not occur in the plankton. On the other hand, in these lakes (and also in many of the shallow lakes) the plankton contains various species of *Zygnema*, *Spirogyra* and *Mougeotia*, principally in the late spring and summer. They are usually the slender species of these genera and almost invariably sterile. In the smaller alpine lakes several slender species of *Mougeotia* are often abundant and they may form no small part of the phytoplankton[1]. Curious coiled *Mougeotia*-filaments sometimes occur (W. & G. S. W., '09 A; '09 B). It would appear that this coiling is a limnetic character developed to augment the floating capacity of the filament and is strictly comparable with that which occurs in certain species of *Anabæna, Lyngbya* and *Melosira.*

* * * * * * * * *

In making a close study of the phytoplankton of lakes, difficulties of comparison are everywhere met with, even when dealing with lakes in the same area. The constituents of the phytoplankton are not the same in all, and species which occur abundantly in one lake may not occur in any of the others. These differences in constituents are partly territorial and partly local, and are in some measure due to the rigorous conditions which govern the distribution of so many aquatic Algæ. A territorial distinction occurs in those lakes situated in drainage-basins in which the rocks are older than the Carboniferous. Local differences between the lakes of one area situated in similar basins, when they occur, are often the result of contamination of the water (W. & G. S. W., '12).

The varying nature of the plankton of different lakes is to be correlated with the fact that the various groups of Algæ require different physiological conditions for rapid multiplication. For instance, the factors which favour the prolific growth of desmids in the plankton are not those which enable an equally rapid increase in the majority of diatoms; and likewise those factors which favour the great multiplication of one species of diatom are not favourable for a similar increase in another. A careful study of the constituents of the phytoplankton in relation to the lake-basins brings with it the conviction that *the factor of greatest importance in both the qualitative and quantitative distribution of plankton is the amount and nature of the dissolved salts present in the water.* The highest percentage of dissolved salts is found in those lakes which are contaminated from adjacent farms, villages and towns, and such lakes contain a greater quantitative bulk of plankton.

Slightly contaminated lakes contain a greater number of diatoms than

[1] In the alpine lakes of the Pike's Peak Region, Colorado, Shantz ('07) states that species of *Spirogyra* and *Œdogonium* form a large part of the summer plankton. Fragmentary filaments of various species of *Œdogonium* are also very frequent in the summer plankton of the British lakes. The pecies are mostly of moderate size, with filaments 20—35 μ in diameter. Dakin & Latarche ('13) make an extraordinary statement that these *Œdogonium*-records may have been confused with *Tribonema*, which 'is almost identical with *Œdogonium*'! There is, of course, no close resemblance between *Œdogonium* and *Tribonema*; even dead empty cells could not be confused. Moreover, the *Œdogonium*-filaments which find their way into the plankton and there live for some time as adventitious constituents, are of much greater diameter than any known species of *Tribonema*.

uncontaminated lakes, and except in extreme northern latitudes some of them are perennial constituents of the plankton. A few species frequently have enormous maxima; in north temperate regions *Asterionella gracillima*, *Tabellaria fenestrata* var. *asterionelloides* and several species of *Melosira* are the most notable; in the large African lakes *Nitzschia nyassensis* and several species of *Melosira* and *Surirella* are at one season among the dominant constituents of the plankton. The desmid-flora of these lakes, except those containing an abundance of *Surirella*, is usually poor, and few species occur in quantity.

Uncontaminated lakes (or lakes with very pure water) contain fewer diatoms, and such as do occur in the plankton rarely attain even a small maximum. The desmids may be very numerous, depending to a great extent upon the nature of the geological formation, and there is sometimes a rich desmid-plankton. The lakes of the Carnarvonshire mountains are excellent examples, being amongst the least contaminated of all the British lakes. In these lakes there are relatively few diatoms (only 11·1 per cent. out of a total of 162 species) and many desmids (62·4 per cent.), and some of them possess a very rich desmid-plankton.

Lakes which possess a mixed plankton of diatoms and desmids are probably of an intermediate character with regard to the nature and amount of the dissolved salts in the water. The Myxophyceæ are to a great extent absent from the lakes with very pure water, but an examination of the occurrence and distribution of the plankton-species of Blue green Algæ indicates that the factors which control their relative abundance are somewhat different from those which govern the prolific occurrence of diatoms (W. & G. S. W., '12).

There is also a considerable reduction in the amount of *Asterionella* and of the star-dispositions of *Tabellaria*, or even an entire absence of them from lakes with pure water. These two genera of diatoms are almost entirely absent from the great African lakes, most probably owing to too high a temperature of the water (G. S. W., '07), but their absence from certain British lakes appears to be directly concerned with purity of the water. Wastwater in the English Lake District furnishes a good example of a lake from which these star-dispositions of the frustules of diatoms are absent (W. & G. S. W., '12).

The plankton-community as a whole is a very ancient one and this fact is particularly emphasized in the case of those lakes which possess a distinct community of plankton-desmids. The general periodicity of the plankton-constituents (*i.e.* the seasonal changes in the composition of the plankton) is much the same in different parts of the world, *diatoms dominant in the early spring, Green Algæ and Blue-green Algæ attaining their maximum later in the year.*

Neither plankton-desmids nor those which occur in other habitats undergo any seasonal form-variations. In certain of the plankton-diatoms seasonal form-variation does occur, but it is in the colony and not in the individual.

Wesenberg-Lund ('08) in commenting upon the cosmopolitanism of the plankton-community stated that 'freshwater plankton-communities, in contrast to all other communities on land or water, everywhere contain the same types, nearly everywhere the same species.' We now know this statement to be erroneous (W. & G. S. W., '09 B). It does not hold good for the desmid-flora of the plankton and in a less degree it is not true of the diatom-flora. The geographical peculiarities of the desmid-flora are especially well marked in the plankton.

* * * * * * * * *

As an appendix to the general account of limnoplankton it is necessary to mention the NANNOPLANKTON, a name first used by Lohmann ('11) to embrace all those organisms which are so minute as to pass easily through the meshes of the finest plankton-nets. The organisms are best obtained by centrifuging the water and they consist of very minute Green Algæ and Flagellates. Pascher ('11) has investigated a number of the freshwater forms and Scourfield ('11) has written a good account of the use of the centrifuge in pond-life work.

POTAMOPLANKTON.

Slow and moderately slow rivers possess a plankton which differs much from that of lakes. It is more mixed in character and its constituents are largely recruited from the backwaters of the rivers, which are the breeding-grounds of the organisms which become mingled to form the river-plankton. Diatoms are the most abundant organisms of the phytoplankton and as a rule the dominant forms are not those which are so conspicuous in lakes. The most complete investigation of river-plankton yet published is that of the Illinois River by Kofoid ('08), but there are numerous observations of importance on the plankton of other rivers: notably those of Schröder ('97; '99) and Zimmer ('99) on the Oder; Zacharias ('98); Brunnthaler ('00) on the Danube; Zykoff ('00) on the Volga; Fritsch ('02; '03; '05) on the Thames, Trent and Cam; Volk ('03; '06) on the Elbe; and Lemmermann ('07 A and B) on the Weser and the Yang-tse-Kiang.

Climatic conditions have a profound influence on the phytoplankton of rivers. In countries with a comparatively mild winter living plankton occurs all the year round [*vide* Fritsch's observations ('03) on the Thames-plankton]; but, in countries with a severe winter, plankton, at any rate in a living condition, is not found in the winter months (Schröder, '99; Brunnthaler, '00). The periodicity of potamoplankton is also somewhat variable, since this type of plankton is subject to extreme fluctuations in quantity and composition.

The plankton of rivers, as clearly stated by Kofoid, is subject to more catastrophic changes than that of lakes. The quality of the plankton and the number of the constituents depend upon the rate of flow of the stream (Zacharias, '98; Zimmer, '99). It has been repeatedly shown that the more rapid the stream the fewer the individuals in the plankton. In the rivers with a slow current the multiplication of plankton-organisms, provided the conditions are favourable, goes on while they are being carried down in the stream.

Zimmer ('99) suggested that there were three types of plankton-organisms in rivers: (1) *eupotamic* planktonts, which thrive and multiply both in the flowing water of the stream and in the backwaters, ponds, etc.; (2) *tychopotamic* planktonts, which only multiply in still water and when carried into the river-current live for a time, but do not reproduce themselves; (3) *autopotamic* planktonts which have adapted themselves to a life in flowing water.

As previously stated diatoms are the most abundant organisms of potamoplankton, and in discussing this point Schmidle ('02) remarks that the more delicate Algæ and animals withdraw themselves from the society of the siliceous Bacillarieæ owing to the fact that they are incapable of withstanding the buffeting they would be subjected to in the main stream. The Centric diatoms are fairly well represented by the genera *Cyclotella*, *Stephanodiscus* and *Melosira*, and more rarely by *Rhizosolenia* and *Attheya*. Species of the three first-named genera are all common, although it is probable that *Melosira varians* is the most abundant Centric diatom. This diatom, and also others, may have both a spring and an autumn maximum, so that the diatom-phases of the potamoplankton are, as a rule, the most important. The genus *Synedra* is much more important than in lake-plankton and is sometimes entirely dominant. *Fragilaria capucina*, *F. virescens* and *F. crotonensis* are all important constituents in temperate climates. *Asterionella gracillima* also occurs in abundance in many of the rivers of north temperate areas. In contrast to lake-plankton the genus *Nitzschia* is often of great importance.

The Green Algæ are chiefly represented by certain of the Protococcales of which members of the Volvocineæ are especially conspicuous. Various species of *Chlamydomonas* are frequent, *Pandorina Morum*, *Gonium pectorale* and *Eudorina elegans* are fairly general, and both *Pleodorina illinoisensis* and *Volvox aureus* sometimes occur. *Platydorina caudata* is one of the characteristic constituents of the plankton of the Illinois River. Several species of *Pediastrum* and *Scenedesmus* are sometimes abundant, and *Actinastrum Hantzschii*, although never occurring in great quantity, is a typical constituent of river-plankton. Species of *Kirchneriella*, *Ankistrodesmus* and *Micractinium* also occur. *Botryococcus Braunii* sometimes occurs, but never in the quantity in which it may occur in lakes. *Ophiocytium* may also be represented. A few species of desmids are occasionally present, mostly one or two of the pond- or ditch-types of *Closterium* and *Cosmarium*. The Peridinieæ are sometimes represented by large numbers of *Glenodinium* and more rarely by *Ceratium hirundinella* and one or two species of *Peridinium*.

CRYOPLANKTON.

This name is applied to the flora and fauna of perpetual ice and snow. On the great snow-fields and glaciers of polar countries and high mountain ranges there lives an association of organisms which in many respects resembles the plankton-communities of lakes and rivers. The vegetable organisms, although including moss-protonema and a few Fungi, are mainly Algæ and they may occur in such great abundance as to give a distinct colouration to the snow or ice. The algal groups represented in these snow-floras are the Myxophyceæ, Bacillarieæ and Chlorophyceæ, and according to the dominance of certain species it is possible to distinguish between red, yellow, green and brown snow. The chief investigators of snow-floras have been Wittrock ('83), Lagerheim ('83; '92; '94); Chodat ('96; '02; '09); Scherffel ('10) and Fritsch ('12).

Red Snow has been most frequently observed and described in the past, probably owing to the fact that it stands out so conspicuously, the tint varying from a delicate rosy red to a deep blood-red or sometimes a dark brick-red colour. It occurs extensively in the Arctic countries, in the Alps, the Carpathians, the Andes, and to a limited extent in the Antarctic; it has also been observed in the Appenines and the Pyrenees. It is caused chiefly by the rounded resting-cells of *Chlamydomonas nivalis* which contain a large amount of hæmatochrome. These cells colour the upper layer of snow to a depth of several centimetres and layer after layer may become buried. A few diatoms are found in the red snow, and *Glœocapsa sanguinea, Scotiella nivalis, Chionaster nivalis* and *Rhaphidonema nivale* have been observed as subsidiary constituents. The species of diatoms vary with the geographical area, and in both Arctic and Antarctic regions the fragmentary remains of marine diatoms are not infrequent.

Yellow snow is quite distinct from red snow and it may cover extensive areas. In colour it is a pale bright yellow and the algal constituents are more numerous than in the case of red snow. The best account of it has been given by Fritsch ('12) who examined in detail yellow snow collected in the South Orkneys. This association included *Protoderma Brownii, Chlorosphæra antarctica, Scotiella antarctica, S. polyptera, S. nivalis, Protococcus viridis, Chodatella brevispina, Rhaphidonema nivale, Ulothrix subtilis, Mesotænium Endlicherianum, Nostoc minutissimum* and several other Algæ. Some of these Algæ are identical with those found in red or other coloured snow, but others are distinct types. Fritsch points out that most of the members of this yellow snow-flora contain a large amount of a solid fat and that the yellow pigment which gives the colour to the snow is present in this fat. The presence of the fat is probably an adaptation against the intense cold of the habitat and thus functions like hæmatochrome, which is capable of absorbing the heat-rays of the sun.

Green snow differs but little from yellow snow. It has been described from the Alps and from the Arctic. Its constituents are mostly Green Algæ: the zoogonidia of *Chlamydomonas nivalis*, *Ankistrodesmus nivalis* and species of *Mesotænium*. A few members of the Myxophyceæ also occur. On the Glacier d'Argentière *Ankistrodesmus Vireti* is one of the constituents of the green snow-flora (Chodat, '09).

Brown snow has been described by Wittrock and it owes its colour to numberless fine mineral particles with which are mixed *Mesotænium* (*Ancylonema*) *Nordenskioldii* (which contains phycoporphyrin in its cell-sap), several other Green Algæ, various diatoms and a few Myxophyceæ.

Lastly it should be mentioned that Chodat ('02) has referred to black snow (neige noir), the principal Alga in which is *Scotiella nivalis* (Fritsch, '12). Scherffel ('10) also records *Rhaphidonema brevirostre* as a constituent of black snow in the high mountains of Tatrla.

LITERATURE CITED

Apstein, C. ('96). Das Süsswasserplankton. Methode und Resultate der quantitativen Untersuchung. Kiel und Leipzig, 1896.

Bachmann, H. ('07). Vergleichende Studien über das Phytoplankton von Seen Schottlands und der Schweiz. Archiv für Hydrobiologie und Planktonkunde, iii, 1907.

Bachmann, H. ('11). Das Phytoplankton des Süsswassers mit besonderer Berücksichtigung des Vierwaldstättersees. Jena, 1911.

Brunnthaler, J. ('00). Plankton-Studien, i. Das Phytoplankton des Donaustromes bei Wien. Verhandl. der k. k. zool.-botan. Ges. in Wien, l, 1900.

Cedergren, G. R. ('13). Bidrag till Kännedomen om Sötvattensalgerna i Sverige. I. Algfloran vid Upsala. Arkiv för Botanik utgif. af K. Sv. Vet.-Akad. xiii, no. 4. Stockholm, 1913.

Chodat, R. ('96). Flore des neiges du col des Ecandies. Bull. de l' Herb. Boiss. iv, 1896

Chodat, R. ('02). Algues Vertes de la Suisse. Berne, 1902.

Chodat, R. ('09). Sur la Neige verte du Glacier d'Argentière. Bull. de la Soc. botan. de Genève, 2me sér. i, 1909.

Dakin, W. J. & Latarche, Margaret ('13). The Plankton of Lough Neagh. Proc. Roy. Irish Acad. xxx, sect. B, no. 3, 1913.

Delf, E. M. ('15). The Algal Vegetation of some Ponds on Hampstead Heath. New Phytologist, xiv, 1915.

Entz, G. ('04). In Result. der wiss. Erforschung des Balatonsees, ii, Budapest, 1904.

Fritsch, F. E. ('02). Preliminary Report on the Phytoplankton of the Thames. Ann. Bot. xvi, 1902.

FRITSCH, F. E. ('03). Further Observations on the Phytoplankton of the River Thames. Ann. Bot. xvii, 1903.

FRITSCH, F. E. ('05). The Plankton of some English Rivers. Ann. Bot. xix, 1905.

FRITSCH, F. E. ('06). Problems in Aquatic Biology, with special reference to the Study of Algal Periodicity. New Phytologist, v, no. 7, July, 1906.

FRITSCH, F. E. ('07 A). The Subaerial and Freshwater Algal Flora of the Tropics. Ann. Bot. xxi, April, 1907.

FRITSCH, F. E. ('07 B). A General Consideration of the Subaërial and Fresh-water Algal Flora of Ceylon. A Contribution to the Study of Tropical Algal Ecology. Part I.—Subaërial Algæ and Algæ of the Inland Fresh-waters. Proc. Roy. Soc. B. vol. 79, 1907.

FRITSCH, F. E. ('07 C). The Rôle of Algal Growth in the Colonization of New Ground and in the Determination of Scenery. Geographical Journal, Nov. 1907.

FRITSCH, F. E. ('12). Freshwater Algæ in the South Orkneys by Mr R. N. Rudmose Brown, of the Scottish National Antarctic Expedition, 1902–04. Journ. Linn. Soc. Bot. xl, 1912.

FRITSCH, F. E. & RICH, F. ('07). Studies on the Occurrence and Reproduction of British Freshwater Algæ in Nature. I. Preliminary Observations on Spirogyra. Ann. Bot. xxi, no. lxxxiii, July, 1907.

FRITSCH, F. E. & RICH, F. ('09). Studies on the Occurrence and Reproduction of British Freshwater Algæ in Nature. II. A five years' observation of the Fish Pond, Abbot's Leigh, near Bristol. Proc. Bristol Nat. Soc. fourth ser. vol. ii, part ii, 1909.

FRITSCH, F. E. & RICH, F. ('13). Studies upon the Occurrence and Reproduction of British Freshwater Algæ in Nature. III. A four years' observation of a freshwater pond. Ann. de Biol. lacustre, Bruxelles, vi, 1913.

HOLMBOE, J. ('99). Undersögelser over Norske Feeskvandsdiatoméer. I. Diatoméer fra Indsjöer i dat Sydlige Norge. Archiv for Mathem. og Naturvidenskab. xxi, no. 8, 1899.

KOFOID, C. A. ('97). Plankton Studies. I. Methods and Apparatus in Use in Plankton Investigations at the Biological Experiment Station of the University of Illinois. Bull. Ill. State Lab. Nat. Hist. v, 1897.

KOFOID, C. A. ('08). The Plankton of the Illinois River. Part II. Constituent Organisms and their Seasonal Distribution. Bull. Ill. State Lab., Urbana, Illinois, xiii, Article i, May, 1908.

LAGERHEIM, G. ('83). Bidrag till kännedomen om Snöfloran i Luleå Lappmark. Botaniska Notiser, 1883.

LAGERHEIM, G. ('92). Die Schneeflora des Pichincha. Ber. Deutsch. Botan. Ges. x, 1892.

LAGERHEIM, G. ('94). Ein Beitrag zur Schneeflora Spitzbergens. Nuova Notarisia, 1894.

LEMMERMANN, E. ('03). Das Phytoplankton einiger Plöner Seen. Forschungsb. Biol. Stat. Plön, x, 1903.

LEMMERMANN, E. ('04). Das Plankton schwedischer Gewässer. Arkiv for Botanik utgifv. af K. Sven. Vet.-Akad. ii, no. 2, 1904.

LEMMERMANN, E. ('07 A). Das Plankton der Weser bei Bremen. Archiv für Hydrobiologie u. Planktonkunde, ii, 1907.

LEMMERMANN, E. ('07 B). Das Plankton des Jang-tse-kiang (China). Archiv für Hydrobiologie u. Planktonkunde, ii, 1907.

LEMMERMANN, E. ('08). Beiträge zur Kenntnis der Planktonalgen XXIII—XXV. Archiv für Hydrobiologie und Planktonkunde, iii, 1908.

LEMMERMANN, E. ('10). Peridiniales in Kryptogamenflora der Mark Brandenburg. Algen I. Juli, 1910.

LOHMANN, H. ('11). Ueber das Nannoplankton und die Zentrifugierung kleinster Wasserproben zur Gewinnung desselben in lebendem Zustande. Int. Revue d. ges. Hydrobiologie u. Hydrographie, iv, 1911.

MARSH, C. DWIGHT ('03). The Plankton of Lake Winnebago and Green Lake. Bull. Wisconsin Geol. and Nat. Hist. Survey, no. xii, Madison, Wis. 1903.

MÜLLER, O. ('04). Bacillariaceen aus dem Nyassalande und einigen benachbarten Gebieten. Zweite Folge. Engler's Botan. Jahrbüch. xxxiv, 1904.

OSTENFELD, C. H. ('08). Phytoplankton aus dem Victoria Nyanza. Engler's Botan. Jahrbüch. xli, 1908.

OSTENFELD, C. H. ('09). Notes on the Phytoplankton of Victoria Nyanza, East Africa. Bull. Mus. Comp. Zool. at Harvard College, lii, no. 10, 1909.

OSTENFELD, C. H. & WESENBERG-LUND, C. ('06). A Regular Fortnightly Exploration of the Plankton of the two Icelandic Lakes, Thingvallavatn and Myvatn. Proc. Roy. Soc. Edinburgh, xxv, part xii, 1906.

PASCHER, A. ('11). Über Nannoplanktonten des Süsswassers. Ber. Deutsch. Botan. Ges. xxix, 1911.

SCHERFFEL, A. ('10). Rhaphidonema brevirostre nov. spec., egyúttal odelék a Magas-Tátra nivális flórájához. Beiblatt zu den Botanikai Közlemények, Heft 2, 1910.

SCHMIDLE, W. ('02). Das Chloro- und Cyanophyceenplankton des Nyassa und einiger anderer innerafrikanischer Seen. Engler's Botan. Jahrbüch. xxxiii, 1902.

SCHRÖDER, B. ('97). Ueber das Plankton der Oder. Ber. Deutsch. Bot. Ges. xv, 1897.

SCHRÖDER, B. ('99). Das pflanzliche Plankton der Oder. Forschungsber. a. d. Biol. Station zu Plön, vii, 1899.

SCHRÖDER, B. ('00). Das Pflanzenplankton preussischer Seen. Westpreuss. Bot.-Zool. Verein, Danzig, 1900.

SCOURFIELD, D. J. ('11). The Use of the Centrifuge in Pond-life Work. Journ. Quekett Micr. Club, xi, 1911.

SHANTZ, H. L. ('07). A Biological Study of the Lakes of the Pike's Peak Region—Preliminary Report. Trans. Amer. Microscop. Soc. xxvii, March, 1907.

SNOW, JULIA W. ('03). The Plankton Algæ of Lake Erie. U.S. Fish Commission Bulletin, 1902 (Washington, 1903).

TRANSEAU, E. N. ('13). The Periodicity of Algæ in Illinois Trans. Amer. Microscop. Soc. xxxii, Jan. 1913.

VIRIEUX, J. ('13). Plancton du lac Victoria Nyanza. Résultats scientif. Voyage de Ch. Alluard et R. Jeannel en Afrique Orientale 1911-1912. Paris, 1913.

VOLK, R. ('03; '06). Hamburgische Elb-Untersuchung. Mitth. a. d. Naturhist. Mus. in Hamburg. I: xix, 1903; VIII: xxiii, 1906.

WELWITSCH, F. ('68). In Journ. of Travel and Nat. Hist. i, 1868.

WESENBERG-LUND, C. ('04). Studier over de Danske Söers Plankton. Kjöbenhavn, 1904.

WESENBERG-LUND, C. ('08). Plankton Investigations of the Danish Lakes. Copenhagen, 1908.

WEST, G. S. (G. S. W. '07). Report on the Freshwater Algæ, including Phytoplankton, of the Third Tanganyika Expedition conducted by Dr W. A. Cunnington, 1904-1905. Journ. Linn. Soc. Bot. xxxviii, Oct. 1907.

WEST, G. S. (G. S. W. '09A). A Biological Investigation of the Peridinieæ of Sutton Park, Warwickshire. New Phytologist, viii, 1909.

WEST, G. S. (G. S. W. '09B). The Algæ of the Yan Yean Reservoir, Victoria: a Biological and Œcological Study. Journ. Linn. Soc. Bot. xxxix, 1909.

WEST, G. S. (G. S. W. '12). Freshwater Algæ of the Percy Sladen Memorial Expedition in South-West Africa, 1908-11. Ann. South Afric. Mus. ix, part ii, 1912.

West, G. S. & Starkey, Clara B. ('15). A Contribution to the Cytology and Life-History of *Zygnema ericetorum* (Kütz.) Hansg., with some Remarks on the 'Genus' *Zygogonium*. New Phytologist, xiv, 1915.

West, W. & West, G. S. (W. & G. S. W., '94). On some Freshwater Algæ from the West Indies. Journ. Linn. Soc. Bot. xxx, 1894.

West, W. & West, G. S. (W. & G. S. W., '97). Welwitsch's African Freshwater Algæ. Journ. Bot. 1897.

West, W. & West, G. S. (W. & G. S. W., '99). A further Contribution to the Freshwater Algæ of the West Indies. Journ. Linn. Soc. Bot. xxxiv, 1899.

West, W. & West, G. S. (W. & G. S. W., '01). The Alga-flora of Yorkshire. Trans. Yorks. Nat. Union, vol. v, 1900–1901.

West, W. & West, G. S. (W. & G. S. W., '02). A Contribution to the Freshwater Algæ of the North of Ireland. Trans. Roy. Irish Acad. xxxii, sect. B, part i, 1902.

West, W. & West, G. S. (W. & G. S. W., '05). A further Contribution to the Freshwater Plankton of the Scottish Lochs. Trans. Roy. Soc. Edin. xli, part iii, 1905.

West, W. & West, G. S. (W. & G. S. W., '06). A Comparative Study of the Plankton of some Irish Lakes. Trans. Roy. Irish Acad. xxxiii, sect. B, part ii, 1906.

West, W. & West, G. S. (W. & G. S. W., '09 a). The Phytoplankton of the English Lake District. The Naturalist, March—Sept. 1909.

West, W. & West, G. S. (W. & G. S. W., '09 b). The British Freshwater Phytoplankton, with Special Reference to the Desmid-plankton and the Distribution of British Desmids. Proc. Roy. Soc. B, vol. 81, 1909.

West, W. & West, G. S. (W. & G. S. W., '11). The Freshwater Algæ in Reports on Sci. Investig. Brit. Antarctic Expedit. 1907–9, vol. i, part vii, Dec. 1911.

West, W. & West, G. S. (W. & G. S. W., '12). On the Periodicity of the Phytoplankton of some British Lakes. Journ. Linn. Soc. Bot. xl, May, 1912.

Wittrock, V. B. ('83). Om Snöns och Isens Flora, Särskeldt i de Arktiska Trakterna. Stockholm, 1883.

Woloszynska, J. ('12). Das Phytoplankton einiger javanischer Seen, mit Berücksichtigung des Sawa-Planktons. Bull. de l'Acad. Sci. Cracovie, sér. B, sci. nat. June, 1912.

Woloszynska, J. ('14). Studien über das Phytoplankton des Victoriasees. Hedwigia, lv, 1914.

Zacharias, O. ('98). Das Potamoplankton. Zoolog. Anzeiger, xxi, 1898.

Zimmer, C. ('99). Das tiersche Plankton der Oder. Forschungsber. a. d. Biol. Station zu Plön, vii, 1899.

Zykoff, W. ('00). Das Potamoplankton der Wolga bei Saratow. Zoolog. Anzeiger, xxiii, 1900.

ADDENDA

SINCE the foregoing pages were printed off a number of papers have appeared to which it is desirable to make some reference.

In the **Myxophyceæ** the cytology of the Chroococcaceæ has been very carefully studied by Miss Acton (*Ann. Bot.* xxviii, 1914) who states that 'there is a gradual transition in the structure of the cell from an almost undifferentiated condition in the lower types to a somewhat specialized one, of which *Chroococcus macrococcus* represents the highest type examined, and *Merismopedia elegans* an intermediate stage.' In *Chroococcus macrococcus* there is a definite incipient nucleus and cytoplasm. Miss Acton suggests that the evolution of the nucleus and cytoplasm has taken place along the following lines: 'The excess of food material elaborated by the pigment was first stored in the plasmatic microsomes as a carbohydrate—cyanophycin. As more and more material was elaborated the reserve in the central region became more complex, and the protein metachromatin granules were formed. In time, the accumulation of nucleo-protein became restricted to a very limited area in the cell, so as to ensure its equal distribution on division, and this restriction only occurred on division, as in *Merismopedia elegans*. In this way part of the cell became physiologically and morphologically separated on account of its function in connection with division.' This area is called by the present author the 'incipient nucleus' (consult p. 7). Miss Acton goes on to state that 'at a later stage the "nucleus" became stable and was always present, as in *Chroococcus macrococcus*.'

The cytology of *Glaucocystis Nostochinearum* has been investigated by Griffiths (*Ann. Bot.* xxix, 1915) who states that '*Glaucocystis* is probably a member of the Cyanophyceæ owing to the presence of an "open" nucleus at one stage; the tendency of cytoplasmic division to take place independently of nuclear division; and to the presence of phycocyanin in the chromoplast. The very high differentiation of the nucleus in the dividing stage; the elaborate chromoplast to which the phycocyanin is confined; the formation of daughter-cells very similar to those of *Oocystis*; and the cellulose character of the cell-wall, are features which separate *Glaucocystis* from all the rest of the Cyanophyceæ, and probably justify its being placed in an entirely separate group of that division.' Griffiths' conclusions give unqualified support to the suggestion made by the present author in 1904 that the Myxophyceæ should be divided into the Glaucocystideæ and the Archiplastideæ (consult p. 40).

* * * * * * * * * *

In the **Protococcales** quite a number of new genera have been recently described. In the Volvocaceæ *Isococcus* Fritsch (in *New Phytologist*, xiii, 1914), *Platymonas* G. S. West (in *Journ. Bot.* liv, 1916) and *Dysmorphococcus* Takeda (in *Ann. Bot.* xxx, 1916) have been discovered, *Platymonas* being a marine genus and the other two freshwater genera. There is also an interesting paper by Grove (in *New Phytologist*, xiv, 1915) on *Pleodorina illinoiensis*, in which a number of new facts are recorded concerning the morphology and reproduction of this cœnobiate type.

Dispora Printz (in *Vidensk. Skrifter*, Kristiania, 1913) is a genus closely allied to *Crucigenia*. Several genera have been described from Victoria Nyanza by Woloszynska (in *Hedwigia*, lv, 1914): *Schmidleia* Wolosz., *Schrœderiella* Wolosz., *Victoriella* Wolosz., and *Peniococcus* Wolosz. All are to be referred to the Selenastreæ of this work, but *Victoriella* is identical with *Tetradesmus* Smith (*vide* G. S. West in *Journ. Bot.* liii, 1915, p. 83) and it is doubtful whether *Peniococcus* can rightly be separated from *Desmatractum* W. & G. S. West. *Peniococcus Nyanzæ* would be better placed as *Desmatractum Nyanzæ.*

The genus *Quadrigula* has been founded by Printz (in *Kgl. Norske Vidensk. Selsk. Skrifter*, Trondhjem, 1915) to include two Algæ previously placed in *Ankistrodesmus.* This genus is also one of the Selenastreæ and probably quite a valid one.

Another genus of the Selenastreæ has just been founded by Teiling (in *Svensk Botanisk Tidskrift*, x, 1916) under the name of *Tetrallantos.* The colony is very similar to that of *Schmidleia*, the differences being specific rather than generic, and it would probably be more correct to place *Tetrallantos Lagerheimii* as *Schmidleia Lagerheimii.*

The most interesting of all these new genera of the Protococcales is *Phytomorula* Kofoid (in *Univ. California Publ.* vi, 1914), a member of the Cœlastreæ consisting of a small *compressed* cœnobium of 16 cells. It occurred in the plankton of a reservoir at Berkeley, California, and the cœnobium is 'of exceptional regularity and remarkable resemblance to a lenticular egg with equal cleavage, in a sixteen-cell stage.'

Concerning the genus *Centrosphæra*, an investigation has just been completed in the botanical laboratory of Birmingham University by Miss B. M. Bristol. The results confirm the view expressed by the present author on p. 212 that *Endosphæra* Klebs and *Scotinosphæra* Klebs cannot be separated from *Chlorochytrium* Cohn; but, in addition, there is clear evidence that *Centrosphæra* Borzi must also be placed as a synonym of *Chlorochytrium*. It is hoped that the results of this investigation will shortly be published.

* * * * * * * * * *

Little addition has been made to our knowledge of the **Ulotrichales** during the past two years. Printz has described (*l.c.* 1916) two new genera from the Siberian side of the Ural Mountains: *Epibolium* Printz, which is allied to *Chloroclonium*, and *Lochmium* Printz, which stands near to *Microthamnion*. Miss Acton has also described a new species of *Gomontia* (*G. Ægagropilæ*) inhabiting the floating balls of *Cladophora* (*Ægagropila*) *holsatica*. The investigation of this Alga has shown that *Tellamia* Batters should most probably be regarded as synonymous with *Endoderma* Lagerh. and that *Foreliella* Chodat is really synonymous with *Gomontia* Born. & Flah. The present author is therefore in error in following Wille, and placing *Foreliella perforans* Chodat as a species of *Tellamia* (*vide* p. 305), since this Alga should be placed as *Gomontia perforans* (Chodat) Acton (in *New Phytologist*, xv, 1916).

INDEX

CAMBRIDGE: PRINTED BY J. B. PEACE, M.A., AT THE UNIVERSITY PRESS.

Printed in the United States
By Bookmasters